Lecture Notes in Applied and Computational Mechanics

Volume 12

Series Editors

Prof. Dr.-Ing. Friedrich Pfeiffer
Prof. Dr.-Ing. Peter Wriggers

Springer

Berlin
Heidelberg
New York
Hong Kong
London
Milan
Paris
Tokyo

Analysis and Simulation of Multifield Problems

Wolfgang Wendland
Messoud Efendiev (Eds.)

Springer

Professor Wolfgang Wendland
Professor Messoud Efendiev
Universität Stuttgart
Lehrstuhl für Angewandte Mathematik
Institut für Angewandte Analysis und Numerische Simulation
Pfaffenwaldring 57
D-70569 Stuttgart
Germany

e-mail: wendland@mathematik.uni-stuttgart.de
e-mail: efendiev@mathematik.uni-stuttgart.de

With 131 Figures

Cataloging-in-Publication Data applied for
Bibliographic information published by Die Deutsche Bibliothek
Die Deutsche Bibliothek lists this publication in the Deutsche Nationalbibliografie;
detailed bibliographic data is available in the Internet at <http://dnb.dd

ISBN 3-540-00696-6 Springer Verlag Berlin Heidelberg New York

Springer-Verlag Berlin Heidelberg New York
a member of BertelsmannSpringer Science+Business Media GmbH

http://www.springer.de

Printed in Germany

Cover design: design & production GmbH, Heidelberg
Typesetting: Digital data supplied by editors

Printed on acid-free paper 62/3020Rw - 5 4 3 2 1 0

Preface

The analysis and simulation of multifield problems became recently one of the most actual and vivid areas of research. Of course, it is clear that complex technical and physical phenomena modelled by the interaction of different fields require a corresponding complex mathematical modelling and numerical simulation. Although the corresponding subproblems often are quite well understood separately, however, their interaction and coupling creates not only new difficulties but also a completely new level and quality of the interacting coupled field problems. In order to tackle multifield problems, the German Research Foundation DFG started in 1995 the Collaborative Research Center on Multifield Problems (Sonderforschungsbereich Mehrfeldprobleme, SFB 404) at the University of Stuttgart where engineers from applied mechanics and mathematicians work together in multidisciplinary interaction on multifield problems from continuum mechanics. Whereas during the first years the models of multifield interaction involved mostly the coupling of different fields within their common space-time regions of definition as well as transmission problems connecting different field models, the focus moved recently more to the interactions between different spacial, time and state scales. The International Conference on Multifield Problems (April 8 – 10, 2002) at the University of Stuttgart was organized by the members of the Collaborative Research Center. (The program of the conference can be found in the appendix.)

The organizers of the conference are very happy that three of the keynote speakers are willing to publish their contributions in this volume. (The lectures of the other five keynote speakers have already been published elsewhere.) In addition, we have selected 38 further contributions to our conference and this whole collection will present a rather colourful picture of corresponding ongoing research.

Keynote speakers

An inherent difficulty in the treatment of multifield problems is the fact that the subproblems usually are well understood and even corresponding robust and reliable software is available, whereas the coupling of these subproblems in general cannot be done in a naive manner. Hermann Matthies and Jan Steindorf present a fundamental approach to this problem by viewing the strong coupling of these subproblems as a block Gauß-Seidel procedure. Since without particular care such a procedure will not converge in general, they give a modification in terms of a block Newton formulation which then allows to use the available subproblem solvers appropriately. As an example fluid-structure coupling is treated.

Many of current important applications of multifield problems involve multiphase flows either in porous media or in dispersed particle fluid mixtures. Donald Drew shows how to develop a rigorous mechanical and mathematical model for two-phase flows of particle or bubble suspensions involving averaging based on the conservations of mass and momentum. He shows how the additional constitutive laws can be obtained by an appropriate analysis of the microscale behaviour of ensembles and obtains the dispersion tensor, an appropriate diffusion model and the Reynolds stress tensor which allow to define the constitutive laws needed for closing the system of equations.

Mary Wheeler, Shuyu Sun, Owen Eslinger and Béatrice Rivière present the complex model and corresponding numerical simulation of reactive flows through porous media. These multifield problems are very complicated since they involve not only multiphase flows but also chemical reactions through time, space and phase regions where the corresponding physical and time scales by drastically different magnitudes whereas the physical properties of the porous materials admit strongly variable permeabilities. The numerical method used is based on discontinuous Galerkin approximation which needs to be locally conservative and to avoid high numerical diffusion. This contribution shows that it is nowadays possible to simulate the transport of radionucleids, chemicals and biological species in order to control water pollution in large geographical areas.

Part I – Material Modelling and Multiscale Problems
is devoted to the mathematical and mechanical modelling of composite materials and materials showing nonlinear behaviour. Here we find essentially three different types of models: Variational models based on nonconvex energy functionals describing phase transitions as in the contribution by Marcel Arndt, the abstract approach to rate-independent processes based on two energy functionals as presented by Messoud Efendiev and the modelling of elastoplasticity without and with hardening via energy minimization as in the article by Klaus Hackl, Alexander Mielke and Dirk Mittenhuber. Tomáš Roubíček presents the functional analysis with measure-valued solutions to relaxed convexification of the nonconvex energy functional associated with laminated composites. The second type of problems consists of two-scale models (which are related to the first type). The corresponding contributions are due to Claus Bayreuther and Cristian Miehe who treat the homogenization by employing appropriate transfer operators to a multigrid formulation to couple heterogeneous linear elastic microstructures with the macroscopic nonlinear behaviour. Romana Piat and Eckart Schnack use an experimentally determined stochastic distributional function describing the geometrical properties of mesoscopic ellipsoids and cracks within pyrolytic carbon composites in order to obtain macroscopic material laws by homogenization. The third type of material modelling involves interior variables. Klaus Hackl, Alexander Mielke and Dirk Mittenhuber present a strain-based mathematical model of materials without and with hardening where the energy functional does not need to be smooth in the state space, and they are able to model elastoplasticity and the flow rules for single crystal deformation with switching slip systems. Sergey Lurie, Petr Belov and Dmitrij Volkov-Bogorodsky use a mul-

tiscale variational formulation and an extended Cosserat model for two-scale formulations of materials with a periodic distribution of inclusions having simple geometry. Slav Dimitrov and Eckart Schnack present a two-scale damage model based on additional interior variables in order to simulate the stress-based evolution of materials containing a big number of cracks.

Part II – Numerical Methods
In this section, the contributions deal with flow problems, elliptic and free boundary value problems and with inverse problems. Raimund Bürger and Kenneth Karlsen present a new version of the Enquist-Osher finite difference method which now can be applied to nonstandard conservation laws with discontinuous flux functions. Cristian Coclici, Ivan Sofronov and Wolfgang Wendland present a nonlocal nonreflecting boundary condition based on the discretization of the Dirichlet to Neumann map subject to a farfield given by the solution to the linearized compressible Euler equations. Christian Wieners defines a complete family of p2-p1 Taylor-Hood finite elements on tetrahedrons, pyramids, prisms and hexahedrons which now can be used within three-dimensional saddlepoint discretizations involving adaptively refined 3D-meshes. He shows that for his elements the inf-sup stability conditions are satisfied. Andrzej Nowakowski, Wanwilai Kraipech and Tom Dyakowsky present numerical experiments for the three-dimensional simulation of Navier-Stokes flows in hydrocyclones and show that the p2-p0 Taylor-Hood elements will provide robust approximations whereas the p2-p1 elements fail. Iuliu Sorin Pop shows how transient Darcy flows can be approximated by the use of a lumped upwind box scheme applied to an Euler implicit time discretization of the Krichhoff transform and regularized Darcy flow. He shows asymptotic convergence of the method and presents corresponding numerical simulations. Elliptic problems are treated by Günther Of and Olaf Steinbach who use a multipole method for solving stabilized boundary element Galerkin equations with the hypersingular boundary integral operator. The corresponding discrete systems are solved with an appropriately preconditioned conjugate gradient iteration. This method can be used to solve Neumann problems which will be one of the basic blocks in 3D domain decomposition. Victor Ryaben'kii is using finite difference potentials which can be seen as to be a discrete version of potential methods. He applies this approach to the solution of the coupling of different elliptic equations which again is needed in domain decomposition methods. A highly nonlinear inverse problem is treated by Bernd Scholz and Wolfgang Ehlers who use a least-squares fit of the solution to the micropolar model of a Cosserat medium with experimental biaxial tests in order to identify the physical constants characterizing the Cosserat medium. Tom-Alexander Langhoff and Eckart Schnack simulate the growth of the solid pyrolytic carbon layer with a free boundary in pores by solving corresponding reaction diffusion equations.

Part III – Solid Mechanics
In this section one finds contributions to contact and crack problems and

to the asymptotic singularity behaviour of stress fields in the vicinity of corners and edges. One contribution is devoted to the material behaviour of fine grained structures on the microscale. Jiří Jarušek and Christof Eck give a survey on friction laws under which the incremental and quasistatic formulations of contact problems with generalized Coulomb friction as well as the corresponding transient formulation with viscoelastic material behaviour allow a mathematically rigorous existence proof. Gwideon Szefer and D. Kedzior show results for contact with materials that have negative Poissons ratio where the corresponding stress field at the contact zone is significantly different from fields observed in traditional materials. Bin Hu, Werner Schiehlen and Peter Eberhard show that the simulation of dynamic elastic contact problems needs three different time scales so that the elastodynamic boundary approach for the wave propagation can be combined with flexible multiple body systems for structural vibrations and with the rigid multibody systems for the rigid body motion. Victor Kovtunenko shows how the quasistatic modelling of crack propagation including nonpenetration within the crack can be formulated in terms of variational inequalities based on the Griffith criterion which allow the simulation of the growing crack geometry. Ch. Zhang presents transient dynamic crack analysis in piezoelectric materials by employing a time-domain boundary integral equation approach where the time integration is based on Lubich's convolution quadrature. Atanas Dimitrov and Eckart Schnack use Kondratiev's singularity analysis for three-dimensional structures and present a complete algorithm to compute stress concentrations and stress intensity distributions at corners and edges including those that appear at surface penetrating cracks. Dorothee Knees is able to characterize the corresponding stress intensity distributions for linear elastic structures composed of polyhedral subdomains occupied by different linear elastic materials, by some geometric ordering so that global bounds for the corrresponding stress intensity factors can be found which are basic for the preconditioning of corresponding finite element computations. Alain Miranville and Messoud Efendiev extend the Cahn-Hilliard theory to fine grained structures and obtain new results for corresponding material properties.

Part IV – Porous Media and Fluid – Structure Interaction

In this section we collect the contributions to multiphase flows in porous media and polydisperse particle flows as well as the coupling of vibrations and wave phenomena. Małgorzata Peszyńska is using a heterogeneous domain decomposition in order to couple reactive transport models of two-phase and three-phase flows with applications to crude oil exploration. Hartmut Jakobs, Andreas Bielinski, Holger Class and Rainer Helmig analyze the influence of interface conditions on the performance of numerical algorithms for the simulation of multiphase flows through porous media with heterogeneities. Wolfgang Ehlers, Ayhan Acartürk and Bernd Markert present a finite element simulation of an electrochemical multiphase formulation of electrolytes and fluids in porous media modelling the behaviour of swelling tissues. Martin Schanz and Dobromil Pryl use a time-dependent boundary integral equation method to compare two different models of wave propagation phenomena in poro-elastic media. Stefan Berres, Raimund Bürger and Elmer Tory investigate the deterministic macroscale mathematical modelling of polydisperse sedimentation by using the theory of mixtures; whereas Elmer Tory, R.A. Ford and M. Bargiel show how to use steady state parameters of a Markov model for the stochastic n particle simulation of the same kind of polydisperse fluid-particle mixtures. Dieter Dinkler, Björn Hübner and Elmar Walhorn use a discontinuous Galerkin method to simulate the fluid-structure interaction between buildings and strong airflows. Matthias Fischer and Lothar Gaul use a finite element - boundary element coupling procedure to simulate the coupling between a vibrating structure with an exterior acoustic field by using mortar elements on the coupling surfaces. Harmen Schippers, Guus Vos and Hans van Tongeren investigate the influence of the structure vibrations generated by aerodynamic loads on the vibrations of antenna appertures and the corresponding perturbations of the electromagnetic fields. Thomas Schwartzkopff and Claus-Dieter Munz use a nonoverlapping domain decomposition with different physical models, namely coupling of the nonlinear Euler equations with the linearized Euler equations for modelling acoustic noise generated by flows with discontinuities as shocks. Decan Ivanovic and Viktor Saljinikov use perturbation analysis to model flows round porous airfoils where injection or ejection influences the separation of the flow and will therefore be useful for the control of laminar-turbulent transition. Boško Rašuo investigates the wall corrections needed to modify wind tunnel experiments appropriately due to the perforated tunnel walls by using two-dimensional stationary flow models and corresponding perturbation analysis. Yuli Lifshitz and David Degani show how a nonsymmetric three-dimensional flow about a conical obstacle can be modelled by an appropriate decomposition into two simplified two-dimensional flows which then are coupled. This allows a significant reduction of the computational effort for the corresponding flow simulations.

The authors and editors express their sincere gratitude to the German Research Foundation which supported this activity significantly and the conference participants from the East. Special thanks are due to Jan Jung who did the technical work in preparing the final form of this book and to Mr. Richardson and his team for all the organizational work in connection with our conference. We also thank the Springer-Verlag for its patience and excellent outfit of this volume.

Stuttgart, January 2003

Messoud Efendiev and Wolfgang Wendland

Contents

Part III. Solid Mechanics

Part IV. Porous Media and Fluid – Structure Interaction

Appendix

List of Coeditors

Wolfgang Ehlers
Institute of Applied Mechanics (CE), Chair II,
University of Stuttgart, Germany

Lothar Gaul
Institut A für Mechanik,
Universität Stuttgart, Germany

Klaus Kirchgässner
Institut für Analysis, Dynamik und Modellierung,
Universität Stuttgart,
Pfaffenwaldring 57,
70569 Stuttgart, Germany

Christian Miehe
University of Stuttgart,
Institute of Applied Mechanics (Civil Engineering),
Pfaffenwaldring 7,
70569 Stuttgart, Germany

Alexander Mielke
Institut für Analysis, Dynamik und Modellierung,
Universität Stuttgart,
Pfaffenwaldring 57,
70569 Stuttgart, Germany

Ekkehard Ramm
Institute of Structural Mechanics,
University of Stuttgart,
Pfaffenwaldring 7,
70550 Stuttgart, Germany

Werner Schiehlen
Institute B of Mechanics,
University of Stuttgart,
Pfaffenwaldring 9,
70550 Stuttgart, Germany

List of Contributors

Ayhan Acartürk
Institute of Applied Mechanics (CE)
University of Stuttgart,
Pfaffenwaldring 7,
70569 Stuttgart, Germany

Marcel Arndt
Universität Bonn, Institut für
Angewandte Mathematik,
Wegelerstr. 6,
53115 Bonn, Germany

M. Bargieł
Institute Of Computer Science,
University of Mining and
Metallurgy, Cracow, Poland

Claus G. Bayreuther
University of Stuttgart,
Institute of Applied Mechanics
(Civil Engineering),
Pfaffenwaldring 7,
70569 Stuttgart, Germany

Petr Belov
Institute of
Applied Mechanics of Russian
Academy of Science,
Leninskij prospect, 32-A,
Moscow 117334, Russia

Stefan Berres
Institute of Mathematics A,
University of Stuttgart,
Stuttgart, Germany

Andreas Bielinski
Stuttgart University, Stuttgart,
Germany

Raimund Bürger
Institute of Mathematics A,
University of Stuttgart, Stuttgart,
Germany

Holger Class
Stuttgart University, Stuttgart,
Germany

Cristian A. Coclici
Department of Mathematics,
University of Kaiserslautern

David Degani
Faculty of Mechanical Engineering,
Technion – Israel Institute of
Technology, Haifa 32000, Israel

Atanas Dimitrov
Institute of Solid Mechanics,
Karlsruhe University,
Kaiserstr. 12, D-76128 Karlsruhe,
Germany

Slav Dimitrov
Institute of Solid Mechanics,
University of Karlsruhe,
Kaiserstrasse 12,
D-76128 Karlsruhe, Germany

Dieter Dinkler
Technische Universität
Braunschweig, Institut für Statik,
Beethovenstr. 51,
38106 Braunschweig, Germany

Donald A. Drew
Center for Multiphase Research
Rensselaer Polytechnic Institute
Troy, NY 12180-3590 USA

Tom Dyakowski
Department of Chemical
Engineering, University of
Manchester
Institute of Science and Technology,
P.O. Box 88, Manchester M60 1QD,
United Kingdom

Peter Eberhard
Institute of Applied Mechanics,
University of Erlangen
Egerlandstr. 5,
D-91058 Erlangen, Germany

Christof Eck
Institute of Applied Mathematics,
University Erlangen-Nürnberg

Messoud A. Efendiev
Mathematisches Institut A,
Universität Stuttgart,
Pfaffenwaldring 57,
70569 Stuttgart, Germany

Wolfgang Ehlers
Institute of Applied Mechanics
(CE), Chair II,
University of Stuttgart, Germany

Owen Eslinger
The Center for Subsurface
Modeling,
Texas Institute for Computational
and Applied Mathematics,
The University of Texas at Austin,
1 University Station C0200,
Austin TX 78712, U.S.A.

Matthias Fischer
Institut A für Mechanik, Universität
Stuttgart, Germany

R.A. Ford
Spielo Gaming International,
Moncton, NB, Canada

Lothar Gaul
Institut A für Mechanik, Universität
Stuttgart, Germany

Klaus Hackl
Fakultät für Bauingenieurwesen,
Lehrstuhl für Allgemeine Mechanik,
Ruhr-Universität Bochum,
Universitätsstraße 150, IA 3/125,
D-44801 Bochum, Germany

Rainer Helmig
Stuttgart University, Stuttgart,
Germany

Bin Hu
Institute B of Mechanics, University
of Stuttgart,
Pfaffenwaldring 9, D-70550
Stuttgart, Germany

Björn Hübner
Technische Universität
Braunschweig, Institut für Statik,
Beethovenstr. 51, 38106
Braunschweig, Germany

Decan J. Ivanovic
Faculty of mechanical Engineering,
University of Montenegro, 81000
Podgorica, Yugoslavia

Hartmut Jakobs
Stuttgart University, Stuttgart,
Germany

Jiří Jarušek
Mathematical Institute,
Academy of Sciences of the Czech
Republic

Kenneth H. Karlsen
Department of Mathematics,
University of Bergen,
Bergen, Norway

D. Kedzior
Cracow University of Technology,
Kraków, Poland

Dorothee Knees
Mathematisches Institut A,
Universität Stuttgart, Germany

Victor A. Kovtunenko
Lavrent'ev Institute of Hydrodynamics, 630090 Novosibirsk, Russia

Wanwilai Kraipech
Department of Chemical Engineering, Srinakharinwirot University, Rangsit-Nakhonnayok Rd, Klong 16, Ongharak, Nakhonnayok, 26120, Thailand

Matthias Lambrecht
University of Stuttgart, Institute of Applied Mechanics (Civil Engineering), Pfaffenwaldring 7, 70569 Stuttgart, Germany.

Tom-Alexander Langhoff
Institute of Solid Mechanics Karlsruhe University, Kaiserstrasse 12,
D-76128 Karlsruhe, Germany

Yuli Lifshitz
Faculty of Mechanical Engineering, Technion – Israel Institute of Technology, Haifa 32000, Israel

Sergey Lurie
Computer Center of Russian Academy of Science,
Vavilova str., 40, GSP-1, Moscow 119991, Russia

Bernd Markert
Institute of Applied Mechanics (CE) University of Stuttgart, Pfaffenwaldring 7,
70569 Stuttgart, Germany

Hermann G. Matthies
Institute of Scientific Computing, Technische Universität Braunschweig, D-38092 Braunschweig, Germany

Christian Miehe
University of Stuttgart, Institute of Applied Mechanics (Civil Engineering), Pfaffenwaldring 7, 70569 Stuttgart, Germany

Alexander Mielke
Institut für Analysis, Dynamik und Modellierung, Universität Stuttgart, Pfaffenwaldring 57, D-70569 Stuttgart, Germany

Alain Miranville
Université de Poitiers, Laboratoire d'Applications des Mathématiques, SP2MI, Téléport 2, Avenue Marie et Pierre Curie, 86962 Chasseneuil Futuroscope Cedex, France

Dirk Mittenhuber
Mathematisches Institut A, 1. Lehrstuhl, Universität Stuttgart, Pfaffenwaldring 57, D-70569 Stuttgart, Germany

Daniel P. Mok
University of Stuttgart, Institute of Structural Mechanics, Pfaffenwaldring 7, D-70550 Stuttgart, Germany

Claus-Dieter Munz
Stuttgart University, Institute for Aero- and Gasdynamics, Pfaffenwaldring 21, 70550 Stuttgart, Germany

Andrzej F. Nowakowski
Department of Chemical Engineering, University of Manchester
Institute of Science and Technology, P.O. Box 88, Manchester M60 1QD, United Kingdom

Günther Of
Institut für Angewandte Analysis und Numerische Simulation,
Universität Stuttgart,
Pfaffenwaldring 57, 70569 Stuttgart, Germany

Małgorzata Peszyńska
University of Texas at Austin,
Austin, TX 78712

Romana Piat
Institute of Solid Mechanics
University of Karlsruhe, Kaiserstr. 12,
D-76128 Karlsruhe, Germany

Iuliu Sorin Pop
Department of Mathematics and Computer Science,
Technical University Eindhoven, P. O. Box 513, 5600 MB Eindhoven,
The Netherlands

Dobromil Pryl
Institute of Applied Mechanics,
Technical University
Braunschweig, D-38023
Braunschweig, P.O. Box 3329,
Germany

Ekkehard Ramm
University of Stuttgart, Institute of Structural Mechanics,
Pfaffenwaldring 7, D-70550
Stuttgart, Germany

Boško Rašuo
Aeronautical Department, Faculty of Mechanical
Engineering, University of Belgrade,
27 Marta 80, 11 000
Belgrade, Yugoslavia

Béatrice Rivière
Department of Mathematics,
301 Thackeray,
University of Pittsburgh,
Pittsburgh, PA 15260, U.S.A.

Tomáš Roubíček
Mathematical Institute, Charles University, Sokolovská 83,
CZ-186 75 Praha 8, Czech Republic,
Institute of Information Theory and Automation, Academy of Sciences,
Pod vodárenskou věží 4, CZ-182 08
Praha 8, Czech Republic.

Victor S. Ryaben'kii
Keldysh Institute of Applied Mathematics, RAS, Moscow

Viktor N. Saljnikov
Faculty of Mechanical Engineering,
University of Belgrade, 11000
Beograd, Yugoslavia

Martin Schanz
Institute of Applied Mechanics,
Technical University
Braunschweig, D-38023
Braunschweig, P.O. Box 3329,
Germany

Werner Schiehlen
Institute B of Mechanics, University of Stuttgart,
Pfaffenwaldring 9, D-70550
Stuttgart, Germany

Harmen Schippers
National Aerospace Laboratory
NLR, P.O. Box 153, 8300 AD
Emmeloord,
The Netherlands

Eckart Schnack
Institute of Solid Mechanics,
University of Karlsruhe,
Kaiserstrasse 12,
D-76128 Karlsruhe, Germany

Bernd Scholz
Institute of Applied Mechanics (CE), Chair II,
University of Stuttgart, Germany

Jan Schotte
University of Stuttgart,
Institute of Applied Mechanics (Civil Engineering),
Pfaffenwaldring 7, 70569 Stuttgart, Germany.

Thomas Schwartzkopff
Stuttgart University, Institute for Aero- and Gasdynamics,
Pfaffenwaldring 21, 70550 Stuttgart, Germany

Ivan L. Sofronov
Keldysh Institute of Applied Mathematics, RAS, Moscow

Olaf Steinbach
Institut für Angewandte Analysis und Numerische Simulation,
Universität Stuttgart,
Pfaffenwaldring 57, 70569 Stuttgart, Germany

Jan Steindorf
Institute of Scientific Computing,
Technische Universität Braunschweig, D-38092 Braunschweig, Germany

Shuyu Sun
The Center for Subsurface Modeling,
Texas Institute for Computational and Applied Mathematics,
The University of Texas at Austin,
1 University Station C0200,
Austin TX 78712, U.S.A.

G. Szefer
Cracow University of Technology,
Kraków, Poland

Hans van Tongeren
National Aerospace Laboratory
NLR, P.O. Box 153, 8300 AD Emmeloord,
The Netherlands

Elmer M. Tory
Professor Emeritus, Department of Mathematics and Computer Science, Mount Allison University,
Sackville, NB, E4L 1E8, Canada

Dmitrij B. Volkov-Bogorodsky
Institute of
Applied Mechanics of Russian Academy of Science, Leninskij prospect, 32-A, Moscow 117334,
Russia

Guus Vos
National Aerospace Laboratory
NLR, P.O. Box 153, 8300 AD Emmeloord,
The Netherlands

Elmar Walhorn
Technische Universität Braunschweig, Institut für Statik,
Beethovenstr. 51, 38106 Braunschweig, Germany

Wolfgang A. Wall
University of Stuttgart, Institute of Structural Mechanics,
Pfaffenwaldring 7, D-70550 Stuttgart, Germany

Wolfgang L. Wendland
Institut für Angewandte Analysis und Numerische Simulation,
Universität Stuttgart,
Pfaffenwaldring 57, 70569 Stuttgart, Germany

Mary F. Wheeler
The Center for Subsurface Modeling,
Texas Institute for Computational and Applied Mathematics,
The University of Texas at Austin,
1 University Station C0200,
Austin TX 78712, U.S.A.

Christian Wieners
Universität Erlangen-Nürnberg,
Institut für Angewandte Mathematik,
Martensstr. 3, D-91058 Erlangen

Ch. Zhang
Hochschule Zittau/Görlitz,
University of Applied Sciences,
02763 Zittau, Germany

Keynote Speakers

A Turbulent Dispersion Model for Particles or Bubbles

Donald A. Drew

Center for Multiphase Research Rensselaer Polytechnic Institute Troy, NY 12180-3590 USA

1 Introduction

The purpose of this paper is to discuss the multiscale aspects of two-phase flow, with the ultimate object being a mechanistically-based model for particle and bubble transport by turbulence. The model is the "standard" two-fluid model for two-phase flow that is derived using ensemble averaging, and consequently that model describes macroscale motions for two-phase flows. The microscale origin of the turbulent dispersion force will be highlighted, and a model for this force derived. Results of application of this model to sediment transport will be shown.

2 Multiphase Balance Equations

The equations of motion can be derived by ensemble averaging. By considering the averaging process, we can also use the method to obtain constitutive equations. This method is described in Drew and Passman (1999).

3 The Microstructure Equations

The exact equations of motion, valid inside each material are

Mass

$$\frac{\partial \rho}{\partial t} + \nabla \bullet \rho \mathbf{v} = 0 \tag{1}$$

Momentum

$$\frac{\partial \rho \mathbf{v}}{\partial t} + \nabla \bullet \rho \mathbf{v}\mathbf{v} = \nabla \bullet \mathbf{T} + \rho \mathbf{g} \tag{2}$$

where ρ is the density, $\mathbf{v}$ is the velocity, $\mathbf{T}$ is the stress tensor, and $\mathbf{g}$ is gravity. We assume that the materials are nonpolar, so that $\mathbf{T} = \mathbf{T}^T$.

At an interface between components, properties are discontinuous, although mass and momentum must be conserved. The jump conditions valid across the interface are

Mass Jump

$$\mathbf{v} \bullet \mathbf{n} = \mathbf{v}_i \bullet \mathbf{n} \tag{3}$$

Momentum Jump

$$-[\mathbf{T}] \bullet \mathbf{n} = \nabla_S \sigma + \sigma \kappa \mathbf{n} \tag{4}$$

where $[\cdot]$ denotes the jump across the interface, $\mathbf{n}$ is the unit normal, σ is the surface tension, and κ is the curvature.

4 Ensemble Averaging

We wish to predict averages over ensembles of flows. The ensemble average is a generalization of adding the values of the variable for each realization, and dividing by the number of observations. We shall refer to a "process" as the set of possible flows that could occur in a device. We refer to a "realization" of the flow as a possible motion that could have happened. Generally, we expect an infinite number of realizations of the flow, consisting of variations of position and velocities of the discrete units and the fluid between them.

The ensemble average allows the interpretation of the phenomena in terms of the *repeatability* of multiphase flows. Any particular exact experiment or realization will not be repeatable, because the individual particles or bubbles will not be in exactly the same place. However, any repetition of the experiment will lead to another member of the ensemble.

Consider the flow of spheres and fluid in a device. This situation is represented schematically in Figure 4.

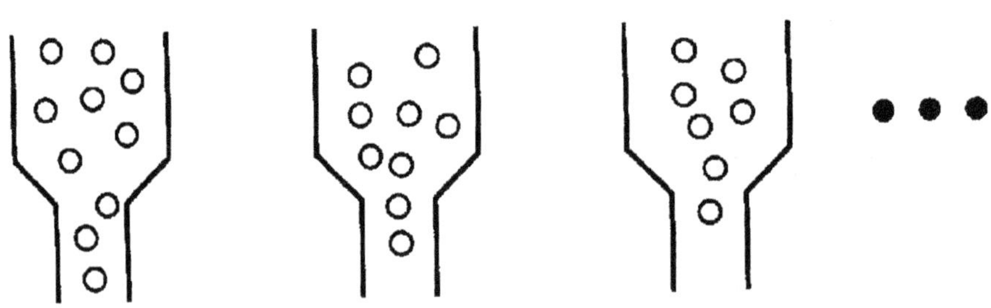

Fig. 1. Realizations of sphere flow in a device

Using the ensemble average as the basis for the theory of multiphase flow has several interesting implications. First, we note that in situations where time and/or volume averaging are appropriate, these averages can be used as "samples" of the ensemble. Data acquisition based on time averaging is much easier than that implied by ensemble averaging. Second, ensemble averaging does not require that a control volume contain a large number of particles in any given realization.

An *ensemble* is a set of motions possible in a system. A realization consists of the values of the velocity, the stress, the density, and the position of the interfaces for the duration of the motion, over all spatial points that are occupied by the material during its motion. Suppose f is one of these fields. We denote the realization by μ, and denote the dependence of the field on the realization by $f(\mathbf{x}, t; \mu)$. We refer to the set of all realizations μ, denoted by $\mathcal{E}$, as the *ensemble*. If the measure is given by a density $dm(\mu)$, we define $\overline{f}$, the *ensemble average* of , by

$$\overline{f}(\mathbf{x}, t) = \int_{\mathcal{E}} f(\mathbf{x}, t; \mu)\, dm(\mu) \tag{5}$$

Phase Indicator Function

It is often desirable to isolate each component mathematically in the microscale description. To do this, we introduce the *phase indicator function*, for phase k

$$X_k(\mathbf{x}, t; \mu) = \begin{cases} 0, & x \notin \text{phase } k \\ 1, & x \in \text{phase } k \end{cases} \tag{6}$$

This function "picks out" phase k, and ignores all other phases and interfaces. This function is used repeatedly in the description of multiphase materials.

Gauss and Leibniz Rules

Applying the ensemble average to the product rule for differentiation gives the *Gauss Rule*

$$\overline{X_k \nabla f} = \nabla \left(\overline{X_k f}\right) - \overline{f_{ki} \nabla X_k} \tag{7}$$

and the *Leibniz Rule*

$$\overline{X_k \frac{\partial f}{\partial t}} = \frac{\partial \overline{X_k f}}{\partial t} - \overline{f_{ki} \frac{\partial X_k}{\partial t}} \tag{8}$$

5 Averaged Equations

Averages of balance equations are obtained by taking the product of the balance equations with X_k, then performing the averaging process. If we neglect phase change, the averaged equations are

Mass

$$\frac{\partial \overline{X_k \rho}}{\partial t} + \nabla \bullet \overline{X_k \rho \mathbf{v}} = 0 \tag{9}$$

Momentum

$$\frac{\partial \overline{X_k \rho \mathbf{v}}}{\partial t} + \nabla \bullet \overline{X_k \rho \mathbf{v}\mathbf{v}} = \nabla \bullet \overline{X_k \mathbf{T}} + \overline{X_k \rho \mathbf{g}} - \overline{\mathbf{T}_{ki} \bullet \nabla X_k} \tag{10}$$

Definition of Average Variables

The exact, or microscopic, situation is defined in terms of the phase indicator function X_k. The average of X_k is called the volume fraction of phase k, and denoted by $\alpha_k = \overline{X_k}$. The density of phase k is defined by $\alpha_k \rho_k = \overline{X_k \rho}$, and the velocity is defined by $\alpha_k \rho_k \mathbf{v}_k = \overline{X_k \rho \mathbf{v}}$. The averaged stress is $\alpha_k \mathbf{T}_k = \overline{X_k \mathbf{T}}$. The stress on the interface acts as a source of momentum. The interfacial momentum source is defined by $\mathbf{M}_k = -\overline{\mathbf{T} \bullet \nabla X_k}$. The effect of velocity fluctuations is accounted for by introducing the fluctuation from the mean. Thus, if

$$\mathbf{v}'_k = \mathbf{v} - \mathbf{v}_k$$

the Reynolds stress is defined by

$$\alpha_k \mathbf{T}_k^{\mathrm{Re}} = -\overline{X_k \rho \mathbf{v}'_k \mathbf{v}'_k}\,. \tag{11}$$

Averaged Equations

The averaged equations governing each phase are

Mass

$$\frac{\partial \alpha_k \rho_k}{\partial t} + \nabla \bullet \alpha_k \rho_k \mathbf{v}_k = 0 \tag{12}$$

Momentum

$$\frac{\partial \alpha_k \rho_k \mathbf{v}_k}{\partial t} + \nabla \bullet \alpha_k \rho_k \mathbf{v}_k \mathbf{v}_k = \nabla \bullet \alpha_k \left(\mathbf{T}_k + \mathbf{T}_k^{\mathrm{Re}} \right) + \alpha_k \rho_k \mathbf{g} + M_k \tag{13}$$

The jump conditions can also be derived by averaging. This process yields:

$$\mathbf{M}_1 + \mathbf{M}_2 = \mathbf{m} \tag{14}$$

where $\mathbf{m}$ is the resultant from surface tension.

6 Constitutive Equations

The equations of balance of mass and momentum for each phase form an underdetermined model for multiphase flow. These equations must be supplemented by constitutive equations specifying the way the materials behave.

Microstructure Considerations

Unlike gases, multiphase flows are characterized by easily observable microstructure. In this section, we discuss some of the connections between microstructure and the averaged equations. We focus on the motions of spheres in a fluid. The velocity field depends on the locations of all the particles, denoted so that $\mathbf{v}(\mathbf{x}, t \,|\, \mathbf{z}_1, \mathbf{z}_2, \ldots)$ is the velocity at spatial point $\mathbf{x}$, at time t, given that the centers of the spheres are at $\mathbf{z}_1, \mathbf{z}_2, \ldots$. If $f(z_1, z_2, \ldots)$ is the probability distribution of the locations of the centers of the spheres, then the ensemble average liquid velocity is

$$\overline{\mathbf{v}}(\mathbf{x}, t) = \int_{Z_1} \int_{Z_2} \cdots \int f(\mathbf{z}_1, \mathbf{z}_2, \ldots) \mathbf{v}(\mathbf{x}, t \,|\, \mathbf{z}_1 \mathbf{z}_2, \ldots) \, d\mathbf{z}_1 d\mathbf{z}_2 \cdots . \tag{15}$$

Solutions for $\mathbf{v}(\mathbf{x}, t \,|\, \mathbf{z}_1 \mathbf{z}_2, \ldots)$ are not readily obtained. However, for special situations, approximate solutions are available. Drew and Passman (1999) give a thorough discussion.

For our purposes here, we shall re-derive the balance equations for a dispersed phase consisting of spheres of a given size and mass, starting from a Boltzmann equation. Comparison of the balance equations for mass and momentum with the two-fluid momentum equation results in a form for the interfacial force that is related to the random motions in the liquid. Let $f(\mathbf{v}, \mathbf{x}, t \,|\, \mathbf{u})$ be the number density of identical spheres in phase space, given the liquid velocity field $\mathbf{u}$. Neglecting the effects of collisions, the Boltzmann equation for f is

$$\frac{\partial f}{\partial t} + \nabla \mathbf{v} f + \frac{\partial}{\partial v} \mathbf{a} f = 0 .$$

Within the framework of this model, the ensemble average of a dispersed phase quantity becomes

$$n \overline{g}_d = \int_{\mathbf{u}} \int_{\mathbf{v}} f(\mathbf{v}, \mathbf{x}, t \,|\, \mathbf{v}) g(\mathbf{u}, \mathbf{v}) \, d\mathbf{v} d\mathbf{u}$$

where n is the number density.

If we multiply the Boltzmann equation by and average, we have

$$\frac{\partial n \overline{\mathbf{v}}_d}{\partial t} + \nabla n \overline{\mathbf{v}}_d \overline{\mathbf{v}}_d = \nabla n \mathbf{T} + \overline{\mathbf{a} f}$$

where the stress $\mathbf{T}$ is defined as $\mathbf{T} = -\overline{\mathbf{v}' \mathbf{v}' f}$. This is a version of the momentum equation. In general, the acceleration of the particle or bubble is given as a function of the velocities, $\mathbf{a} = \mathbf{a}(\mathbf{v}, \mathbf{u})$. If we write $\mathbf{a}(\mathbf{v}, \mathbf{u}) = \mathbf{a}\left(\overline{\mathbf{v}}, \overline{\mathbf{u}}\right) + \mathbf{a}'$, then the momentum equation can be written as

$$\frac{\partial n \overline{\mathbf{v}}_d}{\partial t} + \nabla n \overline{\mathbf{v}}_d \overline{\mathbf{v}}_d = \nabla n \mathbf{T} + n \mathbf{a}\left(\overline{\mathbf{v}}, \overline{\mathbf{u}}\right) + \mathbf{M}_d^{TD}$$

where

$$\mathbf{M}_d^{TD} = \overline{\mathbf{a}' f}$$

is the turbulent dispersion force, which is the resultant (i.e., ensemble average) of the fluctuating forces on the bubble or particle.

If we make specific assumptions about the particle or bubble dynamics, and the nature of the turbulence (as it relates to the dynamics of the dispersed phase), we can derive forms for the constitutive equations for the turbulent dispersion and the particle Reynolds stress. If the carrier fluid velocity field $\mathbf{u}(\mathbf{x}, t)$ is given, the trajectory of any particle or bubble can be obtained by integrating the equation of motion

$$\frac{d\mathbf{v}}{dt} = \mathbf{a}(\mathbf{vu})$$

to obtain a particle trajectory $\mathbf{z} = \zeta\left(t \,\middle|\, \mathbf{v}_0, \mathbf{v}_0, t_0\right)$, where $\mathbf{v}_0, \mathbf{x}_0, t_0$ denote the velocity and position at time t_0. Thus, the probability that the particle is at $\mathbf{x}$ at time t is related to the probability that it was at $\mathbf{x}_0$ at an earlier time t_0 with just the right velocity to follow the necessary trajectory. If that earlier time is sufficiently far in the past, then

$$f(\mathbf{v}, \mathbf{x}, t \,|\, \mathbf{u}) = n(\mathbf{x}_0, t_0)\delta(\mathbf{v}_0 - \mathbf{v}_0^*)$$

where $\mathbf{v}_0^*$ is the velocity necessary to follow the correct trajectory. Expanding $n(\mathbf{x}_0, t_0)$ in a Taylor series, we have $n(\mathbf{x}_0, t_0) \cong n(\mathbf{x}, t) + (\mathbf{x} - \mathbf{x}_0)\nabla n(\mathbf{x}, t) + \cdots$ where we neglect the terms denoted by $\cdots$. Substituting, and noting that the first term is zero gives

$$\mathbf{M}_d^{TD} = \overline{\mathbf{a}'(\mathbf{x} - \mathbf{x}_0)}\nabla n = -\mathbf{D}^{TD}\nabla n$$

where $\mathbf{D}^{TD}$ is the dispersion tensor, and is the correlation between fluctuating acceleration and the (vector) displacement of the particle or bubble.

Next, we assume that the particle momentum equation is

$$\frac{d\mathbf{v}}{dt} = \frac{1}{\tau_p}(\mathbf{u} - \mathbf{v}) + \mathbf{a}''$$

where $\mathbf{a}'' = 0$ for particles, and $\mathbf{a}''$ contains the effects of lift, virtual mass, and liquid pressure gradient for bubbles.

In order to solve this equation for a reasonable turbulence model, we assume that the turbulence consists of "eddies" that have a constant (but random) velocity, and that the residence time in any eddy is random. For bubbles, we also assume that $\mathbf{a}''$ is constant in each eddy. This allows the determination of the trajectories

$$\mathbf{z} = -\tau_p\left(\mathbf{v}_0 - (\mathbf{u} + \tau_p\mathbf{a}'')\right)e^{-t/\tau_p} + (\mathbf{u} + \tau_p\mathbf{a}'')\,t + \mathbf{x}_0$$

From this, we can calculate the dispersion tensor, which is given by Drew [5]

$$\mathbf{D}^{TD} = \overline{\mathbf{v}'\mathbf{v}'}\left[\frac{\tau_f}{\tau_p} + 2\,e^{-2\tau_f/\tau_p}\,\frac{1 - e^{-\tau_f/\tau_p}}{1 - e^{-2\tau_f/\tau_p}}\right]. \tag{16}$$

7 Relation to Diffusion Model

It is important to note that the flux of particle mass is given by $\alpha_s \rho_s \mathbf{v}_s$. That is, there is no explicit model for particle "diffusion" in the particle mass equation. Particle diffusion can be obtained from this model by neglecting inertia and turbulence terms in the particle and fluid momentum equations, so that

$$\begin{aligned} 0 &= -(1-\alpha_s)\nabla p - S\left(\mathbf{v}_f - \mathbf{v}_s\right) + \rho_f C_{TD} k_f \nabla \alpha_d - (1-\alpha_s)\rho_f \mathbf{g} \\ 0 &= -\alpha_s \nabla p + S(\mathbf{v}_f - \mathbf{v}_s) - \rho_f C_{TD} k_f \nabla \alpha_d - \alpha_s \rho_s \mathbf{g} \,. \end{aligned}$$

Eliminating the pressure gradient and solving for the particle velocity field $\mathbf{u}$ gives

$$\mathbf{v}_s = \mathbf{v}_f - \frac{\rho_f C_{TD} k_f}{S} \nabla \alpha_s - \frac{\alpha_s(\rho_s - \rho_f)\mathbf{g}}{S} = \mathbf{v}_T - \frac{\rho_f C_{TD} k_f}{S} \nabla \alpha_s \,.$$

Note that the first and last terms on the right hand side give the settling velocity of the particles in a still fluid. (The buoyant force appears in the particle forces through the pressure gradient.) The middle term is responsible for the particle diffusion. Substituting in the particle mass equation results in

$$\frac{\partial \alpha_s \rho_s}{\partial t} + \nabla \cdot \alpha_s \rho_s \mathbf{v}_T = \nabla \cdot \left(\alpha_s \frac{\rho_f C_{TD} k_f}{S} \nabla \alpha_s \right) .$$

Clearly the quantity $D_p = \alpha_s \frac{\rho_f C_{TD} k_f}{S \rho_s}$ plays the role of particle diffusivity in this approximation to the full model.

This approximation is valid when there is no inertia in the flow and when turbulence effects are negligible. There are many flows where this is not the case. Specifically, the spatial effects due to flow spreading and deceleration when the waterway widens or narrows will not be adequately modeled by the diffusion model.

8 Sediment Transport

Note that $\mathbf{M}_k$ contains the drag force and other forces on the particles due to the motion of the particles through the fluid. A model for $\mathbf{M}_k$ is

$$\mathbf{M}_s = -\mathbf{M}_f = S(\mathbf{v}_f - \mathbf{v}_s) + \mathbf{M}_s^{TD}$$

where $\mathbf{M}_s^{TD}$ is the force causing turbulent dispersion. This particular model neglects lift and virtual mass effects, but they could easily be included in the model.

The Reynolds stress for the fluid is modeled as

$$\mathbf{T}_f^T = -\frac{2}{3}\rho_f k_f \mathbf{I} + \mu_f^{\mathrm{Re}}\left(\nabla \mathbf{v}_f + \mathbf{v}_f \nabla\right)$$

where the eddy viscosity for the fluid is given by

$$\mu_f^T = C_\mu \rho_f \frac{k_f^2}{\varepsilon_f}$$

where k_f is the turbulent kinetic energy, and ε_f is the dissipation.

We assume that the fluid turbulence drives the particle turbulence. This implies that

$$\mathbf{T}_s^{\mathrm{Re}} = C_r \frac{\rho_s}{\rho_f} \mathbf{T}_f^{\mathrm{Re}} .$$

For the turbulent dispersion model, we study two similar, but different, models. Drew [5] assumes that

$$\mathbf{M}_s^{TD} = C_{TD}^D \frac{\rho_f}{\rho_s} \mathbf{T}_s^{\mathrm{Re}} \bullet \nabla \alpha_s$$

where the coefficient C_{TD}^D depends on the Stokes number, $St = \vartheta_s / \vartheta_f$ where ϑ_s and ϑ_f are time scales for the particles and the fluid. Carrica, et al. [3] assume that

$$\mathbf{M}_s^{TD} = C_{TD}^D \frac{\rho_f}{\rho_s} \mathbf{T}_s^{\mathrm{Re}} \bullet \nabla \alpha_s .$$

Here $\nu_f^{\mathrm{Re}} = \mu_f^{\mathrm{Re}} / \rho_f$ is the eddy viscosity, and $Sc = (C_{TD}^C)^{-1}$ is the Schmidt number. Both k_f and ε_f satisfy conservation equations of the form derived by Rodi [9], and modified for two-phase flows by Alajbegovic, et al. [1].

9 Fully Developed Flow

We assume $\alpha_s = \alpha_s(y)$, $k_f = k_f(y)$, $\varepsilon_f = \varepsilon_f(y)$, $\mathbf{v}_k = u_k(y)\mathbf{e}_x$, and $p_f = p_f(y)$. We further assume that the volumetric concentration of sediment is small, and that the eddy viscosity term is smaller (in magnitude) than the turbulent kinetic energy. With these assumptions, the equations become

$$\begin{aligned}
u_s &= u_f + \frac{\rho_s g \sin\theta}{S} \\
\frac{du_f}{dy} &= \frac{g \sin\theta (H - y)}{C_\mu} \frac{\varepsilon_f}{k_f^2} \\
0 &= -\alpha_s \left(C_r \rho_s - \rho_f \right) \frac{2}{3} \frac{dk_f}{dy} - (1 + C_{TD}) C_r \rho_s \frac{2}{3} k_f \frac{d\alpha_s}{dy} \\
&\quad - \alpha_s (\rho_s - \rho_f) g \cos\theta \\
0 &= \frac{d}{dy} \left(\frac{C_\mu}{\sigma_k} \frac{k_f^2}{\varepsilon_f} \frac{dk_f}{dy} \right) + C_\mu \frac{k_f^2}{\varepsilon_f} \left(\frac{du_f}{dy} \right)^2 - \varepsilon_f \\
0 &= \frac{d}{dy} \left(\frac{C_\mu}{\sigma_s} \frac{k_f^2}{\varepsilon_f} \frac{d\varepsilon_f}{dy} \right) + C_1 C_\mu k_f \left(\frac{du_f}{dy} \right)^2 - C_2 \frac{\varepsilon_f^2}{k_f^2}
\end{aligned}$$

where θ is the angle that the bed makes from horizontal.

Finding the volume fraction consists of solving the third equation, with an appropriate boundary condition. We use the data to determine the concentration scale. The solution can be written symbolically as

$$\alpha_s = A \cdot \exp\left(\int B(y)\, dy\right)$$

where A is a constant of integration and

$$B(y) = -\frac{(C_r\rho_s - \rho_f)}{(1 + C_{TD})C_r\rho_s} \frac{1}{k_f} \frac{dk_f}{dy} - \frac{3\,(\rho_s - \rho_f)\, g \cos\theta}{2\,(1 + C_{TD})\, C_r \rho_s K_f}$$

In order to get concentration profiles, we chose the parameter A to fit the numerical solution to the data of Einstein and Chien [6]. We used both least squares (which minimizes the error in the α–direction), and a method that minimized distance from the curve to the data points Aquino [2]. Better fit was obtained from the latter method. These curves are shown in Figure 9. Also shown are curves with the values of C_{TD} chosen to fit the data.

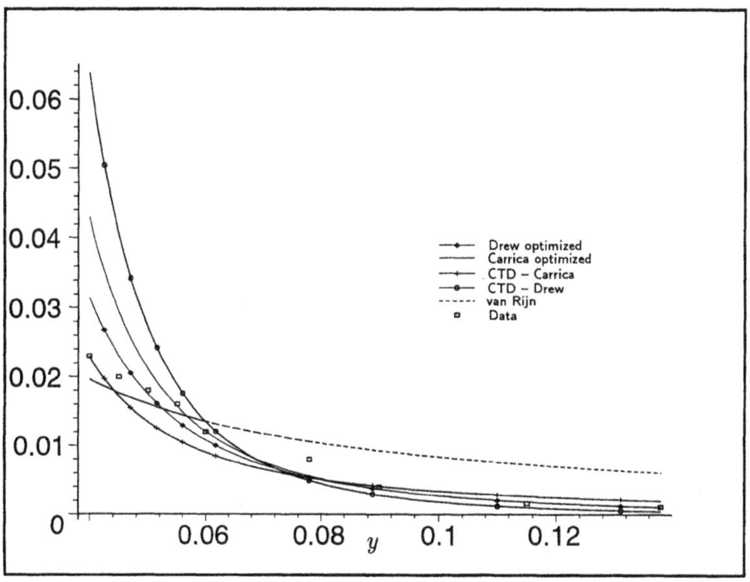

Fig. 2. Comparisons of the theoretical predictions of concentration with that of van Rijn [10] and with data from Einstein and Chien [6]

Conclusion Multiscale dynamics for particle-fluid flows can be described by ensemble averaged equations for mass and momentum balance for each phase. Turbulent dispersion is modeled by considering the microscale flow as the particles responding to constant velocity eddies. The model is capable of predicting the volumetric concentration of suspended sediment at least as well as the theory of van Rijn [10]. Since it can be used in situations with other particle forces, it appears to be superior. The uncertainty in the value of C_{TD} reflects the two aspects of the physics that are not addressed in the models that have led to the theory presented here; namely, that the statistical correlations between particle and fluid turbulence are not known, and that the turbulence near the wall is nonisotropic.

References

1. Alajbegovic, A., Drew. D.A. and Lahey, R.T., Jr. (1999) An Analysis of Phase Distribution and Turbulence in Dispersed Particle/Liquid Flows, Chemical Engineering Communications, Vol-174, 85–133
2. Aquino, L.J. (2001) Modeling Sediment Transport Using a Two-Phase Flow Approach, Ph. D. Thesis, Rensselaer Polytechnic Institute, Troy, NY
3. Carrica, P.M., Drew, D.A., Lahey, R.T., Jr. (1999) A Polydisperse Model for Bubbly Two-Phase Flow Around a Surface Ship, Int. J. of Multiphase Flow, Vol. 25
4. Drew, D.A. and Passman, S.L. (1999) Theory of Multicomponent Flow, Springer Verlag, NY
5. Drew, D.A. (2001) A dispersion model for particles and bubbles, J. Engr. Math. 41, 2/3: 259–274
6. Einstein, H.A. and Chien, N. (1955) Effects of Heavy Sediment Concentration Near the Bed on Velocity and Sediment Distribution, MRD Series 8: 78, Univ. of Calif., Inst. of Eng. and U.S. Army Div., M. River, Corps of Eng., Omaha, Nebraska
7. EPA Report, Phase 2 Report, Further Site Characterization and Analysis, Volume 2D-Revised Baseline Modeling Report, Hudson River PCBs Reassessment RI/FS, Fate and Transport Models, January 2000
8. Lahey, R.T., Jr. and Drew, D.A. (2000) An analysis of two-phase flow and heat transfer using a multidimensional, multifield, two-fluid computational (CFD) model, Proceedings Japan/US Seminar on Two-Phase Flow Dynamics, Santa Barbara, CA June 5–8, 2000
9. Rodi, W., (1984) IAHR/AIRH Monograph
10. van Rijn, L.C. (1983) Sediment Transport, Part I: Bed Load Transport, Journal of Hydraulic Engineering, Vol 110, No. 10, 1983a., Sediment Transport, Part II: Suspended Load Transport, Journal of Hydraulic Engineering, Vol 110, No. 11, 1983b., Sediment Transport, Part III: Bed Forms and Alluvial Roughness, Journal of Hydraulic Engineering, Vol 110, No. 11, 1983c.

Strong Coupling Methods

Hermann G. Matthies and Jan Steindorf

Institute of Scientific Computing, Technische Universität Braunschweig,
D-38092 Braunschweig, Germany

Abstract. We consider the procedure of computing the response of a coupled problem with a partitioned approach. Often we have existing procedures or even software to solve each sub-problem separately, and want to couple both. This setting seems to allow only the so-called weak coupling which is not sufficient for some problems. The so-called strong coupling – a totally implicit formulation – requires iteration in each time step. With the partitioned approach, one simple computational procedure is similar to a block-Gauss-Seidel iteration. We show why this approach may experience difficulties, and how they may be circumvented with block-Newton methods, still in the partitioned framework, by only using the solvers for the two sub-problems. We supply some examples from fluid-structure coupling.

1 Introduction

Our interest in this field started from fluid-structure interaction (FSI) problems [1]. These often show strong interplay between the fluid and the structure [1–7], e.g. in the design of aircraft [8,9] and in many other situations [10].

But apart from FSI-problems, the algorithms may often be used in many other situations where coupled problems arise [11], e.g. thermo-mechanical [12] coupling, soil-pore fluid interaction [13], to name but a few.

There are different ways to tackle a coupled interaction problem, cf. [14–17]. One possibility is to develop new software and solution methods for each of these coupled applications, as undoubtably will happen in some areas. This is referred to as a *monolithical* approach [18], or sometimes as the *direct* method [19]. On the other hand, we shall assume here that the methods and software systems which have been developed for either applications will continue to be used, in our example fluid or structural software. Thefore we consider *partitioned* methods [20,21,14,6,22,23] – also known as *iterative* methods [19] for fluid-structure interaction, i.e. separate solvers are used for the fluid and the structure [24,25].

For stability reasons, often a fully implicit formulation has to be used [7,19]. In this approach, we have to solve a large system of nonlinear equations with the use of the (iterative) solvers for the subsystems. Commonly this is performed with block-Jacobi, block-Gauss-Seidel or related relaxation methods [26]. These simple methods do not always converge [14,27,25,28–31]. We will introduce here a superior approximative block-Newton method [32,25,28–31].

These ideas arise and can be applied in a multitude of other areas such as general multifield/multiphysics phenomena [14,9], co-simulation and multi-numerics [33,34,27,35], domain-decomposition [36], wave-form relaxation [37], and are in line with efforts of parallelisation and modularisation of software. Algorithms derived from Newton's method have been considered [38,39]; here we have to observe the constraint that certain parts of the Jacobian system during the implicit iteration may not be accessible. In the end not the – incremental – equlibrium equation are solved, but an equivalent, numerically better conditioned system [32,25,28–31].

2 Coupled Systems

For the sake of simplicity we shall only consider a global system composed of two subsystems. It is no problem to extend the ideas and methods presented here to more than two coupled subsystems, but this makes harder to convey the basic ideas, which may be seen already in the simplest case of two subsystems.

Often the coupling will introduce additional variables – we shall show this later for the FSI example – and hence algebraic equations in addition to the subsystem equations.

2.1 Pure Differential Coupling

Let us assume that the time evolution of the first subsystem is given by a differential equation of the form

$$\dot{\boldsymbol{x}}_1 = \boldsymbol{f}_1(\boldsymbol{x}_1, \boldsymbol{x}_2) \,, \tag{1}$$

where $\boldsymbol{x}_1 \in \mathcal{X}_1$, $\boldsymbol{x}_2 \in \mathcal{X}_2$ are elements of some Banach-spaces, and the superimposed dot denotes the time derivative. The spaces $\mathcal{X}_1$ and $\mathcal{X}_2$ may be spaces of functions, and $\boldsymbol{f}_1$ a differential operator, so that (1) may describe subsystems modeled by partial differential equations, such as fluid flow. The variables $\boldsymbol{x}_2$ are not further specified in this model, they provide the coupling to the second subsystem, which is described completely analogous to (1):

$$\dot{\boldsymbol{x}}_2 = \boldsymbol{f}_2(\boldsymbol{x}_2, \boldsymbol{x}_1) \,. \tag{2}$$

Here analogous remarks as with (1) apply. The coupled system (1,2) is nothing but a larger evolution system for the combined vector $(\boldsymbol{x}_1, \boldsymbol{x}_2)^T$, and actually may come from splitting a larger system into smaller ones for parallelisation, such as in wave-form relaxation [37]. For all variables we have differential evolution equations, and therefore we refer to this case as pure differential coupling. In a later section we shall describe the case when the coupling is modeled by "algebraic" equations, and the subsystems themselves may be differential algebraic equations (DAEs).

Explicit Coupling: We may assume that both subsystems (1,2) have been discretised – if necessary both in space and – in time with some appropriate method. If the methods for the individual subsystems are both explicit, probably the most natural is to stay explicit also for the coupled system. For the sake of simplicity of exposition we shall assume here that both subsystems have the same time step Δt, otherwise what we say refers to their synchronisation points, and one subsystem may be performing what is called sub-cycling. Let us denote the the approximation to the solution at step n by $\boldsymbol{x}_j^{(n)}$, $(j = 1, 2)$, then the explicit integration algorithms – assumed to be single step for ease of notation – for the two subsystems may be denoted by

$$\boldsymbol{x}_1^{(n)} = \varphi_1(\boldsymbol{x}_1^{(n-1)}, \boldsymbol{y}_2) \, , \tag{3a}$$

$$\boldsymbol{x}_2^{(n)} = \varphi_2(\boldsymbol{x}_2^{(n-1)}, \boldsymbol{y}_1) \, , \tag{3b}$$

where the two functions $\boldsymbol{y}_j(t)$, $(j = 1, 2)$ are assumed given, and dependence on time t or the time step Δt has been suppressed. They are of course to represent the variables of the other subsystem in the time interval of length Δt from t_{n-1} to t_n, but of course these are not known yet! Staying explicit we may extrapolate these by some function $\Psi_j(\boldsymbol{x}_j^{(n-1)})$, $(j = 1, 2)$ of the past values of the approximate solution, in the simplest case one may assume the value at t_{n-1} as constant throughout the interval, i.e. $\boldsymbol{y}_j(t) = \Psi_j(\boldsymbol{x}_j^{(n-1)}) \equiv \boldsymbol{x}_j^{(n-1)}$, $(j = 1, 2)$. This leads to the simplest case of *weak* or *loose* coupling, the so-called *staggering* method [22,40], here with explicit subsystem integrators:

$$\boldsymbol{x}_1^{(n)} = \varphi_1(\boldsymbol{x}_1^{(n-1)}, \Psi_2(\boldsymbol{x}_2^{(n-1)})) = \varphi_1(\boldsymbol{x}_1^{(n-1)}, \boldsymbol{x}_2^{(n-1)}) \, , \tag{4a}$$

$$\boldsymbol{x}_2^{(n)} = \varphi_2(\boldsymbol{x}_2^{(n-1)}, \Psi_1(\boldsymbol{x}_1^{(n-1)})) = \varphi_2(\boldsymbol{x}_2^{(n-1)}, \boldsymbol{x}_1^{(n-1)}) \, . \tag{4b}$$

With this, the method is completely specified. The single system integrators are most likely going to have an associated critical time step, and by coupling the systems in this explicit fashion the critical time step for the subsystem will most likely decrease. The two sub-steps in (4a,4b) can be performed in parallel. If we sacrifice this inherent parallelism, we can formulate a partly implicit method: First perform (4a), to be followed by

$$\boldsymbol{x}_2^{(n)} = \varphi_2(\boldsymbol{x}_2^{(n-1)}, \Psi_1(\boldsymbol{x}_1^{(n-1)})) = \varphi_2(\boldsymbol{x}_2^{(n-1)}, \boldsymbol{x}_1^{(n)}) \, , \tag{5}$$

where the new value $\boldsymbol{x}_1^{(n)} = \Psi_1(\boldsymbol{x}_1^{(n-1)})$ – we may say that we have as extrapolation $\Psi_1(\boldsymbol{x}_1^{(n-1)}) = \varphi_1(\boldsymbol{x}_1^{(n-1)}, \boldsymbol{x}_2^{(n-1)})$ – is already used. This is how the staggering method is mostly applied. Of course the combined method is still explicit, but only "half as much" as before.

In case that the subsystem integrators are implicit, e.g. in order not to be constrained by a critical time step in the subsystem, the new approximations

$x_j^{(n)}$, $(j = 1, 2)$ appear also on the r.h.s. of (6a,6b), meaning that we have to solve the equation system:

$$x_1^{(n)} = \phi_1(x_1^{(n)}, x_1^{(n-1)}, y_2) , \tag{6a}$$

$$x_2^{(n)} = \phi_2(x_2^{(n)}, x_2^{(n-1)}, y_1) . \tag{6b}$$

Here again the two functions $y_j(t)$, $(j = 1, 2)$ are assumed to be given, and we need two extrapolation operators as before

$$y_j = \Psi_j(x_j^{(n-1)}) , \quad (j = 1, 2) . \tag{7}$$

Applying the same simple approximation

$$\Psi_j(x_j^{(n-1)}) \equiv x_j^{(n-1)} , \quad (j = 1, 2) \tag{8}$$

as in (4a,4b), we arrive at

$$x_1^{(n)} = \phi_1(x_1^{(n)}, x_1^{(n-1)}, \Psi_2(x_2^{(n-1)})) = \phi_1(x_1^{(n)}, x_1^{(n-1)}, x_2^{(n-1)}) , \tag{9a}$$

$$x_2^{(n)} = \phi_2(x_2^{(n)}, x_2^{(n-1)}, \Psi_1(x_1^{(n-1)})) = \phi_2(x_2^{(n)}, x_2^{(n-1)}, x_1^{(n-1)}) . \tag{9b}$$

This is a staggering type algorithm for implicit integrators. The two subsystems (9a,9b) may be solved independently of each other, but the method, considered for the system as a whole, is explicit, and will again introduce limitations connected with a critical time step associated to the global system. To circumvent this, generally we have to introduce a global coupling, although there are methods to retain at least the linear stability characteristics in the staggered approach [21,14,41]. We should mention that naturally the partly implicit variant of (4b) is also possible: After performing (9a), replace (9b) by

$$x_2^{(n)} = \phi_2(x_2^{(n)}, x_2^{(n-1)}, \Psi_1(x_1^{(n-1)})) = \phi_2(x_2^{(n)}, x_2^{(n-1)}, x_1^{(n)}) ; \tag{10}$$

where again we may say that we have as extrapolation $\Psi_1(x_1^{(n-1)}) = \varphi_1(x_1^{(n-1)}, x_2^{(n-1)})$.

Implicit Coupling: To alleviate the time step restriction still present in (9a,9b), we may formulate the whole system in an implicit way, extrapolating $y_j(t) = \Psi_j(x_j^{(n-1)}) \equiv x_j^{(n)}$, $(j = 1, 2)$ with the constant – and yet unknown – value of the approximate solution at the end of the time step. Again here some more elaborate approximation is possible, but we restrict ourselves to this simplest and most used case for the sake of brevity. The equation system to be solved at each time step now reads

$$x_1^{(n)} = \phi_1(x_1^{(n)}, x_1^{(n-1)}, \Psi_2(x_2^{(n-1)})) = \phi_1(x_1^{(n)}, x_1^{(n-1)}, x_2^{(n)}) , \tag{11a}$$

$$x_2^{(n)} = \phi_2(x_2^{(n)}, x_2^{(n-1)}, \Psi_1(x_1^{(n-1)})) = \phi_2(x_2^{(n)}, x_2^{(n-1)}, x_1^{(n)}) . \tag{11b}$$

This is a case of *strong* or *tight* coupling, and the results from this are completely equivalent to what would be achieved by a monolithical formulation, although we arrived here through a partitioned approach. The equations (11a,11b) are a coupled system of equations, and unlike (9a,9b) can not be solved independently of each other. Hence, even if we have methods to solve each of (9a,9b) by itself, this does not appear to help us in the solution of the global system, we need a global iteration. But in section 5 we shall see that the methods for the subsystems can be put to good use in the solution of the global system.

Thus although the formulation of the *partitioned* but *strongly coupled* approach is equivalent to a *monolithical* approach, the methods to solve the coupled system will be different: In the monolithical approach we have all the information desired about the subsystems at our disposal, and the solution methods are sometimes termed as *direct* [19], although the actual solution process may – and in many cases will – very well be iterative.

Starting from the partitioned approach, we do not have all the information desired about the subsystems at our disposal, and there has to be some kind of iteration across the subsystems. Therefore this approach is sometimes termed the *iterative* [19] method. The iteration across the subsystems means also that information has to be exchanged, and that, if the subsystems are solved iteratively, we have a nested iteration. This always brings up the question how these iteration processes can be coordinated. In section 5 we will show in what sense the global iteration can control the inner iteration of the individual iterative subsystem solvers.

2.2 Algebraic and Differential Coupling

In the last paragraph we arrived at (11a,11b), a global system of equations to be solved at each time step. It will turn out that the case of two coupled differential algebraic equations (DAEs) is no more difficult conceptually at each time step, so we shall look at this situation next [27].

Assume that the two subsystems are DAEs of index 1, denoted by

$$\dot{\boldsymbol{x}}_1 = \boldsymbol{f}_1(\boldsymbol{x}_1, \boldsymbol{x}_2, \boldsymbol{y}_1, \boldsymbol{z}) \,, \tag{12a}$$

$$\boldsymbol{0} = \boldsymbol{g}_1(\boldsymbol{x}_1, \boldsymbol{x}_2, \boldsymbol{y}_1, \boldsymbol{z}) \,. \tag{12b}$$

The description is similar to (1), only that we have additionally a local – to the first subsystem – algebraic variable $\boldsymbol{y}_1$ and a global one $\boldsymbol{z}$. As with the evolution law $\boldsymbol{f}_1$ in (12a), also the function $\boldsymbol{g}_1$ in the "algebraic" equation (12b) may be a spatial differential operator. Completely analogous, just by switching indices, we model the second subsystem by

$$\dot{\boldsymbol{x}}_2 = \boldsymbol{f}_2(\boldsymbol{x}_2, \boldsymbol{x}_1, \boldsymbol{y}_2, \boldsymbol{z}) \,, \tag{13a}$$

$$\boldsymbol{0} = \boldsymbol{g}_2(\boldsymbol{x}_2, \boldsymbol{x}_1, \boldsymbol{y}_2, \boldsymbol{z}) \,. \tag{13b}$$

The two subsystems are now not only coupled by the differential variables $\boldsymbol{x}_j$, $(j = 1, 2)$ as before, but we additionally allow for an algebraic coupling

$$\mathbf{0} = \boldsymbol{h}(\boldsymbol{x}_1, \boldsymbol{x}_2, \boldsymbol{y}_1, \boldsymbol{y}_2, \boldsymbol{z}) \;, \tag{14}$$

through the global algebraic variable $\boldsymbol{z}$.

As already mentioned, we assume that each subsystem is a DAE of index 1, and that the global system is a DAE of index 1. This means that the matrices

$$D_{y_j}\boldsymbol{g}_j \;, \qquad \begin{bmatrix} D_{y_j}\boldsymbol{g}_j & D_z\boldsymbol{g}_j \\ D_{y_j}\boldsymbol{h} & D_z\boldsymbol{h} \end{bmatrix} , \; (j = 1, 2) \;, \qquad \begin{bmatrix} D_y\boldsymbol{g} & D_z\boldsymbol{g} \\ D_y\boldsymbol{h} & D_z\boldsymbol{h} \end{bmatrix} \tag{15}$$

have to be regular, where D_q is the partial derivative w.r.t. q, and we have set $\boldsymbol{g} = (\boldsymbol{g}_1, \boldsymbol{g}_2)^T$ and $\boldsymbol{y} = (\boldsymbol{y}_1, \boldsymbol{y}_2)^T$.

Upon discretising equations (12a,12b) and (13a,13b) in time with an implicit method, we will again arrive at an implicit equation to be solved for each subsystem. Treating also the differential variable from the other subsystem implicitly like in (11a,11b), and also treating the global algebraic condition (14) implicitly, we arrive at a global system to be solved:

$$\boldsymbol{x}_1^{(n)} = \boldsymbol{\Phi}_1(\boldsymbol{x}_1^{(n)}, \boldsymbol{x}_1^{(n-1)}, \boldsymbol{x}_2^{(n)}, \boldsymbol{y}_1^{(n)}, \boldsymbol{z}^{(n)}) \;, \tag{16a}$$

$$\boldsymbol{x}_2^{(n)} = \boldsymbol{\Phi}_2(\boldsymbol{x}_2^{(n)}, \boldsymbol{x}_2^{(n-1)}, \boldsymbol{x}_1^{(n)}, \boldsymbol{y}_2^{(n)}, \boldsymbol{z}^{(n)}) \;, \tag{16b}$$

$$\mathbf{0} = \boldsymbol{H}(\boldsymbol{x}_1^{(n)}, \boldsymbol{x}_2^{(n)}, \boldsymbol{y}_1^{(n)}, \boldsymbol{y}_2^{(n)}, \boldsymbol{z}^{(n)}) \;. \tag{16c}$$

Again this is a coupled system of equations, but we may only have solvers for each subsystem separately.

3 Fluid-Structure Interaction

Before describing in the abstract setting of the preceeding section 2 how to solve these systems numerically by using the subsystem solvers, we shall take a glance at the FSI problem, and how to formulate it such that it fits the above setting.

3.1 The Fluid

We assume the fluid to be modeled adequately as an incompressible Newtonian fluid, satisfying the appropriate Navier-Stokes equation. As we want to couple this with a moving structure, we have to allow for a moving boundary. We take account of this by formulating the Navier-Stokes equation in an Arbitrary Lagrangean-Eulerian (ALE) [4,42,43] framework in the moving spatial fluid domain Ω_f:

$$\varrho_f\,(\dot{v} + ((v - \dot{\chi}) \cdot \nabla)v) - \operatorname{div}\sigma + \nabla p = r_f, \tag{17a}$$

$$2\sigma = \nu(\nabla v + (\nabla v)^T), \qquad \operatorname{div} v = 0. \tag{17b}$$

The boundary $\partial\Omega_f$ we assume to be divided into three disjoint parts $\partial\Omega_f = \Gamma_v \cup \Gamma_q \cup \Gamma_c$ where on Γ_v the velocity is prescribed, on Γ_q the traction is given, and Γ_c is the coupling boundary, where the coupling conditions will be specified below. Here ϱ_f is the fluid density, v the velocity, σ the viscous stress, χ is the position of the reference ALE-coordinate system and $\dot{\chi}$ its velocity. The fluid shear viscosity is denoted by ν, p is the pressure, r_f the body force in the fluid, and the differential operators are in the spatial frame.

There is no essential difficulty in assuming other fluid models, such as compressible or non-Newtonian ones, the abstract formulation introduced earlier in section 2 would still be fitting. But it is the example of an incompressible fluid which makes it more challenging in one respect, as the incompressibility condition is an "algebraic" constraint.

The movement of the ALE-coordinate system $\chi(t)$ is not specified yet, this will be done after we have considered the discretisation.

3.2 The Structure

The structure we assume to be modeled by a neo-Hookean material, hence the equilibrium equation takes the following form in a Lagrangean framework in a fixed material domain Ω_s:

$$\varrho_s \ddot{u} - \mathrm{DIV}\,(FS) = r_s, \qquad F = \mathrm{GRAD}\, u \tag{18a}$$

$$S = \lambda(\mathrm{tr}\, E)I + 2\mu E, \qquad 2E = (C - I), \qquad C = F^T F \tag{18b}$$

The boundary $\partial\Omega_s$ we again assume divided into three disjoint parts $\partial\Omega_s = \Gamma_u \cup \Gamma_t \cup \Gamma_c$, where on Γ_u the displacements u are prescribed, on Γ_t the tractions, and Γ_c is the coupling boundary with the fluid. The other quantities are the structure density ϱ_s, the displacement gradient F, the second Piola-Kirchhoff stress S, and the body load r_s. The Lamé moduli are denoted by λ and μ, and E is the Lagrange-Green strain, derived from the Cauchy-Green tensor C, and the differential operators are in the material frame.

Again here it would be possible to assume other models for the solid, e.g. non-elastic ones with internal variables such as plasticity or damage, they would still fit into the general setting introduced in section 2.

3.3 The Interface

On the coupling boundary Γ_c let a unique normal n be defined in the spatial frame. Then the coupling conditions may be expressed as requiring the velocities to coincide at the location $\chi(t) = \chi_0 + u(\chi_0, t)$ and the tractions from fluid and structure to balance each other:

$$v(\chi(t), t) = \dot{u}(\chi_0, t), \tag{19a}$$

$$(\sigma - pI) \cdot n = -\frac{1}{J} FSF^T \cdot n\,, \qquad J = \det F\,. \tag{19b}$$

4 FSI as a Coupled DAE

We shall further assume that the fluid – described by (17a,17a), the structure – described by (18a,18a), and the interface – described by (19a,19b), have been discretised; each one may use her favourite discretisation, e.g. finite elements, finite volumes etc., e.g. [44,45].

4.1 Discrete Form of FSI

Up to now we have only hinted at how to move the reference coordinate system in the fluid domain. Several possibilities exist [46–48,7], here we have followed [49] and modeled the connections between the nodes in the fluid domain as elastic springs. This fictitious inertia-free elastic body has displacement loading from the moving structure, on the outer boundaries it is fixed. The traction balance (19b) introduces additional forces on the fluid, and the reaction force on the solid. Hence for the fluid the complete discrete equations of motion are [25]:

$$\boldsymbol{M}_f\dot{\boldsymbol{v}} + \boldsymbol{N}(\boldsymbol{v} - \dot{\boldsymbol{\chi}})\boldsymbol{v} + \boldsymbol{K}_f\boldsymbol{v} + \boldsymbol{B}_f\boldsymbol{p} = \boldsymbol{r}_f + \boldsymbol{T}_f^T\boldsymbol{\tau} , \tag{20a}$$

$$\boldsymbol{B}_f^T\boldsymbol{v} = \boldsymbol{0} , \tag{20b}$$

$$\boldsymbol{K}_g\boldsymbol{\chi} = \boldsymbol{A}\boldsymbol{u} . \tag{20c}$$

The terms in (20a) are the discrete analogues of those in (17a), the term $\boldsymbol{r}_f$ includes the prescribed boundary stresses, and the additional term $\boldsymbol{T}_f^T\boldsymbol{\tau}$ comes from the interaction with the structure, where $\boldsymbol{\tau}$ is the stress on the fluid-structure-interface. In (20b) we may recognise the discrete form of the incompressibility condition, and (20c) describes the movement $\boldsymbol{\chi}(t)$ of the fluid domain, driven by the structure displacements $\boldsymbol{u}$.

In a similar vein we obtain for the structure

$$\boldsymbol{M}_s\ddot{\boldsymbol{u}} + \boldsymbol{K}_s(\boldsymbol{u})\boldsymbol{u} = \boldsymbol{r}_s - \boldsymbol{T}_s^T\boldsymbol{\tau} . \tag{21}$$

To this we add the discrete coupling condition for the velocities,

$$\boldsymbol{T}_f\boldsymbol{v} = \boldsymbol{T}_s\dot{\boldsymbol{u}} . \tag{22}$$

This combined set (20a,20b,20c), and (21,22) is a system of differential algebraic equations (DAEs) in the time variable t of index 2, as can be easily verified. In order to allow an easier numerical treatment and to make them fit the general formulation in subsection 2.2, they will be converted to index 1 by differentiation [25].

There are quite a few items we have glossed over in our quick description, as these are not central to our issue. Nevertheless, for a proper implementation of FSI they have to be dealt with. This includes, but is not limited to, the problem of non-matching grids for fluid, structure, and possibly interface, and the ALE-fluid domain. There are different possibilities to tackle those problems, such as consistent interpolation and mortar elements, to name but a few [50–53]. Another problem is a kind of consistency when moving the ALE-fluid mesh; the numerical time stepping algorithm together with the spatial discretisation has to satisfy a special property, referred to as the geometric conservation law GCL [46,54].

4.2 Index and Order Reduction

First we will reduce the index of the fluid system by differentiation. Differentiating (20b) once gives $\boldsymbol{B}_f^T \dot{\boldsymbol{v}} = \boldsymbol{0}$. As $\boldsymbol{M}_f$ is non-singular, we may solve for $\dot{\boldsymbol{v}}$ from (20a) and insert this into the above relation, giving

$$\boldsymbol{B}_f^T \boldsymbol{M}_f^{-1}(\boldsymbol{r}_f + \boldsymbol{T}_f^T \boldsymbol{\tau} - \boldsymbol{N}_f(\boldsymbol{v} - \boldsymbol{\psi})\boldsymbol{v} - \boldsymbol{K}_f \boldsymbol{v} - \boldsymbol{B}_f \boldsymbol{p}) = \boldsymbol{0} , \tag{23}$$

where we have set $\boldsymbol{\psi} := \dot{\boldsymbol{\chi}}$ for the grid velocity. In the same manner we differentiate equation (20c) for the grid movement

$$\boldsymbol{K}_g \boldsymbol{\psi} - \boldsymbol{A}\boldsymbol{w} = \boldsymbol{0} , \tag{24}$$

where we have at the same time introduced the new variable $\boldsymbol{w} := \dot{\boldsymbol{u}}$, the structural velocities.

With this relation the structural equation (21) can be reduced to first order in a standard manner:

$$\boldsymbol{M}_s \dot{\boldsymbol{w}} + \boldsymbol{K}_s(\boldsymbol{u})\boldsymbol{u} = \boldsymbol{r}_s - \boldsymbol{T}_s^T \boldsymbol{\tau} . \tag{25}$$

4.3 The DAE Correspondence

For our concrete application, the abstract Eqs.(12a,12b) may now be identified with the following terms [25]:

$$\boldsymbol{x}_1 = \begin{bmatrix} \boldsymbol{v} \\ \boldsymbol{\chi} \end{bmatrix} , \quad \boldsymbol{y}_1 = \begin{bmatrix} \boldsymbol{b} \\ \boldsymbol{p} \\ \boldsymbol{\psi} \end{bmatrix} , \quad \boldsymbol{f}_1 = \begin{bmatrix} \boldsymbol{b} \\ \boldsymbol{\psi} \end{bmatrix} , \quad \boldsymbol{z} = \boldsymbol{\tau} \tag{26a}$$

$$\boldsymbol{g}_1 = \begin{bmatrix} \boldsymbol{M}_f \boldsymbol{b} + \boldsymbol{N}_f(\boldsymbol{v} - \boldsymbol{\psi})\boldsymbol{v} + \boldsymbol{K}_f \boldsymbol{v} + \boldsymbol{B}_f \boldsymbol{p} - \boldsymbol{r}_f - \boldsymbol{T}_f^T \boldsymbol{\tau} \\ -\boldsymbol{B}_f^T \boldsymbol{M}_f^{-1}(-\boldsymbol{N}_f(\boldsymbol{v} - \boldsymbol{\psi})\boldsymbol{v} - \boldsymbol{B}_f \boldsymbol{p} - \boldsymbol{K}_f \boldsymbol{v} + \boldsymbol{r}_f + \boldsymbol{T}_f^T \boldsymbol{\tau}) \\ \boldsymbol{K}_g \boldsymbol{\psi} - \boldsymbol{A}\boldsymbol{w} \end{bmatrix} , \tag{26b}$$

so the first subsystem is the fluid with the ALE-grid. In (26a,2) we have introduced the new variable $\boldsymbol{b} := \dot{\boldsymbol{v}}$, the fluid acceleration.

The second subsystem is the structure,

$$\boldsymbol{x}_2 = \begin{bmatrix} \boldsymbol{u} \\ \boldsymbol{w} \end{bmatrix} , \quad \boldsymbol{y}_2 = \boldsymbol{a} , \quad \boldsymbol{f}_2 = \begin{bmatrix} \boldsymbol{w} \\ \boldsymbol{a} \end{bmatrix} , \quad \boldsymbol{z} = \boldsymbol{\tau} \tag{27}$$

$$\boldsymbol{g}_2 = \boldsymbol{M}_s \boldsymbol{a} + \boldsymbol{K}_s(\boldsymbol{u})\boldsymbol{u} - \boldsymbol{r}_s + \boldsymbol{T}_s^T \boldsymbol{\tau}; , \tag{28}$$

where no index reduction, but only a standard order reduction of the differential equation was needed. In (27,2) we have introduced the new variable $\boldsymbol{a} := \dot{\boldsymbol{w}} = \ddot{\boldsymbol{u}}$, the structural acceleration.

The final item to identify is the global coupling condition,

$$\boldsymbol{h} = \boldsymbol{T}_f \boldsymbol{b} - \boldsymbol{T}_s \boldsymbol{a} \,, \tag{29}$$

which expresses the equality of the accelerations at the fluid-structure interface.

Let us remark that in the chosen formulation the differential equation only gives the definition of the various variables, whereas the equilibrium equations have gone into the algebraic conditions. What is still left is to verify the index 1 conditions (15). This is a lengthy calculation [25], but the index 1 conditions are indeed satisfied as long as $\boldsymbol{K}_g$, $\boldsymbol{M}_f$, and $\boldsymbol{M}_s$ are regular — in our case they even can be assumed to be symmetric positive definite.

5 Numerical Procedures for Partitioned Methods

In section 2 we have already tried to give an overview on how this might be possible. We also see that the computational procedures are very dependent on the formulation chosen for the global system. If, as we have done here, the coupled system is formulated as a differential-algebraic equation, we are almost inevitably led to an at least partly implicit formulation, and such mixed time-discretisations have also been investigated [55,56]. For our example this problem already comes within one subsystem, namely the fluid. Here we have an incompressibility constraint, and practically all discretisations have some implicit elements [45].

For our problem we now have

- *the fluid problem* with the components
  1. the Navier-Stokes equation (17a) – a partial differential equation in space and time, or its semi-discrete form (20a), an ordinary differential equation in time.
  2. the incompressibility constraint (17b) – a partial differential equation in space, or its discrete form (20b), an algebraic constraint.
  3. the ALE-grid movement, here only formulated in discrete form as an additional algebraic constraint (20c).
- *the structure problem*, a partial differential equation in space and time, and of second order in time (18a), or in its semi-discrete form (21) a second order ordinary differential equation in time.
- *the coupling condition*, a partial differential equation (19a,19b) on the spatial interface between the fluid and the structural domain, or in its discrete form (22) an algebraic constraint.

The grid movement could very well be taken as a subsystem in its own right, but because it is so intimately connected to the fluid problem, it is customarily treated simultaneously.

Through index and order reduction the above problems have been reduced to the abstract coupled DAE system (26a–28). We assume that at least – as it is otherwise trivial – both subsystems have some implicit parts; something which is true for our example and also in line with common numerical wisdom for the discretisation of DAEs. In our partitioned and modular approach this also means that we have solvers for each subsystem, we write them here in fixed point form, as they may be used during the iterative solution process for each subsystem. This certainly includes also direct solvers, they may be simply regarded as very fast convergent iterative solvers – they only need one iteration.

After discretisation in time, we then arrive at the concrete versions of our abstract Eqs.(16a-16c). The first Eq.(16a) solves for the variables $(\boldsymbol{x}_1, \boldsymbol{y}_1)^T = (\boldsymbol{v}, \boldsymbol{\chi}, \boldsymbol{b}, \boldsymbol{p}, \boldsymbol{\psi})^T$ assuming the others are given, the second (16b) solves for $(\boldsymbol{x}_2, \boldsymbol{y}_2)^T = (\boldsymbol{u}, \boldsymbol{w}, \boldsymbol{a})^T$ assuming the others are given, and the last (16c) solves for $\boldsymbol{z} = \boldsymbol{\tau}$ assuming the others are given. For simplicity, and as it is also customarily done that way, when solving we shall include the discrete coupling condition with one of the other two subsystems, i.e. either with the fluid, meaning that on the coupling interface the velocities are prescribed (Dirichlet data), and the interface tractions are passed from the fluid to the structure (Neumann data); or the other way around. In the first case we shall jointly denote the variables as $\boldsymbol{\xi} := (\boldsymbol{x}_1, \boldsymbol{y}_1, \boldsymbol{z})^T = (\boldsymbol{v}, \boldsymbol{\chi}, \boldsymbol{b}, \boldsymbol{p}, \boldsymbol{\psi}, \boldsymbol{\tau})^T$ and $\boldsymbol{\zeta} := (\boldsymbol{x}_2, \boldsymbol{y}_2)^T = (\boldsymbol{u}, \boldsymbol{w}, \boldsymbol{a})^T$, and in the second case the variable $\boldsymbol{z} = \boldsymbol{\tau}$ is included in $\boldsymbol{\zeta}$ and not in $\boldsymbol{\xi}$.

As we want now to concentrate on how to solve the coupled Eqs.(16a–16c), we shall drop the time step counter, and also suppress the dependence on the previous time steps. We may assume both iterative solution processes are in fixed point form, i.e. the iteration

$$\boldsymbol{\xi}_\kappa = \boldsymbol{F}_1(\boldsymbol{\xi}_{\kappa-1}, \boldsymbol{\zeta}) , \quad \kappa = 1, 2, \dots ; \tag{30}$$

converges for reasonable starting values $\boldsymbol{\xi}_0$ with given $\boldsymbol{\zeta}$. Similarly, the iteration process

$$\boldsymbol{\zeta}_\kappa = \boldsymbol{F}_2(\boldsymbol{\zeta}_{\kappa-1}, \boldsymbol{\xi}) , \quad \kappa = 1, 2, \dots ; \tag{31}$$

converges for reasonable starting values $\boldsymbol{\zeta}_0$ with given $\boldsymbol{\xi}$.

5.1 Non-linear Block-Jacobi

The conceptually simplest way to use these two solvers F_1 and F_2 to solve the combined system is a non-linear block-Jacobi process, or in the context of domain decomposition it would be called an additive or parallel Schwarz procedure [36]. Given $\boldsymbol{\xi}_{\kappa-1}$ and $\boldsymbol{\zeta}_{\kappa-1}$, perform the following iterative step:

$$\boldsymbol{\xi}_{\kappa} = \boldsymbol{F}_1^{\nu_1}(\boldsymbol{\xi}_{\kappa-1}, \boldsymbol{\zeta}_{\kappa-1}) \,, \tag{32a}$$

$$\boldsymbol{\zeta}_{\kappa} = \boldsymbol{F}_2^{\nu_2}(\boldsymbol{\zeta}_{\kappa-1}, \boldsymbol{\xi}_{\kappa-1}) \,; \tag{32b}$$

meaning that in (32a) the iteration (30) has been performed ν_1 times with a fixed $\boldsymbol{\zeta}_{\kappa-1}$, and in (32b) the iteration (31) has been performed ν_2 times with a fixed $\boldsymbol{\xi}_{\kappa-1}$. Very often one takes $\nu_1 = \nu_2 = 1$. The starting values $\boldsymbol{\xi}_0$ and $\boldsymbol{\zeta}_0$ are provided by the extrapolation mapping (7).

5.2 Non-linear Block-Gauss-Seidel

It is well known, that usually the corresponding Gauss-Seidel process converges faster [26], and so we are led to: Given $\boldsymbol{\xi}_{\kappa-1}$ and $\boldsymbol{\zeta}_{\kappa-1}$, do

$$\boldsymbol{\xi}_{\kappa} = \boldsymbol{F}_1^{\nu_1}(\boldsymbol{\xi}_{\kappa-1}, \boldsymbol{\zeta}_{\kappa-1}) \,, \tag{33}$$

and then, with the newly computed $\boldsymbol{\xi}_{\kappa}$, do

$$\boldsymbol{\zeta}_{\kappa} = \boldsymbol{F}_2^{\nu_2}(\boldsymbol{\zeta}_{\kappa-1}, \boldsymbol{\xi}_{\kappa}) \,. \tag{34}$$

This uses the new information as soon as it is available, and in the context of domain decomposition it would be called a multiplicative or serial Schwarz procedure [36]. Let us remark that this is very similar to the staggering variant (9a,10), only that what before was time stepping is now iteration. The starting values $\boldsymbol{\xi}_0$ and $\boldsymbol{\zeta}_0$ are again given by the extrapolation mapping (7).

Naturally, for both (32a,32b) and (33,34) we have to ask the question whether the iterative process converges. In this situation it is useful to note the following result [27,35]:

Theorem 1. *Let*

$$\alpha = \max_{t\in[0,T]} \| \left(D_{y_2}\boldsymbol{g}_2\right)^{-1} D_z\boldsymbol{g}_2 \left(D_{y_1}\boldsymbol{h}\left(D_{y_1}\boldsymbol{g}_1\right)^{-1} D_z\boldsymbol{g}_1\right)^{-1} D_{y_2}\boldsymbol{h}\| \,, \tag{35}$$

and let L be the Lipschitz-constant of the extrapolation (7). Assume that $\alpha < 1$, and that at least κ iterations of the block-Gauss-Seidel scheme (33,34) are performed, such that $L\alpha^{\kappa} < 1$, and that Δt is small enough. Then the global block-Gauss-Seidel method converges, and the global error in the n-th time step

$$\delta^{(n)} = \|\boldsymbol{x}^{(n)} - \boldsymbol{x}(t_n)\| + \|\boldsymbol{y}^{(n)} - \boldsymbol{y}(t_n)\| + \|\boldsymbol{z}^{(n)} - \boldsymbol{z}(t_n)\|$$

is bounded by

$$\delta^{(n)} \leqslant C(\mu^{\max\{0,\kappa-2\}}\delta^{(n)}\boldsymbol{x} + \mu^{\kappa-1}\delta^{(n)}\boldsymbol{y}) + \varepsilon_1^{(n)} + \varepsilon_2^{(n)}. \tag{36}$$

Here $\varepsilon_1^{(n)}$ and $\varepsilon_2^{(n)}$ are the errors incurred by the single system integrators (16a–16c) and $\delta^{(n)}\boldsymbol{y}$ and $\delta^{(n)}\boldsymbol{z}$ are the extrapolation errors, and $\mu = \alpha + O(\Delta t)$.

If $\alpha < 1$ and Δt is small enough such that $\mu < 1$, the iteration will converge. We see that the contractivity constant α is crucial, and the convergence depends strongly on the ordering of the subsystems in the block-Gauss-Seidel solution strategy.

If enough iterations are performed, essentially only the error components from the single system integrators remain. If we have $\varepsilon_1^{(n)} = O((\Delta t)^p)$ and $\varepsilon_2^{(n)} = O((\Delta t)^q)$ as convergence orders for the single system integrators, we obtain $\delta^{(n)} = O((\Delta t)^{\min(p,q)})$, in contrast to the staggering scheme where we only have $O(\Delta t)$.

We now want to summerise for our example of FSI how the solution strategy, the grouping of the systems and algebraic constraint equations, and the order in the Gauss-Seidel process determine the contraction constant α. The computations leading to this are a bit tedious, and details will be given elsewhere.

- Assume that we have the subsystems and ordering
  1. Fluid plus coupling conditions,
  2. and the structure.

  Then the contractivity constant is

  $$\alpha = \|\boldsymbol{M}_s^{-1}\boldsymbol{T}_s^T\tilde{\boldsymbol{M}}_\tau^{-1}\boldsymbol{T}_s\| , \tag{37}$$

  where $\tilde{\boldsymbol{M}}_\tau$ is some sort of mass matrix acting on the coupling boundary.
- Assume the subsystems and ordering
  1. Structure plus coupling conditions,
  2. and the fluid.

  Then the contractivity constant is

  $$\alpha = \|(\ldots)\boldsymbol{M}_f^{-1}\boldsymbol{T}_f^T(\boldsymbol{T}_s\boldsymbol{M}_s^{-1}\boldsymbol{T}_s^T)^{-1}\boldsymbol{T}_f\| , \tag{38}$$

  where $(\ldots)$ is a lengthy but not so interesting expression.
- Assume the ordering and subsystems as
  1. The fluid, and
  2. the structure and coupling conditions.

  Then the contractivity constant is

  $$\alpha = \|(\boldsymbol{T}_s\boldsymbol{M}_s^{-1}\boldsymbol{T}_s^T)^{-1}\tilde{\boldsymbol{M}}_\tau\| , \tag{39}$$

  with the same $\tilde{\boldsymbol{M}}_\tau$ as in (37).
- Finally, assume the following subsystems and ordering
  1. Structure plus coupling conditions,
  2. and the fluid.

  Then the contractivity constant is

  $$\alpha = \|\tilde{\boldsymbol{M}}_\tau^{-1}\boldsymbol{T}_s\boldsymbol{M}_s^{-1}\boldsymbol{T}_s^T\| , \tag{40}$$

  which is the norm of the inverse of the matrix whose norm is taken in (39).

In our view this strong dependence on ordering and grouping in the Gauss-Seidel process calls for strong or tight coupling methods which will converge unconditionally provided the time step is small enough; this then would roughly be a situation similar for the single system implicit integrator. With block-Gauss-Seidel methods there is no easy way of achieving this, although there are possibilities of preconditioning [27].

5.3 Block-Newton

The possibility for such an algorithm can be sought in the group of Newton-like methods, we shall here for the sake of brevity consider the full Newton-Raphson method. It is also in this context the more challenging, as it requires the solution of a global system of linear equations in each iteration, and especially as in the partitioned approach we have no recourse to the cross-coupling parts of the system matrix. With the Newton algorithm we have a robust method, where the convergence behaviour does not depend on

- in which subsystem we include the coupling,
- and in which order we solve for the subsystems.

Usually we would start from (16a–16c), the "equilibrium" equations; but it has been noted [32] that the solutions of (16a–16c) are solutions of the fixed point equations (30,31) and vice versa, but the latter are numerically much better conditioned as they implicitly already have the effect of the single system solvers in them. Hence, we perform the Newton-Raphson iteration on the fixed point equations corresponding to the iteration (30,31), with $\Delta\boldsymbol{\xi}_\kappa := \boldsymbol{\xi}_{\kappa+1} - \boldsymbol{\xi}_\kappa$ and $\Delta\boldsymbol{\zeta}_\kappa := \boldsymbol{\zeta}_{\kappa+1} - \boldsymbol{\zeta}_\kappa$, and the iteration counter κ:

$$\begin{bmatrix} \boldsymbol{I} - D_\xi \boldsymbol{F}_1 & D_\zeta \boldsymbol{F}_1 \\ D_\xi \boldsymbol{F}_2 & \boldsymbol{I} - D_\zeta \boldsymbol{F}_2 \end{bmatrix} \begin{bmatrix} \Delta\boldsymbol{\xi}_\kappa \\ \Delta\boldsymbol{\zeta}_\kappa \end{bmatrix} = - \begin{bmatrix} \boldsymbol{\xi}_\kappa - \boldsymbol{F}_1(\boldsymbol{\xi}_\kappa, \boldsymbol{\zeta}_\kappa) \\ \boldsymbol{\zeta}_\kappa - \boldsymbol{F}_2(\boldsymbol{\zeta}_\kappa, \boldsymbol{\xi}_\kappa) \end{bmatrix} . \tag{41}$$

We only want to use the existing solvers, i.e. the iteration mappings $\boldsymbol{F}_1$ and $\boldsymbol{F}_2$. In particular, we do not have direct access to the partial derivative cross terms in the global system matrix in (41). But if we solve the system (41) by an iterative method [34,25,30,31], all we need is a way to compute the product of the Jacobian matrix in (41) by an arbitrary vector, at least approximately.

6 Solving the Newton System

As we now want to consider just a single iteration, we drop the iteration index κ, and only look at computing the vector $[\Delta\xi, \Delta\zeta]^T$ from the right hand side. For a start, we use – symbolically – block-Gauss elimination [33,25] on the system (41):

$$\Delta\boldsymbol{\xi} = -(\boldsymbol{I} - D_\xi \boldsymbol{F}_1)^{-1}(\xi - \boldsymbol{F}_1(\boldsymbol{\xi}, \zeta)) - \boldsymbol{C}\Delta\boldsymbol{\zeta} \, , \tag{42}$$

with the multiplier matrix

$$C := (I - D_\xi F_1)^{-1}[D_\zeta F_1] . \tag{43}$$

Inserting this into the second equation, we obtain

$$S\Delta\zeta := (I - [D_\zeta F_2] - [D_\xi F_2]C)\,\Delta\zeta = -r , \tag{44}$$

where S is the Schur complement matrix, and

$$r := (\zeta - F_2(\zeta, \xi)) + [D_\xi F_2]q , \tag{45a}$$

where we have set

$$q := -(I - D_\xi F_1)^{-1}(\xi - F_1(\xi, \zeta)) . \tag{45b}$$

Hence we solve (44) for $\Delta\zeta$, and with this then solve (42) for $\Delta\xi$. One step of this Newton-Raphson iteration may now be formulated as follows [25,29–31]:

1. From (45b), solve
$$(I - D_\xi F_1)q = F_1(\xi, \zeta) - \xi \qquad \text{for} \qquad q . \tag{46}$$
2. Solve the Schur-complement system (44) $S\Delta\zeta = -r$ for $\Delta\zeta$ with
$$r = \zeta - (F_2(\zeta, \xi) - [D_\xi F_2]q) \tag{47}$$
from (45a).
3. Compute $\Delta\xi = q - C\Delta\zeta$. This involves application of the multiplier matrix C from (43).

6.1 Solution of a System with a Diagonal Sub-Block

We look at the first step in (46): To solve for q, we use the iterative solver for the first subsystem F_1. The resolution of (46) can also be seen as one Newton-Raphson step for the solution q of the equation

$$\xi - F_1(\xi + q, \zeta) = 0 \tag{48}$$

when ξ and ζ are fixed. But (48) is exactly what the iterative solver F_1 is designed for. Hence with the iterative solver F_1 we obtain

$$q \approx \eta_m - \xi , \tag{49}$$

where η_m is the result of the iterative process

$$\eta_{j+1} = F_1(\eta_j, \zeta) , \quad j = 0, 1, \ldots, m-1 > 0 , \quad \text{with} \quad \eta_0 = \xi , \tag{50}$$

using the solver F_1 of the first subsystem m times.

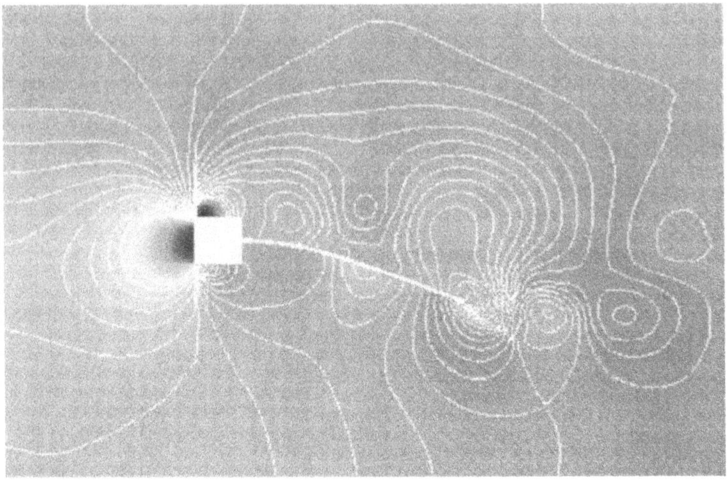

Fig. 1. Vortex Shedding and Pressure Field; colored, see page 369

6.2 Application of the Multiplier Matrix

We shall now next look at the third step, as the techniques used here are needed also in the second step, which is the most involved. In step 3 the application of the multiplier matrix $\boldsymbol{C}$ from (43) is the only point which needs some explanation. Looking at $\boldsymbol{C}$ in (43), we see that it needs the solution of a linear system with a diagonal block $(\boldsymbol{I} - D_\xi \boldsymbol{F}_1)$ – but this is exactly the same task as in the preceeding section 6.1. The other part is to apply the cross-derivative $D_\zeta \boldsymbol{F}_1$. This we do by using finite differences. For some arbitrary vector $\boldsymbol{s}$ we want $[D_\zeta \boldsymbol{F}_1]\boldsymbol{s}$, and with an appropriate small step h we have

$$[D_\zeta \boldsymbol{F}_1]\boldsymbol{s} \approx \frac{1}{h}(\boldsymbol{F}_1(\boldsymbol{\xi}, \boldsymbol{\zeta} + h\boldsymbol{s}) - \boldsymbol{F}_1(\boldsymbol{\xi}, \boldsymbol{\zeta})). \tag{51}$$

This uses the iterative solver $\boldsymbol{F}_1$ for the first subsystem once.

6.3 Solution of the Schur-Complement System

Now we can look at step 2. It involves two sub-steps: Computation of the vector $\boldsymbol{r}$ in (45a,47) and solution of the Schur-complement system (44).

Application of an Off-Diagonal Sub-Block: To apply $[D_\xi \boldsymbol{F}_2]$ in the computation of $\boldsymbol{r}$ in (45a,47), we use finite differences as in the preceeding section 6.2, but in a reversed way. From (47), we have analogously to (51):

$$\boldsymbol{r} = \boldsymbol{\zeta} - (\boldsymbol{F}_2(\boldsymbol{\zeta}, \boldsymbol{\xi}) - [D_\xi \boldsymbol{F}_2]\boldsymbol{q}) \approx \boldsymbol{\zeta} - \boldsymbol{F}_2(\boldsymbol{\zeta}, \boldsymbol{\xi} - \boldsymbol{q}) \, . \tag{52}$$

This uses the iterative solver $\boldsymbol{F}_2$ for the second subsystem once.

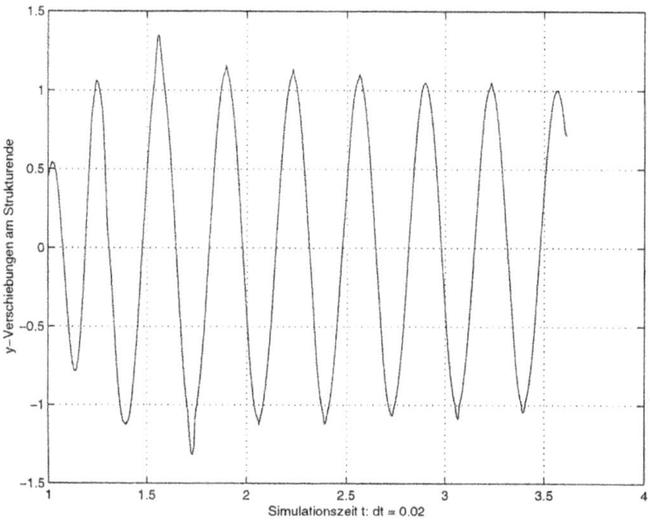

Fig. 2. Displacement Response for Strong Coupling.

Application of the Schur-Complement Matrix: To solve the system with the Schur complement matrix $\boldsymbol{S}$ in (44), we use an iterative Krylov method – here BiCGStab – and we only need the action of $\boldsymbol{S}$ on some arbitrary vector $\boldsymbol{s}$, again approximated via finite differences – in a reversed way as in the preceeding sub-step – with some (small) step-size h:

$$\boldsymbol{S}\boldsymbol{s} = \frac{1}{h}\boldsymbol{S}(h\boldsymbol{s}) = \frac{1}{h}[(\boldsymbol{I} - D_\zeta \boldsymbol{F}_2)(h\boldsymbol{s}) - [D_\xi \boldsymbol{F}_2]\boldsymbol{C}(h\boldsymbol{s})] \tag{53}$$
$$\approx \frac{1}{h}[h\boldsymbol{s} + \boldsymbol{F}_2(\boldsymbol{\zeta} - h\boldsymbol{s}, \boldsymbol{\xi} - \boldsymbol{C}(h\boldsymbol{s})) - \boldsymbol{F}_2(\boldsymbol{\zeta}, \boldsymbol{\xi})] \; .$$

This uses the iterative solver $\boldsymbol{F}_2$ for the second subsystem once. We need in (53) $\boldsymbol{C}(h\boldsymbol{s})$, but this we know how to compute from section 6.2.

Now all steps for the approximative block Newton-Raphson method have been explained. Convergence of these approximative Newton methods has been investigated in [38,39,33,34], and we have the following result:

Theorem 2. *If the single system solvers are quadratically convergent (or enough iterations are made in the approximative steps), the global iteration is also quadratically convergent.*

But in our view even more important than the quadratic convergence is the fact that in this way we have the same convergence characteristics as if we were using the Newton-Raphson method in a monolithic approach, in contrast to theorem 1, where convergence in the Gauss-Seidel process depends on subsystem grouping and ordering. Naturally, the numerical work of the block Newton-method will depend on the subsystem grouping and ordering, as the subsystem solvers are not used in a symmetric way. The computationally

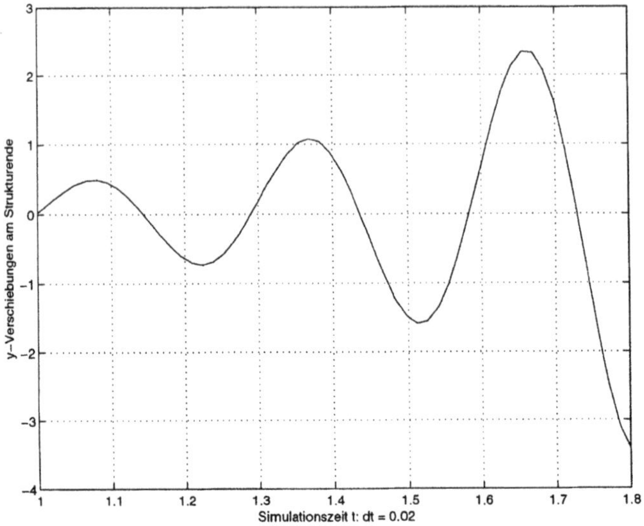

Fig. 3. Displacement Response for Weak Coupling.

most advantageous ordering has to be determined according to the effort involved in using each of the subsystem solvers once.

7 Numerical Examples

The methods presented so far will be demonstrated [25] on a small two-dimensional example shown in 1, introduced in [43]. It is a rigid block with a thin elastic appendage, placed in an incompressible viscous flow coming from the left. The bluff shape of the square block causes separation and vortices, which are transported along the elastic appendage. Although the setup is symmetric, the vortices develop unsymmetrically, alternating from top and bottom. The pressure variations in the vortices interact with the elastic appendage, which in turn starts oscillating. This situation is shown in Fig. 1, together with the pressure field in the flow.

In Fig. 2 the vertical displacement response of the tip of the elastic appendage is shown. The computations were performed with the strong coupling algorithm introduced in sections 2 and 5.3. If we solve this problem with the same time step but with the staggering or so called weak coupling method – keeping the same subsystem solvers as before, we observe the response in Fig. 3, and one immediately recognises the instability of the staggering or weak coupling scheme.

In another test, we compare the block-Gauss-Seidel method as described in section 5.2 with the block-Newton method from section 5.3. For the same configuration as before, we show the number of iterations of either method

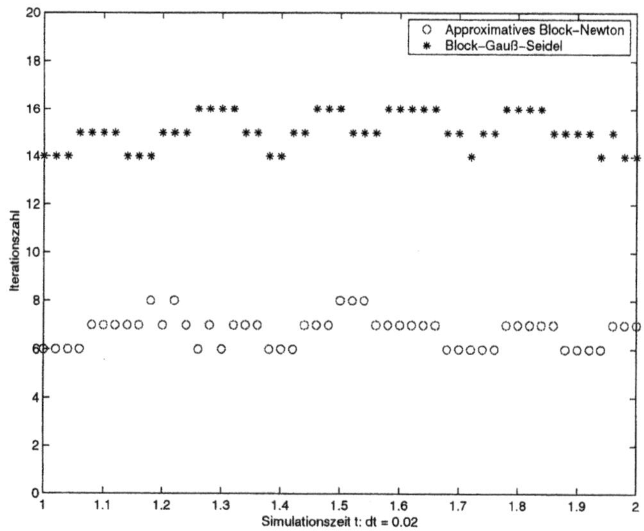

Fig. 4. Iteration Count for block-Gauss-Seidel and block-Newton.

in each time step in 4. The superior convergence characteristic of the block-Newton method is obvious.

As the block-Newton method is more expensive to perform per iteration, this does not necessarily translate into an overall performance gain – of course apart from those cases where the block-Newton converges and the block-Gauss-Seidel method does not. Therefore, as a measure of numerical expenditure, we count the number of times the subsystem solvers have to be called per time step. This is shown in 5, and again the advantage is on the block-Newton side.

8 Conclusion

For coupled but *partitioned* problems, we have proposed a *strong* or *tight* coupling method, which achieves the same results as a monolithical approach. We have formulated this coupling problem as a differential algebraic equation (DAE) following [35,27]. As usually appropriate to the numerical treatment of DAEs, we consider globally implicit methods.

The global system to be solved in each time step is treated with the use of the solvers for the individual subsystems. The non-linear block-Jacobi and non-linear block-Gauss-Seidel methods come very naturally, but they are sub-optimal and not robust. We have discussed the problems which arise with this approach. We propose to solve the global implicit equations with an approximate block-Newton method, following [32–34]. The linear system which arises in each iteration is solved via the subsystem solvers and a Krylov iterative method.

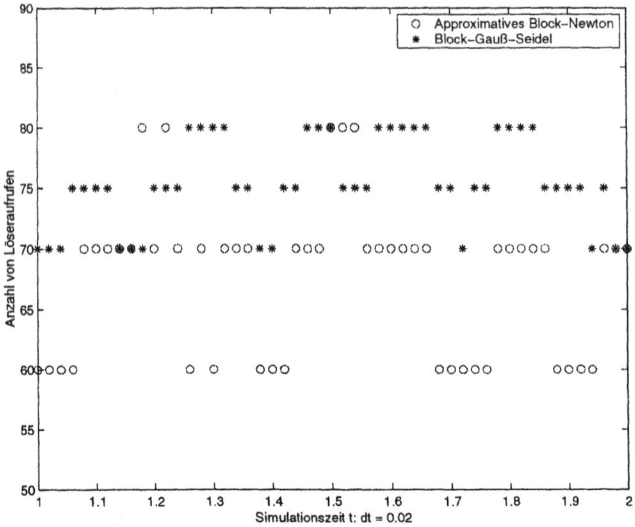

Fig. 5. Count of Subsystem Solver Calls for block-Gauss-Seidel and block-Newton.

This iterative method only requires the possibility to use the sub-problem solvers in an iteration-by-iteration fashion. The computation of the derivatives is approximated by the sub-problem solvers and by finite differencing. These methods are faster both in number of iterations and in numerical effort than the usual staggering methods; additionally they have superior convergence characteristics independent of sub-problem grouping and ordering, unlike the staggering resp. block-Gauss-Seidel method. We have demonstrated the methods on a fluid-structure interaction problem, first fitting it into the general framework, and then applying the proposed procedures. Needless to say that the general ideas about formulating and solving coupling problems are not specific to fluid-structure interaction and can be used in other circumstances as well.

The partitioned approach offers modularity both in the formulation and modeling of physical phenomena, and – a point of paramount practical importance – especially in the implementation of software [57]. This allows the use of existing software packages for the subsystems, and this was how the computations presented here were done. The structural solver was *FEAP* [58,11], and the flow solver used was *FEATFLOW* [59,60], which could be taken as building blocks and only needed a small interface to be able to perform the coupling [25]. Software designed to do the clerical work of managing the communication between different software packets and their meshes [61] has been employed, and was of great help. With this kind of approach, the effort and special knowledge and techniques which has gone into designing and implementing software for specific application areas can be re-used, and the coupled simulation can benefit from this, and things do not have to be designed from scratch.

In closing, we would like to emphasise the fact that it is possible to have the above mentioned advantages of the partitioned approach, and at the same time the superior robustness and convergence characteristics of a monolithical approach – in our view this is the best of both worlds.

References

1. Morand H., Ohayon R. (1995) *Fluid-Structure Interaction.* John Wiley & Sons, Chichester
2. Belytschko T.,Mullen R. (1981) Two-dimensional fluid-structure impact computations with regularization. Comp. Meth. Appl. Mech. Eng. **27**, 139–154
3. Kennedy J. M., Belytschko T. (1982) A survey of computational methods for fluid-structure analysis of reactor safety. Nuclear Engineering and Design **69**, 379–398
4. Donea J., Giuliani S., Halleux J. P., (1982) An arbitrary Lagrangian-Eulerian finite element method for transient fluid-structure interaction. Comp. Meth. Appl. Mech. Eng. **33**, 689–723
5. Bathe K.-J., Zhang H., Wang M. H. (1995) Finite element analysis of incompressible and compressible fluid flows with free surfaces and structural interactions. Computers & Structures **56**, 193–213
6. Piperno S., Farhat C., Larrouturou B. (1995) Partitioned procedures for the transient solution of coupled aeroelastic problems. Comp. Meth. Appl. Mech. Eng. **124**, 79–112
7. Mouro, J., Le Tallec, P. (1998) Fluid Structure Interaction with Large Structural Displacements. Preprint submitted to Elsevier Science
8. Farhat C. (1997) Parallel and distributed solution of coupled nonlinear dynamic aeroelastic response. In: Papadrakakis M. (ed.) *Parallel Solution Methods in Computational Mechanics.* John Wiley & Sons, Chichester
9. Rifai S., Johan Z., Wang W.-P., Grisval J.-P., Hughes T. J. R., Ferencz M. (1999) Multiphysics simulation of flow induced vibrations and aeroelasticity on parallel computing platforms. Comp. Meth. Appl. Mech. Eng. **174**, 393–417
10. Bathe K.-J., Zhang H., Ji S. H. (1999) Finite element analysis of fluid flows fully coupled with structural interactions. Computers & Structures **72**, 1–16
11. Zienkiewicz O. C., Taylor R. L., (2001) *The Finite Element Method.* Vol. 1–3. Butterworth Heinemann, London
12. Argyris J., Doltsinis I. S., Pimenta P., Wüstenberg H. (1982) Thermomechanical response of solids at high strains — natural approach. Comp. Meth. Appl. Mech. Eng. **32**, 3–57
13. Zienkiewicz O. C., Paul D. K., Chan A. H. C. (1988) Unconditionally stable staggered solution procedure for soil-pore fluid interaction problems. Int. J. Num. Meth. Eng. **26**, 1039–1055
14. Park K. C., Felippa C. A. (1983) Partitioned analysis of coupled systems. In: Belytschko T., Hughes T. J. R. (eds.) *Computational Methods in Transient Analysis.* North-Holland, Amsterdam

15. Park K. C., Felippa C. A. (1984) Recent developments in coupled field analysis methods. In: Lewis R. W., Bettes P., Hinton E. (eds.) *Numerical Methods in Coupled Systems.* John Wiley & Sons, Chichester
16. Zienkiewicz O. C. (1984) Coupled problems and their numerical solution. In: Lewis R. W., Bettes P., Hinton E. (eds.) *Numerical Methods in Coupled Systems.* John Wiley & Sons, Chichester
17. Zienkiewicz O. C., Chan A. H. C. (1989) Coupled problems and their numerical solution. In: Doltsinis I. S. (ed.) *Advances in Computational Nonlinear Mechanics.* Springer, Berlin
18. Blom F. J. (1998) A monolithical fluid-structure interaction algorithm applied to the piston problem. Comp. Meth. Appl. Mech. Eng. **167**, 369–391
19. Rugonyi S., Bathe K.-J. (2000) On the analysis of fully-coupled fluid flows with structural interactions – a coupling and condensation procedure. Int. J. of Comp. Civil and Struct. Eng. **1**, 29–41
20. Felippa C. A., Park K. C. (1980) Staggered transient analysis procedures for coupled mechanical systems: formulation. Comp. Meth. Appl. Mech. Eng. **24**, 61–111
21. Park K. C. (1980) Partitioned transient analysis procedures for coupled-field problems: stability analysis. J. Appl. Mech. **47**, 370–376
22. Piperno S. (1997) Explicit/implicit fluid-structure staggered procedures with a structural predictor and fluid subcycling for 2D inviscid aeroelastic simulations. Int. J. Num. Meth. Fluids **25**, 1207–1226
23. Mok D. P., Wall W. A. (2001) Partitioned analysis schemes for the transient interaction of incompressible flows and nonlinear flexible structures. In: Wall W. A., Bletzinger K.-U., Schweizerhof K. (eds.) *Trends in Computational Structural Mechanics.* CIMNE, Barcelona
24. Steindorf J., Matthies H. G. (1999) Efficient partitioned methods for the computation of fluid-structure interaction on parallel computers. In: Topping B. H. V. (ed.) *Proceedings of the Third Euro-Conference on Parallel and Distributed Computing for Computational Mechanics.* Civil-Comp Press, Edinburgh
25. Steindorf J. (2002) Partitionierte Verfahren für Probleme der Fluid-Struktur Wechselwirkung, PhD thesis; Technische Universität Braunschweig.
26. Codina R., Cervera M. (1996) Block-iterative algorithms for nonlinear coupled problems. In: Papadrakakis M., Bugeda G. (eds.) *Advanced Computational Methods in Structural mechanics.* CIMNE, Barcelona
27. Arnold M., Günther M. (2001) Preconditioned dynamic iteration for coupled differential-algebraic systems. BIT Numerical Mathematics **41**, 1–25
28. Matthies H. G., Steindorf J. Efficient iteration schemes for non-linear fluid-structure interaction problems. In: Topping B. H. V. (ed.) *Computational Mechanics: Techniques and Developments.* Civil-Comp Press, Edinburgh
29. Matthies H. G., Steindorf J. (2001) How to make weak couplings strong. In: Bathe K.-J. (ed.) *Computational Fluid and Solid Mechanics.* Elsevier, Amsterdam
30. Matthies H. G., Steindorf J. (2002) Fully coupled fluid-structure interaction using weak coupling. Proc. in Appl. Math. and Mech. **1**(1), 37–38
31. Matthies H. G., Steindorf J. (2002) Partitioned Strong Coupling Algorithms for Fluid-Structure-Interaction. submitted to Computers & Structures
32. Artlich S., Mackens W. (1995) Newton-coupling of fixed point iterations. In: Hackbusch W., Wittum G. (eds.) *Numerical Treatment of Coupled Systems.* Vieweg, Braunschweig

33. Mackens W., Menck J., Voss H. (1998) Numerical system synthesis: Concepts for coupling subsystem solvers. Technical Report, Technische Universität Hamburg-Harburg
34. Mackens W., Menck J., Voss H. (1999) Numerical coupling of subsystems. Zeitschr. für Angew. Math. und Mech. **79**(S3), S871–S872
35. Arnold M. (2000) Constraint partitioning in dynamic iteration methods. submitted to Zeitschr. für Angew. Math. und Mech.
36. Smith B., Bjorstad P., Gropp W. (1996) *Domain Decomposition.* Cambridge University Press, Cambridge
37. Burrage K. (1995) *Parallel and Sequential Methods for Ordinary Differential Equations.* Clarendon Press, Oxford
38. Chan T. F. (1985) An approximate Newton method for coupled nonlinear systems. SIAM J. Numer. Anal. **22**(5), 904–913
39. Miekkala U., Nevanlinna O. (1987) An approximate Newton method for coupled nonlinear systems. SIAM J. Sci. Stat. Comput. **8**, 459–482
40. Farhat C., Lesoinne M. (2000) Two efficient staggered algorithms for the serial and parallel solution of three-dimensional transient aeroelastic problems. Comp. Meth. Appl. Mech. Eng. **182**, 499–515
41. Farhat C., Park K. C., Dubois-Pelerin Y. (1991) An uncoditionally stable staggered algorithm for transient finite element analysis of coupled thermoelastic problems. Comp. Meth. Appl. Mech. Eng. **85**, 349–365
42. Donea J. (1983) Arbitrary Lagrangian-Eulerian finite element methods. In: Belytschko T., Hughes T. J. R. (eds.) *Computational Methods in Transient Analysis.* North-Holland, Amsterdam
43. Wall W. A., Ramm E. (1998) Fluid-structure interaction based upon a stabilized (ALE) finite element method. In: *Proceedings of the 4th World Congress on Computational Mechanics — New Trends and Applications.* CIMNE, Barcelona
44. Bathe K.-J. (1996) *Finite Element Procedures.* Prentice-Hall, Englewood Cliffs, NJ
45. Ferziger J. H., Perić M. (1996) *Computational Methods for Fluid Dynamics.* Springer, Berlin
46. Thomas P. D., Lombard C. K. (1979) Geometric conservation law and its application to flow computations on moving grids. AIAA Journal **17**(10), 1030–1037
47. Dubini G., Pietrabissi R., Montevecchi F. M. (1995) Fluid-structure interaction in bio-fluid mechanics. Med. Eng. Phys. **17**(2), 609–617.
48. Grandmont C., Guimet V., Maday Y. (1997) Numerical analysis of some decoupling techniques for the approximation of the unsteady fluid-structure interaction. In: *Proceedings of the Second European Conference on Numerical Mathematics.* Heidelberg
49. Batina J. T. (1990) Unsteady Euler airfoil solutions using unstructured dynamic meshes. AIAA Journal **28**(8), 1381–1388
50. Cebral J. R., Löhner R. (1997) Conservative load projection and tracking for fluid-structure problems. AIAA Journal **35**(4), 687–692
51. Farhat C., Lesoinne M., Le Tallec P. (1998) Load and motion transfer algorithms for fluid-structure interaction problems with non-matching discrete interfaces: Momentum and energy conservation, optimal discretization and application to aeroelasticity. Comp. Meth. Appl. Mech. Eng. **157**, 95–114
52. Maman N., Farhat C. (1995) Matching fluid and structure meshes for aeroelastic computations: a parallel approach. Computers & Structures **54**, 779–785

53. Wohlmuth B. (2001) *Discretization Methods and Iterative Solvers Based on Domain Decomposition.* Lect. Notes in Computational Science and Engrg., Springer, Berlin
54. Le Tallec P., Mani S. (1999) Conservation laws for fluid-structure interactions. Technical Report CEREMADE, Université de Paris Dauphine
55. Belytschko T., Mullen R. (1978) Stability of explicit-implicit mesh partitions in time integration. Int. J. Num. Meth. Eng. **12**, 1575–1586
56. Farhat C., Lesoinne M., Maman N. (1995) Mixed explicit/implicit time integration of coupled aeroelastic problems: three-field formulation, geometric conservation law and distributed solution. Int. J. Num. Meth. Fluids **21**, 807–835
57. Farhat C., Lesoinne M., Stern P. (1997) High performance solution of three-dimensional nonlinear elastic problems via parallel parttitoned algorithms: methodolgy and preliminary results. Adv. in Engrg. Software **28**, 43–61
58. Taylor R. L. (1998) FEAP – A Finite Element Analysis Program, User Manual Version 6.3; Dept. of Civil and Environmental Engrg., University of California at Berkeley, CA
59. Turek S., Becker C. (1999) FEATFLOW User Manual Release 1.2; Institut für Angewandte Mathematik, Universität Heidelberg
60. Turek S., (1999) *Efficient Solvers for Incompressible Flow Problems: An Algorithmic Approach in View of Computational Aspects.* Lect. Notes in Computational Science and Engrg. Vol. 6. Springer, Berlin
61. Münch M. (1998) GRISSLi Kopplungsinterface, Handbuch für die Anwendungsprogrammierung. IBM High Performance Support Center, Heidelberg; Institut für Algorithmen und Wissenschaftliches Rechnen der GMD, St. Augustin; Institut für Parallele und Verteilte Höchstleistungsrechner, Universität Stuttgart

Discontinuous Galerkin Method for Modeling Flow and Reactive Transport in Porous Media

Mary F. Wheeler[1], Shuyu Sun[1], Owen Eslinger[1], and Béatrice Rivière[2]

[1] The Center for Subsurface Modeling, Texas Institute for Computational and Applied Mathematics, The University of Texas at Austin, 1 University Station C0200, Austin TX 78712, U.S.A.
[2] Department of Mathematics, 301 Thackeray, University of Pittsburgh, Pittsburgh, PA 15260, U.S.A.

Abstract. An important example of an intradomain (i.e. within a specified physical system) multifield problems in the management of water resources is the transport of radionuclides, chemicals, and/or biological species through soil and aquifers. The simulation of such processes with chemical interactions is inherently ill-conditioned due to widely varying scales. Species can be present at trace amounts, equilibrium constants and other parameters can vary over perhaps 50-100 orders of maginitude, the time period of interest can involve millions of years, and the porous media can involve many orders of variablility in permeability.

In this paper we analyze a discontinuous Galerkin method for flow and reactive transport in porous media and apply it to the simulation of a far field nuclear waste management problem. The problem is characterized by large discontinous jumps in permeability, effective porosity, and diffusivity; and by the need to model small levels of concentration of the radioactive constituents. Theoretical derivation shows that optimal *a priori* error estimates in the energy norm can be obtained for concentration and pressure. Numerical computation of a benchmark case demonstrates the importance of the locally conservative property and low numerical diffusion for DG. Methods to optimize the numerical performance of DG such as the use of a slope limiter are also discussed.

1 Introduction and Problem Description

1.1 Introduction

Fluid flows at and below the earth's surface are the cause and the cure for problems of water and soil pollution. Energy production, petroleum and natural gas or nuclear, and its consequences such as the disposal of CO_2 (sequestration by injection into aquifers or oceans) or storage of nuclear wastes in salt domes or underground repositories impacts flows in the earth's subsurface. Length scales of practical and economic interest range from tens of meters to kilometers. However, large scale behavior depends critically upon physics at much smaller length scales, from meters (lithography and fractures in the subsurface) to centimeters (depositional variations) to microns (pore throats in soil and sedimentary rock, fracture apertures, and microbial and geochemical activity). A wide disparity in time scales also exists, from nearly instantaneous chemical reactions to daily tidal movements to millenia-long water infiltration. Moreover, different physical processes may occur simultaneously in different parts of the domain (e.g., single phase flow within an aquifer, multiphase flow in the vadose zone above the aquifer, and shallow water transport in a river or wetland in contact with the porous medium).

Chemical transport is modeled by mass conservation systems which account for advection, diffusion, and reactions of chemical species. These processes must be modeled accurately, and for reliability, all mass in the system should be accounted for (i.e. mass should not be lost due to poor use of numerics). Solutions are often characterized by steep advection fronts due to advection-dominated flow as well as chemical reactions. Schemes which produce nonphysical oscillations and which are not based on mass conservation principles can lead to very erroneous results.

The discontinuous Galerkin (DG) method has become widely used in the last five years for at least four reasons [1,4,6,5]: 1) the inherent flexibility allows construction of more general meshes and a degree of non-uniformity that is not permitted by the more conventional finite element methods; 2) it conserves mass locally on any element; 3) it has, in general, less numerical diffusion and provides more accurate local approximations for problems with rough coefficients; 4) it is easy to implement. DG methods for elliptic problems have been analyzed by Rivière, Wheeler and Girault [6] for the nonsymmetric case with and without penalty. The symmetric case was analyzed by Wheeler in [8]. Currently, DG methods applied to Stokes and Navier-Stokes problems are being studied and analyzed [3].

The Center for Subsurface Modeling at the University of Texas at Austin has developed a 2D DG finite element code [5] to simulate subsurface flow and chemical transport as well as miscible displacement. This 2D DG code is well suited for use with the ANDRA-Couplex1 case [9], a far field nuclear waste management model. The test case presented is characterized by large discontinous jumps in permeability and diffusivity and the transport of very small concentrations. The addition of mechanical dispersion into the model is also easily handled by the code.

This paper consists of four additional sections. The DG algorithm scheme and notation will be given in section 2. We present theoretic results on *a priori* error estimate in section 3. In section 4, numerical experiments are presented using the Andra-Couplex1 [9] test case. Conclusions are described in the last section.

1.2 ANDRA-Couplex1 Test Case Description

We consider a nuclear waste repository in a heterogenuous porous media $\Omega = (0m, 25000m) \times (0m, 695m)$ [9]. The repository lies at a depth of 450m inside a clay layer. Above and below the clay layer there are layers of limestone, marl and dogger limestone as shown in Figure 1(a). Water flows through the entire porous medium, and once the radioactive materials leak from the repository they are diffused and convected by the groundwater flow. Two elements, Plutonium and Iodine, leak from the repository, located in the rectangle defined by the corners $(18440m, 244m)$ and $(21680m, 250m)$. These elements leak in small amounts over time periods that are short compared with the total simulation time. The rate of release can be seen in Figure 1(b).

The following properties apply to the transport of the contaminants in the subsurface:

- the permeabilities of the marl, limestone, clay, and dogger layers are 3.1536e−5 m/yr, 6.3072 m/yr, 3.1536e−6 m/yr, and 25.2288 m/yr respectively,
- the molecular diffusivity of the marl, limestone, and dogger layers is 5.0e−4 m^2/yr for Iodine and Plutonium, and in the clay layer is 9.48e−7 m^2/yr for Iodine and 4.42e−4 m^2/yr for Plutonium,
- the effective porosity is 0.001 for Iodine and 0.2 for Plutonium in the clay layer and 0.1 elsewhere for both,
- the half-life of Plutonium is 3.76e5 $years$ and 1.57e7 $years$ for Iodine,
- the retardation factor is $10e5$ for Plutonium in the clay layer and 1 everywhere else for Plutonium and 1 everywhere for Iodine,
- mechanical dispersion is applied in the limestone and dogger limestone only with $\alpha_t = 1m$ and $\alpha_l = 50m$ and 0 elsewhere.

Fig. 1. ANDRA-Couplex1 Domain

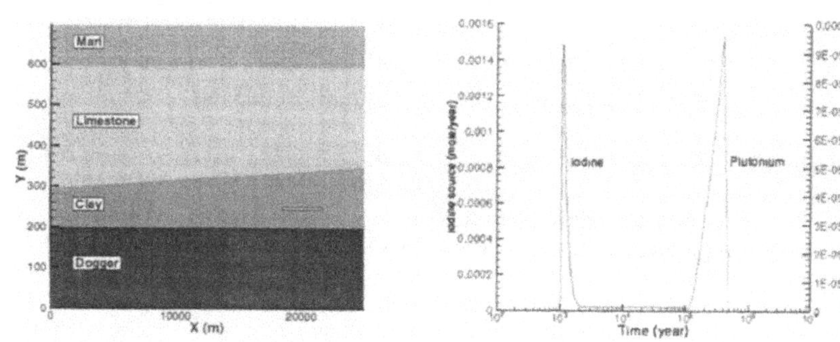

(a) Layers (b) Contaminant Release

2 Governing Equations and Discontinuous Galerkin Scheme

2.1 Governing Equations

Let Ω denote a bounded polygonal domain in $\mathcal{R}^d$, $(d = 2,3)$ and let J denote the time interval $(0, T_f]$. The classical equations governing the flow and transport in porous media are as follows:

- Flow equation

$$-\nabla \cdot (K \nabla p) \equiv \nabla \cdot \mathbf{u} = q \qquad (x,t) \in \Omega \times J \tag{1}$$

- Transport equation

$$\frac{\partial \phi_\alpha c_\alpha}{\partial t} + \nabla \cdot (\mathbf{u} c_\alpha - \mathbf{D}_\alpha(\mathbf{u}) \nabla c_\alpha) = q c_\alpha^* + r(c_\alpha) \qquad (x,t) \in \Omega \times J \tag{2}$$

- Dispersion/diffusion tensor

$$\mathbf{D}_\alpha(\mathbf{u}) = d_{\alpha,m}\mathbf{I} + |\mathbf{u}| \left\{ d_{\alpha,l}\mathbf{E}(\mathbf{u}) + d_{\alpha,t}\left(\mathbf{I} - \mathbf{E}(\mathbf{u})\right) \right\} \tag{3}$$

where $d_m = \phi \tau D_m$ and $\mathbf{E}(\mathbf{u})$ is the tensor that projects onto the $\mathbf{u}$ direction, whose (i,j) component is $(\mathbf{E}(\mathbf{u}))_{i.j} = \frac{u_i u_j}{|\mathbf{u}|^2}$. The imposed external total flow rate q is a sum of sources (injection) and sinks (extraction), c^* is the injected concentration c_w if $q > 0$ and is the resident concentration c if $q < 0$. The flow equation (1) can be obtained by the mass conservation of the whole fluid mixture and Darcy's law. The transport equation (2) is a mass conservation of the species. It governs the convection-diffusion transport process and gives the concentration profile for a given velocity field. It is parabolic but normally convection-dominated. We write $\partial \Omega = \overline{\Gamma}_D \cup \overline{\Gamma}_N = \overline{\Gamma}_{in} \cup \overline{\Gamma}_{out}$, where Γ_D is the Dirichlet boundary and Γ_N is the Neumann boundary for flow subproblem and $\Gamma_D \cap \Gamma_N = \emptyset$; Γ_{in} is the inflow boundary and Γ_{out} is the outflow/noflow boundary condition, defined as follows.

$$\Gamma_{in} = \{x \in \partial\Omega : \ \mathbf{u}(t) \cdot \nu < 0\}, \quad \Gamma_{out} = \{x \in \partial\Omega : \ \mathbf{u}(t) \cdot \nu \geqslant 0\}$$

where ν denotes the unit outward normal vector to $\partial\Omega$. We assume the following boundary conditions:

$$p = p_B \qquad (x,t) \in \Gamma_D \times J \tag{4}$$

$$\mathbf{u} \cdot \nu = u_B \qquad (x,t) \in \Gamma_N \times J \tag{5}$$

$$(\mathbf{u} c_\alpha - \mathbf{D}_\alpha \nabla c_\alpha) \cdot \nu = c_{\alpha,B} \mathbf{u} \cdot \nu \qquad t \in J,\ x \in \Gamma_{in} \tag{6}$$

$$(-\mathbf{D} \nabla c_\alpha) \cdot \nu = 0 \qquad t \in J,\ x \in \Gamma_{out}. \tag{7}$$

The initial concentration is specified by $c_\alpha(x,0) = c_{\alpha,0}(x), \quad x \in \Omega$.

2.2 Discontinuous Galerkin Scheme

We consider the DG scheme for flow and for transport with penalty. Let $\mathcal{T}_h$ be a nondegenerate quasi-uniform subdivision of Ω with triangles or quadrilaterals if $d = 2$ or tetrahedrons if $d = 3$. We assume that the permeability is uniformly positive definite and uniformly bounded above; porosity ϕ is time-independent and is uniformly bounded above and below by positive numbers; no element crosses the boundaries of Γ_D, Γ_N, Γ_{in}, or Γ_{out}. The set of all interior edges (for 2 dimensional domain) or faces (for 3 dimensional domain) for $\mathcal{T}_h$ are denoted by E_h. On each edge or face $e \in E_h$, a unit normal vector ν_e is fixed. The set of all edges or faces on Γ_{out} and on Γ_{in} for $\mathcal{T}_h$ are denoted by $E_{h,out}$ and $E_{h,in}$, respectively, for which the normal vector ν_e coincides with the outward unit normal vector. For $s \geqslant 0$, define,

$$H^s(\mathcal{T}_h) = \left\{\phi \in L^2(\Omega) : \ \phi|_R \in H^s(R), \ R \in \mathcal{T}_h\right\}. \tag{8}$$

The average and jump for $\phi \in H^s(\mathcal{T}_h)$, $s > 1/2$ with $R_i, R_j \in \mathcal{T}_h$, $e = \partial R_i \cap \partial R_j \in E_h$ and ν_e exterior to R_i are defined by

$$\{\phi\} = \frac{1}{2}\left(\phi|_{R_i}\right)\Big|_e + \frac{1}{2}\left(\phi|_{R_j}\right)\Big|_e, \qquad [\phi] = \left(\phi|_{R_i}\right)\Big|_e - \left(\phi|_{R_j}\right)\Big|_e. \tag{9}$$

The usual Sobolev norm on Ω is denoted by $\|\cdot\|_{m,\Omega}$. The broken norms are defined, for positive integer m, as

$$|||\phi|||_m^2 = \sum_{R\in\mathcal{T}_h} \|\phi\|_{m,R}^2 . \tag{10}$$

The finite element space is taken to be

$$\mathcal{D}_r(\mathcal{T}_h) \equiv \left\{\phi \in L^2(\Omega) : \ \phi|_R \in P_r(R), \ R \in \mathcal{T}_h\right\}, \tag{11}$$

where $P_r(R)$ denotes the space of polynomials of (total) degree less than or equal to r on R. The inner product in $\left(L^2(\Omega)\right)^d$ or $L^2(\Omega)$ is indicated by $(\cdot,\cdot)$ and the inner product in boundary function space $L^2(\Gamma)$ is indicated by $(\cdot,\cdot)_\Gamma$, with Γ being edge/face in $\mathcal{R}^{d-1}$. For flow, we introduce the bilinear form and the linear functional,

$$\begin{aligned}
a(p,\psi) = &\sum_{R\in\mathcal{T}_h}\int_R K\nabla p \nabla\psi \\
&- \sum_{e\in E_h}\int_e \{K\nabla p\cdot\nu_e\}[\psi] + \sum_{e\in E_h}\int_e \{K\nabla\psi\cdot\nu_e\}[p] \\
&- \sum_{e\in\Gamma_D}\int_e K\nabla p\cdot\nu_e\psi + \sum_{e\in\Gamma_D}\int_e K\nabla\psi\cdot\nu_e p
\end{aligned}$$

$$l(\psi) = (q,\psi) + \sum_{e\in\Gamma_D}\int_e K\nabla\psi\cdot\nu_e p_B - \sum_{e\in\Gamma_N}\int_e \psi u_B .$$

Let I_S be the index set of all species in the flowing phase, and N_S be the number of species. For transport, we define the bilinear form

$$\begin{aligned} B(\mathbf{c},\mathbf{w};\mathbf{u}) = \sum_{\alpha\in I_S} \Bigg\{ & \sum_{R\in\mathcal{T}_h} \int_R (\mathbf{D}_\alpha(\mathbf{u})\nabla c_\alpha - c_\alpha \mathbf{u})\cdot\nabla w_\alpha \\ & - \sum_{e\in E_h} \int_e \{\mathbf{D}_\alpha(\mathbf{u})\nabla c_\alpha \cdot \nu_e\}\,[w_\alpha] \\ & + \sum_{e\in E_h} \int_e \{\mathbf{D}_\alpha(\mathbf{u})\nabla w_\alpha \cdot \nu_e\}\,[c_\alpha] + \sum_{e\in E_h} \int_e c_\alpha^* \mathbf{u}\cdot\nu_e\,[w_\alpha] \\ & + \sum_{e\in E_{h,out}} \int_e c_\alpha \mathbf{u}\cdot\nu_e w_\alpha - \int_\Omega c_\alpha q^- w_\alpha + J_0^{\sigma,\beta}(c_\alpha, w_\alpha) \Bigg\}, \end{aligned} \tag{12}$$

We note that, for single species, it becomes

$$\begin{aligned} B(c,\psi;\mathbf{u}) = & \sum_{R\in\mathcal{T}_h} \int_R (\mathbf{D}(\mathbf{u})\nabla c - c\mathbf{u})\cdot\nabla\psi - \sum_{e\in E_h} \int_e \{\mathbf{D}(\mathbf{u})\nabla c\cdot\nu_e\}\,[\psi] \\ & + \sum_{e\in E_h} \int_e \{\mathbf{D}(\mathbf{u})\nabla\psi\cdot\nu_e\}\,[c] + \sum_{e\in E_h} \int_e c^* \mathbf{u}\cdot\nu_e\,[\psi] \\ & + \sum_{e\in E_{h,out}} \int_e c\mathbf{u}\cdot\nu_e\psi - \int_\Omega c q^- \psi + J_0^{\sigma,\beta}(c,\psi), \end{aligned}$$

where $c^*|_e$ is the upwind value of concentration,

$$c^*|_e = \begin{cases} c|_{R_1} & \text{if } \mathbf{u}\cdot\nu_e \geqslant 0 \\ c|_{R_2} & \text{if } \mathbf{u}\cdot\nu_e < 0 \end{cases}$$

for $e = \partial R_1 \cap \partial R_2$ and ν_e is the outward unit normal vector to R_1. q^+ is the injection part of source term and q^- is the extraction part of source term, $q^+ = \max(q,0)$, $q^- = \min(q,0)$. $J_0^{\sigma,\beta}(c,\psi)$ is the interior penalty term,

$$J_0^{\sigma,\beta}(c,\psi) = \sum_{e\in E_h} \frac{\sigma_e}{h_e^\beta} \int_e [c]\,[\psi], \tag{13}$$

where σ is a discrete positive function that takes constant value σ_e on the edge or face e, and is bounded below by $\sigma_* > 0$ and above by σ^*, h_e denotes the size of e and $\beta \geqslant 0$ is a real number.

The linear functional $L(\mathbf{w})$ is given by

$$\begin{aligned} & L(\mathbf{w};\mathbf{u},\mathbf{c}) \\ & = \sum_{\alpha\in I_S} \left\{ \int_\Omega c_{\alpha,w} q^+ w_\alpha - \sum_{e\in E_{h,in}} \int_e c_{\alpha,B} \mathbf{u}\cdot\nu_e w_\alpha + \int_\Omega r(\mathcal{M}(\mathbf{c})) \right\}. \end{aligned} \tag{14}$$

We note that, for single species, it becomes

$$L(\psi; \mathbf{u}, c) = \int_{\Omega} c_w q^+ \psi - \sum_{e \in E_{h,in}} \int_e c_B \mathbf{u} \cdot \nu_e \psi + \int_{\Omega} r(\mathcal{M}(c)).$$

The semidiscrete DG scheme for approximating (1) and (2) is as follows. We seek $p_h \in L^\infty (J, \mathcal{D}_k (\mathcal{T}_h))$, $c_h \in L^\infty (J, \mathcal{D}_r (\mathcal{T}_h))$ satisfying,

$$a(p_h, v) = l(v) \qquad \forall \mathbf{v} \in \mathcal{D}_k (\mathcal{T}_h) \quad \forall t \in J \tag{15}$$

$$\sum_{\alpha \in I_S} \left(\phi_\alpha \frac{\partial c_{\alpha,h}}{\partial t}, \psi w_\alpha \right) + B(\mathbf{c}_h, \mathbf{w}; \mathbf{u}_h^M) = L(\mathbf{w}; \mathbf{u}_h^M) \tag{16}$$

$$\forall \mathbf{w} \in (\mathcal{D}_r (\mathcal{T}_h))^{N_S} \quad \forall t \in J$$

$$(\mathbf{c}_h, \mathbf{w}) = (c_0, \psi) \qquad \forall \mathbf{w} \in (\mathcal{D}_r (\mathcal{T}_h))^{N_S} \quad t = 0 \tag{17}$$

where the $\mathbf{u}_h^M = \mathcal{M}(\mathbf{u}_h)$ is defined by the "cut-off" operator $\mathcal{M}$,

$$\mathcal{M}(\mathbf{u})(x) = \min(|\mathbf{u}(x)|, M) \frac{\mathbf{u}(x)}{|\mathbf{u}(x)|}. \tag{18}$$

Here, M is a fixed positive real number and $|\mathbf{u}| = |\mathbf{u}|_2 = \sqrt{\sum_{i=1}^d (\mathbf{u})_i^2}$. The velocity $\mathbf{u}_h$ is computed from the DG approximation for the pressure using Darcy's law followed by averaging on the internal egde/face.

3 Error Estimates for DG Approximation for Reactive Transport Problem

3.1 Notation

For simplicity of discussion, we consider only single species in this section, but the result also applies to the case of multiple species. Throughout this paper, C denotes a generic constant whose value may change with different occurrences. Let $\widehat{P}_h$ be the L^2-projection of $H^s(\mathcal{T}_h)$ to $\mathcal{D}_r (\mathcal{T}_h)$ defined by,

$$\left(\widehat{P}_h c - c, \psi \right) = 0, \qquad \forall \psi \in \mathcal{D}_r (\mathcal{T}_h). \tag{19}$$

It is well known that

$$\|\widehat{P}_h c - c\|_0 \leqslant C h^j \|c\|_j \qquad 0 \leqslant j \leqslant r + 1. \tag{20}$$

Before we present the result, we define some notation for the convenience of discussion. Define the interpolation errors for concentration as $E_c^I = \widehat{P}_h c - c$. Define the finite element solution error for concentration as $E_c = c - c_h$. Define the auxiliary error for concentration as $E_c^A = E_c^I + E_c = \widehat{P}_h c - c_h$. Similarly define the notations for velocity and pressure.

3.2 *a priori* Error Estimate

For brevity in our paper we assume no flow boundary conditions for the flow problem. We recall the DG error estimate result for elliptic problems derived in [6].

Theorem 31 *Assume that $K \in W^{1,\infty}(\mathcal{T}_h)$ and K is symmetric positive definite in $\overline{\Omega}$ unifromly with respect to x, then there is a constant C independent of h, r, p such that for $s \geqslant 2$ and $r \geqslant 2$*

$$|||K^{1/2}\nabla(p - p_h)|||_0 \leqslant C(K)\frac{h^{\mu-1}}{r^{s-\frac{5}{2}}}|||p|||_s$$

where $\mu = \min(r+1, s)$.

We recall several lemmas that were established in [7].

Lemma 32 (Property of operator $\mathcal{M}$) *The "cut-off" operator $\mathcal{M}$ defined as in equation (18) is uniformly Lipschitz continuous and*

$$||\mathcal{M}(\mathbf{u}) - \mathcal{M}(\mathbf{v})||_{(L^\infty(\Omega))^d} \leqslant ||\mathbf{u} - \mathbf{v}||_{(L^\infty(\Omega))^d} \tag{21}$$

A similar property of operator $\mathcal{M}$ holds for scalar c.

Lemma 33 (Uniform positive definiteness of $\mathbf{D}(\mathbf{u})$) *Let $\mathbf{D}(\mathbf{u})$ be defined as in equation (3), where $d_m(x) \geqslant 0$, $\alpha_l(x) \geqslant 0$ and $\alpha_t(x) \geqslant 0$ are nonnegative functions of $x \in \Omega$.*

Then

$$\mathbf{D}(\mathbf{u})\nabla c \cdot \nabla c \geqslant (d_m + \min(\alpha_l, \alpha_t)|\mathbf{u}|)|\nabla c|^2 \tag{22}$$

In particular, if $d_m(x) \geqslant d_{m,*} > 0$ uniformly in the domain Ω, then $\mathbf{D}(\mathbf{u})$ is uniformly positive definite and for all $x \in \Omega$, we have,

$$\mathbf{D}(\mathbf{u})\nabla c \cdot \nabla c \geqslant d_{m,*}|\nabla c|^2 \tag{23}$$

Lemma 34 (Uniform Lipschitz continuity of $\mathbf{D}(\mathbf{u})$) *Let $\mathbf{D}(\mathbf{u})$ be defined as in equation (3), where $d_m(x) \geqslant 0$, $\alpha_l(x) \geqslant 0$ and $\alpha_t(x) \geqslant 0$ are nonnegative of domain $x \in \Omega$, and the dispersivity α_l and α_t is uniformly bounded, i.e. $\alpha_l(x) \leqslant \alpha_l^*$ and $\alpha_t(x) \leqslant \alpha_t^*$.*

Then

$$||\mathbf{D}(\mathbf{u}) - \mathbf{D}(\mathbf{v})||_{(L^2(\Omega))^{d\times d}} \leqslant k_D\,||\mathbf{u} - \mathbf{v}||_{(L^2(\Omega))^d} \tag{24}$$

where $k_D = (4\alpha_t^* + 3\alpha_l^*)\,d^{3/2}$ is a fixed number ($d = 2$ or 3 is the dimension of domain Ω).

It is clear that if $(\mathbf{p}, \mathbf{u}, c)$ is the solution of equations (1) and (2) with boundary conditions (4) through (7) and the initial condition, then it satisfies the following formulation for any $t \in J$.

$$\left(\phi\frac{\partial c}{\partial t}, \psi\right) + B(c, \psi; \mathbf{u}) = L(\psi; \mathbf{u}) \qquad \forall \psi \in \mathcal{D}_r(\mathcal{T}_h)$$

Denote $\widehat{c} = \widehat{P_h}c$. Notice that $[\widehat{c} - c] = [\widehat{c}]$ on any interior edge/face $e \in E_h$ and that $\mathbf{u}^M = \mathbf{u}$ if we picked M large enough, then the above equation can be written as, $\forall \psi \in \mathcal{D}_r(\mathcal{T}_h)$,

$$\begin{aligned}
&\left(\phi\frac{\partial \widehat{c}}{\partial t}, \psi\right) + \sum_{R\in\mathcal{T}_h}\int_R (\mathbf{D}(\mathbf{u}_h^M)\nabla\widehat{c})\cdot\nabla\psi + J_0^{\sigma,\beta}(\widehat{c}, \psi) \\
&= \sum_{R\in\mathcal{T}_h}\int_R \widehat{c}\mathbf{u}_h^M\cdot\nabla\psi + \sum_{e\in E_h}\int_e \{\mathbf{D}(\mathbf{u}_h^M)\nabla\widehat{c}\cdot\nu_e\}[\psi] \\
&\quad - \sum_{e\in E_h}\int_e \{\mathbf{D}(\mathbf{u}_h^M)\nabla\psi\cdot\nu_e\}[\widehat{c}] - \sum_{e\in E_h}\int_e \widehat{c}^*\mathbf{u}_h^M\cdot\nu_e[\psi] \\
&\quad + \int_\Omega \widehat{c}q^-\psi + \int_\Omega c_w q^+\psi + \left(\phi\frac{\partial \widehat{c} - c}{\partial t}, \psi\right) \\
&\quad + \sum_{R\in\mathcal{T}_h}\int_R (\mathbf{D}(\mathbf{u}_h^M) - \mathbf{D}(\mathbf{u}^M))\nabla\widehat{c}\cdot\nabla\psi - \int_\Omega (\widehat{c} - c)q^-\psi \\
&\quad + \sum_{R\in\mathcal{T}_h}\int_R \mathbf{D}(\mathbf{u}^M)\nabla(\widehat{c} - c)\cdot\nabla\psi + J_0^{\sigma,\beta}(\widehat{c} - c, \psi) \\
&\quad - \sum_{R\in\mathcal{T}_h}\int_R \widehat{c}\left(\mathbf{u}_h^M - \mathbf{u}^M\right)\cdot\nabla\psi - \sum_{R\in\mathcal{T}_h}\int_R (\widehat{c} - c)\,\mathbf{u}^M\cdot\nabla\psi \\
&\quad - \sum_{e\in E_h}\int_e \{(\mathbf{D}(\mathbf{u}_h^M) - \mathbf{D}(\mathbf{u}^M))\nabla\widehat{c}\cdot\nu_e\}[\psi] \\
&\quad - \sum_{e\in E_h}\int_e \{\mathbf{D}(\mathbf{u}^M)\nabla(\widehat{c} - c)\cdot\nu_e\}[\psi] \\
&\quad + \sum_{e\in E_h}\int_e \{\mathbf{D}(\mathbf{u}_h^M)\nabla\psi\cdot\nu_e\}[\widehat{c} - c] \\
&\quad + \sum_{e\in E_h}\int_e \widehat{c}^*\left(\mathbf{u}_h^M - \mathbf{u}^M\right)\cdot\nu_e[\psi] + \sum_{e\in E_h}\int_e (\widehat{c} - c)^*\,\mathbf{u}^M\cdot\nu_e[\psi].
\end{aligned}$$

Subtracting the above equation by equation (17) and setting $\psi = E_c^A$, we have,

$$\left(\phi\frac{\partial E_c^A}{\partial t}, E_c^A\right) + \sum_{R\in\mathcal{T}_h}\int_R \left(\mathbf{D}(\mathbf{u}_h^M)\nabla E_c^A\right)\cdot\nabla E_c^A + J_0^{\sigma,\beta}\left(E_c^A, E_c^A\right) \tag{25}$$
$$= \sum_{R\in\mathcal{T}_h}\int_R E_c^A \mathbf{u}_h^M\cdot\nabla E_c^A - \sum_{e\in E_h}\int_e \left(E_c^A\right)^* \mathbf{u}_h^M\cdot\nu_e\left[E_c^A\right]$$
$$+ \int_\Omega E_c^A q^- E_c^A + \sum_{i=1}^{13} T_i$$

where

$$T_1 = \left(\phi\frac{\partial\widehat{c}-c}{\partial t}, E_c^A\right),$$
$$T_2 = \sum_{R\in\mathcal{T}_h}\int_R \left(\mathbf{D}(\mathbf{u}_h^M) - \mathbf{D}(\mathbf{u}^M)\right)\nabla\widehat{c}\cdot\nabla E_c^A,$$
$$T_3 = \sum_{R\in\mathcal{T}_h}\int_R \mathbf{D}(\mathbf{u}^M)\nabla\left(\widehat{c}-c\right)\cdot\nabla E_c^A,$$
$$T_4 = J_0^{\sigma,\beta}\left(\widehat{c}-c, E_c^A\right),$$
$$T_5 = -\sum_{R\in\mathcal{T}_h}\int_R \widehat{c}\left(\mathbf{u}_h^M - \mathbf{u}^M\right)\cdot\nabla E_c^A,$$
$$T_6 = -\sum_{R\in\mathcal{T}_h}\int_R \left(\widehat{c}-c\right)\mathbf{u}^M\cdot\nabla E_c^A,$$
$$T_7 = -\sum_{e\in E_h}\int_e \left\{\left(\mathbf{D}(\mathbf{u}_h^M) - \mathbf{D}(\mathbf{u}^M)\right)\nabla\widehat{c}\cdot\nu_e\right\}\left[E_c^A\right],$$
$$T_8 = -\sum_{e\in E_h}\int_e \left\{\mathbf{D}(\mathbf{u}^M)\nabla\left(\widehat{c}-c\right)\cdot\nu_e\right\}\left[E_c^A\right],$$
$$T_9 = \sum_{e\in E_h}\int_e \left\{\mathbf{D}(\mathbf{u}_h^M)\nabla E_c^A\cdot\nu_e\right\}\left[\widehat{c}-c\right],$$
$$T_{10} = \sum_{e\in E_h}\int_e \widehat{c}^*\left(\mathbf{u}_h^M - \mathbf{u}^M\right)\cdot\nu_e\left[E_c^A\right],$$
$$T_{11} = \sum_{e\in E_h}\int_e \left(\widehat{c}-c\right)^*\mathbf{u}^M\cdot\nu_e\left[E_c^A\right],$$
$$T_{12} = -\int_\Omega \left(\widehat{c}-c\right)q^- E_c^A,$$

and

$$T_{13} = \int_\Omega \left(r(\mathcal{M}(c)) - r(\mathcal{M}(c_h))\right)E_c^A.$$

Theorem 35 (Error estimate for transport) *Let $(\mathbf{u}, p, c)$ be solutions to (1) and (2) with boundary conditions (4) through (7) and initial condition. Assume that no flow boundary conditions are imposed. Further assume $\mathbf{u}$ and c satisfy*

$$\mathbf{u} \in C_B\left(\overline{\Omega} \times (0, T_f]\right) \bigcap L^\infty\left((0, T_f], H^{l+\frac{1}{2}}(\Omega)\right) \tag{26}$$

$$c \in C_B\left((0, T_f], W^{1,\infty}(\Omega)\right) \bigcap L^2\left((0, T_f], H^m(\Omega)\right) \tag{27}$$

and

$$\frac{\partial c}{\partial t} \in L^2\left((0, T_f], H^n(\Omega)\right). \tag{28}$$

We also assume $q^- \in L^\infty(\Omega)$. Assume M is chosen sufficiently large such that $M \geqslant \|\mathbf{u}\|_{(L^\infty(\Omega))^d}$ and $M \geqslant \|c\|_{L^\infty(\Omega)}$ for all $t \in (0, T_f]$. Then, there exists a constant $C > 0$ independent of mesh and an $h_0 > 0$ such that if $h \leqslant h_0$ and $\tau \in (0, T_f]$ then,

$$\begin{aligned} &|||E_c|||_0^2(\tau) + \int_0^\tau |||\nabla E_c|||_0^2(t)dt \\ &\leqslant C\int_0^\tau |||E_c|||_0^2 + C\int_0^\tau \|E_{\mathbf{u}}\|_{(L^2(\Omega))^d}^2 + Ch^{\min(2k,2l,2r,2m-2,2n)} \end{aligned} \tag{29}$$

where $k \geqslant 2$, $r \geqslant 1$ are the order of discontinuous space for flow and transport subproblems, respectively, defined in the above notation subsection; l, m, n describe the regularity order of solution $\mathbf{u}, c, \partial c/\partial t$ respectively, as defined in equations (26), (27) and (28); and the optimal $\beta = 1$ is chosen.

Proof. Using Lemma 33, we bound the left hand side of (25) by,

$$\begin{aligned} &\left(\phi\frac{\partial E_c^A}{\partial t}, E_c^A\right) + \sum_{R\in\mathcal{T}_h}\int_R \left(\mathbf{D}(\mathbf{u}_h^M)\nabla E_c^A\right)\cdot\nabla E_c^A + J_0^{\sigma,\beta}\left(E_c^A, E_c^A\right) \\ &\geqslant \frac{1}{2}\frac{\partial}{\partial t}|||\sqrt{\phi}E_c^A|||_0^2 + d_{m,*}|||\nabla E_c^A|||_0^2 + J_0^{\sigma,\beta}\left(E_c^A, E_c^A\right) \end{aligned}$$

Moreover, we have

$$\begin{aligned} \sum_{R\in\mathcal{T}_h}\int_R E_c^A\mathbf{u}_h^M\cdot\nabla E_c^A &\leqslant M\sum_{R\in\mathcal{T}_h}\left\|E_c^A\right\|_{L^2(R)}\left\|\nabla E_c^A\right\|_{(L^2(R))^d} \\ &\leqslant \frac{C}{\varepsilon}|||E_c^A|||_0^2 + \varepsilon|||\nabla E_c^A|||_0^2 \end{aligned}$$

where ε is a small positive constant. We bound the second term as follows.

$$\begin{aligned}
-\sum_{e\in E_h}\int_e (E_c^A)^* \mathbf{u}_h^M\cdot\nu_e\,[E_c^A] \\
&\leqslant M\sum_{e\in E_h}\left\|(E_c^A)^*\right\|^2_{L^2(e)}\left\|[E_c^A]\right\|^2_{L^2(e)}\\
&\leqslant M\sum_{e\in E_h}\left(\frac{\varepsilon}{M}\frac{\sigma_e}{h_e^\beta}\left\|[E_c^A]\right\|^2_{L^2(e)}+\frac{Ch_e^\beta}{\varepsilon}\left\|(E_c^A)^*\right\|^2_{L^2(e)}\right)\\
&\leqslant \varepsilon J_0^{\sigma,\beta}(E_c^A,E_c^A)+\frac{C}{\varepsilon}\|E_c^A\|_0^2
\end{aligned}$$

where we have assumed $\beta\geqslant 1$. The third term models extraction wells, and can be bounded as

$$\int_\Omega E_c^A q^- E_c^A \leqslant \|q^-\|_{L^\infty}\sum_{R\in\mathcal{T}_h}\|(E_c^A)\|^2_{L^2(R)}\leqslant C\|E_c^A\|_0^2$$

Now we bound the terms T_1 through T_{13}.

$$\begin{aligned}
T_1 &\leqslant \phi^*\left\|\frac{\partial E_c^I}{\partial t}\right\|_{L^2(\Omega)}\|E_c^A\|_{L^2(\Omega)}\leqslant\frac{\phi^*}{2}\left\|\frac{\partial E_c^I}{\partial t}\right\|^2_{L^2(\Omega)}+\frac{\phi^*}{2}\|E_c^A\|^2_{L^2(\Omega)};\\
T_2 &\leqslant \|\nabla\widehat{c}\|_{(L^\infty(\Omega))^3}\sum_{R\in\mathcal{T}_h}\left\|\mathbf{D}(\mathbf{u}_h^M)-\mathbf{D}(\mathbf{u}^M)\right\|_{(L^2(R))^{d\times d}}\left\|\nabla E_c^A\right\|_{(L^2(R))^d}\\
&\leqslant C\|\nabla c\|_{(L^\infty(\Omega))^3}\sum_{R\in\mathcal{T}_h}\left\|\mathbf{u}_h^M-\mathbf{u}^M\right\|_{(L^2(R))^d}\left\|\nabla E_c^A\right\|_{(L^2(R))^d}\\
&\leqslant \frac{C}{\varepsilon}\|E_\mathbf{u}\|^2_{(L^2(\Omega))^d}+\varepsilon\|\nabla E_c^A\|_0^2;
\end{aligned}$$

where we have used the fact $\|\nabla\widehat{c}\|_{(L^\infty(\Omega))^3}\leqslant C\|\nabla c\|_{(L^\infty(\Omega))^3}\leqslant C$ and Lemmas 32 and 34 for bounding the term T_2. We note that

$$\begin{aligned}
T_3 &\leqslant C\sum_{R\in\mathcal{T}_h}\int_R\nabla E_c^I\cdot\nabla E_c^A\leqslant\frac{C}{\varepsilon}\|\nabla E_c^I\|_0^2+\varepsilon\|\nabla E_c^A\|_0^2;\\
T_4 &\leqslant \sum_{e\in E_h}\frac{\sigma_e}{h_e^\beta}\left(\varepsilon\left\|[E_c^A]\right\|^2_{L^2(e)}+\frac{C}{\varepsilon}\left\|[E_c^I]\right\|^2_{L^2(e)}\right)\\
&\leqslant \varepsilon J_0^{\sigma,\beta}(E_c^A,E_c^A)+\frac{C}{\varepsilon h^\beta}\sum_{e\in E_h}\left\|[E_c^I]\right\|^2_{L^2(e)}\\
&\leqslant \varepsilon J_0^{\sigma,\beta}(E_c^A,E_c^A)+\frac{C}{\varepsilon h^\beta}\sum_{R\in\mathcal{T}_h}\left(h^{-1}\left\|E_c^I\right\|^2_{L^2(R)}+h\left\|\nabla E_c^I\right\|^2_{(L^2(R))^d}\right)\\
&\leqslant \varepsilon J_0^{\sigma,\beta}(E_c^A,E_c^A)+\frac{C}{\varepsilon h^{\beta+1}}\|E_c^I\|_0^2+\frac{C}{\varepsilon h^{\beta-1}}\|\nabla E_c^I\|_0^2.
\end{aligned}$$

T_5 can be bounded similarly as the term T_3, i.e.,

$$T_5 \leqslant \|\widehat{c}\|_{L^\infty(\Omega)} \sum_{R\in\mathcal{T}_h} \left\|\mathbf{u}_h^M - \mathbf{u}^M\right\|_{(L^2(R))^d} \left\|\nabla E_c^A\right\|_{(L^2(R))^d}$$
$$\leqslant \frac{C}{\varepsilon} \|E_{\mathbf{u}}\|^2_{(L^2(\Omega))^d} + \varepsilon \|\!|\nabla E_c^A|\!\|_0^2$$

where we have used the fact $\|\widehat{c}\|_{L^\infty(\Omega)} \leqslant C\,\|c\|_{L^\infty(\Omega)} \leqslant C$.

$$T_6 \leqslant M \sum_{R\in\mathcal{T}_h} \left\|E_c^I\right\|^2_{L^2(R)} \left\|\nabla E_c^A\right\|^2_{(L^2(R))^d} \leqslant \frac{C}{\varepsilon}\|\!|E_c^I|\!\|_0^2 + \varepsilon\|\!|\nabla E_c^A|\!\|_0^2.$$

The term T_7 is bounded as

$$T_7 \leqslant \sum_{e\in E_h} \left(\varepsilon\frac{\sigma_e}{h_e^\beta}\left\|[E_c^A]\right\|^2_{L^2(e)} + \frac{Ch^\beta}{\varepsilon}\left\|\mathbf{D}(\mathbf{u}_h^M) - \mathbf{D}(\mathbf{u}^M)\right\|^2_{(L^2(e))^{d\times d}}\right)$$
$$\leqslant \varepsilon J_0^{\sigma,\beta}\left(E_c^A, E_c^A\right) + \frac{Ch^\beta}{\varepsilon}\sum_{e\in E_h}\|\mathbf{u}_h - \mathbf{u}\|^2_{(L^2(e))^d}$$
$$\leqslant \varepsilon J_0^{\sigma,\beta}\left(E_c^A, E_c^A\right) + \frac{Ch^\beta}{\varepsilon}\sum_{e\in E_h}\|\mathbf{u}_h - \widehat{\mathbf{u}}\|^2_{(L^2(e))^d}$$
$$+\frac{Ch^\beta}{\varepsilon}\sum_{e\in E_h}\|\widehat{\mathbf{u}} - \mathbf{u}_h\|^2_{(L^2(e))^d}$$
$$\leqslant \varepsilon J_0^{\sigma,\beta}\left(E_c^A, E_c^A\right) + \frac{Ch^{\beta-1}}{\varepsilon}\sum_{R\in\mathcal{T}_h}\|\mathbf{u}_h - \widehat{\mathbf{u}}\|^2_{(L^2(R))^d}$$
$$+\frac{C}{\varepsilon}h^{\min(2k+\beta-1,2l+\beta-1)}$$
$$\leqslant \varepsilon J_0^{\sigma,\beta}\left(E_c^A, E_c^A\right) + \frac{Ch^{\beta-1}}{\varepsilon}\sum_{R\in\mathcal{T}_h}\left(\|E_{\mathbf{u}}\|^2_{(L^2(R))^d} + \left\|E_{\mathbf{u}}^I\right\|^2_{(L^2(R))^d}\right)$$
$$+\frac{C}{\varepsilon}h^{\min(2k+\beta-1,2l+\beta-1)}$$
$$\leqslant \varepsilon J_0^{\sigma,\beta}\left(E_c^A, E_c^A\right) + \frac{Ch^{\beta-1}}{\varepsilon}\|E_{\mathbf{u}}\|^2_{(L^2(\Omega))^d} + \frac{C}{\varepsilon}h^{\min(2k+\beta-1,2l+\beta-1)},$$

where we have used the fact$\|\nabla\widehat{c}\|_{(L^\infty(\Omega))^d} \leqslant C\,\|\nabla c\|_{(L^\infty(\Omega))^d} \leqslant C$ and the approximation properties for $\mathbf{u}$. The boundedness of T_8 can be shown by using the penalty term

$$T_8 \leqslant C\sum_{e\in E_h}\|\nabla(\widehat{c} - c)\|_{(L^2(e))^d}\left\|[E_c^A]\right\|_{L^2(e)}$$
$$\leqslant \varepsilon J_0^{\sigma,\beta}\left(E_c^A, E_c^A\right) + \frac{C}{\varepsilon}h^{\beta-1}\|\!|\nabla E_c^I|\!\|_0^2;$$

$$T_9 \leqslant C \sum_{e\in E_h} \left\|\nabla E_c^A\right\|_{(L^2(e))^d} \|[\widehat{c}-c]\|_{L^2(e)}$$
$$\leqslant \varepsilon\|\nabla E_c^A\|_0^2 + \frac{C}{\varepsilon h^2}\|E_c^I\|_0^2 + \frac{C}{\varepsilon}\|\nabla E_c^I\|_0^2.$$

The terms T_{10} and T_{11} are similar to T_7 and are bounded as follows:

$$T_{10} \leqslant \|\widehat{c}\|_{L^\infty(\Omega)} \sum_{e\in E_h} \|\mathbf{u}_h - \mathbf{u}\|^2_{(L^2(e))^d} \left\|\left[E_c^A\right]\right\|_{L^2(e)}$$
$$\leqslant \varepsilon J_0^{\sigma,\beta}\left(E_c^A, E_c^A\right) + \frac{Ch^{\beta-1}}{\varepsilon}\|E_{\mathbf{u}}\|^2_{(L^2(\Omega))^d} + \frac{C}{\varepsilon}h^{\min(2k+\beta-1,2l+\beta-1)};$$
$$T_{11} \leqslant C \sum_{e\in E_h} \left\|\left(E_c^I\right)^*\right\|_{L^2(e)} \left\|\left[E_c^A\right]\right\|_{L^2(e)}$$
$$\leqslant \varepsilon J_0^{\sigma,\beta}\left(E_c^A, E_c^A\right) + \frac{Ch^{\beta-1}}{\varepsilon}\|E_c^I\|_0^2 + \frac{Ch^{\beta+1}}{\varepsilon}\|\nabla E_c^I\|_0^2.$$

The term T_{12} can be bounded similarly as the term $\int_\Omega E_c^A q^- E_c^A$ above,

$$T_{12} \leqslant C\|E_c^A\|_0^2 + C\|E_c^I\|_0^2.$$

The term T_{13} can be bounded using properties of the $\mathcal{M}$ operator, i.e.

$$T_{13} \leqslant C\|E_c^A\|_0^2 + C\|E_c^I\|_0^2.$$

Combining the above bounds for the the error equation (25) and choosing ε sufficiently small, we have, for $h \leqslant \left(\frac{\sigma_*}{3M}\right)^{1/\beta}$,

$$\frac{1}{2}\frac{\partial}{\partial t}\left\|\sqrt{\phi}E_c^A\right\|^2 + \frac{d_{m,*}}{2}\|\nabla E_c^A\|_0^2 + \frac{1}{2}J_0^{\sigma,\beta}\left(E_c^A, E_c^A\right)$$
$$\leqslant C\|E_c^A\|_0^2 + \left(C + Ch^{\beta-1}\right)\|E_{\mathbf{u}}\|^2_{(L^2(\Omega))^d}$$
$$+C\|\partial E_c^I/\partial t\|_0^2 + \left(C + \frac{C}{h^{\beta+1}} + \frac{C}{h^2} + Ch^{\beta-1}\right)\|E_c^I\|_0^2$$
$$+\left(C + \frac{C}{h^{\beta-1}} + Ch^{\beta-1} + Ch^{\beta+1}\right)\|\nabla E_c^I\|_0^2 + Ch^{\min(2k+\beta-1,2l+\beta-1)}.$$

We observe that the optimal choice for β is $\beta = 1$, Thus,

$$\frac{1}{2}\frac{\partial}{\partial t}\|\sqrt{\phi}E_c^A\|_0^2 + \frac{d_{m,*}}{2}\|\nabla E_c^A\|_0^2 + \frac{1}{2}J_0^{\sigma,\beta}\left(E_c^A, E_c^A\right)$$
$$\leqslant C\|E_c^A\|_0^2 + C\,\|E_{\mathbf{u}}\|^2_{(L^2(\Omega))^d}$$
$$+C\|\partial E_c^I/\partial t\|_0^2 + \frac{C}{h^2}\|E_c^I\|_0^2 + C\|\nabla E_c^I\|_0^2 + Ch^{\min(2k,2l)}.$$

Integrating above inequality with respect to time from 0 to τ $(0 \leqslant \tau \leqslant T_f)$ and we obtain,

$$\begin{aligned}
&\|\sqrt{\phi}E_c^A\|_0^2(\tau) + \frac{d_{m,*}}{2}\int_0^\tau \|\nabla E_c^A\|_0^2(t)dt + \frac{1}{2}\int_0^\tau J_0^{\sigma,\beta}\left(E_c^A, E_c^A\right) \\
&\leqslant \|\sqrt{\phi}E_c^A\|_0^2(0) + C\int_0^\tau \|E_c^A\|_0^2 + C\int_0^\tau \|E_{\mathbf{u}}\|^2_{(L^2(\Omega))^d} \\
&\quad + C\int_0^\tau \|\partial E_c^I/\partial t\|_0^2 + \frac{C}{h^2}\int_0^\tau \|E_c^I\|_0^2 + C\int_0^\tau \|\nabla E_c^I\|_0^2 + CT_f h^{\min(2k,2l)}.
\end{aligned}$$

Noting that $\left\|\sqrt{\phi}E_c^A\right\|^2(0) = \left\|\sqrt{\phi}E_c^I\right\|^2(0)$, and that ϕ is uniformly bounded below and above, using the approximation results for $c \in \mathcal{D}_r(\mathcal{T}_h)$, we have,

$$\begin{aligned}
&\|E_c^A\|_0^2(\tau) + \int_0^\tau \|\nabla E_c^A\|_0^2(t)dt + \int_0^\tau J_0^{\sigma,\beta}\left(E_c^A, E_c^A\right) \\
&\leqslant C\int_0^\tau \|E_c^A\|_0^2 + C\int_0^\tau \|E_{\mathbf{u}}\|^2_{(L^2(\Omega))^d} + Ch^{\min(2k,2l,2r,2m-2,2n)}.
\end{aligned}$$

The theorem follows by the triangle inequality. □

The optimal error estimate for concentration in $L^2(H^1)$ is obtained as long as the degree of polynomial space for flow is higher than or equal to the degree of polynomial space for transport.

Theorem 36 *Assume all the assumptions in Theorem 35 hold. Then there exists a constant $C > 0$ independent of finite element mesh size h such that*

$$\|E_c\|_{L^\infty(0,T_f;L^2)} + \|E_c\|_{L^2(0,T_f;H^1)} \leqslant Ch^{\min(k,l,r,m-1,n)}.$$

Proof. Combining the above results for transport and for flow, the theorem follows by the Gronwall's inequality. □

Remark 37 *Additionally, the proof above can be easily extended to get the nearly-optimal in p error estimate for concentration in $L^2(H^1)$. We can also obtain the same error estimate for the symmetric DG with penalty by a similar proof. Optimal error estimate for concentration in $L^2(L^2)$ can be be obtained for the symmetric DG with penalty using a special technique, and will be discussed in a later paper.*

4 Numerical Results

Simulations of the ANDRA-Couplex1 test case were run using the 2D DG Code developed by Rivière [5]. The code was run on Linux-PCs with 1.2GHz-Pentium III processors and 512M RAM. For all simulations and figures, time steps of 100 years were used. Simulations began at 0 years and go to 10M years. The grid consisted of 1920 quadrilateral elements where the layer

Fig. 2. Pressure Field; colored, see page 369

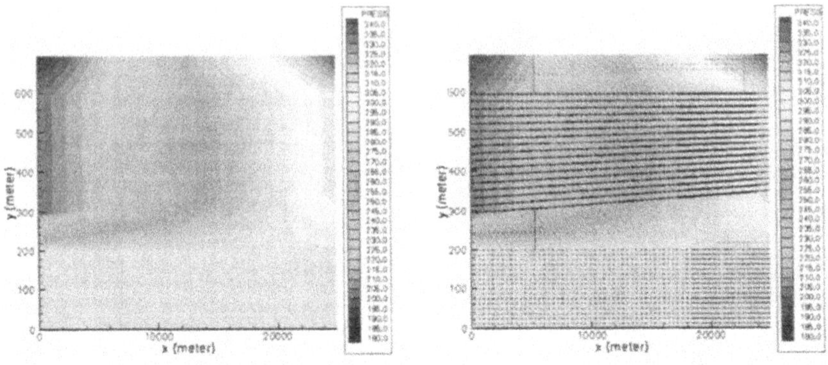

(a) Pressure Only (b) Pressure and Velocity Field

boundaries fall along the edges of the elements. For both flow and transport the complete quadratic basis functions were used, resulting in 11520 unknowns. A sparse linear solver was used for flow, and GMRES with restarts was used for transport.

The resulting pressure field is shown in Figure 2(a), and the velocity field is superimposed in Figure 2(b). Note that magnitude of the velocity varies greatly in the different layers due to the discontinuities in the permeabilities of the layers. In addition, in the clay and marl layers, where permeability is small, transport is dominated by molecular diffusion. In the limestone and dogger limestone layers, where permeability is large, transport is dominated by advection and dispersion.

Figures 3(a,b) are typical concentration results for Iodine at different simulation times. It should be observed that even for small concentrations the DG solution is locally conservative. The low numerical diffusion of the DG method is also important in this benchmark problem because of the long simulation time.

Another nice feature found in numerical experiments is that DG can handle the full dispersion tensor easily and accurately. This feature is also very important to our test case. To show the importance of mechanical dispersion for the Andra-Couplex1 case, we also present the concentration results of Iodine (Figures 4(a,b)) for the same case except without dispersion. The dramatic difference shows the strong influence of dispersion for the Andra-Couplex1 case.

Slope limiting is not covered in the theoretical part of this paper, but is found to be important numerically in this case. It is used to weaken the oscillatory effect of DG without adding small amounts of numerical diffusion.

Fig. 3. Concentration of Iodine with Dispersion; colored, see page 370

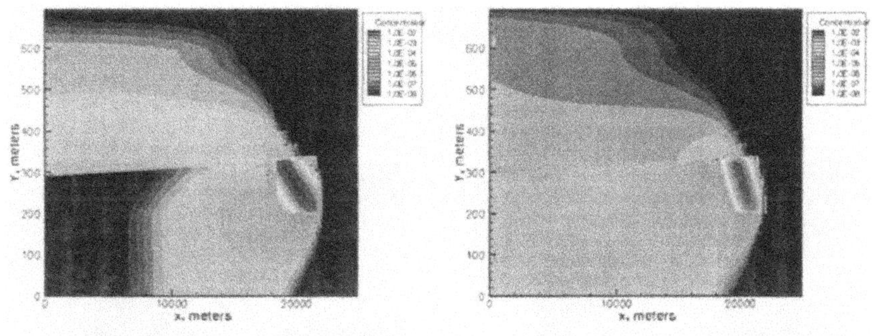

(a) 500k years (b) 1M years

Fig. 4. Concentration of Iodine without Dispersion; colored, see page 370

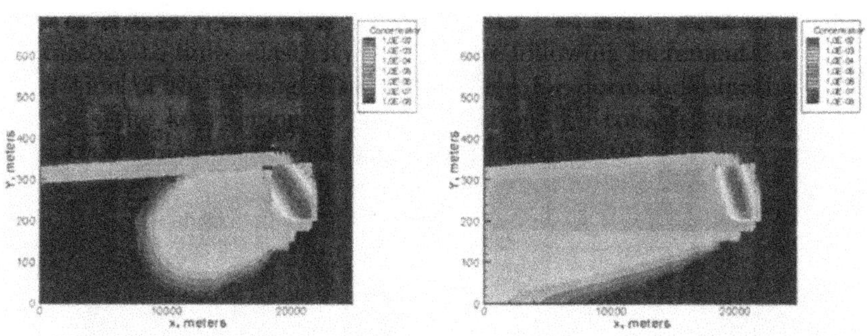

(a) 500k years (b) 1M years

The slope limiter of Cockburn and Shu [2] is implemented with the corrected *minmod* function, $\overline{m}$.

The effect of the slope limiting for early times can be seen in Figures 5(a,b). Note that slope limiting affects the solution in the presence of high gradients of concentration as we expect. However, far from the repository this effect is considerably less pronounced. At later times, when high concentration gradients have diminished, the slope limiter makes little to no noticeable difference in the solution, e.g. Figures 6(a,b). Therefore, in our dispersion cases after $500k$ *years* solutions are shown without the use of slope limiting. All simulations were run out to $10M$ *years* as originally specified [9], but at large times the answer is not interesting and thus the largest time shown is $5M$ *years*.

Fig. 5. Slope Limiting the Concentration of Iodine with Dispersion (Time=$500k$ years); colored, see page 371

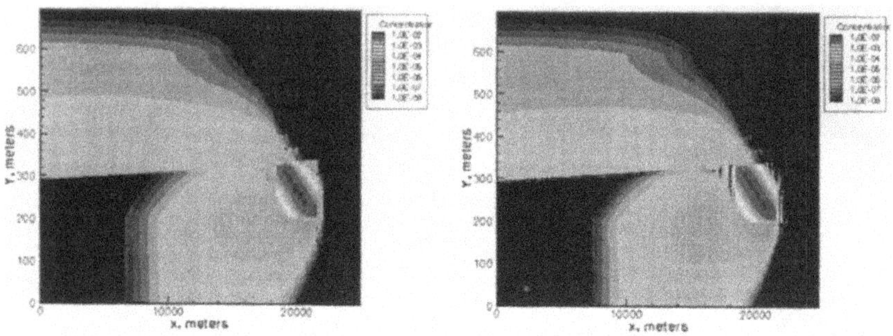

(a) Slope Limited (b) Not Slope Limited

Fig. 6. Concentration of Iodine at Large Times with Dispersion; colored, see page 371

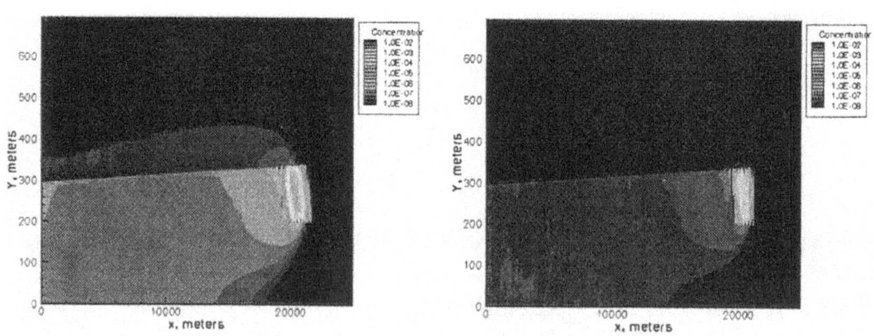

(a) 3M years (b) 5M years

Compared with Iodine, the concentration of Plutonium is localized near the repository as shown in Figures 7(a,b). This is due to the high retardation factor, high effective porosity and shorter half-life of Plutonium, even though the diffusivity of Plutonium is larger in the clay layer.

5 Conclusion and Future Work

A discontinuous Galerkin method for flow and reactive transport in porous media is proposed in this paper. A "cut-off" operator is introduced in the discontinuous Galerkin formulation in order to make the scheme converge. Optimal *a priori* error estimates in the energy norm are derived for concentration. The scheme in this paper can handle fairly large types of geobiochemistry as long as they are locally Lipschitz.

A DG Code for flow and transport has been applied to a nuclear waste management problem. This problem was characterized by large discontinuous jumps in permeability and molecular diffusivity, both of which DG was able to accurately incorporate into the simulation. It has been observed that the local conservation property of DG and low numerical diffusion help limit the accumulation of errors over time. It has also been shown that mechanical dispersion dramatically affects the size and shape of the contaminant plume, and that DG can handle the addition of dispersion easily and accurately. Finally, the ability of DG to make use of slope limiting allows a great deal of flexibility for the problem in which one is interested in very small concentrations.

In the future, we would like to examine the effect of adaptivitiy with respect to the mesh, the orders of approximation, and target functionals. In addition to our observations concerning the numerical diffusion added by penalty terms in the non-symmetric form, we would like to examine the numerical behaviors of the symmetric and non-symmetric forms.

Acknowledgements : Each author was supported in part by the following grants NSF EIA-0121523, ITR 060808-01090729-08, DoD DMS-9873326.

References

1. C. E. Baumann and J. T. Oden. A discontinuous *hp* finite element method for convection-diffusion problems. *Comput. Methods Appl. Mech. Engrg.*, 175(3-4):311–341, 1999.
2. B. Cockburn and C. W. Shu. The Runge-Kutta Discontinuous Galerkin Method for Conservation Laws V. *J. Comput. Phys.*, 141:199-224, 1998.
3. V. Girault, B. Rivière, and M. F. Wheeler. A discontinuous Galerkin method with non-overlapping domain decomposition for the Stokes and Navier-Stokes problems. TICAM report 02-08, Austin 2002, submitted for publication to Mathematics of Computation.
4. J. T. Oden, I. Babuška, and C. E. Baumann. A discontinuous hp finite element method for diffusion problems. *J. Comput. Phys.*, 146:491–516, 1998.
5. B. Rivière. *Discontinuous Galerkin finite element methods for solving the miscible displacement problem in porous media.* PhD thesis, The University of Texas at Austin, 2000.

Fig. 7. Concentration of Plutonium; colored, see page 372

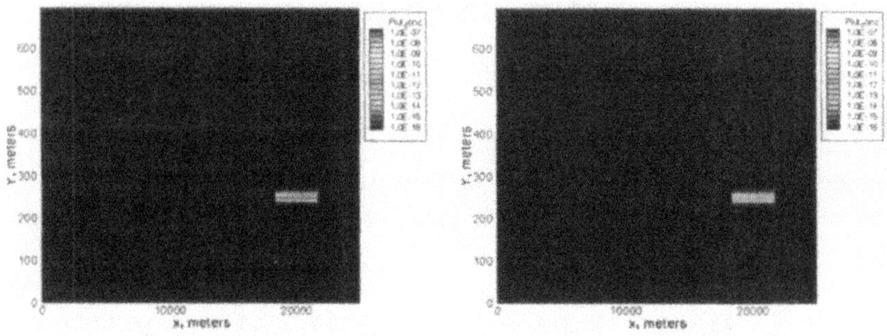

(a) 500k years

(b) 1M years

6. B. Rivière, M. F. Wheeler, and V. Girault. A priori error estimates for finite element methods based on discontinuous approximation spaces for elliptic problems. *SIAM J. Numer. Anal.*, 39(3):902–931, 2001.
7. S. Sun, B. Rivière and M. F. Wheeler. A Combined Mixed Finite Element and Discontinuous Galerkin Method for Miscible Displacement Problem in Porous Media. *International Symposium on Computational and Applied PDEs*, Zhangjiaje, China, 2001.
8. M. F. Wheeler. An Elliptic Collocation-Finite Element Method with Interior Penalties. *SIAM J. Numer. Anal.*, 15(1):152–161, 1978.
9. ANDRA (the French National Radioactive Waste Management Agency). *The Couplex Test Cases.* http://www.andra.fr/couplex/.

Part I

Material Modelling and Multiscale Problems

Modelling and numerical simulation of martensitic transformation

Marcel Arndt

Universität Bonn, Institut für Angewandte Mathematik, Wegelerstr. 6, 53115 Bonn, Germany

Abstract. We consider the time evolution of martensite-austenite phase transformations in shape memory alloys. The process is modelled on a microscopical, continuum mechanical level by partial differential equations. Here the stored energy density is used, which is a rotationally invariant and nonconvex function of the transformation gradient with minima at several wells. The energy dissipation observed in physical experiments is modelled by an additional homogeneous degree-1 dissipation potential for rate-independent phase transformations. Furthermore, viscous and capillarity-like effects are handled by nonlocal terms of higher order.

We discuss the existence of solutions of this model and derive its energetic properties. Its discretization and implementation is described. Special care is taken to handle the nonlocal terms efficiently using wavelet techniques. Finally we present numerical experiments for an Indium-Thallium alloy under various loading regimes.

1 Introduction

There are many crystals such as InTl and CuAlNi alloys that can undergo solid-solid phase transformations, because they admit (under certain circumstances) not only one, but several stable crystal structures. Usually there are one high-temperature phase called the austenite and two or more symmetry-related low-temperature phases called the martensites. Hence, at high temperatures, the crystal purely consists of the austenite phase. At low temperatures it may consist of one pure martensite phase, but also of a mixture of different martensite phases. Close to the transition temperature also mixtures of both austenite and martensite phases can be observed. Typically a mixture occurs in form of microstructure, i.e. fine spatial oscillations like laminates or more complicated structures. Phase transformations may be induced by changes of temperature, outer forces etc.

We use a continuum-mechanical model. A continuum function describes the deformation of the crystal. The phases can then be characterized in terms of the deformation gradients as explained below.

2 The Continuum Model

Let $\Omega \subset \mathbb{R}^n$ be a bounded domain which describes the reference configuration, i.e. the crystal in the pure austenite state. Furthermore let $[0,T]$ be a time interval and define $Q := \Omega \times (0,T)$. The function $u : Q \to \mathbb{R}^n$ describes the deformation of the crystal. Hard-device loading is implemented by prescribing Dirichlet boundary conditions on a part Γ_0 of the boundary $\partial\Omega$.

The model consists of three components: the stored energy describing the elastic properties of the solid, the dissipative mechanism and a capillarity/viscosity part.

2.1 Model components

Let $\varphi : \mathbb{R}^{n\times n} \to \mathbb{R}$ denote the *stored energy density*. The elastic energy is then given by

$$E^{\mathrm{s}}(u) := \int_\Omega \varphi(\nabla u(x))\,\mathrm{d}x. \tag{1}$$

The exact shape of φ depends on the concrete material. Nevertheless some properties are typical in this context:

1. Frame invariance: $\varphi(RF) = \varphi(F)$ for all $F \in \mathrm{SO}(n)$.
2. Crystal symmetry: $\varphi(SFS^T) = \varphi(F)$ for all S from the symmetry group.
3. Multiple well structure: φ has a local minimum at each phase $\mathrm{SO}(n)U_\alpha$, $\alpha = 1,\dots,N$, where N denotes the number of phases.

There is one well $\mathrm{SO}(n)U_1$ for the austenite phase and several wells for the martensite phases. Because the reference configuration Ω is chosen in such a way that it represents the austenite phase, we have $U_1 = \mathrm{Id}$, where Id denotes the identity matrix. The nonconvex structure of φ due to several distinct minima often leads to fine oscillations in the deformation, known as microstructure.

One physical observation is that phase transformations typically dissipate energy. This process is rate-independent and will be modelled phenomenologically. For each subset $\ell \in L \subset 2^{\{1,\dots,N\}}$ of phases we introduce a *phase indicator*

$$\lambda_\ell : \mathbb{R}^{n\times n} \to \mathbb{R}. \tag{2}$$

It takes the constant value $\Lambda_{1,\ell} \in \mathbb{R}$ in a neighbourhood of all phases U_α with $\alpha \in \ell$ and another constant value $\Lambda_{2,\ell} \in \mathbb{R}$ nearby of the other phases. The phase indicators will be later incorporated into the model such that the energy dissipation during the transformation from a phase α to a phase β is

$$\sum_{\ell\in L:\ (\alpha\in\ell\wedge\beta\notin\ell)\vee(\alpha\notin\ell\wedge\beta\in\ell)} |\Lambda_{1,\ell} - \Lambda_{2,\ell}|. \tag{3}$$

The choice of the particular constants gives enough freedom to reflect the actual dissipation energies of the concrete material being modelled.

Furthermore we introduce viscous and capillarity-like terms. Often they are described in terms of second space derivatives of the deformation such as Δu and $\partial_t \Delta u$ respectively. The drawback is that this results in a high regularity of u, which inhibits the sharp interfaces observed in physical experiments. We take this into account by using lower derivatives in the Sobolev-Slobodeckiĭ space $W^{1+\gamma,2}$, also known as fractional order space. Let

$$a : W^{1+\gamma,2}(\Omega, \mathbb{R}^{n\times n}) \to \mathbb{R}, \quad a(z) = \frac{1}{4}\int_\Omega \int_\Omega \frac{\|z(x) - z(y)\|_2^2}{\|x-y\|_2^{n+2\gamma}} \, dx \, dy. \tag{4}$$

We define the viscous and capillarity-like energy by

$$E^{c}(u) := \mu_0 a(\nabla u), \qquad E^{v}(u) := \mu_1 a(\partial_t \nabla u) \tag{5}$$

with coefficients $\mu_0 \geqslant 0, \mu_1 > 0$. A choice of $\gamma \in (0, \frac{1}{2})$ allows sharp interfaces and jumps in the gradient.

2.2 Model formulation

Typically the nonconvexity effects from the stored energy dominate the low velocities of transformation processes. Therefore it is acceptable to simplify the model by neglecting the masses. The dynamics is then governed by

$$\nabla \cdot \Big(\sigma(\nabla u) + \sum_{\ell \in L} \sigma_\ell(\nabla u) + \sigma_a(\mu_1 \partial_t \nabla u + \mu_0 \nabla u) \Big) = 0. \tag{6}$$

Hereby $\sigma := \varphi'$, $\sigma_\ell \in \operatorname{sign}(\partial_t \lambda_\ell(\nabla u))\lambda_\ell'$ and $\sigma_a := a' : W^{\gamma,2} \to W^{\gamma,2^*}$ denote the stresses of the three components of the energy funtional.

Now, we call a tupel $(u, \omega) \in W^{1,2}(W^{1+\gamma,2}) \times L^\infty$ a weak solution of (6), iff for all $v \in L^2(W^{1+\gamma,2})$ with $v|_{\Gamma_0 \times (0,T)} = 0$, $w \in L^\infty$ and $z \in \operatorname{sign} v$:

$$\begin{aligned} \int_0^T \int_\Omega \Big(\sigma(\nabla u) + \sum_{\ell \in L} \omega_\ell \lambda_\ell'(\nabla u) \Big) &: \nabla v \, dx \\ &+ \langle \sigma_a(\mu_1 \partial_t \nabla u + \mu_0 \nabla u), \nabla v \rangle \, dt = 0, \end{aligned} \tag{7a}$$

$$\int_0^T \int_\Omega (\omega_\ell - w) \left((\lambda_\ell'(\nabla u) : \partial_t \nabla u) - z \right) dx \, dt \geqslant 0, \tag{7b}$$

$$u|_{\partial\Omega \times (0,T)} = u_{\mathrm{D}}, \qquad u(\cdot, 0) = u_0. \tag{7c}$$

Here u_{D} denotes Dirichlet boundary conditions and u_0 initial values.

Under moderate assumptions on the problem data, we have

Proposition 1. *The problem* (6) *has a weak solution.*

The proof follows from Proposition 4 below.

2.3 Energetics

We can derive the following energy balance for a weak solution u of (6):

Proposition 2. *For all $t_0 \in [0,T]$ we have*

$$E^s(u|_{t=t_0}) + E^c(u|_{t=t_0}) + \sum_{\ell\in L} \int_\Omega \operatorname*{Var}_{t\in[0,t_0]} \lambda_\ell(\nabla u)\,\mathrm{d}x + 2\mu_1 \int_0^{t_0} a(\partial_t \nabla u)\,\mathrm{d}t$$
$$= E^s(u|_{t=0}) + E^c(u|_{t=0}) + \int_0^{t_0} \langle \sigma_\nu, \partial_t u_{\mathrm{D}} \rangle \,\mathrm{d}t. \quad (8)$$

This states that the energy at time t_0 plus the dissipated energy during time $(0, t_0)$ equals the initial energy plus the energy exchanged over the boundary. The nonlocal character of σ_a makes it necessary to define the boundary stress σ_ν by the relation

$$\langle \sigma_\nu, v|_{\Gamma_0} \rangle := \int_0^T \int_\Omega \left(\sigma(\nabla u) + \sum_{\ell \in L} \sigma_\ell \right) : \nabla v \,\mathrm{d}x$$
$$+ \langle \sigma_a(\mu_1 \partial_t \nabla u + \mu_0 \nabla u), \nabla v \rangle \,\mathrm{d}t. \quad (9)$$

3 Discrete Model

3.1 Discretization

Our space discretization uses finite elements. To this end, we assume that the domain is polygonal. For a given $h > 0$ we divide Ω into a finite number of tetrahedra $\mathcal{T}_h$ with diameter less than h. The finite element spaces are then given by

$$V_h := \{ v_h \in C^0(\overline{\Omega}; \mathbb{R}^n) : v|_T \text{ affine } \forall T \in \mathcal{T}_h \}, \quad (10)$$
$$L_h := \{ \omega_h \in L^\infty(\Omega, \mathbb{R}^L) : v|_T \text{ constant } \forall T \in \mathcal{T}_h \}. \quad (11)$$

We obtain the discrete weak formulation of the problem in the obvious way by replacing the respective spaces in (7) by their discrete counterparts.

The discretization in time is based on finite differences. For a given time step $\tau > 0$, we restrict the problem to the time slices $t = 0, \tau, 2\tau, \ldots, T$. We exploit that our problem stems from a potential and reformulate the discrete weak problem as a time-recursive minimization problem:

For each time step $k = 1, \ldots, T/\tau$ minimize

$$E^k_{h\tau}(u^k_{h\tau}) = \int_\Omega \varphi(\nabla u^k_{h\tau}) + \sum_\ell \left| \lambda_\ell(\nabla u^k_{h\tau}) - \lambda_\ell(\nabla u^{k-1}_{h\tau}) \right|$$
$$+ \tau \mu_1 a \left(\frac{\nabla u^k_{h\tau} - \nabla u^{k-1}_{h\tau}}{\tau} \right) + \mu_0 a(\nabla u^k_{h\tau}) \quad (12)$$

subject to $u^k_{h\tau} \in V_h, u^k_{h\tau}|_{\Gamma 0} = u_D(\cdot, k\tau)$

It is easy to see by a first order necessary condition that each solution of the discrete minimization problem is a solution of the discrete weak problem. Note that it is sufficient for that purpose to find a local minimum of (12) and not a global minimum.

3.2 Existence results

Because V_h is finite-dimensional, we have under the usual coercivity assumption:

Proposition 3. *The discrete minimization problem* (12) *has at least one solution.*

Consider a sequence of solutions with $h \to 0$ and $\tau \to 0$. We can proof an a priori estimate in $W^{1,2}(W^{1+\gamma,2})$ (independent on h and τ) and thus we can choose a convergent subsequence in a weak sense. By using techniques of [2] for the nonlinear terms, we can pass to the limit and obtain

Proposition 4. *Each sequence* $\{(u_{h_n,\tau_n}, \omega_{h_n,\tau_n})\}_{n\in\mathbb{N}}$ *with* $h_n \to 0$ *and* $\tau_n \to 0$ *has a weakly* $\times$ *weakly* convergent subsequence in* $W^{1,2}(W^{1+\gamma,2}) \times L^\infty$. *The limit of each such subsequence is a weak solution of* (6).

This also implies the existence of a weak solution of the continuous problem (7) as stated in Proposition 1. The detailed statements and proofs are lengthy and will not be given here. See [1] for more information.

3.3 Solution

We use a line search method for the solution of the discrete minimization problems (12). The search direction is determined by the steepest descent method or the conjugate gradient method of Fletcher/Reeves. The stepsize is controlled by a modified Armijo method.

There is one difficulty: The double integral for the operator a leads to a complexity of $O(N^2)$, where N denotes the number of unknowns. For 3-dimensional computations this means an $O(h^{-6})$ complexity, which is unacceptable and thus has to be overcome by advanced numerical techniques. The key point for this is a further approximation of (12) by means of wavelets and norm equivalences.

Consider the wavelet representation of u_h for a given mother wavelet ψ:

$$u_h(x) = \sum_l \sum_j u_h^{l,j} \psi^{l,j}(x), \tag{13}$$

where $u_h^{l,j}$ denotes the coefficients; l is the level and j the translation. It is well-known that the following norm equivalence holds:

$$||u_h||^2_{W^{1+\gamma,2}} + a(\nabla u) \cong \sum_l 4^{(1+\gamma)l} \sum_j |u_h^{l,j}|^2 ||\Psi^{l,j}||^2_{L^2}. \tag{14}$$

With the help of a Poincaré-like inequality, we get rid of the leading term $||u_h||^2_{W^{1+\gamma,2}}$ by omitting the lowest level in the level summation. We then replace the non-local operator a in the computations by the right-hand-side of (14). The norm equivalence says that this retains the qualitative properties, hence this strategy is acceptable. The new term can be computed with complexity $O(N)$ instead of $O(N^2)$, which allows an efficient implementation.

4 Numerical Results

We implemented the full model and tested it in different situations. The stored energy density is taken from Ericksen/James (see e.g. [4, formula (2.19)]) to model a realistic 3-dimensional InTl alloy. The reference configuration is a cube. Two opposite sides of it are loaded cyclically by the Dirichlet boundary conditions. We applied tensile, shear or torsion loads and various combinations of the other parameters and let the crystal evolve.

Figure 1 shows the deformation of the crystal at different time steps for the tensile experiment. The different phases are indicated by different gray shades. One can clearly spot the pure austenite in the first cube and a laminate microstructure in the second cube. The overall ratio of the phases is shown in the phase fraction diagram (Figure 2). It proofs that the crystal is in the pure austenite phase nearby of the reference configuration, whereas during pulling a pure martensite phase dominates and during pushing an equal mixture of the remaining two martensite phases comes up, corresponding to the microstructure.

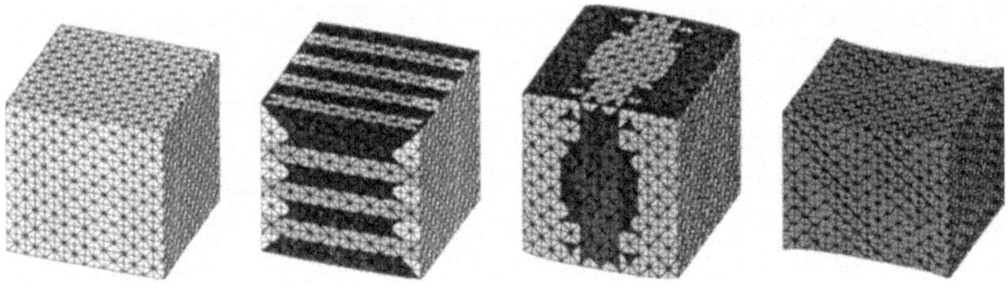

Fig. 1. Deformed specimen under tensile loading of two opposite sides. Different gray shades indicate different phases.

For the shear experiment, Figure 3 shows the deformation with a complex distribution of the different phases at two different time steps. Furthermore the X-Y-component of the stress/strain diagram is displayed. It is of high physical interest for the investigation of the hysteresis effects. The shape of the hysteresis loop is controlled by the viscous and the dissipative mechanism. The fourth diagram in Figure 3 shows the relative error in the error balance. The error stays well below 0.4%, which is a good hint that the minimization routine works reliably.

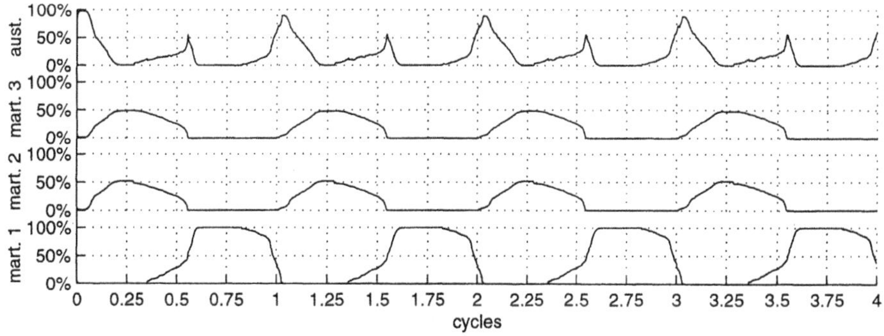

Fig. 2. Phase fraction diagram for the tensile experiment.

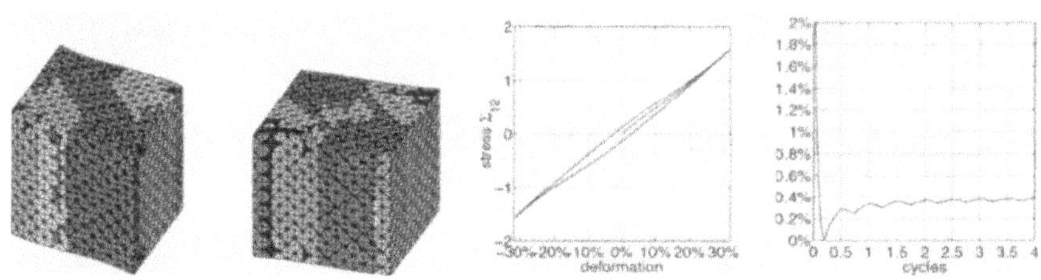

Fig. 3. Shear experiment: deformed specimen at times $t = 0.2$ and $t = 0.56$, stress/strain diagram (X-Y-component), relative error in energy balance.

The numerical results show clearly that the model works well and that its implementation is efficient. Further details are given in [1].

Acknowledgement. This is joint work with Michael Griebel (University of Bonn) and Tomáš Roubíček (University of Prague).

References

1. Arndt, M., Griebel, M., Roubicek, T. (2002) Modelling and numerical simulation of martensitic transformation in shape memory alloys. Preprint.
2. Colli, P., Visintin, A. (1990) On a class of doubly nonlinear evolution equations. Commun. Partial Diff. Equations 15, 737-756.
3. Klouček, P., Luskin, M. (1994) The computation of the dynamics of the martensitic transformation. Continuum Mech. Thermodyn. 6, 209–240.
4. Luskin, M. (1996) On the computation of crystalline microstructure. Acta Numerica 5, pp. 191-257.
5. Plecháč, P., Roubíček, T. (2002) Visco-elasto-plastic model for martensitic phase transformation in shape-memory alloys. Math. Meth. in the Appl. Sci., in print.

Coupling of Homogenization Techniques with Multigrid Solvers for Unstructured Meshes

Claus G. Bayreuther and Christian Miehe

University of Stuttgart, Institute of Applied Mechanics (Civil Engineering), Pfaffenwaldring 7, 70569 Stuttgart, Germany

Abstract. We discuss different concepts for the construction of transfer operators with regard to the application of multigrid methods to an efficient solving of strongly heterogeneous, linear elastic materials. The micro-structure is assumed to be resolved sufficiently well. The key contribution is the formulation of new physically motivated approaches for linear-elastic composites with arbitrary micro-structures. This task is achieved by means of a numerical homogenization concept based on a variational principle. The efficiency of the new approaches will be compared with alternative techniques in the scope of a representative model problem.

1 Introduction

Multigrid methods allow the efficient solution of large systems of equations, arising from finite element discretizations of complex boundary value problems. Extensive reviews of the basic multigrid concept are provided in [2], [3], [8] and [14]. We can differentiate between two classes of transfer operators. Geometrical transfer operators are based on a geometrical coarsening process of the involved sequence of grids, see [4] and [5]. Algebraic transfer operators are obtained by an algebraic coarsening, where an artificial hierarchy of coarse-grid problems is constructed based on the structure of the fine-grid matrix, see [12] and [13]. A current research topic is the combination of homogenization and multigrid solving for an efficient solution of boundary value problems of heterogeneous materials, see e.g. [1], [6], [7] and [11]. In this paper we discuss conceptual approaches to the construction of transfer operators for micro-heterogeneous materials. The new contribution is a transfer for unstructured meshes coupled with patch-wise homogenization. The homogenization concept bases on a variational principle, see [9] and [10]. The efficiency of the new transfer strategy will be demonstrated by means of a representative model problem.

2 Boundary Value Problem

Let $\boldsymbol{u} : \mathcal{B} \rightarrow \mathbb{R}^3$ denote the displacement field and $\boldsymbol{\varepsilon} = \text{sym}[\nabla \boldsymbol{u}]$ the strain field of the considered continuum $\mathcal{B} \subset \mathbb{R}^3$. The boundary value problem of linear elastostatics is governed by the field equation $\text{div}\boldsymbol{\sigma} + \bar{\boldsymbol{f}} = \boldsymbol{0}$ and by the boundary conditions $\boldsymbol{u} = \bar{\boldsymbol{u}}$ on $\partial\mathcal{B}_u$ for the displacements and $\boldsymbol{\sigma}\boldsymbol{n} = \bar{\boldsymbol{t}}$

on $\partial\mathcal{B}_t$ for the surface tractions. A finite element solution of this boundary value problem is based on the weak form of the equilibrium equation

$$G = \int_{\mathcal{B}} \delta\boldsymbol{\varepsilon}^T \boldsymbol{\sigma}\, dV - \int_{\mathcal{B}} \delta\boldsymbol{u}^T \bar{\boldsymbol{f}}\, dV - \int_{\partial\mathcal{B}} \delta\boldsymbol{u}^T \bar{\boldsymbol{t}}\, dA = 0 \tag{1}$$

where $\delta\boldsymbol{u}$ represents virtual displacements and $\delta\boldsymbol{\varepsilon} = \text{sym}[\nabla\delta\boldsymbol{u}]$. The stress-strain relation is governed by the constitutive equation $\boldsymbol{\sigma} = \mathbb{C} : \boldsymbol{\varepsilon}$. The finite element discretization of the displacements $\boldsymbol{u} = \sum N_i\, \boldsymbol{d}_i$, $\delta\boldsymbol{u} = \sum N_i\, \delta\boldsymbol{d}_i$ and strains $\boldsymbol{\varepsilon} = \sum \boldsymbol{B}_i\, \boldsymbol{d}_i$, $\delta\boldsymbol{\varepsilon} = \sum \boldsymbol{B}_i\, \delta\boldsymbol{d}_i$ in the weak form (1) on a local element domain $\mathcal{B}^e \subset \mathcal{B}$ in terms of the ansatz functions N_i, the associated gradients $\boldsymbol{B}_i$ and the nodal displacements $\boldsymbol{d}_i$ yields a linear algebraic system

$$\boldsymbol{K}\boldsymbol{d} = \boldsymbol{f} \tag{2}$$

where $\boldsymbol{K}$ and $\boldsymbol{f}$ represent the stiffness matrix and the right-hand side

$$\boldsymbol{K} = \mathop{\mathsf{A}}_{e=1}^{n_{elem}} \int_{\mathcal{B}^e} \boldsymbol{B}^T \mathbb{C} \boldsymbol{B}\, dV \quad \text{and} \quad \boldsymbol{f} = \mathop{\mathsf{A}}_{e=1}^{n_{elem}} \left\{ \int_{\mathcal{B}^e} \boldsymbol{N}^T \bar{\boldsymbol{f}}\, dV + \int_{\partial\mathcal{B}^e} \boldsymbol{N}^T \bar{\boldsymbol{t}}\, dA \right\}$$

The matrices $\boldsymbol{N}$ and $\boldsymbol{B}$ are element assemblies of nodal ansatz functions N_i and associated gradients $\boldsymbol{B}_i$. We now apply multigrid methodes to the efficient solving of (2), characterizing the fine-grid problem $\boldsymbol{K}^f \boldsymbol{d}^f = \boldsymbol{f}^f$. Let $\mathcal{S}^C \subset \mathcal{S}^f$ be the coarse- and fine-grid ansatz space. Then, the prolongation operator $\boldsymbol{P}^f{}_C \in \mathbb{R}^{n_f \times n_C}$ and the restriction operator $\boldsymbol{R}^C{}_f \in \mathbb{R}^{n_C \times n_f}$ define the mappings $\boldsymbol{P}^f{}_C : \mathcal{S}^C \to \mathcal{S}^f$ and $\boldsymbol{R}^C{}_f : \mathcal{S}^f \to \mathcal{S}^C$. On each level we pre-smooth the high-frequency residuals via Block-Backward Gauß-Seidel and post-smooth via Block-Forward Gauß-Seidel. On the coarse grid we solve the associated coarse-grid problem $\boldsymbol{K}^C \boldsymbol{d}^C = \boldsymbol{f}^C$ in terms of the coarse-grid operator $\boldsymbol{K}^C = \boldsymbol{R}^C{}_f \boldsymbol{K}^f \boldsymbol{P}^f{}_C$ and the restricted defect $\boldsymbol{f}^C = \boldsymbol{R}^C{}_f(\boldsymbol{f}^f - \boldsymbol{K}^f \boldsymbol{d}^f)$. The Galerkin ansatz for $\boldsymbol{K}^C$ implies, that $\boldsymbol{R}^C{}_f$ is the transpose of $\boldsymbol{P}^f{}_C$.

3 A Discrete Homogenization Concept

In this section, we summarize the basic features of a numerical homogenization technique for heterogeneous, linear elastic materials according to [9,10]. The deformation of the micro-structure $\mathcal{B}_p \subset \mathcal{B}$ is assumed to be driven by the macroscopic strain $\bar{\boldsymbol{\varepsilon}}$. Thus, we assume the displacement field on the micro-structure in the form $\boldsymbol{u} = \bar{\boldsymbol{\varepsilon}}\boldsymbol{x} + \tilde{\boldsymbol{w}}$ in terms of a linear part $\bar{\boldsymbol{\varepsilon}}\boldsymbol{x}$ and a superimposed fluctuation field $\tilde{\boldsymbol{w}}$. As depicted in Fig. 1, the micro-to-macro transition is performed by one of three possible constraints on $\partial\mathcal{B}_p$

$$\text{(a)}\ \tilde{\boldsymbol{w}} = \boldsymbol{0} \quad \text{or} \quad \text{(b)}\ \tilde{\boldsymbol{w}}^- = \tilde{\boldsymbol{w}}^+ \quad \text{or} \quad \text{(c)} \int_{\partial\mathcal{B}_p} \text{sym}(\tilde{\boldsymbol{w}} \otimes \boldsymbol{n})\, dA = \boldsymbol{0} \tag{3}$$

linear or periodic boundary displacements or a weak constraint, yielding homogeneous boundary stresses. In (3c) vector $\boldsymbol{n}$ is the outward normal on

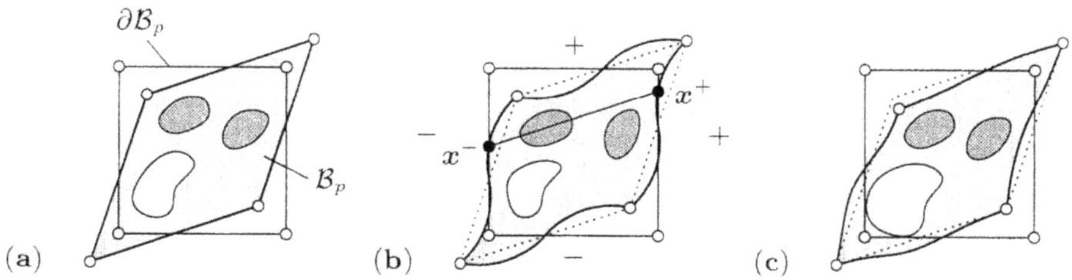

Fig. 1. Deformed micro-structure applied to (**a**) linear boundary displacements, (**b**) periodic boundary conditions and (**c**) homogeneous boundary stresses

$\partial\mathcal{B}_p$. We will neglect constraint (3b) due to our interest in non-periodic micro-structures $\mathcal{B}_p$. The micro-to-macro transition is realized by a variational condition which determines the effective free energy $\bar{\psi}(\bar{\varepsilon})$. We start with a discrete formulation of the homogenization condition

$$\bar{\psi}(\bar{\varepsilon}) = \begin{cases} \min\limits_{\tilde{\boldsymbol{d}}\in\mathcal{V}_0} W(\bar{\varepsilon}, \tilde{\boldsymbol{d}}) & \text{, associated with (3a)} \\ \min\limits_{\tilde{\boldsymbol{d}}} \max\limits_{\bar{\boldsymbol{\sigma}}} \left[W(\bar{\varepsilon}, \tilde{\boldsymbol{d}}) + C(\tilde{\boldsymbol{d}}, \bar{\boldsymbol{\sigma}}) \right] & \text{, associated with (3c)} \end{cases} \tag{4}$$

with $\mathcal{V}_0 = \{u_i \in H^1(\mathcal{B}_p) | u_i = 0 \text{ on } \partial\mathcal{B}_p\}$ and the abbreviations

$$W(\bar{\varepsilon}, \tilde{\boldsymbol{d}}) = \frac{1}{|\mathcal{B}_p|} \sum_{e=1}^{n_{elem}} \int_{\mathcal{B}_p^e} \psi(\bar{\varepsilon} + \boldsymbol{B}\tilde{\boldsymbol{d}})\, dV \ , \ C(\tilde{\boldsymbol{d}}, \bar{\boldsymbol{\sigma}}) = -\frac{\bar{\boldsymbol{\sigma}}}{|\mathcal{B}_p|} \sum_{e=1}^{n_{elem}^s} \int_{\partial\mathcal{B}_p^e} \boldsymbol{B}^s dA\, \tilde{\boldsymbol{d}} \tag{5}$$

Here, we introduced the finite element discretizations $\tilde{\boldsymbol{w}} = \boldsymbol{N}\tilde{\boldsymbol{d}}$, $\tilde{\varepsilon} = \boldsymbol{B}\tilde{\boldsymbol{d}}$ on the element domain $\mathcal{B}_p^e \subset \mathcal{B}_p$ and $\text{sym}(\tilde{\boldsymbol{w}} \otimes \boldsymbol{n}) = \boldsymbol{B}^s\tilde{\boldsymbol{d}}$ on the surface element domain $\partial\mathcal{B}_p^e \subset \partial\mathcal{B}_p$. For given strains $\bar{\varepsilon}$ variation of (4) yields the relation

$$\tilde{\boldsymbol{d}} = -\boldsymbol{F}_p W_{,\tilde{\boldsymbol{d}}\bar{\varepsilon}}\, \bar{\varepsilon} =: \boldsymbol{H}\bar{\varepsilon} \tag{6}$$

for the fluctuation displacements in terms of patch flexibility matrix $\boldsymbol{F}_p$, which we extract from the relations

$$\boldsymbol{F}_p = W_{,\tilde{\boldsymbol{d}}\tilde{\boldsymbol{d}}} \quad \text{or} \quad \begin{bmatrix} \boldsymbol{F}_p & \bullet \\ \bullet & \bullet \end{bmatrix} = \begin{bmatrix} W_{,\tilde{\boldsymbol{d}}\tilde{\boldsymbol{d}}} & C_{,\tilde{\boldsymbol{d}}\bar{\boldsymbol{\sigma}}} \\ C_{,\bar{\boldsymbol{\sigma}}\tilde{\boldsymbol{d}}} & \boldsymbol{0} \end{bmatrix}^{-1} \tag{7}$$

in the case of constraint (3a) or (3c) on $\partial\mathcal{B}_p$, respectively. In the equilibrium state we obtain the effective tangent in the closed form

$$\bar{\mathbb{C}} = W_{,\bar{\varepsilon}\bar{\varepsilon}} - W_{,\bar{\varepsilon}\tilde{\boldsymbol{d}}} \boldsymbol{F}_p W_{,\tilde{\boldsymbol{d}}\bar{\varepsilon}} \tag{8}$$

4 Transfer Operators for Heterogeneous Structures

We discuss different concepts for the construction of transfer operators, applicable to heterogeneous materials. The micro-structure is resolved sufficiently well. For the specification of the following transfer strategies we investigate two grids with accompanying conforming hierarchical ansatz spaces $\mathcal{S}^C \subset \mathcal{S}^f$. The transfer operators act on a priori defined element patches $\mathcal{B}_p \subset \mathcal{B}$.

4.1 Transfer via Injection

In the simplest way, we perform grid transfer via injection

$$P^a{}_B = R^B{}_a = \begin{cases} 1 & \text{, for corresponding degrees of freedom } a \text{ and } B \\ 0 & \text{, else} \end{cases} \tag{9}$$

Injection does not incorporate any geometrical or algebraic information.

4.2 Geometrical Transfer

The geometrical transfer is the most common transfer type in literature. The finite element discretizations of the displacement fields on the fine and coarse grid are given by $u_a^f(\boldsymbol{x}^f) = \sum_i N_i^f(\boldsymbol{x}^f) d_{ia}^f$ and $u_A^C(\boldsymbol{x}^C) = \sum_I N_I^C(\boldsymbol{x}^C) d_{IA}^C$ with nodal degrees of freedom $a = 1,2$ and $A = 1,2$, respectively. The restriction and prolongation operators are obtained from the relations

$$N_I^C(\boldsymbol{x}^C) = \sum_i \sum_{a,A} R^{IA}{}_{ia}\, \delta_{aA}\, N_i^f(\boldsymbol{x}^f) \quad \text{and} \quad \boldsymbol{P}^f{}_C = (\boldsymbol{R}^C{}_f)^T \tag{10}$$

by the evaluation of the ansatz functions at the fine-grid nodes.

4.3 Transfer Operators Associated with Local Schur Complement

Here, the components of the transfer operators represent displacements at fine-grid nodes due to unit displacements $\boldsymbol{d}^{C,f}$ applied to coarse-grid nodes. This leads to the block decomposition of the patch matrix $\boldsymbol{K}_p^f$ and $\boldsymbol{d}^f$ into fine- and coarse-grid components in the form

$$\begin{bmatrix} \boldsymbol{K}^{CC} & \boldsymbol{K}^{Cf} \\ \boldsymbol{K}^{fC} & \boldsymbol{K}^{ff} \end{bmatrix} \begin{bmatrix} \boldsymbol{d}^{C,f} \\ \boldsymbol{d}^{f,f} \end{bmatrix} = \boldsymbol{0} \tag{11}$$

On the finest grid the patch matrixc $\boldsymbol{K}_p^f$ is assembled with moduli $\mathbb{C} = \psi_{\varepsilon\varepsilon}(\varepsilon)$, on coarse grids e.g. with Voigt moduli $\bar{\mathbb{C}} = \frac{1}{|B_p|}\int_{B_p} \mathbb{C}\, dV$. From (11) we extract the fine-grid displacements by $\boldsymbol{d}^{f,f} = -(\boldsymbol{K}^{ff})^{-1}\boldsymbol{K}^{fC}\boldsymbol{d}^{C,f}$. Based on this result and the transfer conditions $\boldsymbol{d}^C = \boldsymbol{d}^{C,f}$ and $\boldsymbol{d}^f = \boldsymbol{P}^f{}_C \boldsymbol{d}^C$ we obtain

$$\boldsymbol{P}^f{}_C = \{\boldsymbol{R}^C{}_f\}^T := \begin{bmatrix} \boldsymbol{I}^C \\ -(\boldsymbol{K}^{ff})^{-1}\boldsymbol{K}^{fC} \end{bmatrix} \tag{12}$$

The coarse-grid operator $\boldsymbol{K}_p^C$, obtained by transfer operators (12), is identical to the local Schur-complement of $\boldsymbol{K}_p^f$ with respect to $\boldsymbol{K}^{ff}$.

4.4 Homogenization Based Transfer Operators

We propose new physically motivated transfer operators, applicable to unstructured meshes. The key idea is the introduction of the results of the homogenization theory, in order to take into account the influence of the heterogeneity of the micro-structure. We consider a patch of finite elements on the fine grid, characterizing a part of the heterogeneous micro-structure. Similar to the work [1] we split the grid transfer operators into a linear part, associated with the linear displacement field

$$\bar{d}^f_{ia} = \sum_I \sum_A N^C_I(\boldsymbol{x}^f_i)\,\delta_{aA}\,d^C_{IA} \tag{13}$$

and a part, associated with the fluctuation displacement field

$$\tilde{d}^f_{ia} = \sum_K H_{(ia)K}\,\bar{\varepsilon}_K \tag{14}$$

The representation (14) is obtained from (6) as a result of our homogenization analysis. We relate the macroscopic strains $\bar{\varepsilon}$ to the coarse-grid displacements in terms of the patch volume averages of the coarse-grid B-matrices. These approximations lead to the required transfer relation $d^f_{ia} = P^{ia}{}_{IA} d^C_{iA}$ within a patch, which yields the explicit expression for the transfer operators

$$P^{ia}{}_{IA} = R^{IA}{}_{ia} := \begin{cases} 1\,, \quad \text{if } \boldsymbol{x}^C_I = \boldsymbol{x}^f_i \\ \{N^C_I(\boldsymbol{x}^f_i)\,\delta_{aA} + \sum_K H_{(ia)K}\,\bar{B}^C_{I(KA)}\}\frac{1}{2} + P^{ia}{}_{IA}\,, \\ \text{only in the case of constraint (3c),} \\ \text{if } \boldsymbol{x}_i \text{ lies on overlapping patch boundaries} \\ N^C_I(\boldsymbol{x}^f_i)\,\delta_{aA} + \sum_K H_{(ia)K}\,\bar{B}^C_{I(KA)}\,, \quad \text{else} \end{cases} \tag{15}$$

In general, the Galerkin representation $\boldsymbol{R}^C{}_f \boldsymbol{K}^f_p \boldsymbol{P}^f{}_C$ is not equal to the coarse-grid matrix $\boldsymbol{K}^C_p(\bar{\mathbb{C}})$, obtained with overall moduli $\bar{\mathbb{C}}$, see also [7].

5 Numerical Examples

We now apply the discussed techniques to a model problem with different micro-structures, see Fig. 2. Due to the symmetry we discretize only a quarter of the specimen with dimension 10×10 and stiffness ratio $1:100$ of the heterogeneities. We consider a four-grid method with $V(3,3)$-cycles. Pre-smoothing is performed by Block-Backward Gauß-Seidel and post-smoothing by Block-Forward Gauß-Seidel. On the four grids we deal with 22300/5148/1150/222 unknowns. In contrast to the geometrical (geom), the Schur complement associated transfer (schur) and injection (inject) the homogenization based strategies associated with (3c) (hom1) and (3a) (hom2) yield a distinct decreasing of the residual Euclidean norm, if anisotropy effects play a key role.

6 Conclusion

In the scope of linear elastic materials with heterogeneous micro-structures different transfer strategies have been discussed. The key contribution was the suggestion of new homogenization based transfer operators, applicable to unstructured meshes. The numerical investigations indicate a good convergence behaviour of the proposed methods, when compared with other approaches.

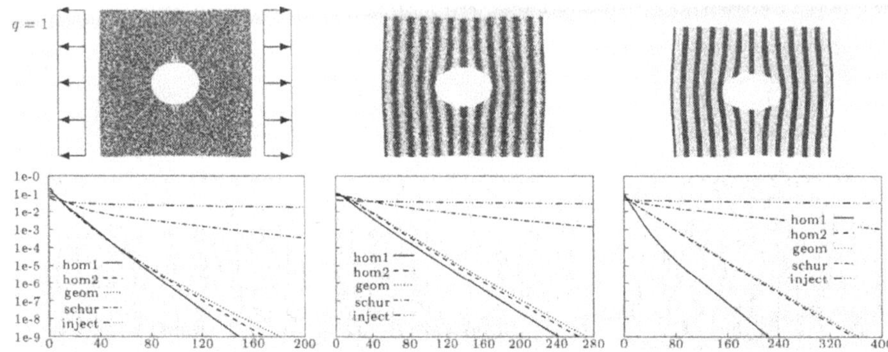

Fig. 2. Tension test of a specimen with a hole: The diagrams depict the residual Euclidian norm, subjected to the numbers of V-cycles. Anisotropy effects due to aggregations of the heterogeneities compound the efficiency of the homogenization based transfer operators in contrast to alternative techniques

References

1. Bayreuther C.G., Miehe C., Schröder J. (2000) Aspects Of Homogenization Techniques And Multigrid Solving. In: Multifield Problems: state of the art A.-M. Sändig, W. Schiehlen, W.L. Wendland, eds., Springer, 88–95
2. Brandt A. (1982) Guide to multigrid development. In: Multigrid Methods, W. Hackbusch, U. Trottenberg (eds.), Lecture Notes in Mathematics 960: 220–312, Springer
3. Briggs W.L. (1987) A multigrid tutorial. SIAM, Philadelphia
4. Dendy J.E. (1982) Black box multi-grid. J. Comput. Physics 48: 366–386
5. de Zeeuw P.M. (1990) Matrix-dependent prolongations and restrictions in a black box multigrid solver. J. Comput. Appl. Math. 33: 1–27
6. Fish J., Belsky V. (1995) Multigrid method for periodic heterogeneous media. Part 1: Convergence studies for one-dimensional case. Comput. Methods Appl. Mech. Engrg. 126: 1–16
7. Fish J., Belsky V. (1995) Multigrid method for periodic heterogeneous media. Part 2: Multiscale modeling and quality control in multidimensional case. Comput. Methods Appl. Mech. Engrg. 126: 17–38
8. Hackbusch W. (1985) Multigrid Methods and Applications, Springer
9. Miehe C. (2002) Strain-Driven Homogenization of Inelastic Microstructures and Composites Based on an Incremental Variational Formulation. International Journal for Numerical Methods in Engineering 55:1285–1322
10. Miehe C., Schotte J., Lambrecht M. (2002) Homogenization of Inelastic Solid Materials at Finite Strains Based on Incremental Minimization Principles. Journal of the Mechanics and Physics of Solids 50:2123–2167
11. Neuss N. (1995) Homogenisierung und Mehrgitter. University of Heidelberg
12. Ruge J.W., Stüben K. (1987) Algebraic multigrid. In: Multigrid Methods, S.F. McCormick, ed., SIAM, Philadelphia
13. Vaněk P., Mandel J., Brezina M. (1996) Algebraic multigrid on unstructured meshes. Technical report, Center for Computational Mathematics, University of Colorado at Denver
14. Wesseling P. (1992) An introduction to multigrid methods. Wiley, Chichester

Cutting-plane integration algorithm for ductile materials with damage

Slav Dimitrov and Eckart Schnack

Institute of Solid Mechanics, University of Karlsruhe,
Kaiserstrasse 12,
D-76128 Karlsruhe, Germany

Abstract. The objective of this contribution is to present a characterization of variational structure and the discrete equations modelling ductile damage in elastic plastic coupons with concave regions on the boundary.

1 Introduction

In this contribution we present a characterization of the variational structure and the discrete equations modelling the isotropic degradation of material stiffness for 2-dimensional elastic-plastic coupons with concave regions on the boundary.

Following the "classical" ideology in damage modelling (cf. [1,2,5]) we develop a *stress based framework* capable of accommodating isotropic elastic response and linear isotropic plastic response. Basic features of damage model are: (a) the damage is introduced trough the notion of *equivalent stress* concept and *strain equivalence principle* [2]; (b) the self-similarity property (see [1]) is used in utilizing the same hardening model for both, mesoscopic and macroscopic scales. For description on macroscopic level we utilize a set of primal variational principles in terms of stresses and stress-like hardening variables inscribed in a rate-independent associative evolutional framework with linear isotropic response into the infinitesimal deformation range. These principles lead in a natural manner again to predictor-corrector structure realized as cutting-plane algorithmic scheme (see [3,4] and [6] for survey). The cutting-plane algorithm is based on the steepest decent strategy without implicit treatment of the governing equations and account for the damage via mesoscale damage calibration function.

2 Continuum formulation

On the basis of the kinematic split $\varepsilon = \varepsilon^e + \varepsilon^d + \varepsilon^p$ we develop in the following the constitutive formulation of our two-level framework. On mesoscopic level we calculate from the respective governing equations the relation $\tilde{\mathbf{C}} = \tilde{\mathbf{C}}(||\hat{\varepsilon}^d||)$, where $||\hat{\varepsilon}^d||$ is the averaged over the RVE norm of the damage strain. Assuming further the coupling condition $||\hat{\varepsilon}^d|| = ||\varepsilon^p||$ we incorporate this dependence in the global macroscopic variational framework.

2.1 Mesoscopic level

Let $\Theta \subset \mathbb{R}^2$ be the RVE, and $\mathfrak{T} \subset \mathbb{R}_+$ be the time interval of interest. Assume that displacements $\mathbf{u}$ and stresses $\boldsymbol{\sigma}_\mu$ are specified on portions of the boundary,

$$\mathbf{u} = \bar{\mathbf{u}} \text{ on } \Gamma_{\mathrm{u}}^{\Theta}, \quad \boldsymbol{\sigma}_\mu \mathbf{n} = \bar{\mathfrak{t}} \text{ on } \Gamma_{\mathrm{t}}^{\Theta} \tag{1}$$

$$\overline{\Gamma_{\mathrm{u}}^{\Theta}} \cup \overline{\Gamma^{\Theta}}_{\mathrm{t}} = \overline{\partial\Theta}, \Gamma_{\mathrm{u}}^{\Theta} \cap \Gamma_{\mathrm{t}}^{\Theta} = \emptyset \tag{2}$$

We suppose further that body forces vanish on Θ.

Incrementally increased load leads to progressive reduction of material stiffness, treated as stress relaxation process. Stress relaxation is constrained from the yield function $f(\boldsymbol{\sigma})$, and can be measured by the quadratic norm of microscale deviatoric stresses $||s_\mu||$. Consider isotropic linear strain-hardening defined in terms of $f(\boldsymbol{\sigma})$ by, $f(\boldsymbol{\sigma}) = \sigma_Y + H\alpha$, where: σ_Y is the yield stress; H is the hardening modulus; α is the norm of inelastic deviatoric strain. The norm, of microscale deviatoric stress reads,

$$||s_\mu|| = \sqrt{\frac{3}{2}}\sqrt{\mathrm{dev}[\boldsymbol{\sigma}_\mu] : \mathrm{dev}[\boldsymbol{\sigma}_\mu]} = \sqrt{\frac{3}{2}}||\mathbf{P}_{dev}\boldsymbol{\sigma}_\mu||, \tag{3}$$

where $\mathbf{P}_{dev} = \mathbf{I} - 1/3[\mathbf{I} \otimes \mathbf{I}]$. Then the constraints enforced by $f(\boldsymbol{\sigma})$ are,

$$||s_\mu|| = f(\boldsymbol{\sigma}), \quad \text{if } \mathfrak{a} \in (0,1) \tag{4}$$

$$||s_\mu|| < f(\boldsymbol{\sigma}), \quad \text{if } \mathfrak{a} = 1 \tag{5}$$

From both the principle of strain equivalence and balance of linear momentum we formulate,

WEAK NAVIER PROBLEM FOR DUCTILE DAMAGE: For particular $t \in \mathfrak{T}$ find $\mathbf{u} \in \{\mathbf{u} \in \mathbb{W}^{1,2}(\Theta) : \mathbf{u} = \bar{\mathbf{u}} \text{ on } \Gamma_{\mathrm{u}}^{\Theta}\}$ such that:

$$\langle \mathfrak{a}(\mathbf{u})\mathfrak{C}_\mu : \nabla^s \mathbf{u}, \eta\rangle + \langle \mathrm{t}, \eta\rangle_{\Gamma_{\mathrm{t}}^{\Theta}} = 0 \tag{6}$$

$$\begin{matrix} 0 < \mathfrak{a}(\mathbf{u}) < 1 \text{ if } ||s_\mu|| = f(\boldsymbol{\sigma}) \text{ at } \mathfrak{x} \\ 0 < \mathfrak{a}(\mathbf{u}) = 1 \text{ if } ||s_\mu|| < f(\boldsymbol{\sigma}) \text{ at } \mathfrak{x} \end{matrix} \tag{7}$$

are satisfied.

where $\mathfrak{C}_\mu$ is the elasticity moduli tensor for material without defects, and $\mathfrak{a}(\mathbf{u})$ is a scalar damage variable. Here $\mathbb{W}^{1,2}(\Theta)$ is the standard SOBOLEV space with closure $\mathbb{W}_0^{1,2}(\Theta)$. Solving the weak NAVIER problem with unilateral constraints one obtains $\boldsymbol{\sigma}_\mu$ and ε_μ with averaged values,

$$\langle \sigma \rangle = \frac{1}{|\Theta|}\int_\Theta \sigma_\mu d\Theta, \quad \langle \varepsilon \rangle = \frac{1}{|\Theta|}\int_\Theta \varepsilon_\mu d\Theta, \tag{8}$$

The effective elasticity moduli tensor then is: $\langle \sigma \rangle \otimes \langle \varepsilon \rangle^{-1} = \widetilde{\mathbf{C}}$ where $\widetilde{\mathbf{C}}$ is the macroscopic damage response tensor. In one-dimensional case $\widetilde{\mathbf{C}}$ has a meaning of secant modulus.

2.2 Macroscopic level

Let the displacements $\mathbf{u}$ and stresses $\boldsymbol{\sigma}$ be specified on the boundary of $\bar{\Omega}$, i.e.

$$\mathbf{u} = \bar{\mathbf{u}} \text{ on } \Gamma_{\mathbf{u}}^{\Omega}, \quad \boldsymbol{\sigma}_{\mu}\mathbf{n} = \bar{\mathbf{t}} \text{ on } \Gamma_{\mathbf{t}}^{\Omega} \tag{9}$$

$$\overline{\Gamma_{\mathbf{u}}^{\Omega}} \cup \overline{\Gamma^{\Omega}}_{\mathbf{t}} = \partial\Omega, \Gamma_{\mathbf{u}}^{\Omega} \cap \Gamma_{\mathbf{t}}^{\Omega} = \emptyset \tag{10}$$

Then freezing the plastic flow ($\dot{\boldsymbol{\varepsilon}}_p = 0$) and denoting the rate of trial elastic stress with $\dot{\boldsymbol{\sigma}}^{tiral}$ one can formulate the initial boundary value problem on macroscopic level as:

DUAL VARIATIONAL PROBLEM: For all $t \in [0, T]$ find $\mathbf{u} \in \mathbb{W}^{1,2}(\Omega)$ and $\boldsymbol{\sigma} \in \mathbb{E}$ such that:

$$\begin{cases} \langle \boldsymbol{\sigma}, \boldsymbol{\eta} \rangle - \langle \bar{\mathbf{t}}, \boldsymbol{\eta} \rangle_{\Gamma_t} = 0, \ \forall \boldsymbol{\eta} \in \mathbb{W}_0^{1,2}(\Omega) \\ \langle\langle \dot{\boldsymbol{\sigma}}^{trial} - \dot{\boldsymbol{\sigma}}, \boldsymbol{\tau} - \boldsymbol{\sigma} \rangle\rangle_{\widetilde{\mathbb{C}}} \leqslant 0, \ \forall \boldsymbol{\tau} \in \mathbb{E} \end{cases} \tag{11}$$

with $\boldsymbol{\tau}$ subject to initial condition: $\langle\langle \boldsymbol{\sigma}|_{t=0}, \boldsymbol{\tau} \rangle\rangle = \langle\langle \boldsymbol{\sigma}_0, \boldsymbol{\tau} \rangle\rangle, \forall \boldsymbol{\tau} \in \mathbb{E}$ and inner product: $\langle\langle \boldsymbol{\sigma}, \boldsymbol{\tau} \rangle\rangle = \int_{\Omega} \boldsymbol{\sigma} : \widetilde{\mathbf{C}}^{-1} \boldsymbol{\xi} d\Omega$

3 Discrete formulation

3.1 Mesoscopic level

The weak NAVIER problem (6) can be reformulated,

$$\mathcal{K}_t(\widetilde{\mathfrak{C}}_{\mu}) = \begin{cases} \nabla \cdot (\widetilde{\mathfrak{C}}_{\mu} : \nabla^s \mathbf{u}) = 0, \\ 0 < \mathfrak{a}(\mathbf{u}) < 1 \text{ if } ||s_{\mu}|| = f(\boldsymbol{\sigma}) \text{ at } \mathfrak{x}, \\ 0 < \mathfrak{a}(\mathbf{u}) = 1 \text{ if } ||s_{\mu}|| < f(\boldsymbol{\sigma}) \text{ at } \mathfrak{x} \end{cases} \tag{12}$$

where $\nabla(\cdot) = \frac{\partial(\cdot)}{\partial \mathfrak{x}}$ is the spatial gradient. The step-forward operator $\mathcal{K}_t(\cdot)$ exhibits iterative structure, with representation:

$$\begin{cases} \dot{\sigma}_{\mu} = \dot{\widetilde{\mathfrak{C}}}_{\mu}\varepsilon_{\mu} \\ \dot{\widetilde{\mathfrak{C}}}_{\mu} = \dot{\mathfrak{a}}(\mathbf{u})\mathfrak{C}_{\mu} \end{cases} = \underbrace{\begin{cases} \dot{\sigma}_{\mu} = \dot{\widetilde{\mathfrak{C}}}_{\mu}\varepsilon_{\mu} \\ \dot{\widetilde{\mathfrak{C}}}_{\mu} = \mathbf{0} \end{cases}}_{\textit{Elastic predictor}} + \underbrace{\begin{cases} \dot{\sigma}_{\mu} = \mathbf{0} \\ \dot{\widetilde{\mathfrak{C}}}_{\mu} = \dot{\mathfrak{a}}(\mathbf{u})\mathfrak{C}_{\mu} \end{cases}}_{\textit{Damage corrector}} \tag{13}$$

where, *damage corrector* is constrained with the inequalities in (12), and $\dot{(\cdot)}$ is time derivative. Because the microscopic strain remains fixed during the iterations it is simple to calculate the magnitude of reduction for share modulus $\widetilde{\mathfrak{G}}_{\mu}$ (ductile damage does not affect bulk modulus), i.e.

$$\widetilde{\mathfrak{G}}_{\mu} = \frac{||s_{\mu}||}{f(\sigma)} \mathfrak{G}_{\mu} \tag{14}$$

Taking this in mind, we are able to compose a numerical procedure for calculation of the macroscopic $\tilde{\mathbf{C}}$. The numerical solution of the weak NAVIER problem involves the following steps (cf. [7]):

(a) Time discretisation of the interval of interest $[0,T] = \bigcup_{n=1}^{N} [t_n, t_{n+1}]$
(b) For given time increment $[t_n, t_{n+1}]$ load RVE with initial elastic response
(c) Compute a trial solution $\mathbf{u}^{trial}$ from the equations of equilibrium
(d) Compute $||s_{\mu}^{trial}||$ at all integration points
(e) Reduce eigenvalues of elasticity moduli tensor where: $||s_{\mu}^{trial}|| > f(\sigma)$
(f) Average the stress and strain fields over the RVE, compute the effective elasticity moduli tensor and store the result

3.2 Macroscopic level

On macroscopic level we utilize a cutting plane scheme (see [3,4]). Once more let us consider the time discretisation $[0,T] = \bigcup_{n=1}^{N} [t_n, t_{n+1}]$. Assume that the local state of the body at point $\mathbf{x}$ and current time t_n is completely defined. The latter means that the triple $\{\varepsilon_n, \varepsilon_n^p, \alpha_n\}$ is known. Here $\alpha_n = ||\varepsilon_n^p||$. The problem to be addressed is that of stress update from, this initial values. For linear isotropic hardening this can be realized in the following form (for details see [6]):

1 Compute trial elastic state: $\mathbf{e}_{n+1} = \varepsilon_{n+1} - \frac{1}{3}(\mathrm{Tr}[\varepsilon_{n+1}]\mathbf{1})$

2 Set initial values: $k = 0,\ \varepsilon_{n+1}^{p^{(0)}} = \varepsilon_n^p,\ \Delta\gamma_{n+1}^{(0)} = 0,\ \alpha_{n+1}^{(0)} = \alpha_n$
3 Compute stress and yield function:

$$\mathbf{s}_{n+1}^{(k)} = 2\tilde{\mu}(\mathbf{e}_{n+1} - \varepsilon_{n+1}^{p^{(k)}}) \tag{15}$$

$$f_{n+1}^{(k)} = ||\mathbf{s}_{n+1}^{(k)}|| - \sqrt{\frac{2}{3}}[\sigma_Y + H\alpha_{n+1}^{(k)}]; \tag{16}$$

4 Check yield condition:
IF $f_{n+1}^{(k)} \leqslant$ TOL THEN update initial values

ELSE:

5 Compute the increment to plastic consistency parameter:

$$\Delta^2\gamma_{n+1}^{(k)} = \frac{f_{n+1}^{(k)}}{[\mathbf{s}_{n+1}^{(k)}/||\mathbf{s}_{n+1}^{(k)}||] : \widetilde{\mathbf{C}}_{n+1}^{ep} : [\mathbf{s}_{n+1}^{(k)}/||\mathbf{s}_{n+1}^{(k)}||]} \tag{17}$$

6 Update state variables and consistency parameter:

$$\varepsilon_{n+1}^{p^{(k+1)}} = \varepsilon_{n+1}^{p^{(k)}} + \Delta^2\gamma_{n+1}^{(k)} \frac{\mathbf{s}_{n+1}^{(k)}}{||\mathbf{s}_{n+1}^{(k)}||} \tag{18}$$

$$\Delta\gamma_{n+1}^{(k+1)} = \Delta\gamma_{n+1}^{(k)} + \Delta^2\gamma_{n+1}^{(k)} \tag{19}$$

$$\alpha_{n+1}^{(k+1)} = \alpha_{n+1}^{(k)} + \sqrt{\frac{2}{3}}\Delta\gamma_{n+1}^{(k+1)} \tag{20}$$

set $k \longleftarrow k+1$ and GOTO 3.

In this realization the effective algorithmic elastoplastic tangent moduli $\widetilde{\mathbf{C}}^{ep}$ are established on the basis of damage response function obtained on mesoscopic level (i.e. $\tilde{\mu} = \tilde{\mu}(\alpha)$, $\tilde{\kappa} = \kappa$).

4 Numerical example

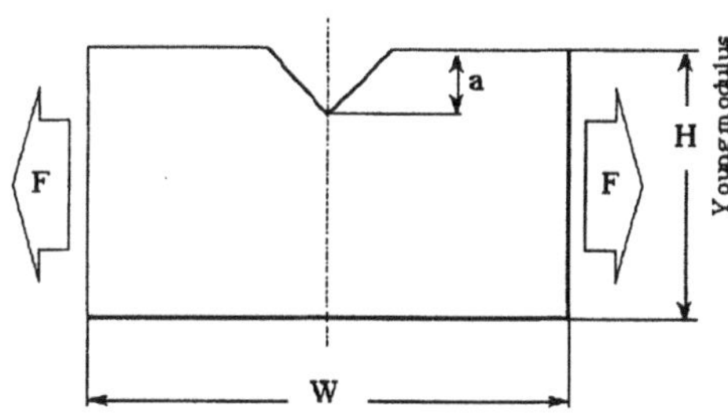

Fig. 1. Geometry

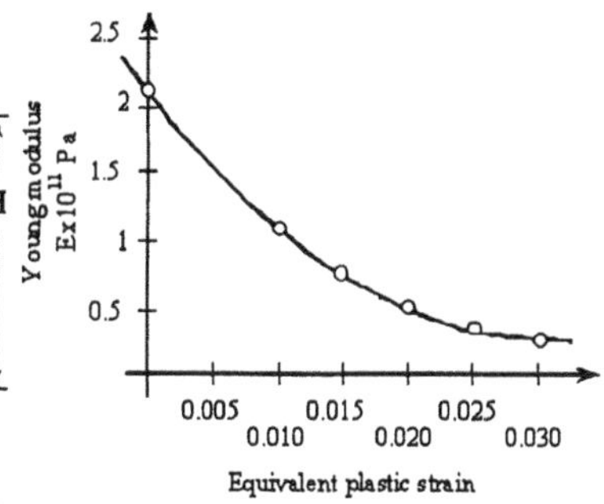

Fig. 2. Damage response function

The problem considered is of extension of a strip with V-notch. The geometry is shown on Figure 1. The following geometrical proportions, $a : H : W = 2 : 20 : 125$ and unit thickness are assumed. The calculations are performed by imposing uniform stress control on the right boundary. For obvious symmetry conditions only one-half of the specimen is analyzed and results are symmetrically expanded after. A von Mises yield condition for elastic perfectly plastic material is considered. The elastic constants and yield stress are: $E = 2,1.10^{11}\, Pa, \nu = 0.3, \sigma_Y^I = 400$ MPa. The calibrated numerically and after fitted, damage response function is shown on Figure 2. The spread of the damage zone at the end of loading is shown on Figure 3, but polar distribution of effective Young modulus close to the V-notch tip on Figure 4.

5 Conclusions

In this article we represent a two-scale constitutive model capable of accommodating progressive reduction of material stiffness under conditions of proportional loading and bilinear isotropic hardening response on macroscopic level. A stress based framework for ductile damage has been employed accounting for isotropic distribution and evolution of micro-failures. The induced by the damage stress relaxation was implicitly represent in displacement based weak formulation with unilateral constraint on share modulus. The additive structure of this formulation has been successively used for establishing the proper methodology for numerical solution.

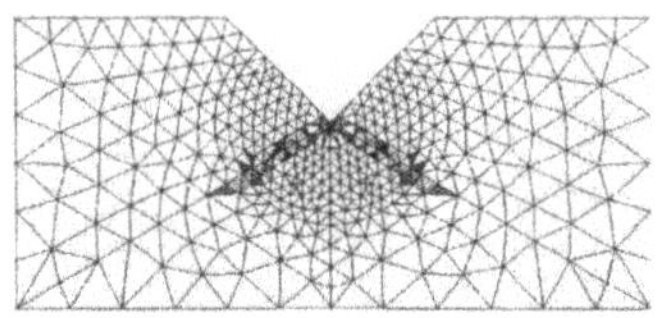

Fig. 3. Damage spread at the notch

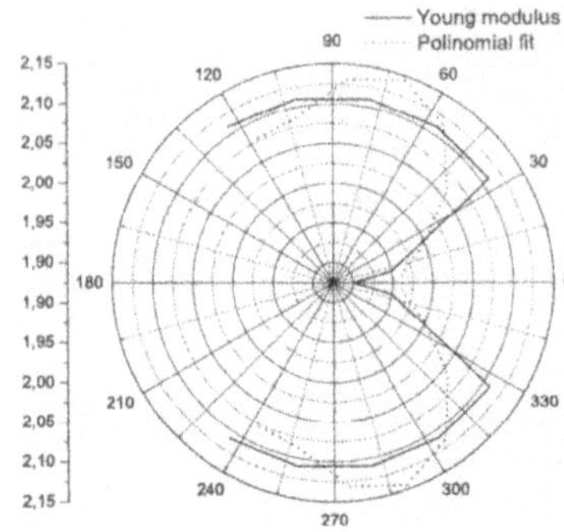

Fig. 4. Polar distribution of the effective Young modulus

Acknowledgements

The authors are gratefully indebted to German Science Foundation (DFG) for the financial support under the grant number Schn 245/19-3.

References

1. Krajcinovic, D. (1996) Damage mechanics, Elsevier Science, Amsterdam
2. Lemaitre J., Chaboche, J. -L. (1990) Mechanics of Solid Materials, Cambridge University Press, Cambridge
3. Simó, J. C., Ortiz M. (1985) Unified approach to finite deformation elastoplastic analysis based on the use of hyperelastic constitutive relations. Comput. Methods Appl. Mech. Engrg., **49**, 221–245.
4. Ortiz M., Simó, J. C. (1986) Analysis of a new class of integration algorithms for elastoplastic constitutive relations. Int. J. Numer. Methods Engrg., **23**, 353–366.
5. Simó,J. C., Ju J. W. (1987) Strain and stress based continuum damage models, Part I: Formulation. Int. J. Solid Structures, **23**, 821–840.
6. Simó, J. C., Hughes, T. J. R. (1998) Computational inelasticity, Springer, New-York.
7. Zohdi, T., Feucht, M. et al (1998) A description of macroscopic damage trough microstructural relaxation, Int. J. Numer. Methods Engrg., **43**, 493–506.

Upper and lower semicontinuity of a stable set for rate-independent processes

Messoud A. Efendiev

Mathematisches Institut A, Universität Stuttgart, Pfaffenwaldring 57,
70569 Stuttgart, Germany

Abstract. This paper deals with a general approach to rate-independent processes, which may have hysteretic behaviour. Such processes play an important role in mechanics, engineering and arises whenever the systems under consideration excited quasistatically. This approach is based on an energy functional $I(t,\cdot)$ and a dissipation functional Δ. We study the upper and lower semicontinuity of a stable set for (I_ε, Δ), $\varepsilon \geqslant 0$, and the uniqueness question.

Let $z \in X$, X a Banach space with dual space X^*, is the variable describing the process, and $\dot{z}$ is the time derivative. The central feature of rate-independent processes is that a solution $z : [0,T] \to X$ remains a solution if the time is rescaled. This leads to a dissipation functional $\Delta : X \to [0,\infty)$ which is homogenous of degree 1, with $\Delta(\alpha v) = \alpha\Delta(v)$, for $\alpha \geqslant 0$ and $v \in X$. Let $\mathbb{P}$ be a closed set in X. Statement of the problem:

Problem 1 *Find $z : [0,T] \to X$, so that for a.a. $t \in [0,T]$*

$$I(t, z(t)) \leqslant I(t,y) + \Delta(y - z(t)), \tag{S}$$

for all $y \in \mathbb{P}$, and

$$I(t,z(t)) + \int_s^t \Delta(dz) \leqslant I(s,z(s)) + \int_s^t \partial_t I(\sigma, z(\sigma))d\sigma, \tag{E}$$

for all $0 \leqslant s \leqslant t \leqslant T$.

As was shown in [1,2], for a proof of solvability of (S),(E), the set of stable points at time t,

$$S(t) = \{z \in \mathbb{P} \,|\, I(t,z) \leqslant I(t,y) + \Delta(y-z) \text{ for all } y \in \mathbb{P}\},$$

plays an important role. In [1] we presented new sufficient conditions on (I,Δ), which leads to existence of at least one solution of (S),(E), which was based on the compactness of $S(t)$. Here, we address to the following questions:

1) Upper and lower semicontinuity of the stable set $S(t)$, that is, in which space $W \subset X$ is valid:

$$\lim_{\varepsilon \to 0} \text{distsymm}_W (S_\varepsilon(t), S_0(t)) = 0, \tag{1}$$

where $S_\varepsilon(t)$ is a stable set for a given family of (I_ε, Δ), $\varepsilon \geqslant 0$. Here, $\mathrm{distsymm}_W(\cdot,\cdot)$ denotes the symmetric Hausdorff distance, which is defined by

$$\mathrm{distsymm}(A,B) := \max\{\mathrm{dist}(A,B), \mathrm{dist}(B,A)\},$$

for $A, B \subset W$ with $\mathrm{dist}(A,B) := \sup_{x\in A} \inf_{y\in B} \|x-y\|$

2) Uniqueness for (S),(E).

In order to answer 1) and 2), we start with some preliminaries.

Condition 1 *Assume that $I_\varepsilon(t,\cdot) \in C^1([0,T],X)$ and Δ is convex and homogenous of degree 1, which satisfies*

$$C_\Delta \|v\|_X \leqslant \Delta(v) \leqslant C_\Delta^{-1} \|v\|_X. \tag{2}$$

Proposition 1 *Assume that a pair (I_ε, Δ), $\varepsilon \geqslant 0$, satisfies Condition 1 and $\mathbb{P} = X$. Then*

a)

$$S_\varepsilon(t) \subset (\mathrm{D}_z I_\varepsilon)^{-1}(-\partial\Delta(0)), \tag{3}$$

b) if $I_\varepsilon(t,\cdot)$ is convex, then

$$S_\varepsilon(t) = (\mathrm{D}_z I_\varepsilon)^{-1}(-\partial\Delta(0)), \tag{4}$$

where by ∂ and D_z are denoted the subdifferential of the convex functional Δ and the Frèchet derivative of $I_\varepsilon(t,\cdot)$ with respect to z respectively.

A proof of these facts is not difficult and we refer readers to [1,2].

Remark 1 *In physics, the set $\partial\Delta(0) \subset X^*$ is called the set of non-hysteretic response.*

For simplicity of presentation, we restrict ourselves here to a class of functionals arising in "ferromagnetic" models. Let us recall that, for "ferromagnetic" models (see [1]), one can choose $\mathbb{P} = H^1(\Omega, \mathbb{R}^n)$, Ω is a bounded domain in $\mathbb{R}^n$, and

$$I_\varepsilon(t,z) = \begin{cases} \displaystyle\int_\Omega \left[\frac{\varepsilon^2}{2}|\nabla_x z|^2 + \varphi(z) - H_\ell \cdot z\right] dx \\ \qquad \displaystyle -\langle g(t), z\rangle + \frac{1}{2}\int_{\mathbb{R}^n} |\nabla_x u|^2\, dx\,, & z \in H^1(\Omega, \mathbb{R}^n)\,, \\ \infty\,, & \text{else,} \end{cases} \tag{5}$$

with $\Delta_x u = \mathrm{div}(\chi_\Omega z)$, $\varepsilon \geqslant 0$. Here, Δ_x is the Laplacian and χ_Ω is the indicator function of Ω. We use $\langle\cdot,\cdot\rangle$ for the duality pairing on $Y^* \times Y$, where Y is a given B-space and assume that $g \in C^1([0,T], \mathbb{P}^*)$.

As a candidate for $\Delta : X \to [0,\infty)$, where $X = L^1(\Omega,\mathbb{R}^n)$ we choose either

$$\Delta(v) = \max\left\{\langle v,w\rangle \,\middle|\, w \in F^* \subset X^*\right\}, \tag{6}$$

where F^* is a closed, bounded and convex set containing zero in the interior or

$$\Delta(v) = \int_\Omega |v|\,dx\,, \tag{7}$$

where $v \in L^1(\Omega,\mathbb{R}^n)$ and $n \leqslant 3$. In the sequel, we assume that $H_\ell \in L^2(\Omega,\mathbb{R}^n)$ and $0 < \alpha \leqslant \varphi''(z) < C$. Then due to Proposition 1 we have

$$\begin{aligned} S_\varepsilon(t) &= \left(\mathrm{D}_z\, I_\varepsilon(t,z)\right)^{-1}\left(-\,\partial\Delta(0)\right),\\ S_0(t) &= \left(\mathrm{D}_z\, I_0(t,z)\right)^{-1}\left(-\,\partial\Delta(0)\right). \end{aligned} \tag{8}$$

For notational convenience, we denote $V_\varepsilon(z) := \mathrm{D}_z\, I_\varepsilon(t,z)$, $\varepsilon \geqslant 0$, for each fixed t. Moreover we will use indifferently the standard notation of Sobolev spaces, that is $L^2(\Omega)$, $H^1(\Omega)$, and $\overset{\circ}{H}{}^1(\Omega)$, for scalar-valued functions as well as for vector-values functions. Let (I_ε, Δ) be the family of functionals defined by (5) and (6) or (5) and (7). Then holds:

Theorem 1 *Let Condition 1 be satisfied. Moreover we assume that for $n \leqslant 3$*

$$0 < \alpha \leqslant \varphi''(z) < C \quad \textit{and} \quad H_\ell \in L^2(\Omega,\mathbb{R}^n). \tag{$*$}$$

Then

a) if $F^* := \partial\Delta(0)$ is compact in $L^2(\Omega,\mathbb{R}^n)$, we have

$$\operatorname{distsymm}_{L^2}\left(S_\varepsilon(t), S_0(t)\right) \to 0 \quad \textit{as} \quad \varepsilon \to 0 \tag{9}$$

and

b) if F^* is not compact in $L^2(\Omega,\mathbb{R}^n)$, then

$$\operatorname{distsymm}_{L^2}\left(S_\varepsilon(t), S_0(t)\right) \not\to 0 \quad \textit{as} \quad \varepsilon \to 0\,. \tag{10}$$

Proof. It is not difficult to see that, from $(*)$ it follows that $V_0 : L^2(\Omega,\mathbb{R}^n) \to L^2(\Omega,\mathbb{R}^n)$ and

$$\gamma_1\|z_1 - z_2\|_{L^2} \leqslant \|V_0(z_1) - V_0(z_2)\|_{L^2} \leqslant \gamma_2\|z_1 - z_2\|_{L^2}\,, \tag{11}$$

with some positive constants γ_1 and γ_2. Hence the sets $V_0^{-1}(F^*)$ and F^*, $F^* \subset L^2(\Omega,\mathbb{R}^n)$ are compact or noncompact in $L^2(\Omega,\mathbb{R}^n)$ simultaneously. First we start proving b). Indeed, assume that $F^* = \partial\Delta(0)$ is not compact in $L^2(\Omega, \mathbb{R}^n)$. This is fulfilled, for example when $\Delta : L^1(\Omega, \mathbb{R}^n) \to [0, \infty)$ is defined by $\Delta(v) = \|v\|_{L^1}$. Indeed in this case we have

$$F^* := \partial\Delta(0) = \{w \in L^\infty(\Omega, \mathbb{R}^n) \,|\, \|w\|_{L^\infty} \leqslant 1\}. \tag{12}$$

Assume on the contrary, that is

$$\text{distsymm}_{L^2}(S_\varepsilon(t), S_0(t)) \to 0 \quad \text{as} \quad \varepsilon \to 0\,.$$

Let us recall that $\mathrm{D}_z\, I : H^1(\Omega, \mathbb{R}^n) \to (H^1(\Omega, \mathbb{R}^n))^*$ is a proper map (see [1]) and $S_\varepsilon(t) = (\mathrm{D}_z\, I_\varepsilon)^{-1}(\partial\Delta(0))$. Since $\partial\Delta(0)$ is compact in $(H^1(\Omega, \mathbb{R}^n))^*$, we have that $S_\varepsilon(t)$ is compact in $L^2(\Omega, \mathbb{R}^n)$. Then, it follows from the definition of the symmetric distance, that $S_0(t)$ is also compact in L^2, which is a contradiction. This proves b).

Next we prove a). Let us recall, that we would like to show that

$$\text{distsymm}_{L^2}(V_\varepsilon^{-1}(F^*), V_0^{-1}(F^*)) \to 0 \quad \text{as} \quad \varepsilon \to 0\,.$$

According to our notational convention,

$$\begin{aligned} V_0(z) &:= \varphi'(z) + \widetilde{\mathrm{D}}z - H_\ell - g\,, \\ V_\varepsilon(z) &:= -\varepsilon\Delta_x z + V_0(z)\,. \end{aligned} \tag{13}$$

Assume on the contrary: Then $\exists\varepsilon_n \to 0$ and $z_n \in S_{\varepsilon_n}(t) = V_{\varepsilon_n}^{-1}(F^*)$ and $z_n' \in S_0(t) = V_0^{-1}(F^*)$, such that

a) $\forall w \in S_0(t)$, $\|z_n - w\| \geqslant \delta_0 > 0$, or
b) $\forall w' \in S_{\varepsilon_n}(t)$, $\|z_n' - w'\| \geqslant \delta_0 > 0$.

Since, $z_n \in V_{\varepsilon_n}^{-1}(F^*) \Rightarrow \exists h_n \in F^* \Subset L^2$, such that

$$-\varepsilon_n\Delta_x z_n + \varphi'(z_n) + \widetilde{\mathrm{D}}z_n = h_n + H_\ell + g\,.$$

Here $\widetilde{\mathrm{D}} : L^2(\Omega, \mathbb{R}^n) \to L^2(\Omega, \mathbb{R}^n)$ is the self-adjoint non-negative operator defined by

$$\widetilde{\mathrm{D}}z := \Pi_\Omega \nabla_x(\Delta_x)^{-1}\mathrm{div}(\widetilde{\Pi}_\Omega z) \tag{14}$$

(for more details of the properties of $\widetilde{\mathrm{D}}$ see [1]). In (14) by Π_Ω and $\widetilde{\Pi}_\Omega$ are denoted the restriction and extension by zero outside of Ω operators respectively. Multiplying both sides of the last equality to z_n and integrating, we obtain

$$\varepsilon_n^2\|\nabla_x z_n\|^2 + (\varphi'(z_n), z_n) + (\widetilde{\mathrm{D}}z_n, z_n) = (h_n, z_n) + (g, z_n) + (H_\ell, z_n).$$

The assumption on φ and Young's inequality yield $\|z_n\|_{L^2} \leqslant C_*$, independent of n. Hence, there exists $z_0 \in L^2$, such that $z_n \rightharpoonup z_0$ (here we use the same letters for the subsequence!). Moreover, we have

$$(V_\varepsilon(z_1) - V_\varepsilon(z_2), z_1 - z_2) \geqslant \alpha\|z_1 - z_2\|_{L^2}^2.$$

Hence, it follows that, for all $w \in H^1$, we have

$$\alpha\|z_n - w\|_{L^2}^2 + \left(V_{\varepsilon_n}(w), z_n - w\right) \leqslant (h_n, z_n - w)$$

or

$$\alpha\left(\|z_n\|_{L^2}^2 - 2(z_n, w) + \|w\|_{L^2}^2\right) + \left(V_{\varepsilon_n}(w), z_n - w\right) \leqslant (h_n, z_n - w).$$

Taking into account that $z_n \rightharpoonup z_0$, we derive from the last inequality

$$\alpha\left(\|z_0\|_M^2 - 2(z_0, w) + \|w\|_{L^2}^2\right) + \left(V_0(w), z_0 - w\right) \leqslant (h_0, z_0 - w), \tag{15}$$

for all $w \in H_0^1(\Omega)$. Here $\|z_0\|_M^2 := \limsup_{n\to\infty} \|z_n\|_{L^2}^2$. Approximating $w \in L^2$ by $w_n \in H_0^1(\Omega)$, that is $\|w_n - w\|_{L^2} \to 0$, we derive that (15) remains valid also for any $w \in L^2(\Omega)$. From (15) it follows that

$$\left(V_0(w), z_0 - w\right) \leqslant (h_0, z_0 - w), \tag{16}$$

for all $w \in L^2(\Omega)$. Since V_0 is maximal monotone, (16) implies that

$$V_0(z_0) = h_0 .$$

Interesting in (15) $w = z_0$, we obtain $\|z_0\|_M^2 \leqslant \|z_0\|_{L^2}^2$. Thus

$$\|z_0\|_{L^2}^2 \leqslant \liminf_{n\to\infty} \|z_n\|_{L^2}^2 \leqslant \limsup_{n\to\infty} \|z_n\|_{L^2}^2 \leqslant \|z_0\|_{L^2}^2,$$

which implies that $\|z_n\| \to \|z_0\|$. Hence we obtain $z_n \to z_0$ strongly in L^2.

Next we show that this fact leads to a contradiction of a) and b) respectively. Indeed, since $z_n \to z_0$, $z_n \in V_{\varepsilon_n}^{-1}(F^*)$ and $z_0 \in V_0^{-1}(F^*)$, we obtain $\|z_n - z_0\| \to 0$ which is contradict to a).

Now we also show $z_n \to z_0$ strongly in L^2 contradicts to b). Indeed, $z_n \in V_{\varepsilon_n}^{-1}(F^*)$, $z_0 \in V_0^{-1}(F^*)$. Then b) implies that $\exists z_n' \in V_0^{-1}(F^*)$ such that $\|z_n' - w'\| \geqslant \delta_0 > 0$ for all $w' \in V_{\varepsilon_n}^{-1}(F^*)$. From $z_n' \in V_0^{-1}(F^*)$ it follows that $\exists h_n' \in F^*$

$$V_0(z_n') = h_n' \in F^*.$$

We denote $z_n'' = V_{\varepsilon_n}^{-1}(h_n')$. Since $h_n' \in F^*$ and F^* is compact, we obtain that sue to the arguments used above $z_n'' \to z_0' \in L^2$ strongly and $V_0(z_0') = h_0'$, $h_n' \to h_0'$ in L^2. Moreover $C\|z_n' - z_0'\| \leqslant \|V_0(z_n') - V_0(z_0')\| = \|h_n' - h_0'\| \to 0$. Thus $z_n' \to z_0'$. Hence $\|z_n'' - z_n'\| \to 0$ which obviously contradicts to $\|z_n' - w'\| \geqslant \delta_0$ for all $w' \in V_{\varepsilon_n}^{-1}(F^*)$. This proves Theorem 1. □

Uniqueness

As before, we deal with the (S),(E) problem in the case, when $I(t,\cdot)$ and $\Delta(\cdot)$ has the form (5) and (6) or (5) and (7). We aim at proving uniqueness of the (S),(E) problem. Let us recall that in [1] and [2] under general conditions on $I(t,\cdot)$ and $\Delta(\cdot)$ it was proved the existence of at least one solution of the (S),(E) problem, which works for nonconvex functionals as well. However, it is easy to construct examples with $I(t,\cdot)$ (not strictly) convex such that the solution is not unique. Here, we present new sufficient conditions on (I,Δ), providing uniqueness of (S),(E) (see also [2]).

Condition 2 *Suppose that,* $I \in C^1([0,T],\mathbb{P})$ *and* $\Delta : X \to [0,\infty)$ *are given by* (5),(6) *or* (5),(7) *respectively. Moreover, we assume that the "anisotropy" energy* φ *and its gradient* $f = \nabla_v \varphi(v)$ *satisfy*

a) $\varphi(v) \geqslant -C + \alpha |v|^{p+3}$,
b) $|f''(v)| \leqslant C(1+|v|^p)$,
c) $f'(v) \geqslant \alpha(1+|v|^{p+1})$,

where C, α, p *are some constants with* $C > 0$, $\alpha > 0$, $p \geqslant 0$.

Theorem 2 *Let Condition 2 be satisfied. The the (S),(E) problem admits a unique solution.*

Proof. Let z_1 and z_2 be two solutions with $z_1(0) = z_2(0) = z_0$. First we start with some formal analysis. Indeed, (S),(E) can be rewritten:

$$J(z(t)) - \langle g(t), z(t)\rangle + \int_s^t \Delta(\dot z)\, d\sigma$$

$$\leqslant J(z(s)) - \langle g(s), z(s)\rangle - \int_s^t \langle \dot g(\sigma), z(\sigma)\rangle\, d\sigma$$

$$J(z(t)) - J(z(s)) - \big(\langle g(t), z(t)\rangle - \langle g(s), z(s)\rangle\big) + \int_s^t \Delta(\dot z)\, d\sigma$$

$$\leqslant - \int_s^t \langle \dot g(\sigma), z(\sigma)\rangle\, d\sigma$$

$$\int_s^t \left[\frac{d}{d\sigma} J(z(\sigma)) - \frac{d}{d\sigma}\langle g(t), z(t)\rangle + \langle \dot g(\sigma), z(\sigma)\rangle\right] d\sigma + \int_s^t \Delta(\dot z)\, d\sigma \leqslant 0\,.$$

Dividing $\frac{1}{t-s}$ and taking the limit, we obtain

$$\langle J'(z(t)), \dot z(t)\rangle - \langle g(t), \dot z(t)\rangle + \Delta(\dot z(t)) \leqslant 0\,. \tag{17}$$

On the other hand (S) implies

$$g - J'(z(t)) \in \partial\Delta(0)\,. \tag{18}$$

Hence, from (17) and (18) it follows that

$$g - J'\big(z(t)\big) \in \partial\Delta(\dot z)\,. \tag{19}$$

Taking into account that $\partial\Delta : X \to X^*$ is maximal monotone, we have

$$\langle J'(z_2) - J'(z_1), \dot z_2 - \dot z_1\rangle \leqslant 0\,. \tag{20}$$

Let $w := z_2 - z_1$. Obviously, $w(0) = 0$ and (20) can be rewritten as

$$\langle J'(z_1 + w) - J'(z_1), \dot w\rangle \leqslant 0\,, \tag{21}$$

We will show that (21) yields $w \equiv 0$. To this end, let us recall that

$$J'(z) = -\varepsilon\Delta_x z + f(z) + \widetilde{\mathrm{D}}z\,. \tag{22}$$

We prove Theorem 2 in two stages. In the first stage, we assume that $\varepsilon = 0$ and $\widetilde{\mathrm{D}} = 0$. Then (21) for $J'(z)$ takes the form

$$\big(f(z_1 + w) - f(z_1), \partial_t w\big)_{L^2(\Omega)} \leqslant 0\,.$$

Taking Condition 2 into account we obtain

$$\begin{aligned} &\big(f(z_1 + w) - f(z_1), \partial_t w\big)_{L^2(\Omega)} \\ &\quad = \partial_t\Big[\big(\varphi(z_1 + w) - \varphi(z_1) - f(z_1)w, 1\big)_{L^2(\Omega)}\Big] \\ &\quad\quad - \big(\partial_t z_1, f(z_1 + w) - f(z_1) - f'(z_1)w\big)_{L^2(\Omega)}\,. \end{aligned} \tag{23}$$

Hence,

$$\begin{aligned} &\partial_t\left[\big(\varphi(z_1 + w) - \varphi(z_1) - f(z_1)w, 1\big)_{L^2}\right] \\ &\quad \leqslant \big(\partial_t z_1, f(z_1 + w) - f(z_1) - f'(z_1)w\big)\,. \end{aligned} \tag{24}$$

Let $y(t) := \big(\varphi(z_1 + w) - \varphi(z_1) - f(z_1)w, 1\big)_{L^2}$, $a(t) := \|\partial_t z_1\|_{L^\infty}$. Due to Condition 2 we have

$$\|f(z_1 + w) - f(z_1) - f'(z_1)w\|_{L^1} \leqslant Cy(t) \quad \text{and} \quad y(t) \geqslant \alpha\|w\|^2_{L^2}\,, \tag{25}$$

where C and α are some positive constants.

From (24) and (25) we obtain

$$\partial_t y(t) \leqslant Ca(t)y(t)\,, \tag{26}$$

and as a result, (26) yields

$$\alpha\|w(t)\|^2_{L^2} \leqslant y(t) \leqslant C_1 y(0) = 0\,. \tag{27}$$

Hence, we obtain $w = 0$. This proves Theorem 2 for the case that $\varepsilon = 0$ and $\widetilde{\mathrm{D}} = 0$. It remains to prove Theorem 2 for $\varepsilon \neq 0$ and $\widetilde{\mathrm{D}} \neq 0$. To this end, we define $y_1(t) := -\varepsilon(\Delta_x w, w) + (\widetilde{\mathrm{D}}w, w)$. In the same manner as before, we show that

$$\partial_t\big[y_1(t) + y(t)\big] \leqslant Cy(t)\cdot a(t)\,. \tag{28}$$

Since $y_1(t) \geqslant 0$ we have

$$\partial_t\big[y_1(t) + y(t)\big] \leqslant Ca(t)\big[y_1(t) + y(t)\big]\,, \tag{29}$$

which leads to

$$\alpha\|w(t)\|^2_{L^2} \leqslant y(t) \leqslant y_1(t) + y(t) \leqslant 0\,. \tag{30}$$

Consequently, it follows from (30) that $w \equiv 0$. This proves Theorem 2. □

Remark 2 *Our approach allows us to prove uniqueness results for the following class of doubly nonlinear problems of the form*

$$\begin{aligned}&\partial_t u\cdot|\partial_t u|^q = a\Delta_x u - u|u|^p + C(x)u\\ &u|_{\partial\Omega} = 0\,,\quad q > p\,,\quad u = (u^1,\dots,u^k)\,,\quad a = a^* > 0\,.\end{aligned}$$

or

$$\begin{aligned}&b(x)\partial_t u + \partial_t u|\partial_t u|^q = a\Delta_x u - u|u|^p + C(x)u\\ &u|_{\partial\Omega} = 0\,,\quad a = a^* > 0 \quad \textit{and} \quad b(x) \geqslant 0 \quad \textit{for all} \quad x \in \Omega\end{aligned}$$

In the forthcoming paper, we will present the long-time behaviour of solutions of the doubly nonlinear equations mentioned above.

References

1. Efendiev, M.A. (2001) On compactness of the stable set for for rate-independent processes. Preprint SFB 404, N17
2. Mielke, A., Theil, F. (2001) On rate-independent hysteresis models. Preprint SFB 404, N7

Dissipation distances in multiplicative elastoplasticity

Klaus Hackl[1], Alexander Mielke[2], and Dirk Mittenhuber[2]

[1] Fakultät für Bauingenieurwesen, Lehrstuhl für Allgemeine Mechanik, Ruhr-Universität Bochum, Universitätsstraße 150, IA 3/125, D-44801 Bochum, GERMANY

[2] Mathematisches Institut A, 1. Lehrstuhl, Universität Stuttgart, Pfaffenwaldring 57, D-70569 Stuttgart, GERMANY

Abstract. We study finite-strain elastoplasticity in a new formulation proposed in [8,1,7]. This theory does not need smoothness and is based on energy minimization techniques. In particular, it gives rise to robust algorithms. It is based on two scalar constitutive functions: an elastic potential and a dissipation potential which give rise to an energy functional and a dissipation distance.

Here we study these dissipation distances in some detail and present situations where they are quite explicitly available. These include isotropic plasticity of Prandtl-Reuß type and examples from two-dimensional single-crystal plasticity. We put special emphasis on the geometric nonlinearities arising from the underlying matrix groups which lead to optimization problems on Lie groups.

1 Introduction

In the recent papers [8,1,7] a new energetic formulation for finite strain elasto–plasticity was proposed. It is based on computational algorithms used in engineering, cf. [12,13,3,5,6]. This theory is based on the elastic potential ψ and the dissipation potential Δ as the underlying constitutive functions:

$$\psi = \widehat{\psi}(\mathbf{F},\mathbf{P},p) \text{ and } \Delta = \widehat{\Delta}(\mathbf{P},p,\dot{\mathbf{P}},\dot{p}) \geqslant 0$$

where $\mathbf{F} = \mathrm{D}\varphi = \mathbf{F}_{\mathrm{el}}\mathbf{F}_{\mathrm{pl}}$ is the total deformation gradient, $\mathbf{P} = \mathbf{F}_{\mathrm{pl}}^{-1}$ the inverse plastic deformation and $p \in \mathbb{R}^m$ denote the hardening variables. The plastic tensor $\mathbf{P}$ is usually assumed to have determinant 1, i.e. $\mathbf{P}$ is an element of the special linear group $\mathrm{SL}(d) = \{\, \mathbf{P} \in \mathbb{R}^{d\times d} \mid \det \mathbf{P} = 1 \,\}$. Consequently, $\widehat{\Delta}$ is defined on the tangent bundle of the manifold $\mathrm{SL}(d)\times\mathbb{R}^m$. The axiom of plastic indifference implies

$$\widehat{\psi}(\mathbf{F},\mathbf{P},p) = \widetilde{\psi}(\mathbf{FP},p), \qquad \widehat{\Delta}(\mathbf{P},p,\dot{\mathbf{P}},\dot{p}) = \widetilde{\Delta}(p,\mathbf{P}^{-1}\dot{\mathbf{P}},\dot{p}).$$

This means that the underlying mathematical structure is that of the Lie group $\mathrm{SL}(d)$, a fact which was first emphasized in [8]. Rate–independency is expressed by the fact, that $\widehat{\Delta}$ is homogeneous of degree 1 in the rate $(\dot{\mathbf{P}},\dot{p})$, see (Sy3) in Section 2.1.

The full global energetic formulation relies heavily on the (global) distance $\widehat{D}(x,\cdot,\cdot)$, called dissipation distance, which is generated via the infinitesimal metric $\widehat{\Delta}(x,\cdot,\cdot)$ on $\mathrm{SL}(d)\times\mathbb{R}^m$ as follows: the distance $\widehat{D}((\mathbf{P}_0,p_0),(\mathbf{P}_1,p_1))$ is the infimum of the dissipation $\int_0^1 \widehat{\Delta}(\mathbf{P}(s),p(s),\dot{\mathbf{P}}(s),\dot{p}(s))\,\mathrm{d}s$ over all paths $(\mathbf{P},p)\in \mathrm{C}^1([0,1],\mathrm{SL}(d)\times\mathbb{R}^m)$ with $(\mathbf{P}(j),p(j))=(\mathbf{P}_j,p_j)$ for $j=0,1$. Consider now a body $\Omega\subset\mathbb{R}^d$, a deformation $\varphi:\Omega\mapsto\mathbb{R}^d$ as well as internal states $(\mathbf{P}_j,p_j):\Omega\mapsto\mathrm{SL}(d)\times\mathbb{R}^n$, then integration over Ω gives the total energies

$$\begin{aligned}\mathcal{E}(t,\varphi,\mathbf{P},p)&=\textstyle\int_\Omega \widehat{\psi}(\mathrm{D}\varphi(x),\mathbf{P}(x),p(x))\,\mathrm{d}x-\langle \ell(t),\varphi\rangle,\\ \mathcal{D}((\mathbf{P}_0,p_0),(\mathbf{P}_1,p_1))&=\textstyle\int_\Omega \widehat{D}((\mathbf{P}_0(x),p_0(x)),(\mathbf{P}_1(x),p_1(x)))\,\mathrm{d}x,\end{aligned}$$

where $\ell(t)$ denotes the external loading depending on the process time $t\in[0,T]$. A triple $(\varphi,\mathbf{P},p):[0,T]\times\Omega\mapsto\mathbb{R}^d\times\mathrm{SL}(d)\times\mathbb{R}^m$ is called a **solution process** if it satisfies the following stability condition (S) and the energy inequality (E):

(S) [**Stability**] For all $t\in[0,T]$ and all comparison states $(\widetilde{\varphi},\widetilde{\mathbf{P}},\widetilde{p})$ we have

$$\mathcal{E}(t,\varphi(t),\mathbf{P}(t),p(t))\leqslant \mathcal{E}(t,\widetilde{\varphi},\widetilde{\mathbf{P}},\widetilde{p})+\mathcal{D}((\mathbf{P}(t),p(t)),(\widetilde{\mathbf{P}},\widetilde{p})).$$

(E) [**Energy inequality**] For all $0\leqslant t_1<t_2\leqslant T$ we have

$$\begin{aligned}&\mathcal{E}(t_2,\varphi(t_2),\mathbf{P}(t_2),p(t_2))+\mathrm{Diss}((\mathbf{P},p);[t_1,t_2])\\ \leqslant\ &\mathcal{E}(t_1,\varphi(t_1),\mathbf{P}(t_1),p(t_1))-\textstyle\int_{t_1}^{t_2}\langle\dot{\ell}(s),\varphi(s)\rangle\,\mathrm{d}s,\end{aligned}$$

where $\mathrm{Diss}((\mathbf{P},p);[t_1,t_2])=\int_{t_1}^{t_2}\int_\Omega\widehat{\Delta}(x,(\mathbf{P},p),(\dot{\mathbf{P}},\dot{p}))\,\mathrm{d}x\,\mathrm{d}s$.

Note that (S) & (E) characterize the process completely and that this formulation does not involve any derivatives, neither of $\mathbf{F}=\mathrm{D}\varphi,\mathbf{P}$ and p (with respect to t or x) nor of the constitutive functions $\widehat{\psi}$ and $\widehat{\Delta}$. It is shown in [7] that this formulation is consistent with the usual flow rules of finite plasticity if the solution of (S) & (E) is sufficiently smooth. In fact, using the Legendre transform there is a one–to–one correspondence between $\widehat{\Delta}$ and an associative flow rule for a suitable yield surface.

The purpose of this work is the investigation of the global **dissipation distance** $\widehat{D}$ in cases when we have no hardening or just a scalar variable measuring the total hardening. The major object studied here is a left–invariant metric $\widehat{D}:\mathrm{SL}(d)\times\mathrm{SL}(d)\to[0,\infty]$ generated by $\widetilde{\Delta}:\mathfrak{sl}(d)\to[0,\infty]$ via

$$\widehat{D}(\mathbf{P}_0,\mathbf{P}_1):=\inf\left\{\int_0^1\widetilde{\Delta}(\mathbf{P}(t)^{-1}\dot{\mathbf{P}}(t))\,\mathrm{d}t \;\middle|\; \begin{array}{l}\mathbf{P}(\cdot)\in\mathrm{C}^1([0,1],\mathrm{SL}(d)),\\ \mathbf{P}(0)=\mathbf{P}_0,\ \mathbf{P}(1)=\mathbf{P}_1\end{array}\right\}.$$

In this work we only investigate the two–dimensional case, so we consider SL(2), the three–dimensional Lie group of 2×2 matrices with determinant 1. A few results for the three–dimensional case are given in [8].

For the isotropic case with $d = 2$ we derive an explicit formula for $\widehat{D}$. This case is characterized by

$$\widetilde{\Delta}(\boldsymbol{\xi}) = \Big(\frac{\alpha}{4}|\boldsymbol{\xi}+\boldsymbol{\xi}^{\mathsf{T}}|^2 + \frac{\beta}{4}|\boldsymbol{\xi}-\boldsymbol{\xi}^{\mathsf{T}}|^2\Big)^{1/2}, \text{ where } |\boldsymbol{\eta}|^2 = \operatorname{trace}(\boldsymbol{\eta}^{\mathsf{T}}\boldsymbol{\eta})$$

and $\alpha > 0$, $\beta \in [0, \infty]$ are material parameters. In this case $\widehat{D}(\mathbf{P}_0, \mathbf{P}_1)$ can be shown to depend solely on the two invariants of $\mathbf{Q} = \mathbf{P}_0^{-1}\mathbf{P}_1$, i.e., the trace and the norm of $\mathbf{Q}$.

Second we report on results for the case of single crystal plasticity where $\widetilde{\Delta}$ is characterized via slip systems $\mathbf{S}^\alpha$, $\alpha = 1, \ldots, m$ as follows

$$\widetilde{\Delta}(\boldsymbol{\xi}) = \min\Big\{\sum_{\alpha=1}^{m} \kappa_\alpha\gamma_\alpha \mid \boldsymbol{\xi} = \sum_{\alpha=1}^{m} \gamma_\alpha \mathbf{S}^\alpha \text{ with } \gamma_\alpha \geqslant 0\Big\},$$

with the usual convention that the minimum is $+\infty$ if the set is empty. These cases lead to functions $\widetilde{\Delta}$ which are piecewise linear. As a consequence one has to expect that geodesic curves which minimize the dissipation distance have corners. Physically these corners correspond to switches between different slip systems.

The case of a square lattice leads to the four slip systems $\pm\mathbf{S}^1 = \pm\left(\begin{smallmatrix}0&1\\0&0\end{smallmatrix}\right)$ and $\pm\mathbf{S}^2 = \pm\left(\begin{smallmatrix}0&0\\1&0\end{smallmatrix}\right)$, which is analyzed in detail in [10]. It is shown that for all geodesic curves $\mathbf{P}\colon [0,1] \to \mathrm{SL}(2)$ the invariant rate $\boldsymbol{\xi}(t) = \mathbf{P}(t)^{-1}\dot{\mathbf{P}}(t)$ is piecewise constant and takes values only in $\{\pm\mathbf{S}^1, \pm\mathbf{S}^2, \pm\frac{1}{2}(\mathbf{S}^1+\mathbf{S}^2)\}$. Moreover, the number of switches can be bounded from above by 5.

As is shown in [10], one easily obtains the solution for a parallelogram lattice from the solution for the square lattice. We also address the case of a hexagonal lattice where $N = 6$ slip systems are present.

Finally we conclude with the calculation of the dissipation distance in the case of a scalar hardening variable.

2 Elastoplasticity with hardening parameters

2.1 Constitutive laws

Multiplicative elastoplasticity uses the split $\mathbf{F} = \mathrm{D}\varphi = \mathbf{F}_{\mathrm{el}}\mathbf{F}_{\mathrm{pl}}$, where $\mathbf{F}_{\mathrm{pl}}$ is an internal variable which is assumed to be generated by movements of dislocations and is such that it maps thc crystallographic lattice onto itself. Only the remainder $\mathbf{F}_{\mathrm{el}} = \mathbf{F}\mathbf{F}_{\mathrm{pl}}^{-1}$ is the part which accounts for elastic energy and stresses. To simplify notation we introduce $\mathbf{P} = \mathbf{F}_{\mathrm{pl}}^{-1}$ as an internal variable together with suitable hardening parameters $p \in \mathbb{R}^m$, i.e.

$$z = (\mathbf{P}, p) \in \mathfrak{G}\times\mathbb{R}^m, \qquad \text{with } \mathfrak{G} \subset \mathrm{GL}_+(\mathbb{R}^d).$$

Here $\mathfrak{G}$ is a Lie group contained in $\mathrm{GL}_+(\mathbb{R}^d)$ which might be different from model to model. We let $\mathfrak{g}$ denote the associated Lie algebra. Typically one chooses $\mathfrak{G} = \mathrm{SL}(\mathbb{R}^d) = \{\,\mathbf{F} \mid \det \mathbf{F} = 1\,\}$. However, $\mathfrak{G} = \mathrm{GL}_+(\mathbb{R}^d)$ or, if only one slip system with $|\mathbf{n}| = |\mathbf{d}| = 1$ and $\mathbf{n}\cdot\mathbf{d} = 0$ is active, $\mathfrak{G} = \{\,\mathbf{I}+\alpha\mathbf{d}\otimes\mathbf{n} \,|\, \alpha \in \mathbb{R}\}$ might also be suitable.

The hardening parameters can include isotropic or kinematic hardening, see [7] and below.

For the constitutive function $\widehat{\psi}$ and $\widehat{\Delta}$ we now specify the associated symmetry conditions. They involve the material symmetry group $\mathfrak{S} \subset \mathrm{SO}(\mathbb{R}^d)$ and they are supposed to hold true for all $(x, \mathbf{F}, \mathbf{P}, p) \in \Omega\times\mathrm{GL}_+(\mathbb{R}^d)\times\mathfrak{G}\times\mathbb{R}^m$:

(Sy1) Objectivity (frame indifference):

$$\widehat{\psi}(x, \mathbf{RF}, \mathbf{P}, p) = \widehat{\psi}(x, \mathbf{F}, \mathbf{P}, p) \text{ for all } \mathbf{R} \in \mathrm{SO}(3);$$

(Sy2) Plastic indifference:

$$\widehat{\psi}(x, \mathbf{FG}^{-1}, \mathbf{GP}, p) = \widehat{\psi}(x, \mathbf{F}, \mathbf{P}, p) \text{ and}$$
$$\widehat{\Delta}(x, \mathbf{GP}, p, \mathbf{G}\dot{\mathbf{P}}, \dot{p}) = \widehat{\Delta}(x, \mathbf{P}, p, \dot{\mathbf{P}}, \dot{p}) \text{ for all } \mathbf{G} \in \mathfrak{G};$$

(Sy3) Rate independency:

$$\widehat{\Delta}(x, \mathbf{P}, p, \alpha\dot{\mathbf{P}}, \alpha\dot{p}) = \alpha\widehat{\Delta}(x, \mathbf{P}, p, \dot{\mathbf{P}}, \dot{p}) \text{ for } \alpha \geqslant 0;$$

(Sy4) Material symmetry:

$$\widehat{\psi}(x, \mathbf{F}, \mathbf{PS}, \tau_{\mathbf{S}}p) = \widehat{\psi}(x, \mathbf{F}, \mathbf{P}, p) \text{ and}$$
$$\widehat{\Delta}(x, \mathbf{PS}, \tau_{\mathbf{S}}p, \dot{\mathbf{P}}\mathbf{S}, \tau_{\mathbf{S}}\dot{p}) = \widehat{\Delta}(x, \mathbf{P}, p, \dot{\mathbf{P}}, \dot{p}) \text{ for all } \mathbf{S} \in \mathfrak{S}.$$

Here $\tau_{\mathbf{S}} \in \mathbb{R}^{m\times m}$ denotes a linear representation of the material symmetry group $\mathfrak{S}$ on the hardening parameters in $\mathbb{R}^m$. It satisfies $\tau_{\mathbf{S}_1\mathbf{S}_2} = \tau_{\mathbf{S}_2}\tau_{\mathbf{S}_1}$.

The special assumption for elastoplasticity is the "plastic indifference" (Sy2) which leads to the multiplicative split in $\widehat{\psi}$ as well as to the correct time rates in the flow rules. We find

$$\widehat{\psi}(x, \mathbf{F}, \mathbf{P}, p) = \widetilde{\psi}(x, \mathbf{FP}, p), \qquad \widehat{\Delta}(x, \mathbf{P}, p, \dot{\mathbf{P}}, p) = \widetilde{\Delta}(x, p, \mathbf{P}^{-1}\dot{\mathbf{P}}, \dot{p}). \quad (1)$$

The consequence of the other symmetries will be studied along with the examples treated below.

2.2 Associative flow rules

We will not need the associated flow rules in order to study the energetic formulation of elastoplasticity. However, they give some insight into the underlying structures. In particular, they are very helpful in finding the paths which minimize the dissipation distance between two points. This minimization is closely related to the Pontrjagin Maximum Principle (PMP), which uses the dual stress variables in the same way as the flow rules in elastoplasticity. We define the thermodynamically conjugate variables

$$\mathbf{Q} = -\tfrac{\partial}{\partial \mathbf{P}}\widehat{\psi}(x,\mathbf{F},\mathbf{P},p) = -\mathbf{F}^{\mathsf{T}}\tfrac{\partial}{\partial \mathbf{F}_{\mathrm{el}}}\widetilde{\psi}(x,\mathbf{FP},p) \in \mathrm{T}^*_{\mathbf{P}}\mathfrak{G} = \mathbf{P}^{-\mathsf{T}}\mathfrak{g}^* \subset \mathbb{R}^{d\times d},$$
$$\mathbf{q} = -\tfrac{\partial}{\partial p}\widehat{\psi}(x,\mathbf{F},\mathbf{P},p) = -\tfrac{\partial}{\partial p}\widetilde{\psi}(x,\mathbf{FP},p) \in \mathbb{R}^m_*.$$

Here $\mathfrak{g}^*$ denotes the dual Lie algebra which is the set of linear mappings from $\mathfrak{g}$ into $\mathbb{R}$ and similarly $\mathbb{R}^m_*$ denotes the dual space of $\mathbb{R}^m$. This notation makes the distinction between the primal internal variables $(\mathbf{P},p) \in \mathfrak{G}\times\mathbb{R}^m$ and the dual (thermodynamically conjugate) variables $(\mathbf{P}^{\mathsf{T}}\mathbf{Q},q) \in \mathfrak{g}^*\times\mathbb{R}^m_*$ more transparent.

The elastic domain $\mathbb{Q}(x,\mathbf{P},p)$ associated with $\widehat{\Delta}(x,\mathbf{P},\mathbf{p},\cdot,\cdot)$ is the set of all thermodynamic forces $(\mathbf{Q},q)$ which are not large enough to overcome the dissipational friction:

$$\begin{aligned}\mathbb{Q}(x,\mathbf{P},p) &= [\partial_{(\dot{\mathbf{P}},\dot{p})}\widehat{\Delta}(x,\mathbf{P},p,\cdot,\cdot)](\mathbf{0},0)\\ &= \{\,(\mathbf{Q},q)\mid \mathbf{Q}{:}\mathbf{V}+q{\cdot}v \leqslant \widehat{\Delta}(x,\mathbf{P},p,\mathbf{V},v) \text{ for all } (\mathbf{V},v)\in \mathrm{T}_{\mathbf{P}}\mathfrak{G}\times\mathbb{R}^m\,\}.\end{aligned}$$

Using (Sy2) and the Lie group structure implying $\mathrm{T}_{\mathbf{P}}\mathfrak{G} = \mathbf{P}\mathfrak{g}$ leads to

$$\begin{aligned}\mathbb{Q}(x,\mathbf{P},p) &= \{\,(\mathbf{Q},q)\mid \mathbf{Q}{:}(\mathbf{P}\boldsymbol{\xi})+q{\cdot}v \leqslant \widetilde{\Delta}(x,p,\boldsymbol{\xi},v) \text{ for all } (\boldsymbol{\xi},v)\in \mathfrak{g}\times\mathbb{R}^m\,\}\\ &= \{\,(\mathbf{Q},q)\mid (\mathbf{P}^{\mathsf{T}}\mathbf{Q},q)\in[\partial_{(\boldsymbol{\xi},\dot{p})}\widetilde{\Delta}(x,p,\cdot)](\mathbf{0},0)\,\} \subset \mathrm{T}^*_{\mathbf{P}}\mathfrak{G}\times\mathbb{R}^m_*.\end{aligned}$$

Defining $\widetilde{\mathbb{Q}}(x,p) = \mathbb{Q}(x,\mathbf{1},p) = [\partial_{(\boldsymbol{\xi},\dot{p})}\widetilde{\Delta}(x,p,\cdot,\cdot)](\mathbf{0},0) \subset \mathfrak{g}^*\times\mathbb{R}^m_*$ we find

$$(\mathbf{Q},q)\in\mathbb{Q}(\mathbf{P},p) \iff (\mathbf{P}^{\mathsf{T}}\mathbf{Q},q)\in\widetilde{\mathbb{Q}}(p). \tag{2}$$

The plastically indifferent objects $\widetilde{\Delta}(p,\cdot,\cdot) : \mathfrak{g}\times\mathbb{R}^m \mapsto [0,\infty]$ and $\widetilde{\mathbb{Q}}(p) \subset \mathfrak{g}^*\times\mathbb{R}^m_*$ are in one–to–one correspondence to each other. On the one hand we have $\widetilde{\mathbb{Q}}(p) = [\partial_{(\boldsymbol{\xi},\dot{p})}\widetilde{\Delta}(p,\cdot,\cdot)](\mathbf{0},0)$. On the other hand, for given convex $\widetilde{\mathbb{Q}}(p)$ the function $\widetilde{\Delta}(p,\cdot)$ is obtained by Legendre transformation of $\mathcal{X}_{\widetilde{\mathbb{Q}}(p)}$, i.e.

$$\widetilde{\Delta}(p,\boldsymbol{\xi},v) = \sup_{(\boldsymbol{\eta},q)\in\widetilde{\mathbb{Q}}(p)} (\boldsymbol{\eta}{:}\boldsymbol{\xi}+v{\cdot}q) = \sup_{(\boldsymbol{\eta},q)\in\mathfrak{g}^*\times\mathbb{R}^m_*} [\boldsymbol{\eta}{:}\boldsymbol{\xi}+v{\cdot}q-\mathcal{X}_{\widetilde{\mathbb{Q}}(p)}(\boldsymbol{\eta},q)]. \tag{3}$$

The flow rule in the thermodynamically conjugate space now takes the form $(\mathbf{Q},q)\in[\partial_{(\dot{\mathbf{P}},\dot{p})}\widehat{\Delta}(\mathbf{P},p,\cdot,\cdot)](\dot{\mathbf{P}},\dot{p})$. Using the above transformation for $\mathbf{P}$ this is equivalent to

$$(\mathbf{P}^{\mathsf{T}}\mathbf{Q},q)\in[\partial_{(\boldsymbol{\xi},\dot{p})}\widetilde{\Delta}(p,\cdot,\cdot)](\mathbf{P}^{-1}\dot{\mathbf{P}},\dot{p}) \subset \mathfrak{g}^*\times\mathbb{R}^m_*. \tag{4}$$

Via the Legendre transform we obtain the formulation in the internal variable space:

$$(\mathbf{P}^{-1}\dot{\mathbf{P}},\dot{p}) \in \partial\mathcal{X}_{\widetilde{\mathbb{Q}}(x,p)}(\mathbf{P}^{\mathsf{T}}\mathbf{Q},q) = \mathrm{N}_{(\mathbf{P}^{\mathsf{T}}\mathbf{Q},q)}\widetilde{\mathbb{Q}}(x,p) \subset \mathfrak{g}\times\mathbb{R}^m, \tag{4a}$$

where $\mathrm{N}_z C$ denotes the outer normal cone at z to the convex set C. This is the well–known associative flow rule of multiplicative elastoplasticity. It contains the "plastically indifferent" plastic rate $\mathbf{P}^{-1}\dot{\mathbf{P}}$ as well as the "plastically indifferent" conjugate force $\mathbf{P}^{\mathsf{T}}\mathbf{Q} = -\mathbf{F}_{\mathrm{el}}^{\mathsf{T}}\frac{\partial}{\partial \mathbf{F}_{\mathrm{el}}}\widetilde{\psi}(x, \mathbf{F}_{\mathrm{el}}, p)$.

3 Dissipation distances in the case without hardening

To facilitate the subsequent discussion we first consider the case without hardening, so $\Delta = \widehat{\Delta}(\mathbf{P}, \dot{\mathbf{P}}) = \widetilde{\Delta}(\mathbf{P}^{-1}\dot{\mathbf{P}})$ is defined via a norm–like function $\widetilde{\Delta} : \mathfrak{g} \mapsto [0, \infty]$. This situation is discussed in [8] in some detail. The dissipation distance satisfies

$$\widehat{D}(\mathbf{P}_0, \mathbf{P}_1) = \widehat{D}(\mathbf{1}, \mathbf{P}_0^{-1}\mathbf{P}_1) =: \widetilde{D}(\mathbf{P}_0^{-1}\mathbf{P}_1),$$

and $\widetilde{D}\colon \mathfrak{G} \mapsto [0, \infty]$ satisfies $\widetilde{D}(\mathbf{1}+\varepsilon\boldsymbol{\xi}) = \varepsilon\widetilde{\Delta}(\boldsymbol{\xi}) + O(\varepsilon^2)$ for $\varepsilon \to 0$. By definition, $\widetilde{D}(\mathrm{e}^{\boldsymbol{\xi}}) \leqslant \widetilde{\Delta}(\boldsymbol{\xi})$ for all $\boldsymbol{\xi} \in \mathfrak{g}$. Hence for $\mathbf{P} = \mathrm{e}^{\boldsymbol{\xi}_1} \cdot \ldots \cdot \mathrm{e}^{\boldsymbol{\xi}_k}$ the triangle inequality yields the estimate $\widetilde{D}(\mathbf{P}) \leqslant \sum_{j=1}^{k} \widetilde{\Delta}(\boldsymbol{\xi}_j)$. It is important to observe that, in general, $\widetilde{D}(\mathrm{e}^{\boldsymbol{\xi}}) < \widetilde{\Delta}(\boldsymbol{\xi})$ which indicates that the matrix exponential curves $t \mapsto \mathrm{e}^{t\boldsymbol{\xi}}$ are **not** the paths of minimal dissipation.

As the dissipation distance is defined via a variational problem, its computation may be fairly complicated. For practical purposes one would like to know situations where $\widehat{D}$ is available at reasonable computational cost. We will present several such cases, in particular in the case of single crystal plasticity.

The definition of the dissipation distance $\widehat{D}$ still contains some redundancy because we assume rate independence (Sy3). A simple reparametrization argument shows that it suffices to consider only curves $\mathbf{P}\colon [0,1] \to \mathfrak{G}$ for which $\widetilde{\Delta}(\mathbf{P}^{-1}\dot{\mathbf{P}})$ is constant. Therefore, if we set $\mathbb{U} = \{\, \boldsymbol{\xi} \in \mathfrak{g} \,|\, \widetilde{\Delta}(\boldsymbol{\xi}) \leqslant 1 \,\}$, we immediately obtain the following characterization:

$$\widehat{D}(\mathbf{P}_0, \mathbf{P}_1) = \inf \left\{ T > 0 \;\middle|\; \begin{array}{l} \text{there exists } \mathbf{P} \in C^1([0,T], \mathfrak{G}) \text{ such that} \\ \mathbf{P}^{-1}\dot{\mathbf{P}} \in \mathbb{U},\ \mathbf{P}(0) = \mathbf{P}_0,\ \mathbf{P}(T) = \mathbf{P}_1 \end{array} \right\}.$$

Thus $\widehat{D}$ is characterized via the solution of a time–optimal control problem. The underlying ODE is $\dot{\mathbf{P}}(t) = \mathbf{P}(t)\boldsymbol{\xi}(t)$, $\boldsymbol{\xi}(\cdot) \in \mathbb{U}$ a.e., with time being the cost functional. The advantage of this point of view is that standard results and tools from optimal control theory can be applied immediately. For example, if $\mathbb{U} \subseteq \mathfrak{g}$ is compact convex, then distance minimizing paths $t \mapsto \mathbf{P}(t)$ always exist within the class of absolutely continuous functions, i.e., $\mathbf{P}^{-1}\dot{\mathbf{P}}$ is L^∞, but not necessarily C^0. Thus shortest paths may have corners, cf. [8] for an example. Typically, corners appear when the boundary of $\mathbb{U}$ is not strictly convex. In particular, this always happens in the case of single crystal plasticity where $\widetilde{\Delta}$ is piecewise linear, so $\mathbb{U}$ is a convex polyhedron.

We note that $\mathbb{U}$ need not have interior points, it suffices that $\mathbb{U}$ generates $\mathfrak{g}$ as a Lie algebra.

The **Pontrjagin Maximum Principle (PMP)** as a first order necessary condition for optimality is a powerful tool for finding shortest paths. For systems on Lie groups it takes a particularly simple form, cf. [4,9], for example. In some sense the (PMP) **is** the flow rule, but our point of view will give additional geometric insight.

3.1 The (PMP) and its relation to the flow rule

Given $\mathbb{U} \subseteq \mathfrak{g}$ we define the **optimal Hamiltonian** $\mathcal{H}\colon \mathfrak{g}^* \to \mathbb{R}$, as follows:

$$\mathcal{H}(\boldsymbol{\eta}) = \min_{\boldsymbol{\xi}\in\mathbb{U}}(\boldsymbol{\eta}{:}\boldsymbol{\xi}).$$

Let $\widetilde{\mathbb{Q}} = \{\boldsymbol{\eta} \in \mathfrak{g}^* \mid \mathcal{H}(\boldsymbol{\eta}) \geqslant -1\}$. If $\mathbb{U}$ is convex and $\mathbb{U} = -\mathbb{U}$, then $\widetilde{\mathbb{Q}}$ is simply the polar of $\mathbb{U}$.

Now let I be an interval and assume that $\mathbf{P}\colon I \to \mathfrak{G}$ is a length minimizing path. Set $\boldsymbol{\xi} = \mathbf{P}^{-1}\dot{\mathbf{P}} \in L^\infty(I, \mathbb{U})$. Then the (PMP) yields an absolutely continuous curve $\boldsymbol{\eta}(t) \in \mathfrak{g}^*$ with the following properties:

(0) Nontriviality: $\boldsymbol{\eta} \neq 0$ in I,
(1) Adjoint equation: $\mathrm{Ad}(\mathbf{P}^{-1}(t))^*\boldsymbol{\eta}(t) \equiv \mathrm{const} \in \mathfrak{g}^*$,
(2) Minimizing condition: $\boldsymbol{\eta}(t){:}\boldsymbol{\xi}(t) = \min\left\{\boldsymbol{\eta}(t){:}\widetilde{\boldsymbol{\xi}} \mid \widetilde{\boldsymbol{\xi}} \in \mathbb{U}\right\}$,
(3) Constant Hamiltonian: $\mathcal{H}(\boldsymbol{\eta}(t)) \equiv \mathrm{const}$, either -1, or 0.

These conditions relate to the classical flow rules as follows: by (3) the curve $\boldsymbol{\eta}(t)$ evolves on a level set of $\mathcal{H}$, either $\{\mathcal{H} = 0\}$, or $\{\mathcal{H} = -1\} = \partial\widetilde{\mathbb{Q}}$. The latter is the **yield surface**, this becomes even more evident if one compares (3) with Eqn. (2). Although $\mathcal{H}(\boldsymbol{\eta}) \equiv 0$ is possible, this occurs only under very degenerate circumstances—it never occurs, for example, if $\mathbb{U}$ contains a zero-neighborhood. The **flow rule** (cf. Eqn. (4a)) is encoded in (1) and (2). The latter implies that $\boldsymbol{\xi}(t) \in \partial\mathcal{H}(\boldsymbol{\eta}(t))$. We stated the adjoint equation (1) already in integrated form using the adjoint action of $\mathfrak{G}$ on $\mathfrak{g}$, resp., the induced action on $\mathfrak{g}^*$. For $\mathfrak{G} = \mathrm{SL}(d)$ and $\mathfrak{g} = \mathfrak{sl}(d)$ the adjoint action is simply conjugation: $\mathrm{Ad}(\mathbf{P})\xi = \mathbf{P}\boldsymbol{\xi}\mathbf{P}^{-1}$. Identifying $\mathfrak{sl}(d)^*$ with $\mathfrak{sl}(d)$ via the trace form $\boldsymbol{\eta}{:}\boldsymbol{\xi} = \mathrm{tr}(\boldsymbol{\eta}^\mathsf{T}\boldsymbol{\xi})$ one may also write $\mathrm{Ad}(\mathbf{P}^{-1})^*\boldsymbol{\eta} = \mathbf{P}^{-\mathsf{T}}\boldsymbol{\eta}\mathbf{P}^\mathsf{T}$.

3.2 The isotropic two-dimensional case

This case relates to isotropic plasticity of Prandtl–Reuß type using the von Mises flow rule. The plastic tensor $\mathbf{P}$ lies in $\mathfrak{G} = \mathrm{SL}(d)$ and the material symmetry–group is $\mathfrak{S} = \mathrm{O}(d)$, see (Sy4). From (Sy2) and (Sy4) one obtains that $\widetilde{\Delta}(\boldsymbol{\xi}) = \widetilde{\Delta}(\mathbf{R}\boldsymbol{\xi}\mathbf{R}^\mathsf{T})$ for all $\boldsymbol{\xi} \in \mathfrak{sl}(d)$, $\mathbf{R} \in \mathrm{SO}(d)$. If one considers Riemannian metrics then $\widetilde{\Delta}(\boldsymbol{\xi})$ can be put into the general form

$$\widetilde{\Delta}(\boldsymbol{\xi}) = (\alpha|\boldsymbol{\xi}_{\mathrm{sym}}|^2+\beta|\boldsymbol{\xi}_{\mathrm{anti}}|^2)^{1/2} \text{ with } \boldsymbol{\xi}_{\mathrm{sym}} = \tfrac{1}{2}(\boldsymbol{\xi}+\boldsymbol{\xi}^\mathsf{T}),\ \boldsymbol{\xi}_{\mathrm{anti}} = \tfrac{1}{2}(\boldsymbol{\xi}-\boldsymbol{\xi}^\mathsf{T}).$$

(5)

It is shown in [8] that a curve is a shortest path if and only if it has the form

$$\mathbf{P}(t) = \mathbf{P}(0)M_{\beta/\alpha}(t\boldsymbol{\xi}) \text{ where } M_\delta(\boldsymbol{\xi}) = \mathrm{e}^{\boldsymbol{\xi}_{\mathrm{sym}}-\delta\boldsymbol{\xi}_{\mathrm{anti}}}\mathrm{e}^{(1+\delta)\boldsymbol{\xi}_{\mathrm{anti}}}.$$

As a consequence, the dissipation distance associated with $\widetilde{\Delta}$ from (5) reads

$$\widetilde{D}(\mathbf{P}) = \min\{\, \widetilde{\Delta}(\boldsymbol{\xi}) \mid \mathbf{P} = M_{\beta/\alpha}(\boldsymbol{\xi})\,\}.$$

For symmetric, positive definite matrices $\mathbf{P} = \mathbf{P}^{\mathsf{T}} > 0$ we find

$$\widetilde{D}(\mathbf{P}) = \sqrt{\alpha}\,|\log \mathbf{P}| = \sqrt{\alpha}\,(\log \mathbf{P}\colon \log \mathbf{P})^{1/2}.$$

Using the polar decomposition $\mathbf{P} = \mathbf{RU}$ with $\mathbf{R} \in \mathrm{SO}(d)$ and $\mathbf{U} = \mathbf{U}^{\mathsf{T}} > 0$ together with the triangle inequality we arrive at the explicit estimate

$$\min\{\widetilde{D}(\mathbf{R}), \widetilde{D}(\mathbf{U})\} \leqslant \widetilde{D}(\mathbf{P}) \leqslant \widetilde{D}(\mathbf{R}) + \widetilde{D}(\mathbf{U}),$$

where $\widetilde{D}(\mathbf{U}) = \frac{1}{2}\sqrt{\alpha}\,|\log(\mathbf{U}^{\mathsf{T}}\mathbf{U})| = \frac{1}{2}\sqrt{\alpha}\,|\log(\mathbf{P}^{\mathsf{T}}\mathbf{P})|$. Thus we see that for large shears $\mathbf{P} = \mathbf{1} + \gamma\,\mathbf{n}\otimes\mathbf{m}$ the dissipation distance $\widetilde{D}(\mathbf{P})$ grows at most like $\log|\gamma|$ for $\gamma \to \infty$.

An important role in isotropic plasticity plays the case of zero plastic spin. This is realized by $\alpha = 1$ and $\beta = \infty$ in (5), or more precisely, on $\mathfrak{sl}(d) = \mathrm{T}_{\mathbf{1}}(\mathrm{SL}(d))$ we set

$$\widetilde{\Delta}_{\text{no spin}}(\boldsymbol{\xi}) = \begin{cases} |\boldsymbol{\xi}_{\text{sym}}| & \text{if } \boldsymbol{\xi}_{\text{anti}} = \mathbf{0}, \\ \infty & \text{otherwise.} \end{cases}$$

The associated geodesic curves are $\mathbf{P}(t) = \mathbf{P}(0)\mathrm{e}^{t(\boldsymbol{\sigma}-\boldsymbol{\omega})}\mathrm{e}^{t\boldsymbol{\omega}}$ with $\boldsymbol{\sigma} = \boldsymbol{\sigma}^{\mathsf{T}}$ and $\boldsymbol{\omega} = -\boldsymbol{\omega}^{\mathsf{T}}$. Note that $\mathbf{P}(t)^{-1}\dot{\mathbf{P}}(t) = \mathrm{e}^{-t\boldsymbol{\omega}}\boldsymbol{\sigma}\mathrm{e}^{t\boldsymbol{\omega}}$ is not constant but gives a constant and finite $\Delta = \widetilde{\Delta}(\mathbf{P}(t)^{-1}\dot{\mathbf{P}}(t)) = |\boldsymbol{\sigma}|$. In particular, $\mathbf{P}(0)^{-1}\mathbf{P}(t)$ can reach every matrix in $\mathrm{SL}(d)$, not just symmetric ones. We find

$$\widetilde{D}_{\text{no spin}}(\mathbf{P}) = \min\left\{ |\boldsymbol{\sigma}| \;\middle|\; \begin{array}{c} \boldsymbol{\sigma} = \boldsymbol{\sigma}^{\mathsf{T}} \text{ and there exists } \boldsymbol{\omega} = -\boldsymbol{\omega}^{\mathsf{T}} \\ \text{such that } \mathbf{P} = \mathrm{e}^{\boldsymbol{\sigma}-\boldsymbol{\omega}}\mathrm{e}^{\boldsymbol{\omega}} \end{array} \right\}. \tag{6}$$

For $d = 2$ the dissipation distances $\widehat{D}_{\text{no spin}}$ can be calculated more explicitly. For $\mathbf{P} \in \mathrm{SL}(2)$ we have

$$\widetilde{D}_{\text{no spin}}(\mathbf{P}) = \min\{\rho \geqslant 0 \mid \text{there exists } \gamma \in \mathbb{R} \text{ such that } N(\rho,\gamma) = \mathbf{P}{:}\mathbf{P} \text{ and } T(\rho,\gamma) = \operatorname{tr}\mathbf{P}\}$$

where the functions N and T are defined via

$$C(t) = \begin{cases} \cosh\sqrt{t} & \text{for } t \geqslant 0, \\ \cos\sqrt{-t} & \text{for } t \leqslant 0, \end{cases} \quad\text{and}\quad S(t) = \begin{cases} \frac{\sinh\sqrt{t}}{\sqrt{t}} & \text{for } t > 0 \\ 1 & \text{for } t = 0 \\ \frac{\sin\sqrt{-t}}{\sqrt{-t}} & \text{for } t < 0 \end{cases}$$

as $T(\rho,\gamma) = 2(\cos\gamma\, C(\rho^2-\gamma^2) + \gamma\sin\gamma\, S(\rho^2-\gamma^2))$, and
$N(\rho,\gamma) = 2(1 + 2\rho^2[S(\rho^2-\gamma^2)]^2)$.
It actually suffices to consider $\gamma \in [0, \sqrt{\pi^2+\rho^2}\,]$ in the minimum defining $\widetilde{D}_{\text{no spin}}$. For instance, for a rotation $\mathbf{R} = \begin{pmatrix} \cos\phi & \sin\phi \\ -\sin\phi & \cos\phi \end{pmatrix}$ with $\phi \in [-\pi, \pi]$, we get $\widetilde{D}(\mathbf{R}) = \sqrt{|\phi|(2\pi+|\phi|)}$. In particular, for $\phi = \pi$ we have $\widetilde{D}(-\mathbf{1}) = \sqrt{3}\,\pi$ which is obtained with $\rho = \sqrt{3}\,\pi$ and $\gamma = 2\pi$ and the geodesic curve $\mathbf{P}(t) = \mathrm{e}^{t(\boldsymbol{\sigma}-\boldsymbol{\omega})}\mathrm{e}^{t\boldsymbol{\omega}}$ with $\boldsymbol{\sigma} = \begin{pmatrix} \rho & 0 \\ 0 & -\rho \end{pmatrix}$ and $\boldsymbol{\omega} = \begin{pmatrix} 0 & \gamma \\ -\gamma & 0 \end{pmatrix}$.

3.3 Single–crystal plasticity

In single–crystal plasticity the plastic flow occurs through plastic slip induced by movements of dislocations. Let $\mathbf{S}^\alpha = \mathbf{d}^\alpha \otimes \mathbf{n}^\alpha$, $\alpha = 1, \dots, m$, be the m slip systems where $\mathbf{n}^\alpha$ is the unit normal to the α-th slip plane and $\mathbf{d}^\alpha$ is the slip direction with $|\mathbf{d}^\alpha| = 1$ and $\mathbf{d}^\alpha \cdot \mathbf{n}^\alpha = 0$. All plastic flow has the form

$$\dot{\mathbf{P}} = \mathbf{P} \textstyle\sum_{\alpha=1}^m \nu_\alpha \mathbf{S}^\alpha$$

where the slip rates ν_α are taken to be positive. This means we formally distinguish the slip systems $\mathbf{S}^\alpha$ and $-\mathbf{S}^\alpha$.

The crystal symmetry group $\mathfrak{S} \subset \mathrm{O}(d)$ is discrete and associates a permutation $\pi_\mathbf{R} \in \mathrm{Perm}(m)$ to each $\mathbf{R} \in \mathfrak{S}$ such that $\pi_{\mathbf{R}\widehat{\mathbf{R}}} = \pi_\mathbf{R} \circ \pi_{\widehat{\mathbf{R}}}$ (composition of permutations) and

$$(\mathbf{R}\mathbf{d}^\alpha, \mathbf{R}\mathbf{n}^\alpha) = (\mathbf{d}^{\pi_\mathbf{R}(\alpha)}, \mathbf{n}^{\pi_\mathbf{R}(\alpha)}) \quad \Longleftrightarrow \quad \mathbf{S}^{\pi_\mathbf{R}(\alpha)} = \mathbf{R}\mathbf{S}^\alpha\mathbf{R}^\top.$$

The set of all slip systems $\{\, \mathbf{S}^\alpha \mid \alpha = 1, \dots, m \,\}$ determines the associated Lie algebra $\mathfrak{g}$ (and hence the Lie group $\mathfrak{G} \subset \mathrm{GL}_+(d)$) as the smallest Lie algebra containing all slip systems:

$$S := \mathrm{span}\{\mathbf{S}_\alpha \mid \alpha = 1, \dots, m\} \subset \mathfrak{g} = \mathrm{T}_\mathbf{1}\mathfrak{G}.$$

Note that $\mathfrak{g}$ may be strictly bigger than S, as is seen in Example 1 where $\dim S = 2 < \dim \mathfrak{g} = 3$. From $\mathbf{d}^\alpha \cdot \mathbf{n}^\alpha = 0$ we know $\mathrm{tr}(\mathbf{S}^\alpha) = 0$ and hence $\mathfrak{g} \subset \mathfrak{sl}(d)$ and $\mathfrak{G} \subset \mathrm{SL}(d)$.

We now postulate the dissipation metric and then show that it gives rise to the classical single–crystal flow rule for the resolved shear stresses τ_α in each slip system. With $\boldsymbol{\xi} = \mathbf{P}^{-1}\dot{\mathbf{P}}$ the dissipation is

$$\widetilde{\Delta}(\boldsymbol{\xi}) = \min\{\, \textstyle\sum_{\alpha=1}^m \kappa_\alpha \gamma_\alpha \mid \gamma_\alpha \geqslant 0,\ \boldsymbol{\xi} = \sum_{\alpha=1}^m \gamma_\alpha \mathbf{S}^\alpha \,\}$$

where $\kappa_\alpha > 0$ are given threshold parameters, see [12,2]. Since $\widetilde{\Delta}(\mathbf{S}^\alpha) = \kappa_\alpha$, the associated set $\{\widetilde{\Delta} \leqslant 1\}$ is $\mathbb{U} = \mathrm{conv}(\{\kappa_\alpha^{-1}\mathbf{S}^\alpha \mid \alpha = 1 \dots m\})$. Computing the dissipation distance $\widetilde{D}(\mathbf{P})$ can be considered as an optimal factorization problem: find $k \in \mathbb{N}$, $t_1, \dots, t_k \in \mathbb{R}^+$, and $\boldsymbol{\xi}_1 \dots \boldsymbol{\xi}_k \in \{\kappa_\alpha^{-1}\mathbf{S}^\alpha \mid \alpha = 1 \dots m\}$ such that $\mathbf{P} = \mathrm{e}^{t_1\boldsymbol{\xi}_1} \cdots \mathrm{e}^{t_k\boldsymbol{\xi}_k}$ and $\sum_k |t_k|$ is minimal.

The associated elastic domain $\widehat{\mathbb{Q}}(\mathbf{P})$ is formulated in the thermodynamically conjugate variables $\mathbf{Q} = -\partial_{\mathbf{P}}\widehat{\psi}$. The invariant form $\widetilde{\mathbb{Q}}$ using $\boldsymbol{\eta} = \operatorname{dev} \mathbf{P}^{\mathsf{T}}\mathbf{Q}$ is given by

$$\widetilde{\mathbb{Q}} = \{\, \boldsymbol{\eta} \in \mathfrak{sl}(d)^* \mid \kappa_\alpha + \mathbf{S}^\alpha{:}\boldsymbol{\eta} \leqslant 0 \text{ for } \alpha = 1, \ldots, m \,\}.$$

Hence the elastic domain is characterized by one yield condition for each slip system $\mathbf{S}^\alpha$. Denoting by $\tau_\alpha = \mathbf{S}^\alpha{:}\boldsymbol{\eta}$ the resolved shear stress, the slip system $\mathbf{S}^\alpha$ becomes active ($\gamma_\alpha > 0$) only if $\tau_\alpha = \kappa_\alpha > 0$.

Example 1. **[Square lattice in $d = 2$.]** Consider $d = 2$ and

$$\widetilde{\Delta}(\begin{pmatrix} \alpha & \beta \\ \gamma & -\alpha \end{pmatrix}) = \begin{cases} |\beta| + |\gamma| & \text{if } \alpha = 0, \\ \infty & \text{otherwise.} \end{cases}$$

This corresponds to the four slip systems $\{\pm\mathbf{S}^1, \pm\mathbf{S}^2\}$ with $\mathbf{S}^1 = \begin{pmatrix} 0 & 1 \\ 0 & 0 \end{pmatrix}$

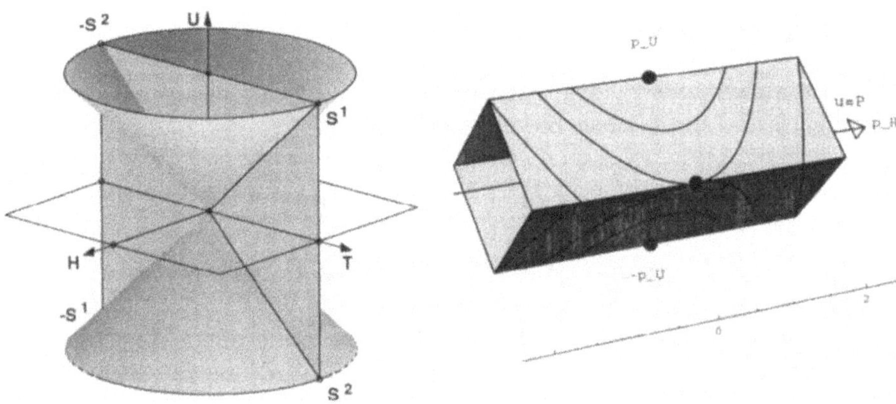

Fig. 1. The set $\mathbb{U}$, the yield surface $\partial\widetilde{\mathbb{Q}}$, and the flow rule for four $2D$-slip systems

and $\mathbf{S}^2 = \begin{pmatrix} 0 & 0 \\ 1 & 0 \end{pmatrix}$. Here we have $\mathbb{U} = \operatorname{conv}(\pm\mathbf{S}^1, \pm\mathbf{S}^2) \subsetneq \mathfrak{sl}(2)$. The Lie algebra $\mathfrak{sl}(2)$ is a 3-dimensional vector space. For visualization purposes we use the basis $\mathsf{H} = \begin{pmatrix} 1 & 0 \\ 0 & -1 \end{pmatrix}$, $\mathsf{T} = \mathbf{S}^1 + \mathbf{S}^2$ and $\mathsf{U} = \mathbf{S}^1 - \mathbf{S}^2$. The set $\{\boldsymbol{\xi} \in \mathfrak{sl}(2) \mid \det(\boldsymbol{\xi}) = 0,\ \boldsymbol{\xi} \neq 0\}$ of all possible slip systems is the boundary of a Lorentzian double cone. Figure 1 shows this double cone and the set $\mathbb{U}$. The horizontal plane $\mathbb{R}\mathsf{H} + \mathbb{R}\mathsf{T}$ is the set of symmetric matrices (in $\mathfrak{sl}(2)$) while the vertical axis is the set of skew-symmetric matrices. Using a dual basis for $\mathfrak{sl}(2)^*$ we depicted the yield surface $\partial\widetilde{\mathbb{Q}}$ (which is a cylinder over a diamond square) and some integral curves of the flow rule. A careful analysis of the information provided by the (PMP) shows that for all geodesic curves $t \mapsto \mathbf{P}(t)$ the curve $\boldsymbol{\xi}(t) = \mathbf{P}(t)^{-1}\dot{\mathbf{P}}(t) \in \mathbb{U}$ is piecewise constant, and

$$\boldsymbol{\xi}(t) \in \left\{ \pm\mathbf{S}^1, \pm\mathbf{S}^2, \pm\frac{1}{2}(\mathbf{S}^1 + \mathbf{S}^2), \pm\frac{1}{2}(\mathbf{S}^1 - \mathbf{S}^2) \right\}.$$

Actually one obtains much more detailed information about the switching behavior of $\boldsymbol{\xi}(t)$. We say that $\boldsymbol{\xi}(t)$ is a **bang-bang control** if it switches only between the vertices of $\mathbb{U}$. Otherwise, we call $\boldsymbol{\xi}(t)$ a **singular control**. From the (PMP) we obtain two types of bang-bang controls:

alternating: $\boldsymbol{\xi}(t)$ alternates between $\mathbf{S}^1, -\mathbf{S}^2$ (or between $-\mathbf{S}^1, \mathbf{S}^2$), and the time τ between successive switches is constant, $\tau \in (0, 2\sqrt{2})$.

cycling: $\boldsymbol{\xi}(t)$ switches cyclically from vertex to vertex (of $\mathbb{U}$), either clockwise or anti-clockwise. Again the time between successive switches is constant, $\tau \in (0, \sqrt{2})$.

There are two types of singular controls:

trivial: $\boldsymbol{\xi}(t) \equiv \pm\frac{1}{2}(\mathbf{S}^1 - \mathbf{S}^2)$ is constant;

singular: $\boldsymbol{\xi}(t)$ switches between $\pm\mathbf{S}^1, \pm\mathbf{S}^2$, and $\pm\frac{1}{2}(\mathbf{S}^1 + \mathbf{S}^2)$, but the switching patterns are more complicated although not arbitrary, see [10] for the details.

These singular controls are difficult to observe in numerical simulations. Depending on the discretization scheme it is most likely, that only the bang-bang-controls are detected.

It is therefore important to emphasize that the singular controls **do** provide geodesics. For example, $\mathbf{P}(t) = \mathrm{e}^{\frac{t}{2}(\mathbf{S}^1+\mathbf{S}^2)}$ is length minimizing for all t. More important, the set of $\mathbf{P}_0 \in \mathrm{SL}(2)$ such that a length-minimizing path from $\mathbf{1}$ to $\mathbf{P}_0$ is either alternating or cycling, is compact and not even a neighborhood of $\mathbf{1}$. It also turns out that the **trivial** controls ($\boldsymbol{\xi} \equiv \pm\frac{1}{2}(\mathbf{S}^1 - \mathbf{S}^2)$ constant) do not generate length minimizing paths. So geodesics have $\boldsymbol{\xi}(t) \in \{\pm\mathbf{S}^1, \pm\mathbf{S}^2, \pm\frac{1}{2}(\mathbf{S}^1 + \mathbf{S}^2)\}$.

For a fully detailed analysis of $\widetilde{D}$ we refer to [10]. There it is shown that geodesics have at most 5 switches and which switching patterns have to be considered. We would like to emphasize that in the present example $\widetilde{D}$ is algorithmically available, and its computation is inexpensive.

Similar results can be obtained for more general slip systems.

Example 2. [**Hexagonal lattice.**] Here we consider 6 slip systems along the sides of an equilateral triangle. A suitable setting is $\mathbb{U} = \mathrm{conv}(\pm\mathbf{S}^1, \pm\mathbf{S}^2, \pm\mathbf{S}^3)$ with $\mathbf{S}^1 = \left(\begin{smallmatrix} 0 & 1 \\ 0 & 0 \end{smallmatrix}\right)$, $\mathbf{S}^2 = \mathbf{R}\mathbf{S}^1\mathbf{R}^\mathsf{T}$, $\mathbf{S}^3 = \mathbf{R}^\mathsf{T}\mathbf{S}^1\mathbf{R}$, where $\mathbf{R} = \left(\begin{smallmatrix} \cos\pi/3 & \sin\pi/3 \\ -\sin\pi/3 & \cos\pi/3 \end{smallmatrix}\right)$. The polytope $\mathbb{U}$ has the combinatorial structure of an octahedron, it is shown in Figure 2. There are 6 vertices, 8 (triangular) faces, and $6 + 8 - 2 = 12$ edges (Euler's formula for polyhedra). The "top" and "bottom" triangles $\pm\,\mathrm{conv}(\mathbf{S}^1, \mathbf{S}^2, \mathbf{S}^3)$ are equilateral while the other triangles are isosceles. The polar $\widetilde{\mathbb{Q}}$ is the dual polytope of $\mathbb{U}$. Therefore its combinatorial structure is that of a cube with 8 vertices, 6 (quadrilateral) faces, and $8 + 6 - 2 = 12$ edges, cf. Figure 2 which also shows some integral curves of the flow rule. The overshooting tips indicate the flow direction and which switches occur. As in

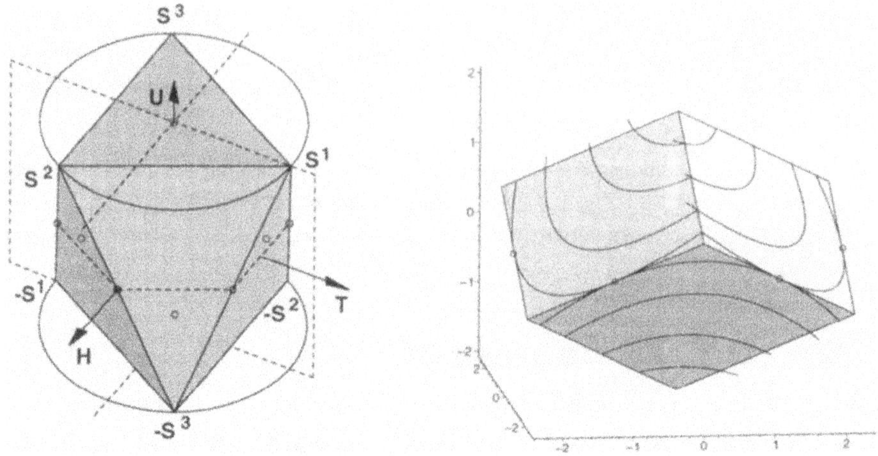

Fig. 2. The set $\mathbb{U}$ and the yield surface $\partial\mathbb{Q}$ for the hexagonal lattice

the previous example one obtains from the (PMP) that $\boldsymbol{\xi}(t) = \mathbf{P}(t)^{-1}\dot{\mathbf{P}}(t)$ is piecewise constant for length-minimizing $\mathbf{P}(t)$. Apart from the 6 possibilities we get from the vertices of $\mathbb{U}$ there are an additional 6 possibilities (each corresponding to a vertex of $\widetilde{\mathbb{Q}}$, resp., a face of $\mathbb{U}$) plus 6 possibilities corresponding to (some) edges of $\mathbb{U}$, resp. $\widetilde{\mathbb{Q}}$. Thus we have a total of 18 possible values for $\boldsymbol{\xi}$. Generically, $\boldsymbol{\xi}(t)$ cycles through the vertices of a single face of $\mathbb{U}$ in a certain order. For example, on the "top" face $\mathrm{conv}(\mathbf{S}^1, \mathbf{S}^2, \mathbf{S}^3)$ the switching sequence is $\mathbf{S}^1 \leadsto \mathbf{S}^3 \leadsto \mathbf{S}^2 \cdots$; on the "lateral" face $\mathrm{conv}(\mathbf{S}^1, -\mathbf{S}^2, -\mathbf{S}^3)$ the switching sequence is $\mathbf{S}^1 \leadsto -\mathbf{S}^2 \leadsto -\mathbf{S}^3 \ldots$. There are trivial cases (i.e. $\boldsymbol{\xi}$ constant) corresponding to the vertices of $\mathbb{Q}$. Nevertheless, it turns out that none of these is optimal. Finally, we also get more complicated singular cases. We will provide the details in the forthcoming paper [11]. Although computation of $\widetilde{D}$ is technically more difficult than for the square lattice, all the obstacles can be overcome.

Considering the previous two examples it is natural to ask whether it is true in general that for arbitrary $2D$-slips systems the length minimizing paths have piecewise constant $\boldsymbol{\xi} = \mathbf{P}^{-1}\dot{\mathbf{P}}$. This is actually true, but it is by no means obvious because it is definitely **not** true for arbitrary polyhedra $\mathbb{U} \subseteq \mathfrak{sl}(2)$. For example, let $\boldsymbol{\xi}_1 = \begin{pmatrix} 1 & 0 \\ 2 & -1 \end{pmatrix}$, $\boldsymbol{\xi}_2 = \begin{pmatrix} 1 & 0 \\ -2 & -1 \end{pmatrix}$, $\boldsymbol{\xi}_3 = -\boldsymbol{\xi}_1^{\mathsf{T}}$, $\boldsymbol{\xi}_4 = -\boldsymbol{\xi}_2^{\mathsf{T}}$, and $\mathbb{U} = \mathrm{conv}(\boldsymbol{\xi}_1, \ldots, \boldsymbol{\xi}_4)$. Since $\mathrm{tr}(\boldsymbol{\xi}_j^{\mathsf{T}}\boldsymbol{\xi}_k) = -2$ for $j \neq k$, we can write $\mathbb{Q} = \mathrm{conv}(\boldsymbol{\eta}_1, \ldots, \boldsymbol{\eta}_4)$ with $\boldsymbol{\eta}_j = \frac{1}{2}\boldsymbol{\xi}_j$. Here it may happen, for instance, that $\boldsymbol{\xi}(t) \in \mathrm{conv}(\boldsymbol{\xi}_1, \boldsymbol{\xi}_2)$ varies arbitrarily. The geometric reason for this is that $\mathbb{R}\boldsymbol{\xi}_1 + \mathbb{R}\boldsymbol{\xi}_2 \subseteq \mathfrak{sl}(2)$ is a 2-dimensional subalgebra. Speaking in geometric terms, degeneracies occur if one of the edges of the polytope $\mathbb{U}$ lies in a 2-dimensional subalgebra of $\mathfrak{sl}(2)$. It is well-known that the 2-dimensional subalgebras in $\mathfrak{sl}(2)$ are precisely the tangent planes of the double cone (depicted in Fig. 1).

Thus for slip systems these degeneracies never occur because for two elements $\mathbf{S}^1, \mathbf{S}^2$ of the double cone the segment $\text{conv}(\mathbf{S}^1, \mathbf{S}^2)$ is tangential to the cone iff $\mathbf{S}^1$ and $\mathbf{S}^2$ are collinear. Moreover, for general slip systems the combinatorial structure of the polytope $\mathbb{U} = \text{conv}(\mathbf{S}^1, \dots, \mathbf{S}^m)$ gives an apriori bound for the number of possible values of $\boldsymbol{\xi}$. If, say, $\mathbb{U}$ has f_0 vertices, f_1 edges and f_2 faces, then the number of possible values of $\boldsymbol{\xi}$ is bounded by $f_0 + f_1 + f_2$. We get at most one singular control for each edge of $\mathbb{Q}$ (i.e., each face of $\mathbb{U}$) and each edge of $\mathbb{Q}$, resp. $\mathbb{U}$. As $f_0 - f_1 + f_2 = 2$ (Euler's polyhedra formula) $f_0 + f_1 + f_2 = 2(f_0 + f_2 - 1)$.

4 Isotropic hardening

Finally we want to address the question of hardening. In general this is a difficult question since many hardening variables may be needed, in particular in the case of latent hardening [12,13]. On the other hand, systems without any hardening are too soft to withstand any nontrivial stresses, since the reduced functionals (cf. [12,1,7]) grow only logarithmic at infinity.

Here we propose a simple model of isotropic hardening using a scalar variable only. In the case of isotropic material this reduces to the classical isotropic Prandtl–Reuß–von Mises model. Thus, we consider the internal variables $(\mathbf{P}, p) \in \mathfrak{G} \times [0, \infty)$ with an infinitesimal dissipation potential $\widehat{\Delta}$ of the form

$$\widehat{\Delta}((\mathbf{P},p),(\dot{\mathbf{P}},\dot{p})) = \begin{cases} \widetilde{\Delta}(\mathbf{P}^{-1}\dot{\mathbf{P}}) & \text{if } \widetilde{\Delta}(\mathbf{P}^{-1}\dot{\mathbf{P}}) \leqslant a(p)\dot{p}, \\ \infty & \text{otherwise,} \end{cases}$$

where $a : [0,\infty[\to]0,\infty[$ is assumed to be continuous. Here $\widetilde{\Delta} : \mathfrak{g} \to [0,\infty]$ is convex and homogeneous of degree 1. By $\widetilde{D} : \mathfrak{G} \to [0,\infty]$ we denote the global dissipation distance associated with $\widetilde{\Delta}$. Moreover, by $\widetilde{\Delta}^* : \mathfrak{g}^* \to [0,\infty]$ we denote its polar function, i.e., $\widetilde{\Delta}^*(\boldsymbol{\eta}) := \max\{\, \boldsymbol{\xi}{:}\boldsymbol{\eta} \mid \widetilde{\Delta}(\boldsymbol{\xi}) \leqslant 1 \,\}$.

The left-invariant elastic domain $\widetilde{\mathbb{Q}}(p)$ is given by

$$\widetilde{\mathbb{Q}}(p) = \{\, (\boldsymbol{\eta}, q) \in \mathfrak{g} \times \mathbb{R} \mid q \leqslant 0,\ \Delta^*(\boldsymbol{\eta}) + q \leqslant 1 + a(p) \,\}.$$

We see that hardening is obtained if $a(p)$ is increasing in p. Denoting by A the primitive function of a (i.e., $A(p) = \int_0^p a(s)\,\mathrm{d}s$) we obtain the following formula for the global dissipation distance:

$$\widehat{D}((\mathbf{P}_0,p_0),(\mathbf{P}_1,p_1)) = \begin{cases} \widetilde{D}(\mathbf{P}_0^{-1}\mathbf{P}_1) + A(p_1) - A(p_0) & \text{if } p_1 \leqslant p_0 + \widetilde{D}(\mathbf{P}_0^{-1}\mathbf{P}_1), \\ \infty & \text{otherwise.} \end{cases}$$

We refer to [7] for more general cases.

References

1. C. Carstensen, K. Hackl, and A. Mielke. Nonconvex potentials and microstructure in finite–strain plasticity. *Proc. Royal Soc. London Ser. A*, 458(2018), 299–317, 2002.
2. M. E. Gurtin. On the plasticity of single crystals: free energy, microforces, plastic-strain gradients. *J. Mech. Phys. Solids*, 48(5), 989–1036, 2000.
3. K. Hackl and U. Hoppe. On the calculation of microstructures for inelastic materials using relaxed energies. *Proc. of IUTAM Symposium on Computational Mechanics of Solid Materials at Large Strains (Aug. 2001)*, to appear.
4. V. Jurdjevic. Non-euclidean elastica. *Am. J. Math*, 117, 93–124, 1995.
5. C. Miehe and M. Lambrecht. Analysis of Microstructure development in shearbands by energy relaxation of incremental stress potentials: large-strain theory for generalized standard solids. Report No. 01-I-10, Inst. Mech., Univ. Stuttgart., 2001.
6. C. Miehe, J. Schotte, and M. Lambrecht. Homogenization of inelastic solid materials at finite strains based on incremental minimization principles. Application to the texture analysis of polycrystals. Report No. 01-I-07, Inst. Mech., Univ. Stuttgart., 2001.
7. A. Mielke. Energetic formulation of multiplicative elastoplasticity using dissipation distances. *J. Mech. Physics Solids*, 2002. Submitted.
8. A. Mielke. Finite elastoplasticity, Lie groups and geodesics on SL(d). In P. Newton, A. Weinstein, and P. Holmes, editors, *Geometry, Dynamics, and Mechanics*. Springer–Verlag, 2002. In press.
9. D. Mittenhuber. Applications of the Maximum Principle to problems in Lie semigroups. In K.H. Hofmann, J.D. Lawson, and E.B. Vinberg, editors, *Semigroups in Algebra, Geometry and Analysis*. de Gruyter, 1995.
10. D. Mittenhuber. The dissipation distance for a 2D single crystal with two symmetric slip systems. In preparation, 2002.
11. D. Mittenhuber. Dissipation distances for 2D single crystals: The hexagonal lattice. In preparation, 2002.
12. M. Ortiz and E. Repetto. Nonconvex energy minimization and dislocation structures in ductile single crystals. *J. Mech. Phys. Solids*, 47(2), 397–462, 1999.
13. M. Ortiz, E. Repetto, and L. Stainier. A theory of subgrain dislocation structures. *J. Mech. Physics Solids*, 48, 2077–2114, 2000.
14. L.S. Pontrjagin, V.G. Boltyanski, R.V. Gamkrelidze, and E.F. Mishchenko. The Mathematical Theory of Optimal Processes. Wiley Interscience, 1962.

Multiscale Modeling in the Mechanics of Materials: Cohesion, Interfacial Interactions, Inclusions and Defects

Sergey Lurie[1], Petr Belov[2], and Dmitrij B. Volkov-Bogorodsky[2]

[1] Computer Center of Russian Academy of Science, Vavilova str., 40, GSP-1, Moscow 119991, Russia
[2] Institute of Applied Mechanics of Russian Academy of Science, Leninskij prospect, 32-A, Moscow 117334, Russia

Abstract. The models of media with multiscale effects are constructed based on the variational formalism. The internal interactions are determined by kinematic constraints of general character. These models take into account the superficial effects and internal interactions similar to interatomic forces of coupling. The models of media with a continuous field of defects of various types are considered. Descriptions of the media with defects, cohesion field models as a special case of the Cosserat models are considered. This paper is devoted to the consideration of disperse composite materials reinforced by nanoparticles. For numerical simulation, the new efficient block analytical-numerical method is developed oriented to the general case of solving problems in complex-shaped 2D and 3D domains. As an example of numerical investigation of the interfacial interactions in heterogeneous media, the medium with inclusions is considered.

1 Kinematic variational modeling of the medium

A variant of the variational formalism is developed. The variant proposed allows us to obtain constitutive equations, the system of equilibrium equations and boundary conditions [1]-[4]. A functional is constructed by taking into account kinematic constraints. The basic idea is that the general character of kinematic constraints determines the internal interactions. An algorithm proposed for the formulation of the model includes the following stages:

- Studying kinematic models, we can state kinematic constraints in media.
- Using the variational principle of possible displacements, we can determine the variation of the work done by internal forces on the kinematic constraints. As a result, the tensor of the Lagrange multipliers describes a spectrum of interactions that corresponds to the internal kinematic constraints.
- Integrating by part, we obtain a linear variational form and a set of arguments.
- Assume that only the reversible processes are considered. Then, the linear variational form is integrable and the potential energy can be formulated. For a linear elastic medium, the quadric form for the density of potential

energy can be obtained. The potential energy is formulated taking into account the tensor's dimensions of the arguments.
- Taking into account the expression of potential energy, we can arrive at the constitutive equations.
- The constitutive equations allow us to obtain particular expressions for the potential energy, which defines the mathematical statement of the model as a whole.

It is worth emphasizing that the proposed variational method needs fixed sets of arguments in the volume and in the superficial parts of potential energy. These sets of arguments are dictated by the introduced kinematic constraints. The correct mathematical statement is realized as a result of the proposed algorithm. By comparison, the set of arguments for the Sedov variational formalism is determined only for the volumetric part of the potential energy; consequently, some additional physical assumptions should be introduced to define the set of arguments for the superficial part of the potential energy.

2 Kinematic models

Let us represent a nonsymmetric distortion tensor referred associated with the Cartesian coordinates as follows:

$$\frac{\partial R_i}{\partial x_j} = \gamma_{ij} + \frac{1}{3}\theta\delta_{ij} - \omega_k \mathcal{E}_{ijk}\,. \tag{1}$$

Here, R_i are the components of the displacement vector, γ_{ij} and θ are the components of the deviator of the deformation tensor and the deviator of the spherical tensor, ω_k are the components of the rotation vector, $\mathcal{E}_{ijk}$ is the Levi–Civita tensor, δ_{ij} is the Kronecker delta, and $\omega_k = -\frac{1}{2}\,\mathcal{E}_{ijk}\,\frac{\partial R_i}{\partial x_j}$.

The conditions of the existence of the quadratures (the conditions of the existence of corresponding line (contour) integral) for the generalized Cauchy relationships (1) can be written in the following from:

$$\frac{\partial}{\partial x_m}\left(\gamma_{in} + \frac{1}{3}\theta\delta_{in} - \omega_k \mathcal{E}_{ink}\right)\mathcal{E}_{nmj} = 0\,. \tag{2}$$

These relationships (2) are known as the Papkovich equations. Similarly, the well-known Saint-Venant equations can be obtained as the necessary and sufficient conditions of continuity for the rotation vector ω_k. The Cauchy relationships (1), the Papkovich equations (2), and the Saint-Venant equations may be treated as kinematic constraints [2]-[4]. All these relationships are used in the algorithm of the model proposed. The quasi-classical model of a medium with a nonsymmetric stress tensor can be formulated by using the nonsymmetric Cauchy relationships (1). Considering the Papkovich equations (2) and the Saint-Venant equations as kinematic constraints, we can obtain a Cosserat model for a medium with nonfree rotations, that is, rotations that depend on other generalized coordinates. In particular, the

Papkovich equations (2) give us the following variational equation:

$$\delta U = \iiint \left\{ \sigma_{ij}\delta\left[\gamma_{ij} + \frac{1}{3}\theta\delta_{ij} - \omega_k \mathcal{E}_{ijk} - \frac{\partial R_i}{\partial x_j}\right] \right. \\ \left. - m_{ij}\delta\left[\frac{\partial}{\partial x_m}\left(\gamma_{in} + \frac{1}{3}\theta\delta_{in} - \frac{\partial\omega_k}{\partial x_m}\mathcal{E}_{ink}\right)\mathcal{E}_{nmj}\right]\right\} dV \,. \tag{3}$$

Here, σ_{ij} and m_{ij} are the components of the stress tensor and the components of the Cosserat stress tensor, respectively. These equations may be used as a base for the mathematical description of the generalized Cosserat model with restricted rotations.

Let us assume that the linear form (3) is integrable. Then, we can represent the potential energy as follows:

$$\delta U = \iiint U_V\left(\frac{\partial\omega_i}{\partial x_j}, \gamma_{ij}, \frac{\partial\theta}{\partial x_n}, \omega_n, R_n, \theta\right) dV \\ + \oiint U_F\left(R_i, \frac{\partial R_i}{\partial x_q} n_q\right) dF \,. \tag{4}$$

Here, n_j are the components of the normal vector of the boundary surface F of the elastic body under consideration.

Thus, for a linear elastic medium, we can obtain the quadric form for the density of the potential energy. The system of constitutive equations can be found with the aid of the Green relationships. The components of generalized static strain and stress state in the body and the superficial forces on the surface can be found by using the formulas:

$$\sigma_{ij} = \partial U_V \Big/ \partial\left(\frac{\partial R_i}{\partial x_j}\right), \quad m_{ij} = \partial U_V \Big/ \partial\left(\frac{\partial\omega_i}{\partial x_j}\right), \quad \mu_i = \partial U_V \Big/ \partial\left(\frac{\partial\theta}{\partial x_i}\right),$$

$$\sigma_i = \partial U_V/\partial R_i, \qquad f_{ij} = \partial U_F/\partial R_i, \qquad m_i = \partial U_F/\partial \dot{R}_i$$

where $\dot{R}_i = (\partial R_i/\partial x_q)\cdot n_q$.

It is worth emphasizing that the coefficients of quadratic terms of U_V in (4) are determined by the deformations γ_{ij}, ω_n, and θ. These coefficients define the system of conventional elastic properties of the medium and have dimensions of stresses. The constant coefficients of quadratic terms which are associated with the vectors R_n, ω_n, and $\frac{\partial\theta}{\partial x_n}$ and the tensor $\frac{\partial\omega_i}{\partial x_j}$ are of the remaining dimensions. The multiscale effects of other type can be determined with the aid of these constants. We further suppose that some of these constants allow us to describe the interactions corresponding to the internal cohesion.

On the other hand, U_F (4) can be represented as

$$U_F = \frac{1}{2}\left\{(AR_iR_j + 2A_1R_i\dot{R}_j + A_2\dot{R}_i\dot{R}_j)n_in_j + BR_iR_j(\delta_{ij} - n_in_j) \right. \\ \left. + 2B_1R_i\dot{R}_j(\delta_{ij} - n_in_j) + B_2\dot{R}_i\dot{R}_j(\delta_{ij} - n_in_j)\right\} . \tag{5}$$

The constants A, A_1 and A_2 in (5) determine the surface effects associated with the normal to the surface of the body, and the constants B, B_1, and B_2 in (5) determine the superficial effects in the tangential plane.

2.1 Kinematic model for the Cosserat pseudocontinuum (the fields of defects)

To describe the models of media with a continuous field of defects, we will consider the conditions of nonintegrability of the Cauchy relationships (1) instead the Papkovich equations (2):

$$\frac{\partial}{\partial x_m}\left(\gamma_{in} + \frac{1}{3}\theta\delta_{in} - \omega_k \mathcal{E}_{ink}\right)\mathcal{E}_{nmj} = \Xi_{ij}\,. \tag{6}$$

The tensor Ξ_{ij} is such that $\frac{\partial \Xi_{ij}}{\partial x_j} = 0$. This tensor determines the field of incompatibility for displacements. A vector of jumps in displacements can be found by using the tensor (6). In other words, we can say that the virtual Burgers vector is represented by Ξ_{ij}. Furthermore, when this vector is written as the flux of the tensor Ξ_{ij} through any surface, this flux gives the virtual Burgers vector of dislocation. That is why the tensor Ξ_{ij} is called the dislocation density tensor.

Using the relationships (6) as kinematic constraints, we can construct the Cosserat pseudocontinuum with free deformations and rotations, that is, with deformations and rotations considered as independent generalized coordinates. In this case, the mathematical statement of the problem is based on the following variational equation:

$$\begin{aligned}\delta U = \iiint \Bigg\{ & \sigma_{ij}\delta\left(\gamma_{ij}^0 + \frac{1}{3}\theta^0\delta_{ij} - \omega_k^0 \mathcal{E}_{ijk} - \frac{\partial R_i^0}{\partial x_j}\right) \\ & + m_{ij}\delta\left[\Xi_{ij} - \frac{\partial}{\partial x_m}\left(\gamma_{in}^{\Xi} + \frac{1}{3}\theta^{\Xi}\delta_{in} - \omega_k^{\Xi}\mathcal{E}_{ink}\right)\mathcal{E}_{nmj}\right]\Bigg\}\, dV\,.\end{aligned} \tag{7}$$

Here, γ_{ij}^{Ξ}, ω_k^{Ξ}, and θ^{Ξ} are defined by the right-hand part of the tensor equation (6) and can be considered as particular solutions of this tensor equation.

Taking into account the definition of the Cosserat continuum, we will define γ_{ij}^0, θ^0, and ω_n^0 as nonfree deformations and γ_{ij}^{Ξ}, θ^{Ξ}, and ω_n^{Ξ} as free deformations.

Similarly, the Cosserat models based on the conditions of nonintegrability for the Papkovich equations was obtained, the disclination density tensor was defied, and the mathematical statement of the problem for the generalized Cosserat pseudocontinuum was formulated in [5].

This process may be continued and the conditions of nonintegrability for the Saint-Venant relationships with respect to the gradient of the volume deformation θ may be obtained [4]. As a result, a new type of defects is modeled. This type of defects is additional to the dislocations and disclinations.

These defects are determined by jumps in $\partial\theta/\partial x_n$ and are described with the aid of the density tensor. Thus, the procedure of modeling proposed gives us the natural classification of models for continuums with defects [5].

3 The model of the cohesion field

One particular case of the Cosserat pseudocontinuum (7) gives us the variant of the cohesion field model [4], [5]:

$$\iiint \left\{ L_{ij} \left[-\frac{l_0^2}{\mu} L_{jk}(\ldots) + \delta_{jk}(\ldots) \right] R_k + X_i \right\} \delta R_i \, dV + \oiint \left[-\widehat{M}_i \delta \frac{\partial R_i}{\partial x_q} n_q + \widehat{Y}_i \delta R_i \right] dF = 0 \,. \tag{8}$$

Here,

$$L_{ij}(\ldots) = (2\mu + \lambda) \frac{\partial^2(\ldots)}{\partial x_i \partial x_j} + (\mu + \chi) \left(\delta_{ij} \Delta(\ldots) - \frac{\partial^2(\ldots)}{\partial x_i \partial x_j} \right)$$

$$\widehat{M}_i = \frac{2(\mu+\chi)^2}{\mu} l_0^2 [n_m n_j \mathcal{E}_{ijn} + (1 - 2\nu_*) n_n n_j \mathcal{E}_{ijm}] \frac{\partial \omega_n}{\partial x_m} - \frac{2(\mu+\lambda)^2}{\mu} l_0^2 \frac{\partial \theta}{\partial x_k} n_k n_i$$

$$\begin{aligned} \widehat{Y}_i = Y_i - \Bigg[& 2\mu\gamma_{ij} + \left(\frac{2\mu}{3} + \lambda \right) \theta \delta_{ij} - 2\chi\omega_k \mathcal{E}_{ijk} \\ & + \frac{2(\mu+\chi)^2}{\mu} l_0^2 \Delta\omega_n \mathcal{E}_{ijn} - \frac{2(\mu+\lambda)^2}{\mu} l_0^2 \Delta\theta \delta_{ij} \Bigg] n_j + (\delta_{qj} - n_q n_j) \\ & \times \frac{\partial}{\partial x_q} \left[-\frac{2(\mu+\chi)^2}{\mu} l_0^2 \left[\frac{\partial \omega_k}{\partial x_p} + (1 - 2\nu_*) \frac{\partial \omega_p}{\partial x_k} \right] n_p \mathcal{E}_{ijk} \right. \\ & \left. + \frac{2(\mu+\lambda)^2}{\mu} l_0^2 \frac{\partial \theta}{\partial x_k} n_k \delta_{ij} \right], \end{aligned}$$

where Δ is the Laplace operator, R_i is the vector of total displacement, μ and λ are the Lame coefficients, ν_* is a dimensionless constant of the model, and l_0^2 is a dimensional constant of the Cosserat pseudocontinuum.

Let us introduce the classical displacement vector associated with the operator of the classical theory of elasticity L_{ij} and the cohesion displacement vector associated with non-classical scale effects:

$$R_i = U_i - u_i, \qquad U_i = \left[-\frac{l_0^2}{\mu} L_{ij}(\ldots) + \delta_{ij}(\ldots) \right] R_j \qquad u_i = -\frac{l_0^2}{\mu} L_{ij}(R_j) \,. \tag{9}$$

Here, U_i is the classical displacement vector and u_i is the cohesion displacement vector.

Then, we can restate the mathematical statement of the problem (8). The variational equation (8) implies the following differential equilibrium equations and boundary conditions for the model under consideration:

$$L_{ij}(U_j) + X_i = 0, \qquad L_{ij}(u_j) - \frac{\mu}{l_0^2}\, u_i + X_i = 0. \tag{10}$$
$$\delta L = \oiint \left\{ \widehat{M}_i\, \delta\dot{U}_i + \widehat{Y}_i\, \delta U_i - \widehat{M}_i\, \delta\dot{u}_i - \widehat{Y}_i\, \delta u_i \right\} dF = 0\,.$$

4 Numerical simulation of media with inclusions by the block analytical-numerical method

The simulation of physical processes in non-classical media with inclusions can be effectively performed by the method based on analytical representations of the solution (so-called block analytical-numerical method [6]-[8]). The general scheme of the method consists in covering the initial region by a set of simple sub-regions called blocks. A solution in every block is approximated by a series in terms of special functions (called multipoles), which best reproduce the analytical properties of the solution and strongly satisfy the problem's operator equations. Comparing with finite element and boundary element methods, the method proposed has the advantage that it gives us the possibility to employ more complicated functions, which can describe analytical properties of the solution near to boundary singularities of the domain (for instance, reentrant corners with small or zero rounding radii [6]). Besides the representation of the solution in analytical form, this method facilitates procedures of calculation and optimization of effective and local characteristics of the heterogeneous media with inclusions.

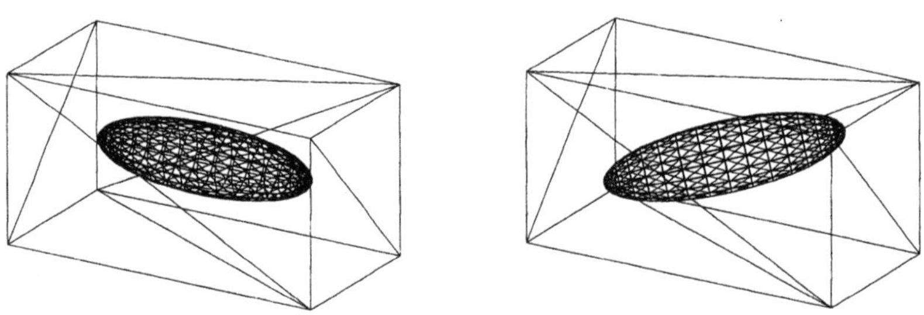

Fig. 1. Various types of inclusions

For various types of inclusions (see Fig. 1), the first step is to construct a block structure covered the initial region. The following method for the construction can be proposed.

For a plane problem, in a domain with an elliptic inclusion (as an example, see Fig. 2), we use a rectangular grid and then additionally split those rectangular cells that are intersecting by the boundary of the inclusion. Then we

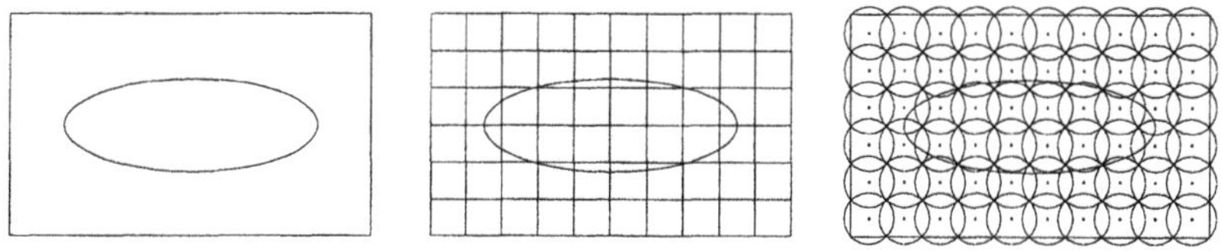

Fig. 2. Construction of block structure

circumscribe a circle about every cell and obtain circular blocks as it shown in Fig. 2.

Let us now consider the double plane problem ($R_1 = R$, $R_2 = R_3 = 0$) for the total displacement vector $R = U - u$ in the cell with constructed block structure. From the above discussion it appears that now the problem can be stated as follows:

$$\mathcal{L}_C \, \mathcal{L} \, R = 0, \quad \mathcal{L} = \mathcal{L}_C\big|_{C=0}, \quad \mathcal{L}_C = (2\,\mu + \lambda)\,\frac{\partial^2}{\partial\, x^2} + \mu\,\frac{\partial^2}{\partial\, y^2} - C\,,$$

$$R = \begin{cases} R_f & \text{in inclusion} \\ R_m & \text{in matrix} \end{cases}, \quad \mu = \frac{E}{2\,(1+\nu)}\,, \quad \lambda = \frac{\nu\, E}{(1-2\,\nu)\,(1+\nu)}\,;$$

these equations should be completed by the following conditions on the boundary of the inclusion and the cell:

$$[R] = \left[\frac{\partial\, R}{\partial\, n}\right] = 0\,, \quad [\widehat{Y}] = 0\,, \quad [\widehat{M}] = 0\,,$$

$$R(\pm\, L,\, y) = \pm 1, \qquad \widehat{M} = 0, \qquad 0 < y < H\,,$$

$$\frac{\partial\, R}{\partial\, n}(x,\, 0) = \frac{\partial\, R}{\partial\, n}(x,\, H) = 0, \qquad \widehat{Y} = 0, \qquad -L < x < L\,,$$

where E is the modulus of elasticity (Young's modulus), ν is Poisson's ratio, $C = 1/l_0^2$ is a dimensional constant of the Cosserat pseudocontinuum, $[R] = R_f - R_m$ is the jump at the inclusion, $\partial/\partial\, n$ is the derivative in the direction of the boundary normal vector with the components n_x and n_y, H is the height of the cell, and L is the half-width of the cell.

For a numerical example, let us assume the following values of parameters: $\nu_f = \nu_m = 0.3$, $E_f/E_m = 3$, and $C_m/\mu_m = C_f/\mu_f = 100$.

It may be proved that an approximate iterative algorithm for the solution of the problem (9) can be used. Here, to construct the approximate solution, we restrict ourselves to the first step in the iterative procedure, which has two stages.

First, the classical problem ($\mathcal{L}\, U = 0$) is solved for the following boundary conditions:

$$[U] = 0\,, \qquad \Big[\mu\, \frac{\partial U}{\partial n} + (\mu + \lambda)\, \frac{\partial U}{\partial x}\, n_x\Big] = 0\,, \tag{11a}$$

$$U(\pm L,\, y) = \pm 1, \qquad 0 < y < H\,, \tag{11b}$$

$$\frac{\partial U}{\partial n}(x,\, 0) = \frac{\partial U}{\partial n}(x,\, H) = 0, \quad -L < x < L\,. \tag{11c}$$

Second, the displacement of the cohesion field u is found as a solution of the non-classical equation ($\mathcal{L}_C\, u = 0$) with the following boundary conditions at the contour of the inclusion and at the boundary of the cell:

$$[u] = \frac{[c^*]}{\langle c^* \rangle}\, \langle u \rangle\,, \qquad \Big[\frac{\partial u}{\partial n} - \frac{\partial U}{\partial n}\Big] = 0\,, \tag{12a}$$

$$u(\pm L,\, y) = 0, \qquad 0 < y < H\,, \tag{12b}$$

$$\frac{\partial u}{\partial n}(x,\, 0) = \frac{\partial u}{\partial n}(x,\, H) = 0, \quad -L < x < L\,. \tag{12c}$$

Here,

$$\langle u \rangle = u_f + u_m, \qquad \langle c^* \rangle = c_f^* + c_m^*, \qquad c_{f,m}^* = C_{f,m}/E_{f,m}\,.$$

In every block, we use a finite series (with sufficient number of multipoles) for the system of the following functions (see [8]):

$$\begin{aligned} \Phi_s^s(x,\, y) &= \Phi_s^s(\sqrt{2\mu + \lambda}\, \hat{x},\, \sqrt{\mu}\, \hat{y}) \\ &= \Gamma(s+1)(\sqrt{C}/2)^{-s}\, I_s(\sqrt{C}\, r)\, \exp(is\, \varphi)\,, \end{aligned}$$

where $\Gamma(s)$ is the Euler gamma-function, $I_s(r)$ is the modified Bessel function, $r = \sqrt{\hat{x}^2 + \hat{y}^2}$, and $\varphi = \arctan\, \hat{y}/\hat{x}$.

The local displacements $U^{(k)}$ and $u^{(k)}$ (in the block with the number k) are coupled to the displacements of adjacent blocks by a particular variant of the least square method for a multi-block system (see [8]); the conditions (11a)-(12c) are enforced by the same method. Some examples of the tension distribution are presented in Figs. 3 and 4. The parameters of the inclusion are taken as follows: the major axis of the ellipse $a = 1.4\,\varepsilon$, the minor axis of the ellipse $b = (0.5,\, 0.3)\,\varepsilon$, where $\varepsilon = (0.9,\, 0.7,\, 0.5)$ and the parameters of the cell are $L = 1$ and $H = 1.2$.

In Fig. 3, the tension patterns are depicted which belongs to the classical model. The first pattern corresponds to $2.4\,\mu_m$, the others correspond to the increment $0.2\,\mu_m$. The corresponding tension patterns associated with the cohesion fields are presented in Fig. 4.

As can be seen, the pattern of distribution differs essentially from the pattern corresponding to the classical model. This is due to the presence of

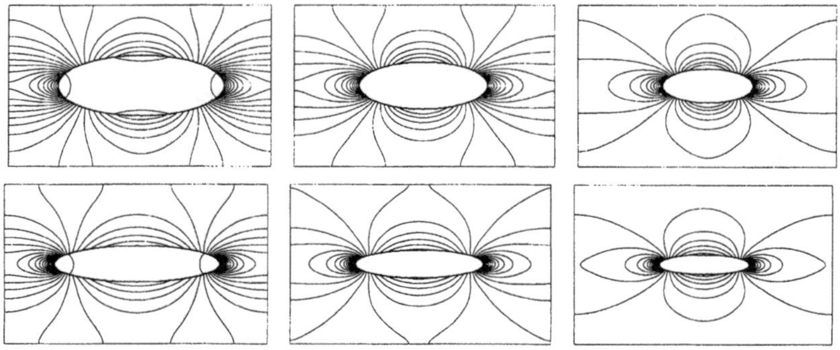

Fig. 3. Examples of the classical tension distributions $\sigma_x(U) = (2\,\mu + \lambda)\,\frac{\partial U}{\partial x}$

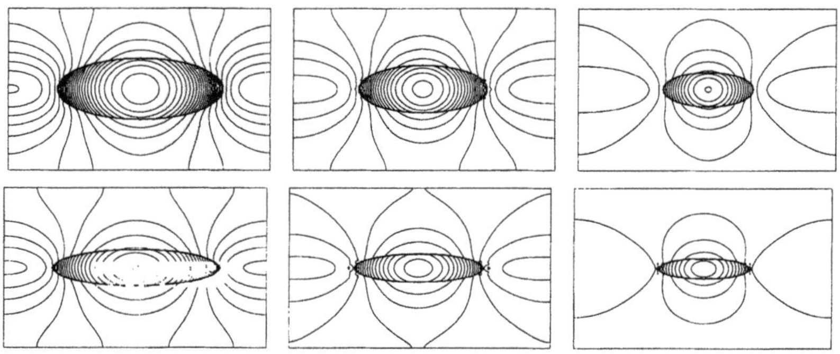

Fig. 4. Examples of tension distributions $\sigma_x(R)$ in the complete problem

the cohesion component u in the complete problem (9), (10). The change in the stress-strain state leads to the change in the balance of the potential energy between the inclusions and the matrix. As a result, the stiffness and strength parameters of a cell as a whole may vary. It seems plausible that an accurate estimation of these effects should be performed for the determination of the parameters of the models.

References

1. Lurie, S.A., Babeshko, V.A., Obraztsov, I.F. (1999) On the thin film theory. Mech. of Compos. Mater. and Design, **5**, No. 2, 48–60 (in Russian)
2. Belov, P.A., Lurie, S.A (1998) Solid mechanics deformation model and analogies in field theory. Mech. of Solids, **3**, 157–166
3. Obraztsov, I.F., Lurie, S.A., Belov, P.A. (2000) On the new cohesion field model. Izvestiya Vuzov. The North-Caucasian Region, **3**, 110–118
4. Lurie, S.A., Belov, P.A. (2000) Mathematical models of the mechanics of solids and physical fields. Sci. Pub. of Comp. Centre of RAS, Moscow (in Russian)
5. Lurie, S.A., Belov, P.A. Babeshko, V.A.(2002) Multiscale effects in continuum mechanics bodies. **8**, No. 1, 71–82 (in Russian)
6. Vlasov, V.I., Volkov, D.B. (1995) The multipole method for Poisson equation in regions with rounded corners. Comp. Math. & Math. Phys. **35**, No. 6, 687–707

7. Vlasov, V.I., Volkov-Bogorodsky, D.B. (1998) Block multipole method for boundary value problems in complex-shaped domains. ZAMM **78**, Sup. 3, 1111–1112
8. Volkov-Bogorodsky, D.B. (2000) Development block analytical-numerical method for problems in mechanic and acoustic. Proc. of the Conf. "Composite materials", IAM RAN, 44–56 (In Russian)

Computational Homogenization of Materials with Microstructures Based on Incremental Variational Formulations

Christian Miehe, Matthias Lambrecht, and Jan Schotte

University of Stuttgart, Institute of Applied Mechanics (Civil Engineering), Pfaffenwaldring 7, 70569 Stuttgart, Germany.

Abstract. The paper presents incremental minimization principles for the broad class of standard dissipative materials at finite strains. Starting with a minimization principle for the local constitutive material response we define a minimization problem for the incremental boundary-value problem of standard dissipative solids. The existence of this principle allows the determination of micro-structure developments in non-stable dissipative materials based on a convexification analysis. Finally, we propose a minimization principle for the boundary-value problem of homogenization in heterogeneous microstructures.

1 Introduction

The paper presents new variational formulations for the homogenization analysis of inelastic solid materials with microstructures undergoing finite strains. The point of departure is a general internal variable formulation for the constitutive response of standard dissipative materials in terms of an energy storage and a dissipation function. Consistent with this type of finite inelasticity we develop an *incremental variational formulation of the local constitutive response*, where a quasi-hyperelastic stress potential is obtained from a local minimization problem. This formulation allows the definition of the *material stability of an inelastic solid based on weak convexity conditions* of the incremental stress potential in analogy to treatments of finite elasticity. Localization phenomena are interpreted as microstructure developments on multiple levels associated with non-convex incremental stress potentials in analogy to elastic phase decomposition problems. These microstructures can be resolved by the relaxation of non-convex energy functionals based on a convexification of the stress potential. The relaxed problem provides a well-posed formulation for a mesh-objective analysis of localizations as close as possible to the non-convex original problem. Furthermore, the existence of quasi-hyperelastic incremental stress potentials allows the extension of homogenization approaches of elasticity to the incremental setting of inelasticity. We develop a new *incremental variational formulation of the homogenization problem* for standard dissipative materials at finite strains, where a quasi-hyperelastic macro-stress potential is obtained from a global minimization problem. Table 1 summarizes the subsequent developments.

Table 1. Minimization principles for standard dissipative materials

(M) Constitutive Model. $\boldsymbol{F} \in GL(3)$ is the local deformation of the material and $\boldsymbol{\mathcal{I}}$ a generalized vector of internal variables. A normal-dissipative set of local material equations has the structure

stress equation $\quad \boldsymbol{P} = \partial_{\boldsymbol{F}}\psi(\boldsymbol{F},\boldsymbol{\mathcal{I}})$

evolution equation $\quad \partial_{\boldsymbol{\mathcal{I}}}\psi(\boldsymbol{F},\boldsymbol{\mathcal{I}}) + \partial_{\dot{\boldsymbol{\mathcal{I}}}}\phi(\dot{\boldsymbol{\mathcal{I}}},\boldsymbol{\mathcal{I}}) = 0 \;,\; \boldsymbol{\mathcal{I}}(0) = \boldsymbol{\mathcal{I}}_0$

defined in terms of energy storage function ψ and dissipation function ϕ.

(V) Incremental Variational Formulation of the Constitutive Model. In a finite time increment $[t_n, t_{n+1}]$, the *minimization problem of constitutive response*

stress equation $\quad \boldsymbol{P}_{n+1} = \partial_{\boldsymbol{F}} W(\boldsymbol{F}_{n+1})$

stress potential $\quad W(\boldsymbol{F}_{n+1}) = \inf_{\boldsymbol{\mathcal{I}}} \int_{t_n}^{t_{n+1}} [\,\dot{\psi} + \phi\,]\, dt \;,\; \boldsymbol{\mathcal{I}}(t_n) = \boldsymbol{\mathcal{I}}_n$

determines the current internal state $\boldsymbol{\mathcal{I}}_{n+1}$ and provides a potential for the stresses at t_{n+1}.

(S) Stability of the Incremental Constitutive Response. In $[t_n, t_{n+1}]$ the material is locally stable at $\boldsymbol{F}_{n+1}$ if the *incremental stress potential* W *is quasi-convex*

quasi-convexity $\quad \frac{1}{|\mathcal{D}|}\int_{\mathcal{D}} W(\boldsymbol{F}_{n+1} + \nabla \boldsymbol{w}(\boldsymbol{y}))\, dV \geqslant W(\boldsymbol{F}_{n+1})$

(R) Relaxation of a Non-Convex Incremental Response. For an instable non-convex incremental response, the *minimization problem of quasi-convexification*

macro-stresses $\quad \boldsymbol{P}_{Q n+1} = \partial_{\boldsymbol{F}} W_Q(\boldsymbol{F}_{n+1})$

convexification $\quad W_Q(\boldsymbol{F}_{n+1}) = \inf_{\boldsymbol{w}\in\mathcal{W}} \frac{1}{|\mathcal{D}|}\int_{\mathcal{D}} W(\boldsymbol{F}_{n+1} + \nabla \boldsymbol{w}(\boldsymbol{y}))\, dV$

determines the current microstructure-fluctuations $\boldsymbol{w}(\boldsymbol{y})$.

(H) Homogenization of Heterogeneous Microstructures. $\bar{\boldsymbol{F}} \in GL(3)$ is the macro-deformation of a heterogeneous microstructure at t_{n+1}. The *minimization problem of homogenization*

macro-stresses $\quad \bar{\boldsymbol{P}}_{n+1} = \partial_{\boldsymbol{F}} \bar{W}(\boldsymbol{F}_{n+1})$

homogenization $\quad \bar{W}(\boldsymbol{F}_{n+1}) = \inf_{\boldsymbol{w}} \frac{1}{|\mathcal{B}|}\int_{\mathcal{B}} W(\boldsymbol{F}_{n+1} + \nabla \boldsymbol{w}; \boldsymbol{X})\, dV$

determines the fluctuation field $\boldsymbol{w}(\boldsymbol{X})$ on the heterogeneous microstructure.

2 Minimization Problem of Standard Dissipative Solids

2.1 Internal Variable Formulation of Inelastic Materials

Let $\boldsymbol{F} \in GL_+(3)$ with $\det[\boldsymbol{F}] > 0$ be the homogeneous deformation of a material at time $t \in \mathcal{R}_+$. Focussing on purely mechanical problems, the homogeneous stress response of the material is physically constrained by the so-called Clausius-Planck inequality for the dissipation of the material

$$\mathcal{D} := \boldsymbol{P} : \dot{\boldsymbol{F}} - \dot{\psi} \geqslant 0 \;, \tag{1}$$

where $\boldsymbol{P}$ denotes the nominal (first Piola) stress tensor. ψ is an objective *energy storage function* that is assumed to depend on the deformation $\boldsymbol{F}$ and a generalized vector $\mathcal{I} \in \mathcal{G}$ of internal variables. $\mathcal{G}$ indicates a vector space $\mathcal{R}^n$ of n scalar functions of internal variables which may be constrained to a manifold, e.g. the Lie group $SL(3)$ of unimodular tensors in isochoric finite inelasticity. Insertion into (1) yields by a standard argument the constitutive equation for the stresses

$$\boldsymbol{P} = \partial_{\boldsymbol{F}}\psi(\boldsymbol{F},\mathcal{I}) \tag{2}$$

and the reduced dissipation inequality

$$\mathcal{D} = \mathcal{F}\cdot\dot{\mathcal{I}} \geqslant 0 \qquad \text{with} \qquad \mathcal{F} := -\partial_{\mathcal{I}}\psi(\boldsymbol{F},\mathcal{I}) \ , \tag{3}$$

where $\mathcal{F} \in \mathcal{R}^n$ is a generalized vector of internal forces conjugate to the internal variables $\mathcal{I}$. A broad spectrum of materials is covered by the so-called standard dissipative materials where the evolution $\dot{\mathcal{I}}$ of the internal variables on $\mathcal{G}$ is governed by a scalar *dissipation function* ϕ, depending on the flux $\dot{\mathcal{I}}$ of the internal variables and the internal variables $\mathcal{I}$ themselves, see for example [25] and [6]. It governs the evolution of $\mathcal{I}$ in time by the constitutive differential equation

$$\partial_{\mathcal{I}}\psi(\boldsymbol{F},\mathcal{I}) + \partial_{\dot{\mathcal{I}}}\phi(\dot{\mathcal{I}},\mathcal{I}) = \boldsymbol{0} \quad \text{with} \quad \mathcal{I}(0) = \mathcal{I}_0 \tag{4}$$

We denote this equation Biot's equation of standard dissipative systems, referring to [2]. The two constitutive equations (2) and (4) determine the stress response of a *smooth* normal-dissipative material in a deformation-driven process where $\boldsymbol{F}$ is prescribed. Plasticity and dry friction are time-independent or non-viscous irreversible processes, governed by *non-smooth* dissipation functions. These functions are positively homogeneous of degree one with respect to the flux $\dot{\mathcal{I}}$, i.e. $\phi(\epsilon\dot{\mathcal{I}},\mathcal{I}) = \epsilon\phi(\dot{\mathcal{I}},\mathcal{I})$ for all $\epsilon \in \mathcal{R}_+$. Such a function has a cone-like graph and is not differentiable at the point $\dot{\mathcal{I}} = \boldsymbol{0}$. A generalization of the above formulation needs a generalization of the differential operator of smooth functions to the notion of a *sub-differential* operator ∂ of non-smooth convex functions as follows, see [17], [6] and and references therein. Using the definition of the sub-differential, in what follows we formally understand Biot's equation (4) to be generalized to $\boldsymbol{0} \in \partial_{\mathcal{I}}\psi(\boldsymbol{F},\mathcal{I}) + \partial_{\dot{\mathcal{I}}}\phi(\dot{\mathcal{I}},\mathcal{I})$.

2.2 Incremental Variational Formulation of Inelastic Materials

We now proceed with the construction of an integrated version of constitutive equations giving a consistent approximation of the continuous differential equation (4) in a finite increment $[t_n, t_{n+1}] \in \mathcal{R}_+$ of time. Following conceptually the works [22] and [4] on plasticity, the key point is the definition of an *incremental stress potential function* W depending of the deformation $\boldsymbol{F}_{n+1} := \boldsymbol{F}(t_{n+1})$ at time t_{n+1} that determines the stresses at t_{n+1} by the quasi-hyperelastic function evaluation

$$\boldsymbol{P}_{n+1} = \partial_{\boldsymbol{F}}W(\boldsymbol{F}_{n+1}) \tag{5}$$

Clearly, this function must cover characteristics of the storage function ψ and the dissipation function ϕ introduced above. To this end, consider the *minimization problem*

$$W(\boldsymbol{F}_{n+1}) = \inf_{\mathcal{I}\in\mathcal{G}} \int_{t_n}^{t_{n+1}} [\dot{\psi} + \phi]\, dt \quad \text{with} \quad \mathcal{I}(t_n) = \mathcal{I}_n \tag{6}$$

for dissipative standard materials proposed in [10]. The principle provides an alternative to the formulation proposed by [22] whose variational principle minimizes the incremental work $\int_{t_n}^{t_{n+1}} \boldsymbol{P} : \dot{\boldsymbol{F}} dt = \int_{t_n}^{t_{n+1}} [\dot{\psi} + \mathcal{F} \cdot \dot{\mathcal{I}}] dt$. Starting with the given initial condition $\mathcal{I}(t_n) = \mathcal{I}_n$, the minimum problem defines an optimal path of the internal variables $\mathcal{I}(t)$ for $t \in [t_n, t_{n+1}]$, including the right boundary value $\mathcal{I}_{n+1} := \mathcal{I}(t_{n+1})$. We refer to [9] and the recent work [16] for a discussion of extremum paths in linear and nonlinear plasticity. The two equations (5) and (6) provide an approximative variational counterpart of the continuous setting (2) and (4) of the constitutive equations in the discrete time step $[t_n, t_{n+1}]$ under consideration. Taking the derivative of the incremental potential function with respect to the deformation $\boldsymbol{F}_{n+1}$, we have

$$\partial_{\boldsymbol{F}} W(\boldsymbol{F}_{n+1}) = \partial_{\boldsymbol{F}} \psi(\boldsymbol{F}_{n+1}, \mathcal{I}_{n+1}) \ , \tag{7}$$

where $\mathcal{I}_{n+1}$ is considered to be given by the minimization problem (6). A comparison with (2) then shows the consistency of the potential equation (5) with the continuous setting.

2.3 Incremental Variational Formulation of Inelastic Solids

The existence of the constitutive minimization problem (6) allows the introduction of an incremental minimization formulation of the boundary-value problem of finite inelasticity for standard dissipative solids. Let $\varphi : \mathcal{B} \times \mathcal{R}_+ \to \mathcal{R}^3$ denote the nonlinear deformation map of an inelastic continuum $\mathcal{B} \subset \mathcal{R}^3$ at a material point $\boldsymbol{X} \in \mathcal{B}$ and time $t \in \mathcal{R}^+$. Then $\boldsymbol{F} := \nabla\varphi$ is the deformation gradient with Jacobian $J := \det[\boldsymbol{F}] > 0$. Now consider a functional of the current deformation field φ_{n+1} at the right boundary of the increment $[t_n, t_{n+1}]$

$$\Pi(\varphi_{n+1}) = \int_{\mathcal{B}} W(\boldsymbol{F}_{n+1})\, dV - \Pi_{ext}(\varphi_{n+1}) \tag{8}$$

with the external part $\Pi_{ext}(\varphi_{n+1}) = \int_{\mathcal{B}} \varphi_{n+1} \cdot \gamma_{n+1}\, dV - \int_{\partial\mathcal{B}_t} \varphi_{n+1} \cdot \boldsymbol{t}_{n+1}\, dA$ of dead body forces $\gamma(\boldsymbol{X}, t)$ in $\mathcal{B}$ and surface tractions $\boldsymbol{t}(\boldsymbol{X}, t)$ on $\partial\mathcal{B}_t$. W is the incremental stress potential function defined in (6). The current deformation map of inelastic standard dissipative materials can then be determined by a *principle of minimum incremental energy for standard dissipative solids*

$$\Pi(\varphi^*_{n+1}) = \inf_{\varphi_{n+1} \in \mathcal{W}} \Pi(\varphi_{n+1}) \tag{9}$$

subject to $\varphi_{n+1} \in \mathcal{W} := \{\varphi \in \mathcal{W}^{1,p}(\mathcal{B}) | \varphi = \bar{\varphi}(\boldsymbol{X}) \text{ on } \partial\mathcal{B}_\varphi\}$ associated with prescribed deformations $\bar{\varphi}$ at $\boldsymbol{X} \in \partial\mathcal{B}_\varphi$. The minimization problem has a structure identical to the principle of minimum potential energy in finite elasticity.

2.4 Incremental Stability of Standard Dissipative Materials

With the incremental minimization problem (9) at hand, existence results of finite elasticity as outlined in [1] and [5] can straightforwardly be applied to the incremental response of standard dissipative materials. The internal part $I(\varphi_{n+1}) := \int_\mathcal{B} W(\boldsymbol{F}_{n+1})\, dV$ of the functional (8) is sequentially weakly lower semicontinuous (s.w.l.s.), if the incremental stress potential $W(\boldsymbol{F}_{n+1})$ defined in (6) is quasiconvex, see [18] and [5]. This is considered to be the key property for the existence of a sufficiently regular minimizer of the variational problem (9). Hence, the quasiconvexity

$$\frac{1}{|D|}\int_D W(\boldsymbol{F}_{n+1} + \nabla \boldsymbol{w}(\boldsymbol{y}))\, dV \geqslant W(\boldsymbol{F}_{n+1}) \tag{10}$$

of the incremental stress potential W is considered to be the fundamental *criterion for a local material stability of the incremental response of standard dissipative solids.* The condition states that for all fluctuations $\boldsymbol{w}$ at $\boldsymbol{y} \in D$ on an arbitrarily chosen part D of the inelastic solid with support $\boldsymbol{w} = \boldsymbol{0}$ on ∂D the homogeneous deformation given by $\boldsymbol{F}_{n+1}$ provides an absolute minimizer of the incremental potential. Thus (10) rules out internal buckling, the development of local fine-scale microstructures and phase decomposition of a homogeneous local deformation state.

2.5 Microstructure Development of Non-Stable Materials

The variational formulation of the incremental boundary value problem for standard dissipative solids outlined above opens up the opportunity to apply the concept of relaxation of non-convex problems where an energy-minimizing *microstructure* develops. A relaxation is associated with a *convexification* of a non-convex function W by constructing its *convex envelopes* W_Q as discussed below, see [8], [5] and [20] for a sound mathematical basis. The convexification procedure determines the developing microstructure. We consider the *quasi-convexified incremental energy functional of standard dissipative solids*

$$\Pi_Q(\varphi_{n+1}) = \int_\mathcal{B} W_Q(\boldsymbol{F}_{n+1})\, dV - \Pi_{ext}(\varphi_{n+1})\;, \tag{11}$$

where the internal part of the relaxed energy functional is obtained by replacing the non-convex integrand by its *quasiconvex envelope.* The current deformation field of the elastic-plastic solid is then determined by the relaxed incremental variational principle

$$\Pi_Q(\varphi^*_{n+1}) = \inf_{\varphi_{n+1}\in\mathcal{W}} \Pi_Q(\varphi_{n+1}) \tag{12}$$

that *minimizes the relaxed incremental potential energy*. This relaxed problem (12) is considered to be a *well-posed problem as close as possible to the ill-posed problem* (9). In (12), the quasi-convexified incremental stress potential is defined by the minimization problem of relaxation with respect to the microscopic fluctuation field $\boldsymbol{w}$

$$W_Q(\boldsymbol{F}_{n+1}) = \inf_{\boldsymbol{w}\in\mathcal{W}_0} \frac{1}{|D|}\int_D W(\boldsymbol{F}_{n+1} + \nabla\boldsymbol{w}(\boldsymbol{y}))\, dV \tag{13}$$

with $\boldsymbol{y}\in D$ subject to the constraint $\boldsymbol{w}\in\mathcal{W}_0 := \{\boldsymbol{w}\in\mathcal{W}^{1,\infty}(D)|\boldsymbol{w}=\boldsymbol{0} \text{ on } \partial D\}$ providing a support on ∂D. The first and second derivatives of this function define relaxed stresses and tangent moduli in the form

$$\boldsymbol{P}_{Qn+1} := \partial_{\boldsymbol{F}} W_Q(\boldsymbol{F}_{n+1}) \quad \text{and} \quad \mathbb{C}_{Qn+1} := \partial^2_{\boldsymbol{F}\boldsymbol{F}} W_Q(\boldsymbol{F}_{n+1}) \ . \tag{14}$$

It determines a micro-fluctuation field $\boldsymbol{w}$ for an arbitrary domain $D\subset\mathcal{R}^3$.

3 Minimization Problem of Homogenization

3.1 Fluctuation Fields on Heterogeneous Microstructures

Let $\varphi : \mathcal{B}\times\mathcal{R}_+ \to \mathcal{R}^3$ denote the deformation map of a heterogeneous microstructure $\mathcal{B}\subset\mathcal{R}^3$ at $\boldsymbol{X}\in\mathcal{B}$ and time $t\in\mathcal{R}_+$. The deformation is assumed to be driven by a prescribed macro-deformation $\bar{\boldsymbol{F}}(t)$

$$\varphi(\boldsymbol{X},t) = \bar{\boldsymbol{F}}(t)\boldsymbol{X} + \boldsymbol{w}(\boldsymbol{X},t) \tag{15}$$

consisting of a linear part $\bar{\boldsymbol{F}}(t)\boldsymbol{X}$ and a superimposed fine-scale fluctuation field $\boldsymbol{w} : \mathcal{B}\times\mathcal{R}_+ \to \mathcal{R}^3$. Following [7], the macro-deformation $\bar{\boldsymbol{F}}(t)$ is assumed to be governed by surface data of the microscopic deformation field $\bar{\boldsymbol{F}}(t) = \frac{1}{|\mathcal{B}|}\int_{\partial\mathcal{B}}\varphi(\boldsymbol{X},t)\otimes\boldsymbol{n}(\boldsymbol{X})\,dA$. Then, as a consequence of the decomposition (15), the fluctuation field $\boldsymbol{w}$ has to satisfy the constraint $\frac{1}{|\mathcal{B}|}\int_{\partial\mathcal{B}}\boldsymbol{w}(\boldsymbol{X},t)\otimes\boldsymbol{n}(\boldsymbol{X})\,dA = \boldsymbol{0}$. This constraint is satisfied for the three classes

$$\left.\begin{aligned}
\bar{C}_D(\boldsymbol{w};\boldsymbol{t}) &:= -\frac{1}{|\mathcal{B}|}\int_{\partial\mathcal{B}} \boldsymbol{t}\cdot\boldsymbol{w}\,dA &&= 0\\
\bar{C}_P(\boldsymbol{w};\boldsymbol{t}^+) &:= -\frac{1}{|\mathcal{B}|}\int_{\partial\mathcal{B}^+} \boldsymbol{t}^+\cdot(\boldsymbol{w}^+ - \boldsymbol{w}^-)\,dA &&= 0\\
\bar{C}_S(\boldsymbol{w};\bar{\boldsymbol{P}}) &:= -\frac{1}{|\mathcal{B}|}\int_{\partial\mathcal{B}} (\bar{\boldsymbol{P}}\boldsymbol{n})\cdot\boldsymbol{w}\,dA &&= 0
\end{aligned}\right\} \tag{16}$$

associated with the surface $\partial\mathcal{B}$ of the micro-structure. The first constraint $\bar{C}_D$ in (16) demands *homogeneous deformation* with zero fluctuations $\boldsymbol{w} = \boldsymbol{0}$ on $\partial\mathcal{B}$, where the traction field $\boldsymbol{t}(\boldsymbol{X}, t)$ is considered as a Lagrangian multiplier. The second constraint $\bar{C}_P$ states a *non-trivial periodicity* $\boldsymbol{w}^+ = \boldsymbol{w}^-$ of the superimposed fluctuation field on $\partial\mathcal{B}$. Here, the boundary $\partial\mathcal{B}$ is understood to be decomposed into two parts $\partial\mathcal{B} = \partial\mathcal{B}^- \cup \partial\mathcal{B}^+$ with outward normals $\boldsymbol{n}^+ = -\boldsymbol{n}^-$ at two associated points $\boldsymbol{X}^- \in \partial\mathcal{B}^-$ and $\boldsymbol{X}^+ \in \partial\mathcal{B}^+$. $\boldsymbol{t}^+(\boldsymbol{X}, t)$ is a Lagrangian multiplier field that characterizes antiperiodic tractions on $\partial\mathcal{B}^+$. The third constraint $\bar{C}_S$ in (16) is associated with *homogeneous stresses* $\boldsymbol{t} = \bar{\boldsymbol{P}}\boldsymbol{n}$ on the boundary $\partial\mathcal{B}$, where the Lagrangian multiplier $\bar{\boldsymbol{P}}(t)$ is the macro-stress dual to the macro-deformation $\bar{\boldsymbol{F}}(t)$.

3.2 Incremental Variational Formulation of Homogenization

Based on the constitutive minimization problem for the incremental response of standard dissipative materials defined in (6) we obtain the current stresses by the quasi-hyperelastic function evaluation (5). Hence, the micro-stress $\boldsymbol{P}_{n+1}$ and micro-moduli $\mathbb{C}_{n+1}$ at a local point $\boldsymbol{X} \in \mathcal{B}$ of the heterogeneous micro-structure at time t_{n+1} are defined by

$$\boldsymbol{P}_{n+1} = \partial_{\boldsymbol{F}} W(\boldsymbol{F}_{n+1}; \boldsymbol{X}) \quad \text{and} \quad \mathbb{C}_{n+1} = \partial^2_{\boldsymbol{F}\boldsymbol{F}} W(\boldsymbol{F}_{n+1}; \boldsymbol{X}) \,. \tag{17}$$

The analogy to finite elasticity induces the following incremental variational formulation of the homogenization problem for normal-dissipative inelastic solids. As the key homogenization condition, we consider the *principle of minimum average incremental energy* proposed in [10]

$$\bar{W}(\bar{\boldsymbol{F}}_{n+1}) = \inf_{\boldsymbol{w}_{n+1}} \frac{1}{|\mathcal{V}|} \int_{\mathcal{B}} W(\bar{\boldsymbol{F}}_{n+1} + \nabla \boldsymbol{w}_{n+1}; \boldsymbol{X}) \, dV \tag{18}$$

for heterogeneous micro-structures with constituents of standard dissipative materials. The minimization problem is understood to be subject to the three alternative constraints defined in (16). The minimization problem determines the fluctuation field $\boldsymbol{w}_{n+1}$ on the micro-structure at discrete time t_{n+1}. The macro-stress potential $\bar{W}$ is assumed to define the macro-stresses $\bar{\boldsymbol{P}}_{n+1}$ and the macro-moduli $\bar{\mathbb{C}}_{n+1}$ by the quasi-hyperelastic function evaluations

$$\bar{\boldsymbol{P}}_{n+1} = \partial_{\bar{\boldsymbol{F}}} \bar{W}(\bar{\boldsymbol{F}}_{n+1}) \quad \text{and} \quad \bar{\mathbb{C}}_{n+1} = \partial^2_{\bar{\boldsymbol{F}}\bar{\boldsymbol{F}}} \bar{W}(\bar{\boldsymbol{F}}_{n+1}) \tag{19}$$

in complete analogy to (17). Thus the minimization problem (18) provides a shift of associated variables from the micro-scale to the macro-scale, often denoted as micro-to-macro transition. It extends the so-called *average variational principle* of nonlinear elasticity, outlined in [19] and [24], to the incremental formulation of finite inelasticity. The attractive feature of the proposed formulation (18) is that the structure of the average variational principle is preserved, i.e. the energy storage function ψ of finite elasticity is replaced by the incremental stress potential W of finite inelasticity.

4 Characteristic Numerical Examples

4.1 Microstructure Development in Single-Slip Plasticity

We apply a two-phase rank-one relaxation analysis developed in [14] and [15] to a model problem of single-slip crystal plasticity and investigate the relaxed and non-relaxed stress response of a specimen in shear. The main goals are the demonstration of the mesh-invariance of the technique and the analysis of the developing microstructures. For details of the constitutive functions, material parameters and geometry we refer to [15]. The evolution of the first-

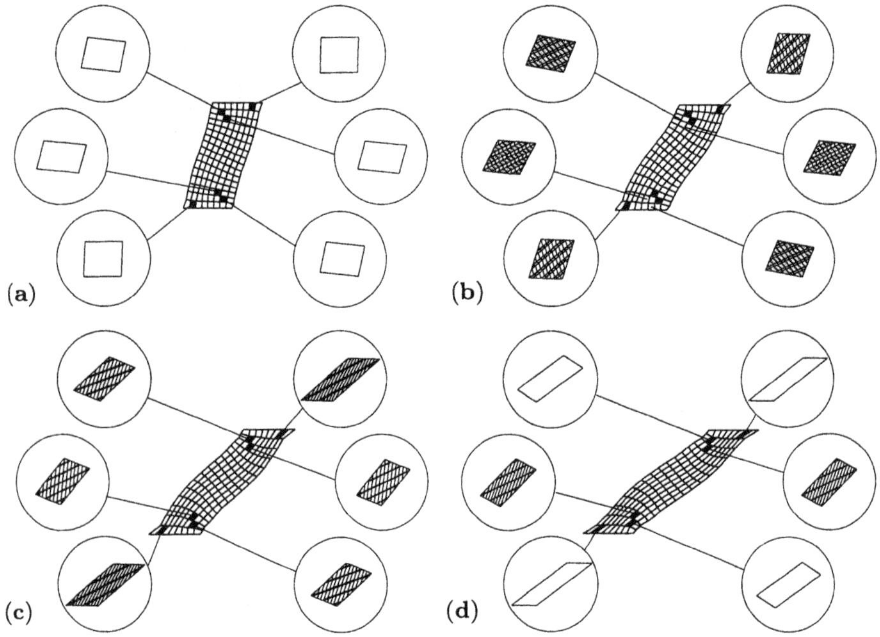

Fig. 1. Specimen in shear. Development of microstructures at the colored integration points for top displacements of (**a**) 10 mm, (**b**) 20 mm, (**c**) 30 mm, (**d**) 40 mm

order laminates is documented in Fig. 1 for a 8×16-element mesh. With increasing deformation the bifurcation of the homogeneous deformation intensifies and the volume fraction $1-\xi$ of the micro-phase $(-)$ augments until a stable homogeneous deformation state is recovered $(\xi = 0)$. The resolution of the microstructures as first-order laminates results from the rank-one convexification of the non-convex incremental stress potentials W, see Fig. 2a. Obviously, the incremental stress potential $W(\boldsymbol{F}_{n+1})$ is greater than the interpolation of potentials $W(\boldsymbol{F}^{+})$ and $W(\boldsymbol{F}^{-})$ corresponding to micro-phases. As a consequence, the homogeneous deformation state is not stable and decomposes into the micro-deformations $\boldsymbol{F}^{\pm}$. The relaxed stress-deformation

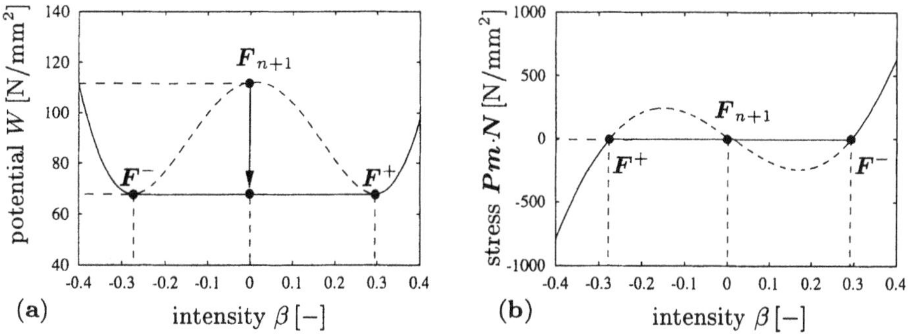

Fig. 2. Specimen in shear. β parameterizes the intensity of the laminate $\boldsymbol{F}^{\pm} = \boldsymbol{F}_{n+1} + \beta^{\pm}\boldsymbol{m} \otimes \boldsymbol{N}$. (**a**) At $\boldsymbol{F}_{n+1}$ the potential is not rank-one convex (dashed line). $\boldsymbol{F}_{n+1}$ decomposes into micro-phases $\boldsymbol{F}^{\pm}$ (solid line). (**b**) The relaxed stress-strain relation characterizes a snap-through Maxwell-line behavior between the phases $\boldsymbol{F}^{\pm}$

relation plotted in Fig. 2b characterizes a snap-through behavior (Maxwell line) between the two micro-phases $\boldsymbol{F}^{\pm}$ due to the constant slope of the rank-one convex envelope. In order to point out the mesh-objectivity of the

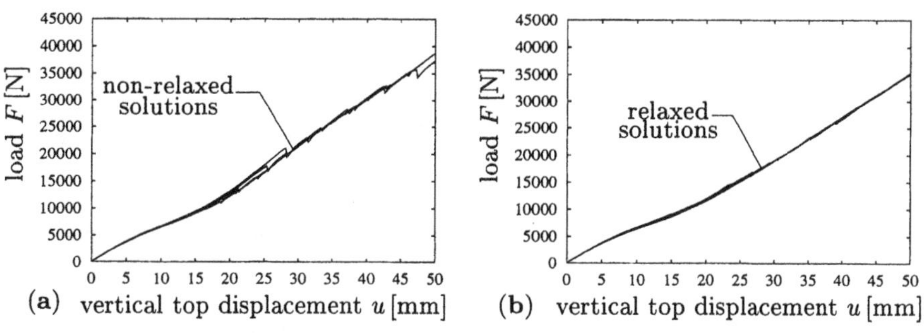

Fig. 3. Specimen in shear. Load-displacement curves for five finite element meshes in terms of (**a**) non-relaxed and (**b**) objective relaxed formulation

relaxed formulation we compare the associated load-deflection curves plotted in Fig. 3 for the five different mesh-sizes considered. The non-relaxed incremental boundary-value problem is ill-posed and the non-convexified stress response not objective. In contrast to the non-relaxed formulation, application of the proposed relaxation technique yields a mesh-invariant response. The load-deflection curves do not depend on the mesh-size, but are identical for all different mesh-densities, see Fig. 3b. Note that the objectivity of the material behavior is obtained without the introduction of an internal length scale parameter.

4.2 Homogenization of a Polycrystalline Microstructure

In the second example we investigate the effect of different boundary constraints on the fluctuation field for the prediction of the texture evolution in polycrystals. For a more detailed description of these analyses we refer to [13]. The micro-structure considered is a copper unit cube discretized with $7^3 = 343$ so-called Q1P0 mixed finite elements. Individual crystal orientations are assigned to each integration point to model a polycrystalline aggregate of 2744 single crystal grains of equal volume. These grains are governed by a model of finite single crystal plasticity described in [13]. The micro-structure is driven by a prescribed macro-deformation of plane strain compression with $\bar{\boldsymbol{F}}(t) = 1(\boldsymbol{e}_1 \otimes \boldsymbol{e}_1) + \exp[vt](\boldsymbol{e}_2 \otimes \boldsymbol{e}_2) + \exp[-vt](\boldsymbol{e}_3 \otimes \boldsymbol{e}_3)$ and $v = 1.0s^{-1}$. During the simulation the specimen is compressed down to 21% of its initial

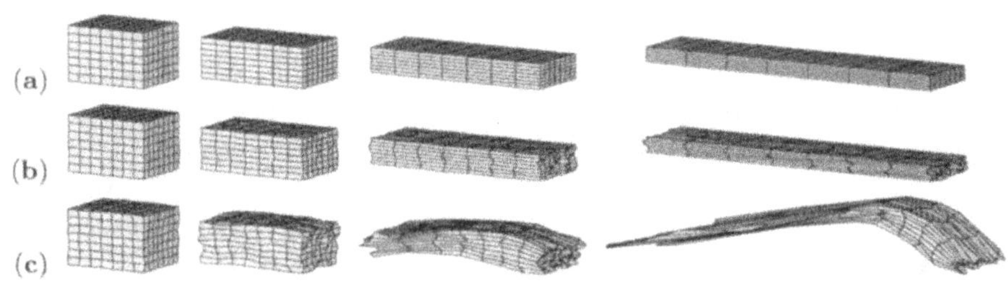

Fig. 4. Plane strain compression test. Deformed micro-structures after a reduction of 18%, 41%, 63% and 79%. (**a**) Homogeneous deformation. (**b**) Periodic deformation. (**c**) Homogeneous stresses on the boundary

height, see Fig. 4 for plot sequences of the deformed geometry for three different fluctuation constraints. The homogeneous deformation of the boundary in Fig. 4a provides the stiffest response, the homogeneous stress boundary condition of Fig. 4c the softest response of the micro-structure, resulting in heavily distorted finite element meshes. The texture development of the micro-structure is compared with experimental data and the prediction of a corresponding Taylor-type analysis. Figure 5 presents pole figures done by equal-area projection, which yield a constant point density in the initial quasi-isotropic state. As expected, the Taylor-type approach in Fig. 5a shows the sharpest orientation distributions due to the stiff constraint of uniform deformation. The simulations in Fig. 5b–c relax this assumption, their pole figures are more diffuse with a somewhat wider elliptical band. Fig. 5d reports on the development of the crystal orientations in case of homogeneous stresses on the boundary of the aggregate. This approach simulates the softest micro-structure and, accordingly, the texture plots are more diffuse. The last pole figure should be ignored, since it results from the heavily distorted mesh of Fig. 4c.

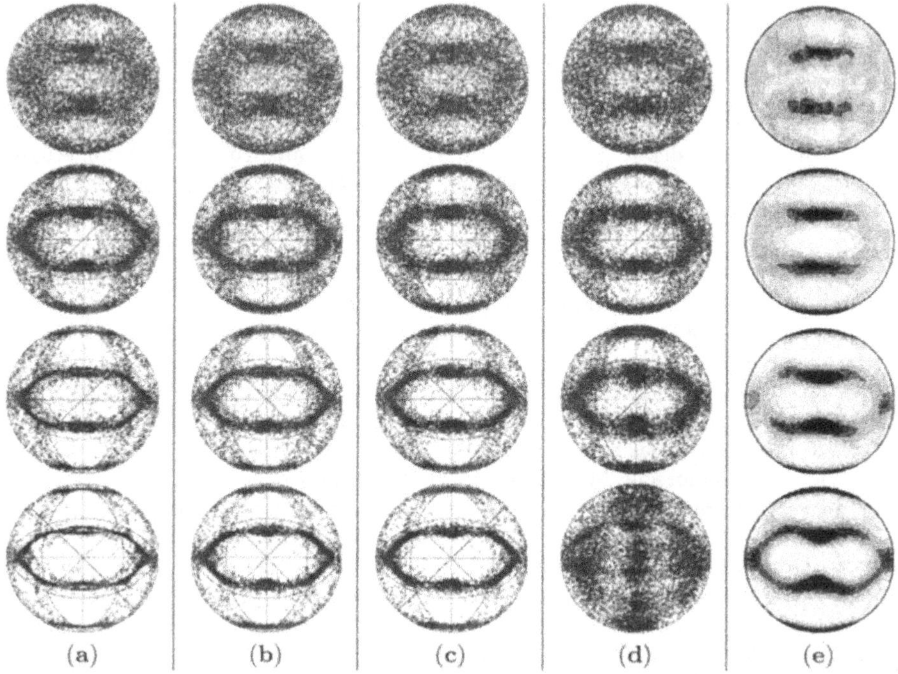

Fig. 5. Plane strain compression test. Sequences of ⟨111⟩-pole figures in the $\boldsymbol{e}_1\boldsymbol{e}_2$-plane corresponding to the deformation states of Figure 4. (**a**) Taylor calculation using 2744 grains. (**b**) Homogeneous deformation. (**c**) Periodic deformation. (**d**) Homogeneous stresses on the boundary. (**e**) Experimental result from [3]

References

1. Ball J.M. (1977) Convexity Conditions and Existence Theorems in Nonlinear Elasticity. Archive of Rational Mechanics and Analysis 63:337–403
2. Biot M.A. (1965) Mechanics of Incremental Deformations. John Wiley & Sons, New York
3. Bronkhorst C.A., Kalidindi S.R., Anand L. (1992) Polycrystalline Plasticity and the Evolution of Crystallographic Texture in f.c.c. Metals. Philosophical Transactions Royal Society London A 341:443–477
4. Carstensen C., Hackl K., Mielke A. (2002) Nonconvex potentials and microstructures in finite-strain plasticity. Royal Society London, Proc. Ser. A 458:299–317
5. Dacorogna B. (1989) Direct Methods in the Calculus of Variations. Springer-Verlag, Berlin
6. Halphen B., Nguyen Q.S. (1975) Sur les Matériaux Standards Généralisés. Journal de Mécanique 40:39–63
7. Hill R. (1972) On Constitutive Macro-Variables for Heterogeneous Solids at Finite Strain. Proceedings of the Royal Society London A 326:131–147
8. Kohn R.V., Strang G. (1983) Explicit Relaxation of a Variational Problem in Optimal Design. Bulletin of the American Mathematical Society 9:211–214

9. Martin J.B. (1975) Plasticity. Fundamentals and General Results. MIT press, Cambridge, Massachusetts
10. Miehe C. (2002) Strain-Driven Homogenization of Inelastic Microstructures and Composites Based on an Incremental Variational Formulation. International Journal for Numerical Methods in Engineering 55:1285–1322
11. Miehe C., Schröder J., Schotte J. (1999) Computational Homogenization Analysis in Finite Plasticity. Simulation of Texture Development in Polycrystalline Materials. Computer Methods in Applied Mechanics and Engineering 171:387–418
12. Miehe C., Schröder J., Becker M. (2002) Computational Homogenization Analysis in Finite Elasticity. Material and Structural Instabilities on the Micro- and Macro-Scales of Periodic Composites and their Interaction. Computer Methods in Applied Mechanics and Engineering 191:4971–5005
13. Miehe C., Schotte J., Lambrecht M. (2002) Homogenization of Inelastic Solid Materials at Finite Strains Based on Incremental Minimization Principles. Journal of the Mechanics and Physics of Solids 50:2123–2167
14. Miehe C, Lambrecht M. (2000) Analysis of Microstructure Development in Shearbands by Energy Relaxation of Incremental Stress Potentials: Large-Strain Theory for Generalized Standard Solids. International Journal for Numerical Methods in Engineering, in press
15. Miehe C., Lambrecht M., Gürses E. (2002) Analysis of Material Instabilities of Inelastic Solids based on Incremental Variational and Relaxation Methods: Application to Single-Slip Crystal Plasticity. Journal of the Mechanics and Physics of Solids, submitted
16. Mielke A. (2002) Finite elastoplasticity, Lie Groups and geodesics on SL(d). In: Newton P., Weinstein A., Holmes P. (Eds.), Geometry, Dynamics, and Mechanics, 61–90, Springer-Verlag
17. Moreau J.J. (1976) Application of Convex Analysis to the Treatment of Elastoplastic Systems. In: Germain P., Nayroles B. (Eds.), Application of Methods of Functional Analysis to Problems in Mechanics. Springer-Verlag, Berlin
18. Morrey C.B. (1952) Quasiconvexity and the Semicontinuity of Multiple Integrands. Pacific Journal of Mathematics 2:25–53
19. Müller S. (1987) Homogenization of Nonconvex Integral Functionals and Cellular Elastic Materials. Archive of Rational Mechanics and Analysis 99:189–212
20. Müller S. (1999) Variational models for microstructure and phase transitions. In: Bethuel F., Huisken G., Müller S., Steffen K., Hildebrandt S., Struwe M. (Eds.), Lectures given at the 2nd Session of the Centro Internazionale Matematico Estivo held in Cetaro, Italy, June 15–22, 1996, Springer, Berlin
21. Nguyen Q.S. (2000) Stability and Nonlinear Solid Mechanics. Wiley & Sons, Chichester
22. Ortiz M., Repetto E.A. (1999) Nonconvex Energy Minimization and Dislocation Structures in Ductile Single Crystals. Journal of the Mechanics and Physics of Solids 47:397–462
23. Ortiz M., Repetto E.A., Stainier L. (2000) A Theory of Subgrain Dislocation Structures. Journal of the Mechanics and Physics of Solids 48:2077–2114
24. Ponte Castañeda P., Suquet P. (1998) Nonlinear Composites. Advances in Applied Mechanics 34:171–303
25. Ziegler H. (1963) Some Extremum Principles in Irreversible Thermodynamics with Application to Continuum Mechanics. In: Sneddon I.N., Hill R. (Eds.), Progress in Solid Mechanics IV. North-Holland Publishing, Amsterdam

Aspects of Material Modelling of Pyrolytic Carbon

Romana Piat and Eckart Schnack

Institute of Solid Mechanics
University of Karlsruhe, Kaiserstr. 12, D-76128 Karlsruhe, Germany

Abstract. Carbon/carbon composites fabricated by chemical vapor infiltration (CVI) consisting of carbon fibers in a pyrolytic carbon (pyro-C) matrix are particularly interesting for industrial application due to their high strength and light weight. One of the difficulties in hierarchical modelling of this composite [1] is that the material constants of pyrolytic carbon are unknown. The pyro-C shows four forms with a different polycrystalline structure: isotropic, low textured (LT), medium textured (MT) and highly textured (HT) [2]. The pyro-C is modelled as a system of ellipsoidal single crystals of graphite embedded in an infinite homogeneous matrix with unknown effective overall parameters of the polycrystal [3]. The local average stress and strain in the embedded crystals are then calculated and used to obtain the overall moduli. A probability density function is introduced to simulate spatial crystal orientations. The different forms of the pyro-C are described by this density function. The calculation of the overall elasticity and compliance tensors for each pyro-C modification is reduced to a system of equations, which are solved numerically. The theoretical results are compared with experimental data. A system of parallel cracks is embedded in this brittle homogenized material. The overall elasticity and compliance tensors for the material with cracks are also calculated.

1 Introduction

Carbon/carbon composites (C/C composites) fabricated by chemical vapor infiltration (CVI) consist of carbon fibers in a pyrolytic carbon (pyro-C) matrix. Especial difficulty in studying the mechanical properties of C/C composites is the fact that the mechanical properties of the different pyro-C modifications unknown. This paper presents: 1) modelling the pyro-C as an aggregate of single graphite crystals with discrete and continuous probability density function describing the distribution of the crystal orientations; 2) modelling of this material with a set of cracks.

2 Material description

The technology of chemical vapor infiltration consists of the synthesis of carbon particles from carbon-hydrate gas and their deposition on carbon fibers preliminarily posed in that environment. The process runs under elevated temperatures and pressures until the carbon particles deposit in form of a

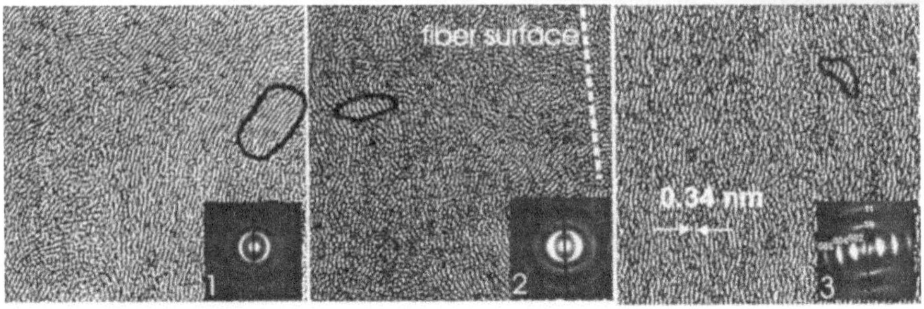

Fig. 1. High-resolution images with select area electron diffraction patterns as insets of 1) low; 2) medium and 3) highly textured pyro-C

pyrolytic carbon matrix filling the space between fibers [4]. On the basis of four different types of preferred orientation (or texture) there are four basic forms of pyro-C: isotropic, low textured, medium textured and highly textured (Figure 1). The difference between the modifications of pyro-C is in the sizes of coherent domains and their orientation. The distribution functions *normal distribution* (Gauss distribution) of the preferred orientation can be determined for each form of pyro-C as:

$$f(\alpha) = \frac{1}{\sigma\sqrt{2\pi}} \exp\left(-\alpha^2/2\,\sigma\right). \tag{1}$$

Experiment allow to identify the parameter σ in the distribution functions of the orientations for every material modification of the pyro-C [4]. These values are: a) for low textured pyro-C $0.7 \leqslant \sigma \leqslant \pi$; b) for medium textured pyro-C: $0.44 \leqslant \sigma \leqslant 0.7$ and c) for highly textured pyro-C: $\sigma \leqslant 0.44$.

3 Material Modelling

The material presents a part of one pyro-C modification (see Figure 1). The modelling of this material as an aggregate of single crystals is proposed for the identification of these unknown material parameters. The pyro-C is modelled as a system of ellipsoidal single crystals of graphite embedded in an infinite homogeneous matrix with unknown effective (overall) elastic moduli of the polycrystal. Each ellipsoidal crystal of graphite is characterized by three angles $\boldsymbol{\alpha}$ $(\alpha_1, \alpha_2, \alpha_3)$ with their orientations to the global coordinate system (see Figure 2). The elasticity and compliance tensors of the crystal of graphite in the global coordinate system are:

$$\boldsymbol{C}^\alpha = \boldsymbol{Q}^{\alpha\,\mathrm{T}} : \boldsymbol{C} : \boldsymbol{Q}^\alpha\,, \quad \boldsymbol{D}^\alpha = (\boldsymbol{C}^\alpha)^{-1}, \tag{2}$$

where $\boldsymbol{Q}^\alpha$ is a transformation matrix for the crystals with the orientation $\boldsymbol{\alpha}$. The unknown overall elasticity $\overline{\boldsymbol{C}}$ for the case n different crystal orientations

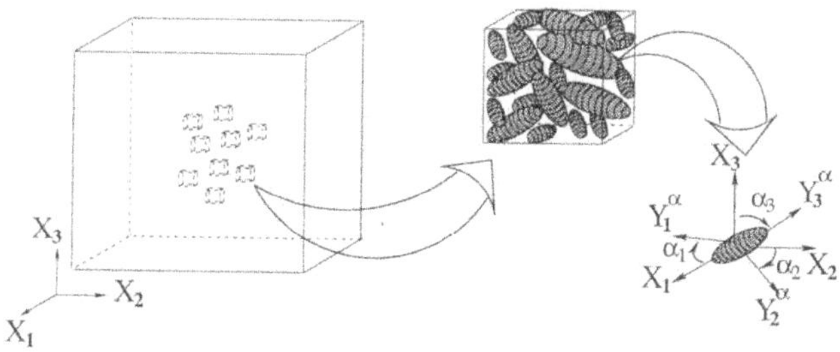

Fig. 2. The RVE the system of ellipsoidal single crystals of graphite with different orientations relative to the global coordinate system

[3] can be identify from :

$$\sum_{\alpha=0}^{n} f_\alpha \left[\boldsymbol{I} + \overline{\boldsymbol{T}}^I : (\overline{\boldsymbol{C}} : \boldsymbol{D}^\alpha - \boldsymbol{I}) \right]^{-1} = \boldsymbol{I} \,, \tag{3}$$

were f_α is the volume fraction of the inclusions with the orientation $\boldsymbol{\alpha}$, $\overline{\boldsymbol{T}}^I$ is the conjugate of Eshelby's tensor $\overline{\boldsymbol{S}}^I$ where the overbar signifies that this tensor corresponds to the yet unknown overall elasticity or compliance tensors $\overline{\boldsymbol{C}}$ and $\overline{\boldsymbol{D}}$. After some transformations the calculation of the unknown components of the tensor $\overline{\boldsymbol{C}}$ is reduced to:

$$\sum_{\alpha=0}^{n} f_\alpha \left\{ \boldsymbol{P} : \overline{\boldsymbol{C}} : \boldsymbol{D}^\alpha - (\boldsymbol{D}^\alpha + \boldsymbol{P}) \right\}^{-1} = \overline{\boldsymbol{C}} \,, \tag{4}$$

with $\boldsymbol{P}$ from $\overline{\boldsymbol{S}}^I = \boldsymbol{P} : \overline{\boldsymbol{C}}$.
For the case that the distribution function (1) is continuous, the upper bound [3] for the overall elasticity can be calculated for each pyro-C modification as:

$$\overline{\boldsymbol{C}} = \lim_{\Delta\alpha_i \to 0} \sum_{\alpha_i} f_{\alpha_i} \boldsymbol{C}^{\alpha_i} = \int_\alpha f(\alpha)\, \boldsymbol{C}^\alpha \, d\alpha \,. \tag{5}$$

4 Material modelling with cracks

A RVE with a system of cracks is presented in Figure 3. The mechanical properties of the material without cracks is described by the effective elasticity $\overline{\boldsymbol{C}}$ of the polycrystalline. It is assumed that the overall strain $\boldsymbol{\varepsilon}^0$ is prescribed at the boundary of the RVE: $\boldsymbol{u}^0 = \boldsymbol{x} \cdot \boldsymbol{\varepsilon}^0$ on ∂V. Each crack is modelled as an elliptical inclusion. The ratio between the smaller b_α and the larger a_α half axes of the ellipsoid is presented by a very small positive number. The volumetrically given principal strain through the crack with the orientation $\boldsymbol{\alpha}$ can be presented as:

$$\overline{\boldsymbol{\varepsilon}}^\alpha = \boldsymbol{J}^\alpha : \boldsymbol{\varepsilon}^0 . \tag{6}$$

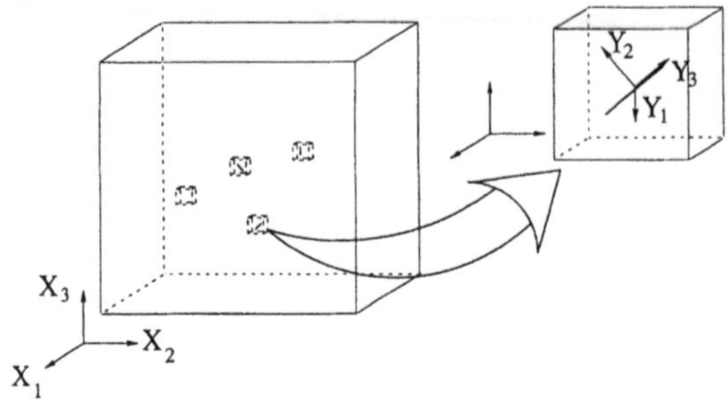

Fig. 3. RVE with a set of cracks

The tensor $\boldsymbol{J}^{\alpha}$ describes the effects of the existence of cracks in the direction $\boldsymbol{\alpha}$. The unknown components J^{α}_{ijkl} for cracks in the direction $\boldsymbol{\alpha}$ can be calculated in terms of Eshelby's tensor $\boldsymbol{S}^{\alpha}$: $\boldsymbol{J}^{\alpha} = (\boldsymbol{I} - \boldsymbol{S}^{\alpha})^{-1}$. The crack density function for the RVE with a set of cracks with n different orientations $\boldsymbol{\alpha}$ and N_{α} cracks in each orientation $\boldsymbol{\alpha}$ and unknown overall elasticity tensor $\widehat{\boldsymbol{C}}$ for a material with a set of cracks then can be presented as:

$$f = \sum_{\alpha=1}^{n} f_{\alpha} = \sum_{\alpha=1}^{n} N_{\alpha}\, a_{\alpha}^{2}\,, \quad \widehat{\boldsymbol{C}} = \overline{\boldsymbol{C}} : (\boldsymbol{I} - \sum_{\alpha=1}^{n} f_{\alpha}\, \boldsymbol{J}^{\alpha}). \tag{7}$$

5 Numerical results

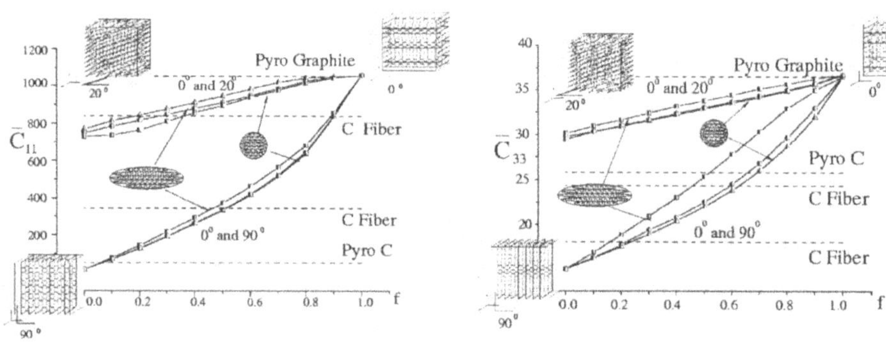

Fig. 4. $\overline{C}_{11}$ and $\overline{C}_{33}$ for different discrete distributions of single crystal orientations

Numerical solutions of Eq.(4) for the discrete distribution of single crystal orientations are presented in Figure 4. The unknown values of the components of the elasticity tensor $\overline{C}_{11}$ and $\overline{C}_{33}$ are calculated for a material which consists of crystals with two orientations, 0° and 90° (or 0° and 20°) to the global coordinate system. The parameter f characterizes the volume fraction of the crystals with the orientation 90° (or 20°). The numerical results are compared with known material constants for pyrolytic graphite and a particular case of C-fibers [5].

After the numerical evaluation of the unknown values of $\overline{C}$ for the case of continuous distribution function $f(\alpha)$ Eq.(1) and Eq.(5) the engineering elasticity constants are calculated and then normalized. The results of the

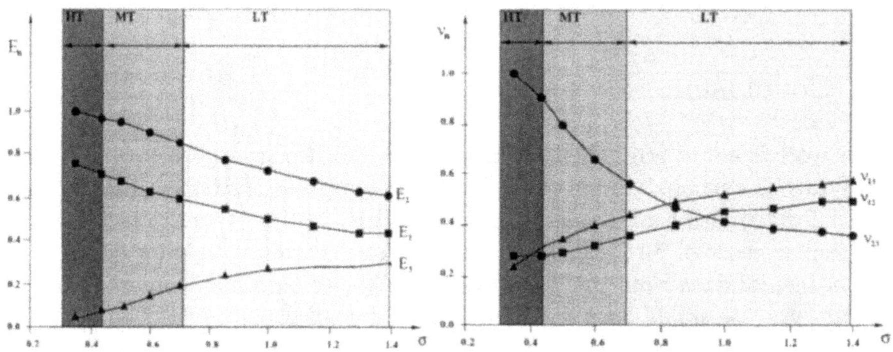

Fig. 5. Normalized a) Young moduli E_1, E_2, E_3 and b)Poisson coefficients ν_{12}, ν_{23} and ν_{13} as a functions of σ

numerical evaluation of the normalized Young moduli as a functions of the texture degree are presented at the Figure 5a. The E_3 moduli in the perpendicular direction to the axis for highly textured pyrolytic carbon is smaller as for another modifications. Normalized values of the Poisson coefficients

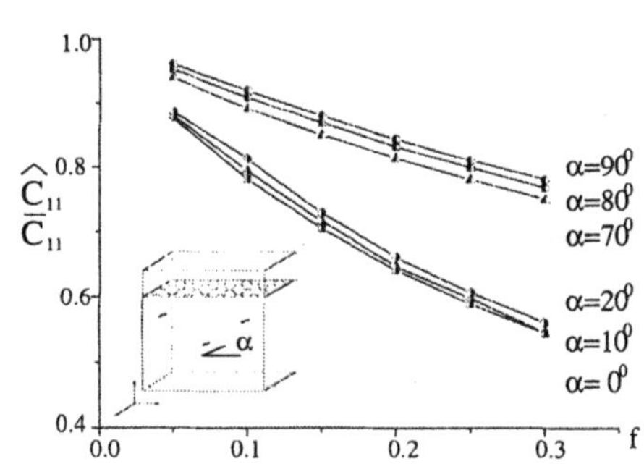

Fig. 6. $\widehat{C}_{11}/C_{11}$ for graphite with different values of the cracks density function and the orientation of cracks

ν_{12}, ν_{23} and ν_{13} as functions of texture degree are presented at the Figure 5b.

The Poisson coefficient ν_{23} is for highly textured pyrolytic carbon very large in comparison to two other coefficients ν_{12} and ν_{13} . For medium textured pyrolytic carbon this difference is smaller and for low textured pyrolytic carbon the difference between all three coefficients is small.
The numerical solution of Eq.(7) for a set of parallel cracks in pure graphite is presented in Figure 6. The ratio of the component $\widehat{C}_{11}$ of the elasticity tensor for graphite with cracks to the component C_{11} of the elasticity tensor for undamaged graphite characterizes the influence of the cracks on the mechanical properties of the material.

6 Conclusion

This work is a first implementation of homogenization methods for modelling of pyrolytic carbon. Moreover: 1) The work proposes a stochastic model for spatial distribution of single graphite crystals; 2) The overall elasticity for particular cases of discrete single crystal distribution are calculated; 3) The experimental data from the real physical structure obtained from experiment [2] are used as input data for modelling; 4) On the basis of the model, the upper bounds of the normalized Young moduli as functions of the texture degree of pyrolytic carbon are calculated for continuous single crystal distribution; 5) The overall properties of graphite with a set of parallel cracks are obtained for different cases of crack orientations and crack density functions.

Acknowledgements

The authors are gratefully indebted to the Deutsche Forschungsgemeinschaft (DFG) for the financial support. We would like to thank our colleague Dr. B. Reznik for constructive discussions and the provided experimental material.

References

1. Piat R., Schnack E., Andrä H. (2001) Entwicklung des hierarchischen Materialmodells von CVI-CFC-Werkstoffen. Proc. Appl. Math. Mech. **1** (www.interscience.wiley.com).
2. Reznik B., Gerthsen D., Hüttinger K.J.(2001) Macro- and nanostructure of the carbon matrix of infiltrated carbon fiber felts. Carbon **39**, 215–229.
3. Nemat-Nasser S., Hori M. (1999) Micromechanics: overall properties of heterogeneous materials. North-Holland, Amsterdam.
4. Reznik B., Hüttinger K.J. (2002) On the terminology for pyrolytic carbon. Carbon **40**, 621–624.
5. Piat R., Schnack E.(2002) Mechanical properties and damage processes of pyrolytic carbon. WCCM V, Vienna, Austria, ISBN 3-9501554-0-6.

Optimal design of laminated composites

Tomáš Roubíček[1,2]

[1] Mathematical Institute, Charles University, Sokolovská 83, CZ-186 75 Praha 8, Czech Republic,
[2] Institute of Information Theory and Automation, Academy of Sciences, Pod vodárenskou věží 4, CZ-182 08 Praha 8, Czech Republic.

Abstract. A relaxed-control-theory approach is applied to optimal design of composite structures whose response is fully determined by volume fractions. This allows us, e.g., to derive optimality conditions in terms of a maximum principle.

1 Introduction

The contribution deals with optimal material design of nonhomogeneous linear media. As an optimal design often fails to exists and minimizing sequences necessarily develops faster and faster oscillations in space, this problem is closely related with *homogenization* theory; we refer to [1,6,8,10].

Here, we want to present briefly a method based on standard techniques developed for *relaxation* in optimal control theory; see [7,15,20]. This theory handles fast oscillations of minimizing sequences by *Young measures* which record only volume fractions of particular materials at a current "macroscopical" point x. This allows to develop a relaxation theory including optimality conditions but, at the same time, represents an essential restriction to situations where the limit composite is fully determined only by volume fractions.

Such situation indeed occurs if the properties of the medium can vary only in a specified direction; we put the coordinate x_1 in this direction, cf. Fig.1. Such *simply laminated* configuration are often used in technology of *composites* because they can be relatively easily manufactured. Those media are also called *stratified* and there is a developed homogenization theory in this case, see McConnell [11], Murat [13], Tartar [17]. Here, in Section 2, we want to demonstrate relaxed-control-theory approach on a very simple example while various generalizations and modifications are only outlined in Section 3.

2 A model problem: isotropic conductive medium

For simplicity, we will consider Ω a cylindrical domain in $\mathbb{R}^n$ with $n \geqslant 2$, say $\Omega := I \times \omega$ with $I \subset \mathbb{R}$ an open interval and $\omega \subset \mathbb{R}^{n-1}$ a Lipschitz domain. Then we will write $x = (x_1, x_2, ..., x_n) \equiv (x_1, x_{2\text{-}n})$ and consider an isotropic linearly-responding material characterized by one coefficient $\sigma = \sigma(x_1)$ and the state response governed by the single 2nd-order elliptic equation

$$\operatorname{div}\,(\sigma(x_1)\nabla y(x)) = f. \tag{1}$$

Depending on particular applications, the variable y (resp. the coefficient σ) can be interpreted as temperature (resp. heat conductivity), electrostatical potential (resp. electrical conductivity or permitivity), or magnetostatical potential (resp. permeability).

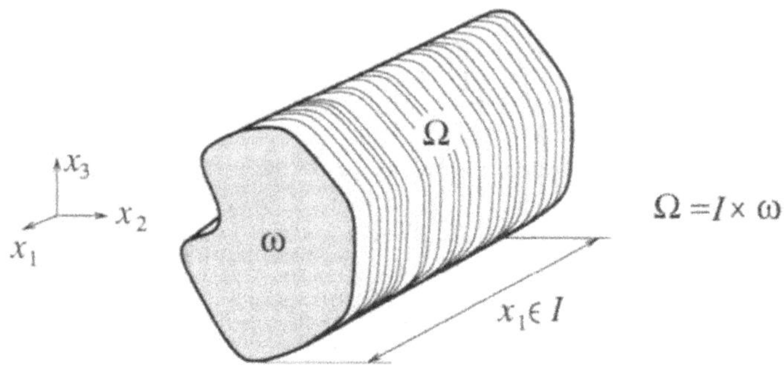

Fig. 1. Schematic situation: a laminated conductive medium.

For further simplicity, we will confine ourselves to Dirichlet boundary conditions and consider the following model constrained optimal-design problem:

$$(\mathfrak{P}) \qquad \left\{\begin{array}{ll} \text{minimize} & J(\sigma, y) := \displaystyle\int_\Omega \varphi(x, y(x), \sigma(x_1))\,\mathrm{d}x \\ \text{subject to} & \operatorname{div}(\sigma(x_1)\nabla y(x)) = f \qquad \text{on } \Omega, \\ & \displaystyle\int_\Omega \psi(x, y(x), \sigma(x_1))\,\mathrm{d}x \leqslant 0, \\ & s_{\min} \leqslant \sigma(x_1) \leqslant s_{\max} \qquad \text{for a.a. } x_1 \in I, \\ & y \in W_0^{1,2}(\Omega), \quad \sigma \in L^\infty(I), \end{array}\right.$$

where the equation (1) with boundary conditions $u|_{\partial\Omega} = 0$ is understood in the weak sense, $\varphi : \Omega \times \mathbb{R} \times S \to \mathbb{R}$, and $\psi : \Omega \times \mathbb{R} \times S \to \mathbb{R}^k$, $\mathbb{R}^k$ considered ordered by an ordering $\succeq$ inducing a cone $\{\lambda \succeq 0\}$ having a non-empty interior in $\mathbb{R}^k$, and $S := [s_{\min}, s_{\max}]$.

We already mentioned in Section 1 that $(\mathfrak{P})$ need not have any solution, and thus we are urged to extend the original problem somehow. The most natural way is, of course, a continuous extension (in suitable topologies, if exist) on a compact set; we call this procedure a *relaxation.* The general method [15] relies on a *convex compact envelope* of $\Sigma_{\mathrm{ad}} = \{\sigma \in L^\infty(I);\ \sigma(x_1) \in S \text{ for a.a. } x_1\}$ sufficiently fine so that all functionals occurring in $(\mathfrak{P})$ can be continuously extended (and the extended functionals will be even *smooth* with respect to the convex structure).

As a suitable convex, compact envelope of Σ_{ad} for the case (1), one can use the classical construction by *Young measures*, i.e. the set of all families $\nu := \{\nu_{x_1}\}_{x_1 \in I}$ of probability Radon measures on S that are weakly measurable in the sense that, for all $v \in C(S)$, $v \bullet \nu$ is Lebesgue measurable where we used the shorthand notation $[v \bullet \nu](x_1) = \int_S v(s)\nu_{x_1}(\mathrm{d}s)$. Due to $L^\infty_{\mathrm{w}}(I;\mathrm{rca}(S)) \cong L^1(I;C(S))^*$ with "rca" standing for Radon measures, we have $\langle \nu, h\rangle = \int_I \int_S h(x_1,s)\nu_{x_1}(\mathrm{d}s)\mathrm{d}x_1$ for any $h \in L^1(I;C(S))$. The set of all Young measures, denoted by $\mathcal{Y}(I;S)$, is indeed convex and compact in $L^\infty_{\mathrm{w}}(I;\mathrm{rca}(\mathbb{R}))$ and the embedding $\delta : \sigma \mapsto \nu := \{\delta_{\sigma(x_1)}\}_{x_1\in I}$ is homeomorphical and $\delta(\Sigma_{\mathrm{ad}})$ is dense in $\mathcal{Y}(I;S)$.

Now, the crucial point is a suitable decomposition of the flux $j(\sigma,\nabla y) := \sigma\nabla y$, if possible at all. In our stratified situation $\sigma = \sigma(x_1)$, this means:

$$j(\sigma,\nabla y) = \left(\frac{1}{\alpha(\sigma)}\frac{\partial y}{\partial x_1}, \beta(\sigma)\frac{\partial y}{\partial x_2}, \beta(\sigma)\frac{\partial y}{\partial x_3}\right), \quad \alpha(x) = \frac{1}{s}, \quad \beta(s) = s. \tag{2}$$

This leads to the *relaxed problem* in the form

$$(\mathfrak{RP}) \qquad \begin{cases} \text{minimize} & \bar{J}(\nu,y) := \langle \nu, \varphi\circ y\rangle \\ \text{subject to} & \operatorname{div}\bar{\jmath}(\nu,\nabla y) = f, \quad \langle \nu, \psi\circ y\rangle \preceq 0, \\ & y \in W^{1,2}_0(\Omega), \quad \nu \in \mathcal{Y}(I;S), \end{cases}$$

where the integrand $\psi\circ y \in L^1(I;C(S;\mathbb{R}^k))$ is defined by $[\psi\circ y](x_1,s) := \int_\omega \psi(x,y(x),s)\,\mathrm{d}x_{2\text{-}n}$, and $\varphi\circ y \in L^1(I;C(S))$ is defined analogously, and

$$\bar{\jmath}(\nu,\nabla y) := \left(\frac{1}{\alpha\bullet\nu}\frac{\partial y}{\partial x_1}, (\beta\bullet\nu)\frac{\partial y}{\partial x_2}, (\beta\bullet\nu)\frac{\partial y}{\partial x_3}\right). \tag{3}$$

The important and deep fact, shown by Murat [13] for a more general case, is that $\nu \mapsto y = y_\nu : \mathcal{Y}(I;S) \to W^{1,2}_0(\Omega)$, y_ν the solution to $\operatorname{div}\bar{\jmath}(\nu,\nabla y) = f$, is a (weak*,weak)-continuous extension of the mapping $\sigma \mapsto y = y_\sigma : \Sigma_{\mathrm{ad}} \to W^{1,2}_0(\Omega)$, where y_σ denotes the weak solution to $\operatorname{div}(\sigma\nabla y) = f$.

Contrary to the original problem $(\mathfrak{P})$, the relaxed problem $(\mathfrak{RP})$ always has a solution, and $\min(\mathfrak{RP}) \leqslant \inf(\mathfrak{P})$. Moreover, the solutions to $(\mathfrak{RP})$ are naturally related with cluster-points of minimizing sequences $\{\sigma_i\}_{i\in\mathrm{N}}$ for $(\mathfrak{P})$ that satisfy the constraints only asymptotically, see [15,16] for details.

To set up the 1st-order *optimality conditions*, we consider multipliers $\lambda_0 \in \mathbb{R}$, $\lambda_1 \in \mathbb{R}^k$, and $w \in W^{1,2}_0(\Omega)^{**} \cong W^{1,2}_0(\Omega)$ and compose the *Lagrangean* for the relaxed problem $(\mathfrak{RP})$:

$$L_{\lambda_0,\lambda_1,w}(y,\nu) = \int_\Omega \left(\nabla w\cdot\bar{\jmath}(\nu,\nabla y) + fw + \int_S \left[\lambda_0\varphi + \lambda_1\cdot\psi\right](x,y,s)\nu_{x_1}(\mathrm{d}s)\right)\mathrm{d}x.$$

Note that $L_{\lambda_0,\lambda_1,w}$ is smooth with respect to ν if restricted on $\mathcal{Y}(I;S)$ because (3) involves analytical nonlinearities α and β. Assuming still a smoothness of $[\varphi,\psi](x,\cdot,s) : \mathbb{R} \to \mathbb{R}^{1+k}$ in the natural mode

$$\left|[\varphi,\psi]'(x,y,s)\right| \leqslant a_1(x) + b|y|^{q-1}, \tag{4}$$

$$\left|[\varphi,\psi]'(x,y,s) - [\varphi,\psi]'(x,\tilde{y},s)\right| \leqslant \left(a_2(x) + b|y|^{q-2} + b|\tilde{y}|^{q-2}\right)|y-\tilde{y}|, \tag{5}$$

where $[\varphi,\psi]' := \partial[\varphi,\psi]/\partial y$, $a_i \in L^{q/(q-i)}(\Omega)$, $i = 1,2$, $b \in \mathbb{R}$, and $2 \leqslant q < 2n/(n-2)$, we have smoothness of $L_{\lambda_0,\lambda_1,w}$ still with respect to $y \in W_0^{1,2}(\Omega)$.

The *adjoint equation* in the weak formulation means standardly that $[L_{\lambda_0,\lambda_1,w}(y,\nu)]'_y(\tilde{y}) = 0$ for any $\tilde{y} \in W_0^{1,2}(\Omega)$, which results here to

$$\operatorname{div} \bar{j}(\nu, \nabla w) = \int_S \frac{\partial(\lambda_0\varphi + \lambda_1\cdot\psi)}{\partial y}(x, y(x), s)\, \nu_{x_1}(\mathrm{d}s), \qquad w|_{\partial\Omega} = 0, \tag{6}$$

while the directional derivative of $L_{\lambda_0,\lambda_1,w}$ with respect to $\tilde{\nu}$ results to $[L_{\lambda_0,\lambda_1,w}(y,\nu)]'_\nu(\tilde{\nu}) = \langle\tilde{\nu}, -h_{y,\nu,\lambda_0,\lambda_1,w}\rangle$ with the *Hamiltonian* $h_{y,\nu,\lambda_0,\lambda_1,w} \in L^1(I;C(S))$ defined by

$$h_{y,\nu,\lambda_0,\lambda_1,w}(x_1,s) := \int_\omega -\frac{\alpha(s)}{[\alpha\bullet\nu](x_1)}\frac{\partial w}{\partial x_1}\frac{\partial y}{\partial x_1} + \beta(s)\sum_{i=2,..n}\frac{\partial w}{\partial x_i}\frac{\partial y}{\partial x_i} \tag{7}$$
$$-\big[\lambda_0\varphi + \lambda_1\cdot\psi\big](x, y(x), s)\,\mathrm{d}x_{2\text{-}n}.$$

Note that, contrary to usual relaxed optimal control problems, the Lagrangean $L_{\lambda_0,\lambda_1,w}$ is *not affine* with respect to ν which causes the Hamiltonian from (7) to depend explicitly on ν.

Proposition 1. (F.John's necessary optimality conditions.) *If $\nu \in \mathcal{Y}(I;S)$ is an optimal design, i.e. a solution to $(\mathfrak{RP})$, then there exist $\lambda_0 \geqslant 0$ and $\lambda_1 \overset{*}{\succeq} 0$, where "$\overset{*}{\succeq}$" denotes the dual ordering to "$\succeq$", not vanishing simultaneously and satisfying the transversality condition $\langle\nu, \lambda_1\cdot(\psi\circ y)\rangle = 0$ and the maximum principle*

$$\int_S h_{y,\nu,\lambda_0,\lambda_1,w}(x_1,s)\nu_{x_1}(\mathrm{d}s) = \max_{s\in S} h_{y,\nu,\lambda_0,\lambda_1,w}(x_1,s) \tag{8}$$

for a.a. $x_1 \in I$ with $h_{y,\nu,\lambda_0,\lambda_1,w}$ from (7) and $w \in W_0^{1,2}(\Omega)$ solving (6).

Note that (8) says precisely that ν_{x_1} can be supported only where the Hamiltonian attains its maximum on S. Also, note that (8) is equivalent to the inequality $[L'_\nu(y,\nu,\lambda_0,\lambda_1,w)](\tilde{\nu}-\nu) \geqslant 0$ for any $\tilde{\nu} \in \mathcal{Y}(I;S)$, which, together with $L'_y(y,\nu,\lambda_0,\lambda_1,w) = 0$ (i.e. (6)), is the usual optimality condition.

Analysis of the maximum principle can be performed in particular cases. In special cases, one can sometimes say more. E.g., let us consider a single constraint

$$\int_I \sigma(x_1)^q \mathrm{d}x_1 \leqslant K, \tag{9}$$

this means $k = 1$ and $\psi(x,r,s) = s^q - K/|I|$. We assume $q > 0$ and $K > s_{\min}^q|I|$. In this case, there is no relaxation gap, i.e. $\min(\mathfrak{RP}) = \inf(\mathfrak{P})$; a construction of a sequence $\{\sigma_i\}_{i\in\mathbb{N}}$ attaining a feasible Young measure and, at the same time, satisfying the constraints $\sigma_i(x_1) \in S$ and $\int_I \sigma_i(x_1)^q \leqslant K$ has been shown by Muñoz and Pedregal [12, Sect.2].

The Mangasarian-Fromowitz *constraint qualification* (CQ) is always satisfied at any feasible ν except $\nu_{x_1} \equiv \delta_{s_{\min}}$; as now ψ does not depend on y and thus $\Psi(\nu) := \int_\Omega\int_S \psi(x,y(x),s)\nu_{x_1}(\mathrm{d}s)\mathrm{d}x$ is linear, this CQ means that $\Psi(\tilde{\nu}) < \Psi(\nu)$ for some $\tilde{\nu} \in \mathcal{Y}(I;S)$, which indeed holds for $\tilde{\nu}_{x_1} \equiv \delta_{s_{\min}}$. Hence:

Proposition 2. (KARUSH-KUHN-TUCKER CONDITIONS.) *If* $\psi(x, y, s) = s^q - K/|I|$ *with* $K > s_{\min}^q|I|$, *one may take* $\lambda_0 = 1$ *in Proposition 1.*

Note that CQ does not hold at ν, $\nu_{x_1} \equiv \delta_{s_{\min}}$, but the constraint (9) is not active at this ν so that $\lambda_1 = 0$ in this case, hence again one can assume $\lambda_0 = 1$. Then, from (8) it follows that there is at least one *at most 4-atomic optimal design* ν, i.e. each ν_x is a convex combination of at most 4 Dirac's masses, cf. [15, Prop.4.2.9]; note that it is less by 1 than the Carathéodory number corresponding to 4 nonlinearities $\alpha(\cdot)$, $\beta(\cdot)$, $\varphi(x, y(x), \cdot)$ and $|\cdot|^q$. This is certainly an information of a practical value.

3 Various generalizations and modifications

One can generalize the above considerations for *anisotropic media*, i.e. $\sigma(x_1)$ in (1) will be a symmetric, positive definite $(n \times n)$-matrix. Instead of (2), the suitable decomposition of the flux $j(\sigma, \nabla y) := \sigma \nabla y$ is now

$$j(\sigma, \nabla y) = \frac{\beta(\sigma) \cdot \nabla y}{\alpha(\sigma)} \beta(\sigma) + \gamma(\sigma) \nabla y \tag{10}$$

with the nonlinearities $(\alpha, \beta, \gamma) : \mathbb{R}^{n \times n} \to \mathbb{R} \times \mathbb{R}^n \times \mathbb{R}^{n \times n}$ defined by

$$\alpha(s) := \frac{1}{s_{11}}, \quad \beta_i(s) := \frac{s_{i1}}{s_{11}}, \quad \gamma_{ij}(s) := \begin{cases} s_{ij} - \dfrac{s_{i1} s_{1j}}{s_{11}} & \text{for } \max(i, j) \geqslant 2, \\ 0 & \text{for } i = 1 \text{ or } j = 1. \end{cases}$$

The relaxation theory for such anisotropic case (without any constraints, i.e. $\psi \equiv 0$) has been developed by Bouchitté at al. [5] even for $S \subset \mathbb{R}^{n \times n}$ unbounded, while the maximum principle (8) needed S bounded. See also Tartar [18] or [19, Sect.9] even for a nonstratified case but, on the other hand, for a finite-number of materials and very special cost functionals only.

One can also consider a 4th-order equation instead of (1):

$$\operatorname{div} \operatorname{div} \big(u(x_1)^3 A \nabla^2 y(x)\big) = f \tag{11}$$

with a scalar parameter $u = u(x_1)$ and a fixed 4th-order tensor A depending on Young's modulus and Poisson's ratio. This represents a certain model for tangential deflection y of a plate with a thickness u. Then we face an optimal *plate design* and the relaxation theory has been performed by Bonnetier and Conca [2,3], Bonnetier and Vogelius [4], or Muñoz and Pedregal [12], see also [14, Sect.6], and Kohn and Vogelius [9], while the maximum principle (8) has been derived in [16] even for distributed constraints $\psi(x, y, u) \preceq 0$ a.e.

This approach seems possible to be extended for *systems* of 2nd-order m equations, i.e. (1) with $y \in W^{1,2}(\Omega; \mathbb{R}^m)$ and $\sigma = \sigma(x_1)$ a 4th-order tensor (e.g. Lamé system in linear elasticity), by using McConnell [11].

Acknowledgments. This research was covered by the grants 201/00/0768 (GA ČR), A 107 5005 (GA AV ČR), and MSM 11320007 (MŠMT ČR). The author is thankful to Ing. Jan Zeman for drawing attention to [11] and for useful discussions.

References

1. Benssousan, A., Lions, J.L., Papanicolau, G.: *Asymptotic Analysis for Periodic Structures.* North-Holland, Amsterdam, 1978.
2. Bonnetier, E., Conca, C.: Relaxation totale d'un problème d'optimisation de plaques. *C.R. Acad. Sci Paris* **317** (1993), 931–936.
3. Bonnetier, E., Conca, C.: Approximation of Young measures by functions and application to an optimal design problem for plates with variable thickness. *Proc. Royal Soc. Edinburgh* **124 A** (1994), 399–422.
4. Bonnetier, E., Vogelius, M.: Relaxation of a compliance functional for a plate optimization problem. In: *Applications of Multiple Scaling in Mechanics.* (P.G.Ciarlet, E.Sánches-Palencia, eds.) Mason, 1987, pp.31–53.
5. Bouchitté, G., Roubíček, T.: Optimal design of stratified media. *Adv. Math. Sci. Appl.* **12** (2002), 135-150.
6. Cherkaev, A.: *Variational methods for structural optimization.* Springer, New York, 2000.
7. Fattorini, H.O.: *Infinite Dimensional Optimization Theory and Optimal Control.* Cambridge Univ. Press, Cambridge, 1999.
8. Jikov, V.V., Kozlov, S.M., Oleinik, O.A.: *Homogenization of Differential Operators and Integral Functionals.* Springer, Berlin, 1994.
9. Kohn, R.V., Vogelius, M.: Thin plates with varying thickness, and their relation to structural optimization. In: *Homogenization and Effective Moduli of Materials and Media.* (J.Ericksen et al. eds.) IMA Vol. 1, Springer, 1986, pp.126–149.
10. Lewiński, T., Telega, J.J.: *Plates, Laminates and Shells.* World Scientific, Singapore, 2000.
11. McConnell, W.H.: On the approximation of elliptic operators with discontinuous coefficients. *Ann. Scuola Norm. Sup. Pisa Cl. Sci. IV* **3** (1976), 121–137.
12. Muñoz, J., Pedregal, P.: On the relaxation of an optimal design problem for plates. *Asymptotic Anal.* **16** (1998), 125–140.
13. Murat, F.: H-convergence. In: *Seminaire d'Analyse Fonctionnelle et Numérique de l'Université d'Alger*, 1978.
14. Pedregal, P.: Optimization, relaxation and Young measures. *Bull. A.M.S.* **36** (1999), 27–58.
15. Roubíček, T.: *Relaxation in Optimization Theory and Variational Calculus.* W. de Gruyter, Berlin, 1997.
16. T.Roubíček: Maximum principle in optimal design of plates with stratified thickness. *Appl. Math. Optim.* (submitted)
17. Tartar, L.: Homogénéisation. Cours Peccot au Collège de France, Paris, 1977.
18. Tartar, L.: Remarks on the homogenization method in optimal design problems. In: *Homogenization and Applications to Material Sciences.* Gakuto Int. Ser. Math. Sci. Appl. **9**, Gakkokotosho, 1995, pp.393–412.
19. Tartar, L.: An introduction to the homogenization method in optimal design. CIME Summer Course, Troia, June 1998.
20. Warga, J.: *Optimal Control of Differential and Functional Equations.* Academic Press, New York, 1972.

Part II

Numerical Methods

Recent advances in conservation laws with discontinuous flux and clarifier-thickener models

Raimund Bürger[1] and Kenneth H. Karlsen[2]

[1] Institute of Mathematics A, University of Stuttgart, Stuttgart, Germany,
E-Mail: buerger@mathematik.uni-stuttgart.de
[2] Department of Mathematics, University of Bergen, Bergen, Norway,
E-Mail: kennethk@mi.uib.no

Abstract. One-dimensional continuous clarification-thickening processes of ideal suspensions can be described by scalar conservation laws with a flux which is discontinuous with respect to the space variables due to the feed mechanism and jumps of the vessel diameter. The authors provide a survey of extensions of the front tracking and the Engquist-Osher finite difference methods to such non-standard conservation laws. Recent analyses show that these methods can be employed for both proving existence of weak solutions and numerical simulations. The (newer) adaptation of the Engquist-Osher upwind scheme is illustrated by numerical examples.

1 Introduction

Kynch's [6] one-dimensional kinematic theory for the settling of ideal solid-liquid suspensions leads to the following conservation law for the solids concentration u as a function of depth x and time t:

$$u_t + \big(q(x,t)u + h(u)\big)_x = 0, \tag{1}$$

where $q(x,t)$ is the local bulk velocity of the mixture and the function h, which has to be determined experimentally [4], reflects the material properties of the suspension. The basic assumptions on h are $h(u) = 0$ for $u \leqslant 0$ and $u \geqslant 1$, $h(u) > 0$ for $0 < u < 1$, $h'(0) > 0$ and $h'(1) \leqslant 0$. We have chosen "1" as the maximum solids concentration and assume h to be sufficiently smooth.

The simple model for continuous sedimentation by Bustos et al. [8] restricts Eq. (1) to an interval, say $x \in [0,1]$, corresponding to a cylindrical vessel, the upper end $x = 0$ is identified with a feed inlet, and the lower $x = 1$ with a discharge outlet. The vessel is assumed to be fed continuously at the inlet and to be discharged continuously through the outlet. No overflow mechanism is provided. The velocity $q = q_{\mathrm{r}}(t)$ is a function of time only, where q_{r} is determined by the discharge control. In [8], Eq. (1) is studied with Dirichlet boundary conditions in an appropriate entropy boundary framework [9].

This model, which was proposed first by Petty [24], has some severe shortcomings. Among them is the lack of a global conservation principle.

It is preferable to replace the Dirichlet boundary conditions by transitions between the transport flux $q(x,t)u$ and the composite flux $q(x,t)u + h(u)$, which leads to a pure initial value problem. Moreover, the feed suspension should enter *between* the overflow outlet at the top and the discharge outlet at the bottom. This gives rise to an upwards-directed volume average velocity $q_l \leqslant 0$ above and a downwards-directed velocity $q_r \geqslant 0$ below the feed level. The feed source itself is modeled by a singular source term. Such clarifier-thickener configurations were proposed by several authors [1,10,23]. Thorough analyses of clarifier-thickener models were presented by Diehl [13].

We here consider vessels with (discontinuously) varying cross-sectional area $0 < S_{\min} \leqslant S(x) \leqslant S_{\max}$, which leads to the initial-value problem

$$u_t + \frac{g(u,x)_x}{S(x)} = 0,\ x \in \mathbf{R},\ t > 0; \quad u(x,0) = u_0(x) \in [0,1],\ x \in \mathbf{R};$$

$$g(u,x) := Q_{\mathrm{R}} u_{\mathrm{F}} + \begin{cases} Q_{\mathrm{L}}(u - u_{\mathrm{F}}) & \text{for } x < -1, \\ Q_{\mathrm{L}}(u - u_{\mathrm{F}}) + S(x)h(u) & \text{for } -1 < x < 0, \\ Q_{\mathrm{R}}(u - u_{\mathrm{F}}) + S(x)h(u) & \text{for } 0 < x < 1, \\ Q_{\mathrm{R}}(u - u_{\mathrm{F}}) & \text{for } x > 1, \end{cases} \tag{2}$$

where $Q_{\mathrm{R}} \geqslant 0$ and $Q_{\mathrm{F}} \geqslant 0$ are the (prescribed) volume rates of suspension leaving the thickening zone and being fed into the unit, respectively, $Q_{\mathrm{L}} = Q_{\mathrm{F}} - Q_{\mathrm{R}}$ is the resulting volume overflow rate, and u_{F} is the solids concentration of the feed suspension. The front tracking method applies to the case $S \equiv S_0 = \text{const.}$, where $Q_{\mathrm{L}} = Q_{\mathrm{L}}/S_0$ and $q_{\mathrm{R}} = Q_{\mathrm{R}}/S_0$.

2 Front tracking methods

The main idea behind by the front tracking method was introduced by Dafermos [12]. To illustrate it, consider the hyperbolic conservation law $u_t + \phi(u)_x = 0$ for $x \in \mathbf{R}$ and $t > 0$ with the initial condition $u(x,0) = u_0(x)$ for $x \in \mathbf{R}$, where u_0 is assumed to be piecewise constant and ϕ is a given flux function. Then the entropy solution can be constructed by a superposition of solutions of Riemann problems, i.e., solutions of the conservation law with initial data consisting of two constant states separated by a simple discontinuity. If ϕ is piecewise linear, each Riemann solution consists exclusively of constant states separated by shocks. When waves from neighboring Riemann problems interact, the interaction only involves constant states and therefore leads to new Riemann problems, and the construction can be continued forward in time. Thus, the method consists of *solving Riemann problems* and *tracking straight-line discontinuities*. In the general case, u_0 is approximated by a step function and ϕ by a piecewise linear function. This way rarefaction waves are approximated by a sequence of small shocks. See [20] for the history and variants of the method. In particular, Holden et al. [19] proved that the construction is well-defined and terminates in a finite number of

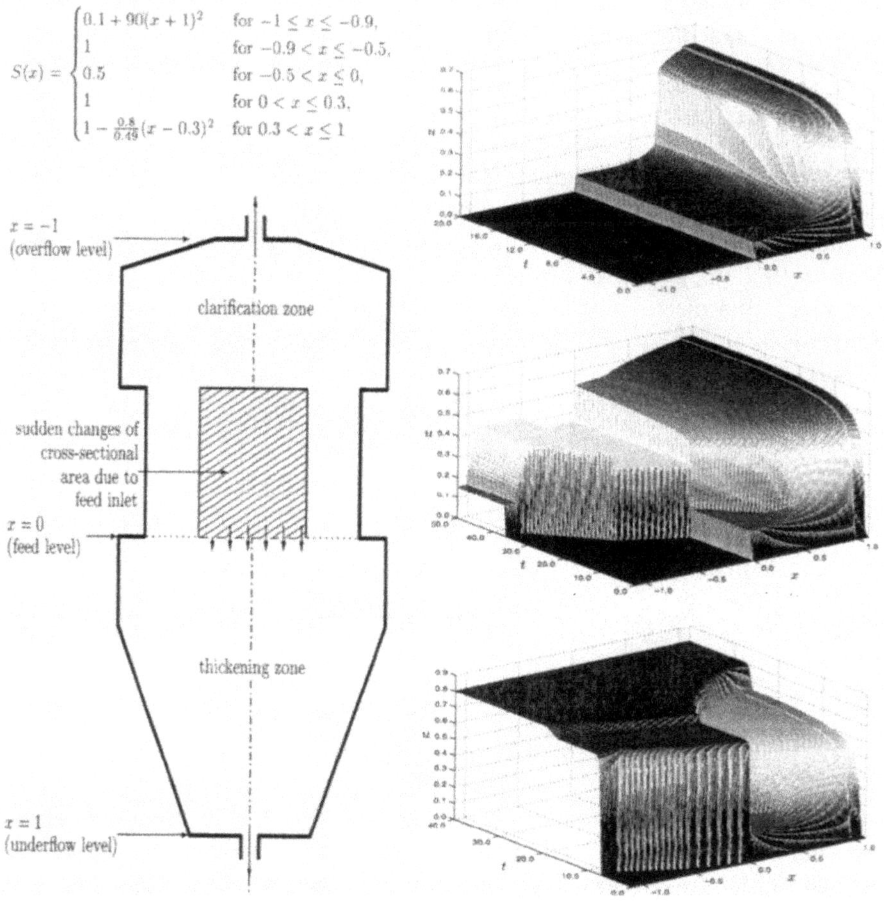

Fig. 1. A clarifier-thickener unit defined by the cross-sectional area function (top left) and simulations of continuous sedimentation processes with $u_F = 0.3$ (right column; top), $u_F = 0.4$ (middle) and $u_F = 0.8$ (bottom); colored, see page 373

steps, even for non-convex flux functions, given a finite number of constant states in $u_0(x)$. Front tracking was later formulated for hyperbolic systems by DiPerna [14], Bressan [2] and Risebro [25]. Variants of the front tracking technique based on [12] have also been applied earlier to batch sedimentation problems [7,22]. However, to apply front tracking to the clarifier-thickener model, one must be able to solve nonstandard Riemann problems involving changes between flux functions depending on x. Such solutions were constructed by Gimse and Risebro [16,17]. In [16] they showed that under some mild conditions, Riemann problems with a discontinuous flux function always have weak solutions. Furthermore, one can always single out a unique weak solution as being the limit of a viscous approximation. A convergence proof for the front tracking method for the clarifier-thickener model is given in [3].

3 The finite difference method

The finite difference algorithm a slight modification to the standard Engquist-Osher (EO) [15] upwind scheme in order to handle the spatial variation of the flux appearing in (2). The flux $g(x,u)$ is considered to depend on a pair $\gamma(x) := (\gamma^1(x), \gamma^2(x))$ of spatially varying parameters such that $g(x,u) = f(\gamma(x),u)$. We discretize the parameter vector γ on a spatial mesh that is staggered against that of u. The result is a scalar finite difference scheme in conservation form: if we choose $\Delta x > 0$, set $x_j := j\Delta x$, and discretize the parameter vector γ, the initial data, and the cross sectional area by

$$\gamma_{j+\frac{1}{2}} := \frac{1}{\Delta x}\int_{x_j}^{x_{j+1}} \gamma(x)dx,\ U_j^0 := \frac{1}{\Delta x}\int_{x_{j-\frac{1}{2}}}^{x_{j+\frac{1}{2}}} u_0(x)dx,\ S_j := \frac{1}{\Delta x}\int_{x_{j-\frac{1}{2}}}^{x_{j+\frac{1}{2}}} S(x)dx,$$

then the resulting scheme for the approximations U_j^n can be written as

$$U_j^{n+1} = U_j^n - \lambda_j \Delta_- f^{\mathrm{EO}}\left(\gamma_{j+\frac{1}{2}}, U_{j+1}^n, U_j^n\right),\ j \in \mathbf{Z},\ n = 1,2,3,\ldots; \tag{3}$$

where $\lambda_j := \Delta t/(S_j \Delta x)$, $\Delta_- V_j := V_j - V_{j-1}$, and the EO flux is

$$f^{\mathrm{EO}}(\gamma, v, u) := \frac{1}{2}\left[f(\gamma,u) + f(\gamma,v) - \int_u^v |f_u(\gamma,w)|dw\right]. \tag{4}$$

By this staggering, we avoid solving the full 2×2 Riemann problem that arises at each cell boundary when the two discretizations are aligned. With our choice of the EO flux (4), the resulting algorithm is a so-called upwind scheme, meaning that the differencing of the flux is biased in the direction of incoming waves, making it possible to resolve shocks without excessive smearing. The choice of the EO flux is also motivated by its close functional relationship to the so-called Kružkov entropy flux $F(\gamma,u) = \operatorname{sgn}(u-c)(f(\gamma,u) - f(\gamma,c))$ and the Temple functional, which enables us to prove convergence of the method. For the constant parameter conservation law, the EO algorithm is well known to converge to the correct entropy solution. In that setting, the classical theory of monotone schemes [11,18] is applicable. In particular, the numerical approximations converge to the unique entropy solution as $\Delta x \Delta t \to 0$, as long as a suitable CFL condition is enforced. For the case of single discontinuous parameter the modified EO scheme was proven to converge for a concave flux in [26] and for a flux with a finite number of critical points in [27]. Karlsen et al. [21] extended the scheme to degenerate parabolic equations with discontinuities in the convective flux, and proved convergence to a weak solution. In fact, in [5] we prove the following theorem.

Theorem 1. *Let the approximate solution u^Δ be given by the scheme* (3). *Let $\Delta = (\Delta t, \Delta x) \to 0$ with $\Delta t/\Delta x = const.$ and assume that the CFL condition $\lambda_j|\max\{f_u(\gamma_{j+\frac{1}{2}}, U_j^n)\} - \min\{f_u(\gamma_{j-\frac{1}{2}}, U_j^n)\}| \leqslant 1$ for all $j \in \mathbf{Z}$ is satisfied. Assume that $\mathrm{TV}(f(\gamma, u_0(\cdot))) < \infty$. Then there exists a subsequence of $\{u^\Delta\}$, still denoted by $\{u^\Delta\}$, and a function u such that $u^\Delta \to u$ in $L^1_{\mathrm{loc}}(\mathbf{R} \times (0,T))$ and a.e. in $\mathbf{R} \times (0,T)$. The limit function u is a weak solution of problem* (2).

4 Numerical example

Consider a clarifier-thickener defined by the cross-sectional area function $S(x)$ given in the upper left corner of Figure 1 and drawn below the formula. We take the typical function $h(u) = u(1-u)^5$, $Q_{\mathrm{R}} = 0.3 = -Q_{\mathrm{L}} = 0.1$, $\Delta x = 1/100$ and $u_{\mathrm{F}} \in \{0.3, 0.4, 0.8\}$. Observe that for u_{F}, continuous sedimentation is restricted to the thickening zone; for $u_{\mathrm{F}} = 0.4$, the sediment breaks into the clarification zone and the system attains a steady state of simultaneous thickening and clarification; and for $u_{\mathrm{F}} = 0.8$, the system is overloaded and after a while no longer able to handle the feed flux (hydraulic forcing).

References

1. Barton, N.G., Li, C.H. and Spencer, S.J., 'Control of a surface of discontinuity in continuous thickeners', *J. Austral. Math. Soc. Ser. B* **33** (1992), 269–289.
2. Bressan, A., 'Global solutions to systems of conservation laws by wave-front tracking', *J. Math. Anal. Appl.* **170** (1992), 414–432.
3. Bürger, R., Karlsen, K.H., Klingenberg, C. and Risebro, N.H., 'A front tracking approach to a model of continuous sedimentation in ideal clarifier-thickener units', *Nonlin. Anal. Real World Appl.*, **4** (2003), 457–481.
4. Bürger, R., Karlsen, K.H., Risebro, N.H. and Towers, J.D., 'Numerical methods for the simulation of continuous sedimentation in ideal clarifier-thickener units', *Int. J. Mineral Process.*, to appear.
5. Bürger, R., Karlsen, K.H., Risebro, N.H. and Towers, J.D., 'Monotone difference approximations for the simulation of clarifier-thickener units', submitted to *Comput. Visual. Sci.* (2002).
6. Bürger, R. and Wendland, W.L., 'Sedimentation and suspension flows: Historical perspective and some recent developments', *J. Eng. Math.* **41** (2001), 101–116.
7. Bustos, M.C. and Concha, F., 'Settling velocities of particulate systems: 10. A numerical method for solving Kynch sedimentation processes', *Int. J. Mineral Process.* **57** (1999), 185–203.
8. Bustos, M.C., Concha, F. and Wendland, W.L., 'Global weak solutions to the problem of continuous sedimentation of an ideal suspension', *Math. Meth. Appl. Sci.* **13** (1990), 1–22.
9. Bustos, M.C., Paiva, F. and Wendland, W.L., 'Entropy boundary conditions in the theory of sedimentation of ideal suspensions', *Math. Meth. Appl. Sci.* **19** (1996), 679–697.

10. Chancelier, J.P., Cohen de Lara, M. and Pacard, F., 'Analysis of a conservation PDE with discontinuous flux: a model of settler', *SIAM J. Appl. Math.* **54** (1994), 954–995.
11. Crandall, M.G. and Majda, A., 'Monotone difference approximations for scalar conservation laws', *Math. Comp.* **34** (1980), 1–21.
12. Dafermos, C.M., 'Polygonal approximations of solutions to the initial value problem for a conservation law', *J. Math. Anal. Appl.* **38** (1972), 33–41.
13. Diehl, S., 'Operating charts for continuous sedimentation I: Control of steady states', *J. Eng. Math.* **41** (2001), 117–144.
14. DiPerna, R.J., 'Global existence of solutions to nonlinear systems of conservation laws', *J. Diff. Eqns.* **20** (1976), 187–212.
15. Engquist, B. and Osher, S., 'Stable and entropy-satisfying approximations for transonic flow calculations', *Math. Comp.* **34** (1980), 45–75.
16. Gimse, T. and Risebro, N.H., 'Riemann problems with a discontinuous lux function', Proc. 3rd Int. Conf. Hyp. Problems, Uppsala, Sweden, 1990.
17. Gimse, T. and Risebro, N.H., 'Solution of the Cauchy problem for a conservation law with a discontinuous flux function', *SIAM J. Math. Anal.* **23** (1992), 635–648.
18. Harten, A., Hyman, J.M. and Lax, P.D., 'On finite difference approximations and entropy conditions for shocks', *Comm. Pure Appl. Math.* **29** (1976), 297–322.
19. Holden, H., Holden, L. and Høegh-Krohn, R., 'A numerical method for first order nonlinear scalar conservation laws in one dimension', *Comp. Math. Appl.* **15** (1988), 595–602.
20. Holden, H. and Risebro, N.H., *Front Tracking for Hyperbolic Conservation Laws*, Springer-Verlag, New York, 2002.
21. Karlsen, K.H., Risebro, N.H. and Towers, J.D., 'On an upwind difference scheme for degenerate parabolic convection-diffusion equations with a discontinuous coefficient', *IMA J. Numer. Anal.* **22** (2002), 623–664.
22. Kunik, M., 'A numerical method for some initial-value problems of one scalar hyperbolic conservation law', *Math. Meth. Appl. Sci.* **15** (1992), 495–509.
23. Lev, O., Rubin, E. and Sheintuch, M., 'Steady state analysis of a continuous clarifier-thickener system', *AIChE J.* **32** (1986), 1516–1525.
24. Petty, C.A., 'Continuous sedimentation of a suspension with a nonconvex flux law', *Chem. Eng. Sci.* **30** (1975), 1451–1458.
25. Risebro, N.H., 'A front-tracking alternative to the random choice method', *Proc. Amer. Math. Soc.* **117** (1993), 1125–1139.
26. Towers, J.D., 'Convergence of a difference scheme for conservation laws with a discontinuous flux', *SIAM J. Numer. Anal.* **38** (2000), 681–698.
27. Towers, J.D., 'A difference scheme for conservation laws with a discontinuous flux—the nonconvex case', *SIAM J. Numer. Anal.* **39** (2001), 1197–1218.

Artificial outlet boundary conditions for steady flows with rotational symmetry

Cristian A. Coclici[1], Ivan L. Sofronov[2], and Wolfgang L. Wendland[3]

[1] Department of Mathematics, University of Kaiserslautern
[2] Keldysh Institute of Applied Mathematics, RAS, Moscow
[3] Mathematical Institute A, University of Stuttgart

Abstract. We present a construction of far–field boundary conditions on the outflow cross-section for axisymmetric two–dimensional compressible flows in a pipe. An unbounded downstream domain is considered, where the original subsonic flow is modelled via steady compressible Euler equations linearized about its mean value. The resulting system for the perturbations leads in particular to an exterior elliptic problem for the pressure. Its solution, calculated by means of Fourier analysis, provides the Dirichlet–to–Neumann map at the open boundary defining artificial boundary conditions in the form of the Steklov–Poincaré operator. We present a method for the approximation of this operator and exemplify our procedure for the case of a compressible plasma flow inside a self–field magnetoplasmadynamic accelerator.

1 Introduction. Motivating CFD–problem

We develop artificial outlet far–field boundary conditions for the numerical treatment of two–dimensional compressible flows with rotational symmetry. The problem under consideration is motivated by a complex practical problem treated in [1–3] and related to compressible viscous plasma flows inside self–field magnetoplasmadynamic (MPD) accelerators. For a complete description of the principle of MPD–thrusters we refer the reader to [4]. Our assumptions in the formulation are related to the local physical properties of the considered flow field: the flow is subsonic and inviscid downstream the outflow boundary.

The original 3D flow–field domain is axisymmetric, and we denote a plane section through the axis by $\Omega \subset \mathbb{R}^2$. The high–enthalpy magneto–plasma flow is supposed axisymmetric, and is modelled by a 2D system of conservation laws *extended* by nonlinear equations describing the flow of the electron component in the plasma, and the electromagnetic field. In order to treat the rather complex original problem efficiently, we decompose the flow–field domain into disjoint model zones, characterized by different physical properties of the flow (Fig. 1). In [1–3], the complete system of the extended compressible Navier–Stokes equations has been employed only within an essentially smaller near–field region $\Omega_1 \subset \Omega$ containing the thruster, being coupled via appropriate transmission conditions across the artificial interface Γ to different simplified models in the complementary far–field region $\Omega \setminus \Omega_1$ which corresponds to a test tank Ω_2 in the laboratory.

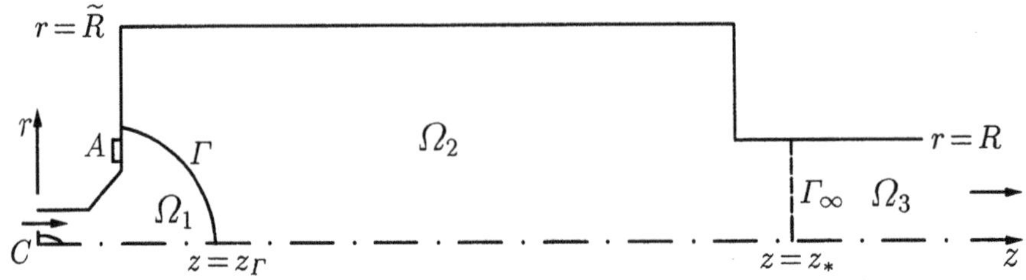

Fig. 1. Decomposition of the computational domain

In these heterogeneous domain decompositions, the flow in the vicinity of the outflow boundary $\Gamma_\infty := \{z_*\} \times [0, R]$ has been modelled by the compressible Euler equations extended for the influence of the electron field; here, $z_* > z_\Gamma$. For the subsonic outflow, *characteristic* boundary conditions with a prescribed pressure at Γ_∞ have originally been employed.

Since the values used for the pressure are obtained from experiments, it becomes desirable to improve the boundary conditions at Γ_∞. Therefore, we consider an infinite continuation of Ω_2, defined by $\Omega_3 := (z_*, \infty) \times (0, R)$, which corresponds to an infinite cylindrical continuation of the tank at the far right. We suppose that also the electron field is negligible in Ω_3, and we start with the classical Euler equations for describing the flow there. The simplified linear model of inviscid subsonic flow shall be used for providing artificial far–field boundary conditions at the outflow boundary Γ_∞.

2 Linearized Euler equations

In [1–3], the steady–state solution of the model systems (extended NS in Ω_1, simplified conservation laws in Ω_2) is obtained iteratively by an alternating method via time stabilization. We employ here a *steady* model of linear Euler equations in Ω_3, which shall be used at every iteration in order to update the values of the flow quantities on Γ_∞. From several related results on the coupling with linear Euler equations, we mention here [5,6] (channel problem), [7,8] (airfoil problem), [9] (extention to 3D case).

In what follows, we use cylindrical coordinates $(z, \cdot, r)$ and collect in $V(z,r) := (\rho, u, v, p)(z,r)$ the primitive variables with the density ρ, the longitudinal velocity u, the radial velocity v, and the pressure p. We begin our analysis in Ω_3 with the steady Euler equations written in the form

$$\begin{aligned} &\rho\Big(\frac{\partial u}{\partial z} + \frac{\partial v}{\partial r} + \frac{v}{r}\Big) + u\frac{\partial \rho}{\partial z} + v\frac{\partial \rho}{\partial r} = 0, \quad \rho u\frac{\partial u}{\partial z} + \rho v\frac{\partial u}{\partial r} + \frac{\partial p}{\partial z} = 0, \\ &\rho u\frac{\partial v}{\partial z} + \rho v\frac{\partial v}{\partial r} + \frac{\partial p}{\partial r} = 0, \quad \kappa\, p\left(\frac{\partial u}{\partial z} + \frac{\partial v}{\partial r} + \frac{v}{r}\right) + u\frac{\partial p}{\partial z} + v\frac{\partial p}{\partial r} = 0. \end{aligned} \tag{1}$$

Our basic assumption is that the flow field in Ω_3 admits a representation in the form of small variations about a variable state $V_0(r) := (\rho_0, u_0, v_0, p_0)(r)$

which depends on r but *not* on z. Assuming $\rho_0, u_0, v_0, p_0 \in C^1[0,R]$ with $\rho_0(r), u_0(r), p_0(r) > 0$ for $r \in [0,R]$, and setting $\frac{\partial}{\partial z}\{\rho, u, v, p\} = 0$ in (1), we get $V_0(r) = (\rho_0(r),\ u_0(r),\ 0,\ p_0)$ for $r \in [0,R]$ where $\rho_0(\cdot)$, $u_0(\cdot)$ and $p_0 > 0$ have to be taken from an additional information on the mean flow in Ω_3. Writing now the primitive variables in the form

$$\rho=\rho_0(r)+\widetilde{\rho}_0(z,r),\ u=u_0(r)+\widetilde{u}_0(z,r),\ v=\widetilde{v}_0(z,r),\ p=p_0+\widetilde{p}_0(z,r),\ (z,r)\in\Omega_3,$$

we obtain for the perturbations $\tilde{V}(z,r) := (\tilde{\rho}, \tilde{u}, \tilde{v}, \tilde{p})(z,r)$ the linear system

$$\rho_0(r)\left[\frac{\partial\widetilde{u}}{\partial z}+\frac{\partial\widetilde{v}}{\partial r}+\frac{\widetilde{v}}{r}\right]+u_0(r)\frac{\partial\widetilde{\rho}}{\partial z}+\rho_0'(r)\widetilde{v}=0,\ (\rho_0u_0)(r)\frac{\partial\widetilde{u}}{\partial z}+(\rho_0u_0')(r)\widetilde{v}+\frac{\partial\widetilde{p}}{\partial z}=0,$$

$$(\rho_0u_0)(r)\frac{\partial\widetilde{v}}{\partial z}+\frac{\partial\widetilde{p}}{\partial r}=0,\quad \kappa\, p_0\left[\frac{\partial\widetilde{u}}{\partial z}+\frac{\partial\widetilde{v}}{\partial r}+\frac{\widetilde{v}}{r}\right]+u_0(r)\frac{\partial\widetilde{p}}{\partial z}=0\quad\text{in}\quad\Omega_3. \tag{2}$$

Difficulties in dealing with (2) in comparison with the planar case [6] consist of the variable coefficients $V_0(r)$, together with the use of cylindrical coordinates.

3 Steklov–Poincaré operator. Numerical treatment

We denote by $c_0(r) := \sqrt{\kappa p_0/\rho_0(r)}$ and $M_0(r) := u_0(r)/c_0(r)$ the *local speed of sound* and the *Mach number* corresponding to the mean flow $V_0(r)$. By using partial differentiation and eliminating $\widetilde{u}$, $\widetilde{v}$ and their partial derivatives in $(2)_{2,3,4}$, we obtain the scalar PDE for the pressure perturbation $\widetilde{p}(z,r)$:

$$\frac{\partial^2\widetilde{p}}{\partial z^2}+\frac{1}{1-M_0^2(r)}\frac{\partial^2\widetilde{p}}{\partial r^2}+\left[\frac{1}{r}-2\frac{M_0'(r)}{M_0(r)}\right]\cdot\frac{1}{1-M_0^2(r)}\frac{\partial\widetilde{p}}{\partial r}=0\quad\text{in}\quad\Omega_3. \tag{3}$$

Note that the coefficients of this equation depend only on $M_0(r)$. Due to the axisymmetry of the flow and to the choice of the boundary $\{r=R\}$, we have the following boundary conditions for $\widetilde{p}$ with respect to r:

$$\left.\frac{\partial\widetilde{p}}{\partial r}\right|_{r=0}=0,\qquad \left.\frac{\partial\widetilde{p}}{\partial r}\right|_{r=R}=0\quad\text{for all}\quad z\geqslant z_*. \tag{4}$$

We require the natural condition of non–perturbed background at infinity:

$$\lim_{z\to\infty}\widetilde{p}_0(z,r)=\text{const.}\quad\text{for all}\quad r\in[0,R]. \tag{5}$$

The divergence form of the *elliptic* equation (3) reads

$$\frac{\partial^2\widetilde{p}}{\partial z^2}+\frac{1}{1-M_0^2(r)}\frac{M_0^2(r)}{r}\frac{\partial}{\partial r}\Big(\frac{r}{M_0^2(r)}\frac{\partial\widetilde{p}}{\partial r}\Big)=0\quad\text{in}\quad\Omega_3, \tag{6}$$

and motivates us to define the operator $\mathcal{L} : C^2[0,R] \to C[0,R]$ by

$$\mathcal{L}f(r):=\frac{1}{1-M_0^2(r)}\frac{M_0^2(r)}{r}\frac{d}{dr}\Big(\frac{r}{M_0^2(r)}\frac{df}{dr}\Big)\quad\text{for}\quad f\in C^2[0,R]. \tag{7}$$

We consider the space $\mathcal{F} := \{ f \in C^2[0,R] : f'(0)=0, f'(R)=0 \}$, and endow it with the scalar product

$$(f,g)_M := \int_0^R r \, \frac{1-M_0^2(r)}{M_0^2(r)} \, f(r)\, g(r)\, dr.$$

Simple calculations show that $(\mathcal{F}, (\cdot,\cdot)_M)$ is a Hilbert space, and that

1. *The operator $\mathcal{L}$ is self-adjoint: $(\mathcal{L}f,g)_M = (f,\mathcal{L}g)_M$ for all $f,g \in \mathcal{F}$.*
2. *For any non-constant function $f \in \mathcal{F}$, there holds: $-(\mathcal{L}f,f)_M > 0$.*

Consequently, the Sturm–Liouville problems

$$(P_k): \quad \begin{cases} \mathcal{L}\varphi_k + \lambda_k^2 \varphi_k = 0 \quad \text{in} \quad (0,R), \\ \varphi_k'(0) = 0, \quad \varphi_k'(R) = 0 \end{cases} \qquad (k = 1,2,\ldots) \tag{8}$$

have real eigenvalues λ_k^2, and the set of solutions $\{\varphi_k\}_{k=1}^{\infty}$ with the norm conditions $\|\varphi_k\|_M = 1$ determine an orthonormal basis of $\mathcal{F}$.

We treat the boundary value problem (3)–(5) by using the Fourier method. Denote by Q the operator mapping f into its Fourier coefficients,

$$Qf := \{a_k\}_{k=1}^{\infty} \quad \text{with} \quad a_k := (f,\varphi_k)_M, \; k = 1,2,\ldots, \tag{9}$$

and consider the inverse operator Q^{-1} defined by

$$Q^{-1}\Big(\{a_k\}_{k=1}^{\infty}\Big)(r) := \sum_{k=1}^{\infty} a_k \varphi_k(r) \quad \text{for} \quad r \in [0,R]. \tag{10}$$

Taking now a Fourier–harmonic term $\widetilde{p}_k(z)\varphi_k(r)$, where $\widetilde{p}_k(z) := (\widetilde{p},\varphi_k)_M$, using (8) and substituting into (6), we obtain

$$\widetilde{p}_k''(z) - \lambda_k^2 \widetilde{p}_k(z) = 0 \quad \text{for} \quad z \in (z_*, \infty). \tag{11}$$

Due to (5), the solution of (11) is given by $\widetilde{p}_k(z) = C_k e^{-\lambda_k z}$ for $z \geqslant z_*$ with $\lambda_k \geqslant 0$ and with $C_k \in \mathbb{R}$, and, consequently, we find the relation

$$\frac{\partial^2 \widetilde{p}_l}{\partial z^2} + \lambda_k \widetilde{p}_k = 0 \quad \text{at} \quad z = z_*. \tag{12}$$

Using the operators Q and Q^{-1}, and also the matrix $\Lambda := \operatorname{diag}\{\lambda_1, \lambda_2, \ldots\}$, we obtain from (12) the Steklov–Poincaré operator for the problem (3)–(5):

$$\frac{\partial \widetilde{p}}{\partial z}(z_*,r) = -\big[Q^{-1}\Lambda\, Q\big](\widetilde{p})\,(z_*,r) \quad \text{for all} \quad r \in [0,R]. \tag{13}$$

Equation (13) is our desired *exact* artificial boundary condition for (3).

A first step in the **numerical approximation** of the Steklov–Poincaré operator (13) is to treat the Sturm–Liouville problem (8) numerically, cf. [10].

We choose $J \in \mathbb{N}$ and introduce on $[0, R]$ the uniform grid $r_j := (j-1/2)h$ with $h = R/J$, $j = \overline{1, J}$. The operator $\mathcal{L}$ is approximated by central differences: For a given function f, we denote $f_j := f(r_j)$ $(j = 1, \ldots, J)$ and define

$$(\mathcal{L}^h f)_j := \frac{M_0^2(r_j)}{r_j(1-M_0^2(r_j))}\left[\frac{r_j+\frac{h}{2}}{M_0^2(r_j+\frac{h}{2})}\,\frac{f_{j+1}-f_j}{h} - \frac{r_j-\frac{h}{2}}{M_0^2(r_j-\frac{h}{2})}\,\frac{f_j-f_{j-1}}{h}\right]\frac{1}{h}$$

for $j = 2, \ldots, J-1$. For $j \in \{1, J\}$, we construct $(\mathcal{L}^h f)_j$ by using the boundary conditions $(8)_2$ in the form $(f_1 - f_0)/h = 0$ and $(f_{J+1} - f_J)/h = 0$.

With the diagonal $J \times J$ matrix

$$\mathcal{D} := \operatorname{diag}\,\{\, d_{j,j} \,\}, \quad d_{j,j} := r_j\left[1 - M_0^2(r_j)\right] / \, M_0^2(r_j) \text{ for } j = 1, \ldots, J,$$

and the symmetric $J \times J$ matrix $\mathcal{A}^h := \{ a_{j-1,j},\ a_{j,j},\ a_{j,j+1} \}$ defined by

$$a_{j-1,j} := \frac{1}{h^2}\frac{r_j - \frac{h}{2}}{M_0^2(r_j - \frac{h}{2})}\ (j = 2, \ldots, J),\ a_{j+1,j} := a_{j,j+1} (j = 1, \ldots, J-1),$$

$a_{j,j} := -\left[a_{j-1,j} + a_{j,j+1}\right]$ $(j = 2, .., J-1)$, $a_{1,1} := -a_{1,2}$ and $a_{J,J} := -a_{J-1,J}$, the matrix of the discrete operator $\mathcal{L}^h$ is $\mathcal{D}^{-1}\mathcal{A}^h$. The eigenvalues λ_k^h and the corresponding eigenfunctions $\varphi_k^h \in \mathbb{R}^J$, $k = 1, \ldots, J$ for this tridiagonal matrix are obtained numerically. Note that the basis depends on the distribution of $M_0(r)$ on Γ_∞.

A second step consists in the approximation of the operators Q, Q^{-1} and Λ. Denoting by $0 = \rho_1 < \rho_2 < \ldots < \rho_n = R$ the mesh on $\Gamma_\infty = \partial\Omega_2 \cap \partial\Omega_3$ coming from Ω_2, we introduce the corresponding piece-wise linear, continuous hat functions $\{w_i(r) | \, i = 1, \ldots, n\}$ satisfying $w_i(\rho_\ell) = \delta_{i\ell}$ for $1 \leqslant i, \ell \leqslant n$. With $\widetilde{p}(z_*, r) = \sum_{i=1}^n \widetilde{p}(z_*, \rho_i)\, w_i(r)$ for $r \in [0, R]$, we have

$$\left(\widetilde{p}(z_*, r), \varphi_k(r)\right)_M = \sum_{i=1}^n \left(w_i, \varphi_k\right)_M \widetilde{p}(z_*, \rho_i),$$

hence the discrete form of the operator Q is given by the matrix

$$\mathbf{M}_Q := \left(\ (w_i, \varphi_k)_M\ \right) \in \mathcal{M}^{J\times n}(\mathbb{R}).$$

We take $\mathbf{M}_\Lambda := \operatorname{diag}\{\lambda_1^h, \ldots, \lambda_J^h\} \in \mathcal{M}^{J\times J}(\mathbb{R})$. Finally, with the matrix

$$\mathbf{M}_{Q^{-1}} := \left(\ \varphi_k(\rho_\ell)\ \right) \in \mathcal{M}^{n\times J}(\mathbb{R}),$$

we obtain the **discrete form of the Steklov–Poincaré operator** (13):

$$\left(\frac{\partial \widetilde{p}}{\partial z}(z_*, \rho_\ell)\right)_{\ell=1,\ldots,n} = -\left[\mathbf{M}_{Q^{-1}}\, \mathbf{M}_\Lambda\, \mathbf{M}_Q\right]\left(\widetilde{p}(z_*, \rho_i)\right)_{i=1,\ldots,n}, \tag{14}$$

or component-wise (note that J is chosen so that $J \ll n$)

$$\frac{\partial \widetilde{p}}{\partial z}(z_*, \rho_\ell) = -\sum_{k=1}^J \varphi_k(\rho_\ell) \cdot \left[\lambda_k \sum_{i=1}^n (w_i, \varphi_k)_M\, \widetilde{p}(z_*, \rho_i)\right] \quad \text{for} \quad \ell = 1, \ldots, n.$$

4 Artificial far–field boundary conditions

Applying the central difference representation for $\frac{\partial \widetilde{p}}{\partial z}$ on Γ_∞ and inserting the values calculated via the equation (14), we update the perturbed pressure $\widetilde{p}$ on Γ_∞, taking herewith into account the influence of the linear model in the far field Ω_3. Let us describe how to update the remaining variables $\widetilde{v}$, $\widetilde{u}$, $\widetilde{\rho}$. By using the differentiation along the interface, we obtain $\frac{\partial \widetilde{p}}{\partial r}$ on Γ_∞ and herewith, by using $(2)_3$, the discrete values of $\frac{\partial \widetilde{v}}{\partial z}$ on Γ_∞. Using the same procedure as for $\widetilde{p}$, we determine the perturbation $\widetilde{v}$ on Γ_∞. We further use the relation $(2)_2$ to determine $\frac{\partial \widetilde{u}}{\partial z}$, and, hence, $\widetilde{u}$ on the interface. Finally, to define $\widetilde{\rho}$ on Γ_∞, the relations $(2)_1$ and $(2)_4$ are combined. Returning to the original variables, the updated values can then be used in the next step of the iterative time–stepping procedure in $\Omega_1 \cup \Omega_2$.

Acknowledgement. The work of C.A. Coclici has been supported by the DFG Project We 659/35–1,2. The work of I.L. Sofronov has been supported by the Programme AMIF of the ESF, SFB-404 at the University of Stuttgart, and grant RFBR No. 01–01–00520.

References

1. Coclici, C.A., Heiermann, J., Auweter–Kurtz, M. and Wendland, W.L. (2000) A heterogeneous domain decomposition for initial– boundary value problems with conservation laws and electromagnetic fields. In: Domain Decomposition Methods in Science and Engineering (Chan, T. ed.), ddm.org, Japan, 281-288
2. Auweter–Kurtz, M., Coclici, C.A., Heiermann, J. and Wendland, W.L. (2001) Heterogeneous domain decomposition methods for compressible magneto–plasma flows. In: Hyperbolic Problems. Theory, Numerics, Applications, ISNM **140** (Freistühler, H., Warnecke, G. eds.), Birkhäuser–Verlag Basel, 89–98
3. Coclici, C.A., Wendland, W.L., Heiermann, J. und Auweter–Kurtz, M.(2002) Artificial boundary conditions for compressible Navier–Stokes equations with electromagnetic fields". Comput. Visual. Sci. **4**, Vol. 3, 157–165
4. Heiermann, J. (2002) Ein Finite–Volumen–Verfahren zur Lösung magnetoplasmadynamischer Erhaltungsgleichungen. Doctoral Thesis (in German), Institute of Space Systems, University of Stuttgart
5. Gustafsson, B.(1982) The choice of numerical boundary conditions for hyperbolic systems. J. Comput. Phys. **48**, 270–283
6. Ferm, L., Gustafsson, B. (1982) A downstream boundary procedure for the Euler equations. Computers and Fluids **10**, 261–276
7. Ferm, L. (1990) Open boundary condition for external flow problems. J. Comput. Phys. **91**, 55–70
8. Coclici, C.A., Sofronov, I.L., Wendland, W.L. (1996) A domain decomposition method and far–field boundary conditions for 2D transonic flow around an airfoil. Pitman Research Notes in Mathematical Series, **379**, p. 58–63
9. Sofronov, I.L., Wendland, W.L. (2001) Exact linear far–field conditions for three–dimensional aerodynamic stationary transonic flows. J. Comp. Appl. Mathematics **136**, 317–335
10. Ryaben'kii, V.S., Sofronov I.L. (1983) Difference spherical functions. Preprint of the Keldysh Inst. Appl. Maths. **75**

Modelling chemical vapour infiltration of pyrolytic carbon

Tom-Alexander Langhoff and Eckart Schnack

Institute of Solid Mechanics
Karlsruhe University, Kaiserstrasse 12,
D-76128 Karlsruhe, Germany

Abstract. The isothermal and isobaric chemical vapour infiltration (CVI) process of pyrolytic carbon for producing carbon reinforced carbon fibres is investigated yielding a moving boundary problem to be solved. The growing surface of the substrate constitutes the free boundary due to steady deposition of solid pyrolytic carbon. A one-dimensional model for studying the infiltration of a single cylindrical pore is set up containing a simplified model of chemical kinetics as well as binary and Knudsen diffusion within the gas phase. This single pore model is integrated in a CVI reactor model additionally describing convection over a plane, porous substrate. Results of simulations reported here provide important insight in strategies to achieve a complete infiltration of a porous substrate.

1 Introduction

An important industrial process for producing carbon reinforced carbon fibres (CFC) is chemical vapour infiltration (CVI). Complex interactions of homogeneous gas phase reactions and heterogeneous surface reactions result in the deposition of solid pyrolytic carbon on the surface of the given porous substrate. Additionally, the deposition of carbon on the surface is significantly influenced by transport processes convection and diffusion [1].

It is the evolution in time and space of the interface between the gas phase and the solid phase that is searched for yielding a moving boundary problem to be solved in modeling CVI. The change of geometry of the substrate surface and also of the region of the gas phase due to deposition of solid material is the crucial point since the spatial distribution of the deposited pyrolytic carbon strongly influences the properties of the composite being produced.

In many existing CVI models, an (over-)simplified description of the reaction kinetics with a single reaction is used [1–4]. The consideration of the whole hydrocarbon chemistry with many hundreds of species represents a real challenge for the modeling procedure of CVI of pyrolytic carbon. A more detailed treatment of modelling CVI of pyrolytic carbon can be found in [5].

In the next chapter, the mathematical model for the infiltration of a single cylindrical pore will be presented. Relying on this model, a CVI reactor model dealing with plane substrates containing several cylindrical model pores is subsequently discussed. The numerical implementation is briefly described. The results of the numerical simulations are finally presented.

2 Mathematical Model

2.1 Single pore model

Assuming ideal cylindrical symmetry of a single pore with initial diameter d_{p}^{0} and fixed length l_0 as shown in Fig. 1, the evolution of the moving boundary $\Gamma(t)$ between the gas phase $\Omega(t)$ and the solid phase $\tilde{\Omega}(t)$ can be studied using a one-dimensional model. The open pore mouth constitutes a fixed

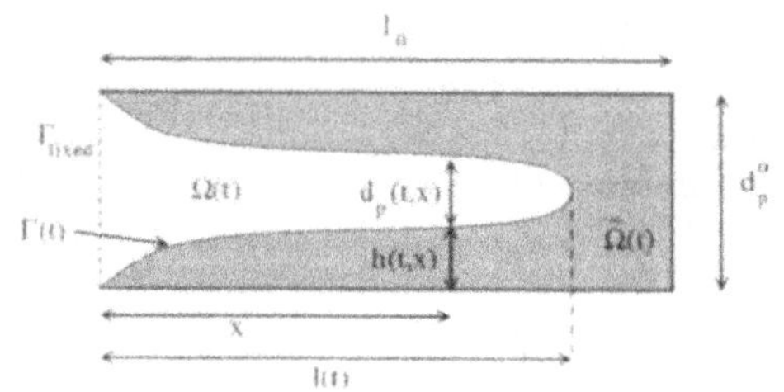

Fig. 1. Schematic view of a single cylindrical pore

boundary Γ_{fixed} whereas the moving boundary $\Gamma(t)$ is determined by the spatially varying pore diameter $d_{\mathrm{p}}(t)\, undx$. Working with N_{gs} different chemical gas phase species and neglecting convection within the pore, the resulting non-linear diffusion-reaction-system for their concentrations $c_i(t)\, undx$, $i = 1, \ldots, N_{\mathrm{gs}}$ for $(t,x) \in [0; t_{\max}] \times \Omega(t)$ reads

$$\frac{\partial}{\partial t} c_i(t,x) = \frac{\partial}{\partial x}\left[D_i(t,x)\frac{\partial}{\partial x} c_i(t)\, undx\right] + r_i^{\mathrm{gas}}(t,x,\mathbf{c}) \tag{1}$$

with $\mathbf{r}^{\mathrm{gas}}$ denoting the vector of reaction rates of the gas phase reactions, $t_{\max}$ the maximal duration of the infiltration process and $\Omega(t)$ denoting the domain of the gas phase within the pore:

$$\Omega(t) \;=\; \{x \in \mathbb{R} : \; 0 \leqslant x \leqslant l(t)\}\,, \qquad t \in (0; t_{\max}]\,. \tag{2}$$

Here, $l(t)$ denotes the actual depth of the pore for $t \in (0; t_{\max}]$ measured from the pore mouth, being restricted by the first point where the pore closes:

$$l(t) \;=\; \begin{cases} l_0 & \forall x \in [0; l_0] : 2h(t,x) < d_{\mathrm{p}}^{0} \\ \min\left\{x \in [0; l_0] : \; 2h(t,x) = d_{\mathrm{p}}^{0}\right\} & \text{otherwise} \end{cases} \tag{3}$$

where $h(t,x)$ represents the height of the pyrolytic carbon layer that has been already deposited. The normal growth velocity v_Γ of this layer is given by

$$v_\Gamma(t,x) \;=\; \frac{M_{\mathrm{C}_\infty}}{\varrho_{\mathrm{C}_\infty}}\,\frac{d_p(t,x)}{4}\, r_{\mathrm{C}_\infty}(t,x)\,, \quad (t,x) \in [0; t_{\max}] \times \Omega(t) \tag{4}$$

with r_{C_∞} denoting the reaction rate of the formation of solid pyrolytic carbon and M_{C_∞} and ϱ_{C_∞} denoting the molecular mass and density of solid pyrolytic carbon, respectively.

Binary diffusion between each gas phase species and methane as well as Knudsen diffusion for each gas phase species are taken into account. The matrix $\mathbf{D}$ of the diffusion coefficients then has diagonal structure and the elements are given by effective diffusion coefficients for each species combining the values of the binary and the Knudsen diffusion coefficient:

$$D_i(t,x)^{-1} = D_i^{\text{bin}^{-1}} + D_i^{\text{K}}(t,x)^{-1}, \qquad i = 1,\ldots,N_{\text{gs}}, \tag{5}$$

where the binary diffusion coefficients are calculated using the Chapman-Enskog-equation and the Knudsen diffusion coefficient is given by

$$D_i^{\text{K}}(t,x) = \frac{d_{\text{p}}(t,x)}{3}\sqrt{\frac{8RT}{\pi M_i}}, \quad (t,x) \in [0;t_{\max}] \times \Omega(t) \tag{6}$$

where R denotes the general gas constant. It is this dependence of the Knudsen diffusion coefficient on the pore diameter that introduces non-linearities into the diffusion-reaction-system (1).

To complete the model, boundary and initial conditions have to be applied: at Γ_{fixed} at the pore mouth, steady species concentrations are prescribed as boundary conditions, while initial conditions for the concentrations are given within the pore.

2.2 CVI reactor model

To describe the infiltration of porous substrates containing several pores a CVI reactor model is developed. Within the plane substrate, the model pores are assumed to be of cylindrical shape and perpendicular to the substrate surface. Domain decomposition of the total gas phase domain distinguishes the domain outside of the pores Ω_{out} from the domains within the different pores. We denote the number of pores by N_{p} and and the domain within pore j by $\Omega_j(t)$, $j = 1,\ldots,N_{\text{p}}$. Outside of the pores deposition of pyrolytic carbon at the reactor walls can be neglected due to the dimensions of the reactor yielding the time independence of the domain Ω_{out}. Neglecting convection within the pores and diffusion in Ω_{out} as one of the transport mechanisms is strongly dominating in each domain, the problem is split into a convection-reaction-system in Ω_{out} and diffusion-reaction-systems withi each pore. We further assume a constant convection velocity in Ω_{out} for all gas phase species. Thus the convection-reaction-system reads for $i = 1,\ldots,N_{\text{gs}}$

$$\frac{\partial}{\partial t}c_i(t,x) = -v\frac{\partial}{\partial x}c_i(t,x) + r_i^{\text{gas}}(t,x,\mathbf{c}), \quad (t,x) \in [0;t_{\max}] \times \Omega_{\text{out}}. \tag{7}$$

For $j = 1,\ldots,N_{\text{p}}$, the diffusion-reaction-system within pore j is the same as (1) except with $\Omega_j(t)$ as gas phase domain. As the pore diameter is very small compared to the length scale outside the pores, flow into or out of the pores as well as the mass loss in Ω_{out} can be neglected. A coupling between these systems is provided by the boundary conditions of the diffusion-reaction-systems: the values of the concentrations determined at the pore mouths yield the boundary values for the concentrations for the diffusion-reaction-systems. Again boundary values for the concentrations at the reactor entrance and initial values within the reactor for the linear convection-reaction-system in Ω_{out} complete the model. Using this model, the influence of the location of the pores within the substrate on the deposition profiles as well as that of the convection of the gas mixture can be studied.

2.3 Numerical solution

For numerically solving the systems in both models, the method of finite differences is applied. Choosing standard central difference quotients to approximate the spatial derivatives and a backward Euler scheme for the time discretisation, implicit schemes are obtained for the concentrations. This leads to big linear systems for the vector of the concentrations at the grid points. Having solved for them by using a restarted version of the GMRES solver with an incomplete LU factorisation as preconditioner, the height of the carbon layer is updated as well as the values of the diffusion coefficients; and the domain of the gas phase within the pores at the next time step is calculated. In case of the CVI reactor model, first the concentrations in Ω_{out} are determined, yielding the boundary conditions for the diffusion-reaction-systems within the pores that are subsequently solved.

3 Results

3.1 Results for the single pore model

The evolution of the moving boundary $\Gamma(t)$ for $T = 1398$ K and $p = 20$ kPa for different infiltration times using an initial pore diameter of $d_{\text{p}}^{0} = 0.001$ m is shown in Fig. 2 for the two pore geometries with pore lengths $l_0 = 0.05$ m and $l_0 = 0.2$ m. The deposition profiles within the pores show very steep gradients next to the pore mouths that monotonically decrease with the depth of the pores. A complete infiltration of the short pore is achieved whereas in the case of the long pore, the maximum deposition occurs within the pore and not at the pore ground (see Fig. 2). This leads to inaccessible pore space and consequently to an incomplete filling of the long pore.

3.2 Results for the CVI reactor model

The values of the rate constants as well as the initial diameters of the pores are the same as in the single pore model. Two different values for the gas ve-

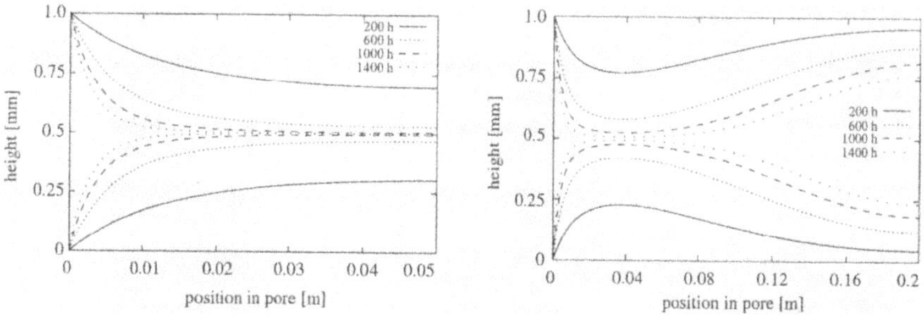

Fig. 2. Deposition profiles within a cylindrical pore of initial length 0.05 m (left) and of initial length 0.2 m (right) at different times

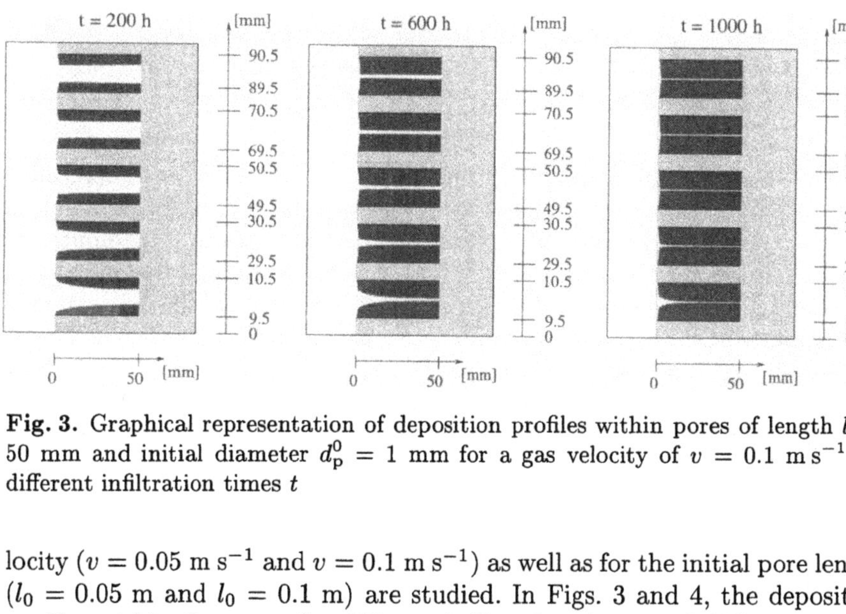

Fig. 3. Graphical representation of deposition profiles within pores of length $l_0 =$ 50 mm and initial diameter $d_\mathrm{p}^0 = 1$ mm for a gas velocity of $v = 0.1\ \mathrm{m\,s^{-1}}$ for different infiltration times t

locity ($v = 0.05\ \mathrm{m\ s^{-1}}$ and $v = 0.1\ \mathrm{m\,s^{-1}}$) as well as for the initial pore length ($l_0 = 0.05$ m and $l_0 = 0.1$ m) are studied. In Figs. 3 and 4, the deposition profiles within the pores for different infiltration times and different combinations of values of the convection velocity and pore length are presented. Obviously, the best situation for complete infiltration in all four cases is given by short pores and a high gas velocity outside of the pores (Fig. 3). Relying on the results of the single pore model, this is reasonable since only small concentrations of the higher hydrocarbons are present at the pore mouths emerging as boundary conditions for the calculations of the concentrations within the pores. The worst case of the situations studied is represented in Fig. 4 leading to incomplete infiltration of all five pores.

4 Conclusions

The surface of a porous substrate used in chemical vapour infiltration of pyrolytic carbon constitutes a temporally and spatially evolving interface

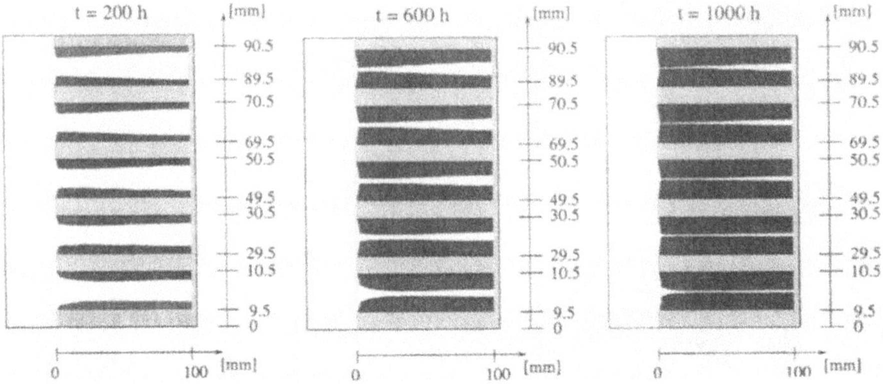

Fig. 4. Graphical representation of deposition profiles within pores of length $l_0 = 100$ mm and initial diameter $d_p^0 = 1$ mm for a gas velocity of $v = 0.05$ m s^{-1} for different infiltration times t

between the surrounding gas phase and the solid phase. In order to determine how the filling of the porous substrate progresses and if complete infiltration can be achieved the position of this moving boundary has to be computed in time and space. Two models are proposed, one for describing the infiltration of a single cylindrical pore, and the other one for the study of the infiltration of several pores within a plane substrate, both using a reduced reaction scheme to describe the chemistry in the gas phase. The position of the pyrolytic carbon layer is explicitly constructed in each time step in both models.

The results of the calculations based on the models presented here provide the basis for the identification of conditions under which complete infiltration of the pores in the substrate is achieved.

References

1. Middleman, S. (1989) The interaction of chemical kinetics and diffusion in the dynamics of chemical vapor infiltration, J. Mat. Res., **4**, 1515–1524.
2. Hou, X., Li, H., Chen, Y., Li, K. (1999) Modeling of chemical vapor infiltration process for fabrication of carbon-carbon composites by finite difference methods, Carbon, **37**, 669–677.
3. McAllister, P., Wolf. E. E. (1991) Modelling of chemical vapor infiltration of carbon in porous carbon substrates, Carbon, **29**, 387–396.
4. Tago, Y. I. T. , Kawase, M., Hashimoto, K. (2001) Numerical simulation of the thermal-gradient chemical vapor infiltration process for production of fiber-reinforced ceramic composite, Chem. Eng. Sci., **56**, 2161–2170.
5. Langhoff, T.-A., Schnack, E. (2001) Modelling chemical vapour infiltration of pyrocarbon as moving boundary problem. Preprint 2001-4, Institute of Solid Mechanics, Karlsruhe University.

Performance of Some Finite Elements in Numerical Simulation of Complex Incompressible Three Dimensional Flow

Andrzej F. Nowakowski[1], Wanwilai Kraipech[2], and Tom Dyakowski[1]

[1] Department of Chemical Engineering, University of Manchester Institute of Science and Technology, P.O. Box 88, Manchester M60 1QD, United Kingdom
[2] Department of Chemical Engineering, Srinakharinwirot University, Rangsit-Nakhonnayok Rd, Klong 16, Ongharak, Nakhonnayok, 26120, Thailand

Abstract. A numerical approach is applied to study the flow in some three-dimensional flow configurations. A physical problem has been discretizied using a finite element method based on mixed approximation of the velocity and pressure space. The effectiveness of the method is confirmed by the computational examples representing the class of internal flows having one inlet and more than one outlet. The selected outflow boundary conditions enable the avoidance of mass split-ratio. The computational examples showed that the choice of function spaces for approximation of velocity and pressure has a critical effect on physically meaningful solutions. It was illustrated that the lack of element mass balance limits the application of the well known Taylor-Hood elements for particular types of flow.

1 Introduction

The finite element approach can offer significant advantages in the solution process of convection dominated flow problems. The major advantages are the method's ability to adopt complicated boundaries and to provide refined local representation of a solution. In addition to this, the finite element method incorporates boundary conditions very efficiently into the numerical formulation. These are desired features of the method in the context of flow simulation in complex physical regimes with various boundary conditions (often of interfacial type) and sophisticated three-dimensional geometry. The good examples of such problems can be found in many multifield chemical engineering applications.

Numerical exploration of the mixed finite element method in the context of multiphase flows (fluid-particles) was the subject of our previous efforts [4], [5]. The work was focused on predicting the highly complicated swirling flow pattern in hydrocyclones. The numerical simulation in this important practical application proved to be a stringent computational challenge, which enabled identification of some problems related to element approximation functions. These are reported in the present contribution.

The analysis of elements in the context of the Navier-Stokes equations in terms of compatibility (or stabilty) of velocity and pressure spaces was widely presented in the literature. We will not attempt to describe the many possibilities but will devote our attention to the "Taylor-Hood" group of elements on a grid of tetrahedrals. There is a complete mathematical framework available (see for instance [1]) providing rigorous derivation of stable element pairs. Thatcher [7], [8] considered different interpolation spaces based on tetrahedral elements. The monograph of Gresho and Sani [2], and references included in this book, are also worth consulting in this respect. One major result is that in general the approximate velocity space should have at least a higher polynomial degree than the corresponding pressure space. Function spaces should satisfy compatibility condition also called Ladyzhenskaya-Babuska-Brezzi (LBB) condition [9]. The failure of the LBB condition leads to numerical instabilities such as spurious pressure oscillations or bad convergence behaviour.

In what follows, the discussion is restricted to incompressible three dimensional flow, for which the computational results are presented. This was the underlying continuous phase for our practical applications.

2 Problem Formulation and Element Choice

The objective is to compute a solution of internal flow problem modelled by the Navier-Stokes equations in three dimensional domain, Ω, with boundary, $\partial\Omega = \Gamma_{\mathrm{Ds}} + \Gamma_{\mathrm{Dw}} + \Gamma_{\mathrm{N}}$. The Γ_{Ds} stands for solid walls and Γ_{Dw} for inlet boundary, while Γ_{N} is used for indicating outflow parts of the boundary. More than one outlet is considered. The underlying equations written in primitive variables and stress-divergence form are:

$$\mathbf{u}_{,t} + (\mathbf{u} \cdot \nabla)\mathbf{u} - \nabla \cdot \tau = 0 \qquad (1)$$

$$\nabla \cdot \mathbf{u} = 0 \qquad (2)$$

where constitutive equations for incompressible fluids are given by:

$$\tau = -p\mathbf{I} + 2\nu\mathbf{D} \qquad \mathbf{D} = (1/2)[\nabla\mathbf{u} + (\nabla\mathbf{u}^T)] \qquad (3)$$

The standard notation is used: $\mathbf{u}[u, v, w]$ is the fluid velocity field, p is the pressure normalized by the constant density, ρ, and ν is kinematic viscosity. The body force vector is neglected, τ is the total stress tensor, $\mathbf{I}$ is the identity tensor and $\mathbf{D}$ is the deformation rate tensor. The initial velocity field, $\mathbf{u}_0$, is assumed to satisfy the incompressibility constraint 2. The stress-divergence form leads to physical Neumann boundary conditions that represent forces, $\mathbf{f}$,

$$\mathbf{f} = \tau \cdot \mathbf{n} = -p\mathbf{n} + 2\nu\mathbf{D} \cdot \mathbf{n} = \hat{\mathbf{f}} \qquad \text{on} \quad \Gamma_{\mathrm{N}} \qquad (4)$$

and Dirichlet boundary condition for velocity field, $\mathbf{u}$, at the inlet and solid walls

$$\mathbf{u} = 0 \quad \text{on} \quad \Gamma_{\mathrm{Dw}} \quad \text{and} \quad \mathbf{u} = \hat{\mathbf{u}} \quad \text{on} \quad \Gamma_{\mathrm{Ds}} \tag{5}$$

where $\mathbf{n}$ is unit vector normal to $\partial\Omega$, $\hat{\mathbf{f}}$ (presumed given) is applied force on the boundary while $\hat{\mathbf{u}}$ is specified velocity profile at the inlet.

Applying the primitive variable finite element method to Galerkin weak form of equations 1 and 2 leads to the following numerical problem:
Find $(\mathbf{u}_h, p_h) \in V_h \times Q_h$ where $V_h \subset V = (H_0^1(\Omega))^3$ and $Q_h \subset Q = L_0^2(\Omega)$ are velocity and pressure spaces depending on mesh parameter h such that

$$\begin{aligned} &\int_\Omega \{\phi(\mathbf{u}_{,t} + (\mathbf{u}\cdot\nabla)\mathbf{u}) + \nu[\nabla\mathbf{u} + (\nabla\mathbf{u}^T)] : \nabla\phi - p\nabla\phi\}\,\mathrm{d}\Omega \\ &\quad = \int_\Gamma \phi(-p\mathbf{n} + 2\nu\mathbf{D}\cdot\mathbf{n})\,\mathrm{d}\Gamma = \int_\Gamma \phi\hat{\mathbf{f}}\,\mathrm{ds} \qquad \forall\,\phi \in V_h \subset V \end{aligned} \tag{6}$$

$$\int_\Omega \xi\,\nabla\cdot\mathbf{u}\,\mathrm{d}\Omega = 0 \qquad \forall\,\xi \in Q_h \subset Q \tag{7}$$

In order to satisfy LBB condition, the Taylor-Hood elements were chosen as an approximation of basis spaces for primitive variables. These elements use the same grid for pressure and velocity with continuous approximation of both variables based on Lagrange interpolation, where the order of approximation is one less for pressure than for the components of velocity. The element based on piecewise continuous quadratic velocity and piecewise continuous linear pressure is implemented in our code. It is referred to as the P2/P1 element. The tetrahedra P2/P0, with the same approximation for velocity but discontinuous constant function as an approximation of pressure field, is also utilized. Only discontinuous pressure element assures an element-level mass balance, which can be easily proved using 7:

$$0 = \int_\Omega \xi\,\nabla\cdot\mathbf{u}\,\mathrm{d}\Omega = \int_{\Omega_e} \nabla\cdot\mathbf{u}_h\,\mathrm{d}\Omega = \int_{\Gamma_e} \mathbf{n}\cdot\mathbf{u}_h \tag{8}$$

The non-linear equation system resulting from the weak formulation of momentum, 6, and continuity, 7 equations, is solved using pressure projection algorithm introduced in [3]. The adopted approach does not require any matrix inversion and uses conjugate gradient type solvers for the solution of linearized equations. The boundary conditions on velocity and forces are explicitly and consistently incorporated in the algorithm.

3 Case Study and Numerical Results

The numerical code was very carefully validated against a comprehensive suite of test cases. These problems were realistic in the sense that they do not scale with the Reynolds number and that meaningful experimental data can be used for comparison purposes.

The first problem considered was cubical cavity [6], for which grid independent results were obtained. In Fig. 1, the centerline velocity profiles for two types of element approximation, P2/P1 and P2/P0, are shown. The grid resolution in these two cases was 32×32×32 (velocity nodes) and Re = 400. The Reynolds number is not high in the sense that Taylor-Görtler-like vorticies are not seen in the cavity, thus permitting steady state flow analysis. Both elements performed well with little difference in obtained field data, due to various order of interpolation spaces and hence accuracy used for pressure approximation.

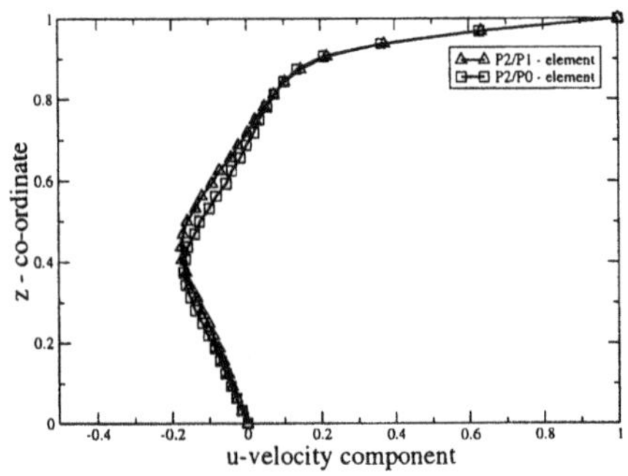

Fig. 1. Profile of streamwise velocity at the midplane of the cavity for Re = 400. Comparison of two types of element approximation P2/P1 and P2/P0

The second test involves a three dimensional cubic container with inlet located at the bottom plane and two outlets placed in two different walls. The boundary conditions are specified for velocity at the inlet and are left open at the outlets, $\hat{\mathbf{f}} = 0$. Fig. 2 presents the results for the P2/P0 element and Re = 100. Although the recirculation region was revealed for two elements P2/P1 and P2/P0 and the overall mass balance is satisfied, the resolved cross-section of computational domain show the difference.

Finally, the numerical code was applied to investigate the flow in a 50 mm hydrocyclone – a piece of separation equipment for solid-liquid systems, which is used to extract particles from a continuous fluid. The unstructured grid comprising profiled head entry region and outer part of the inlet feeder orifice is presented in Fig. 3. No stress boundary conditions are prescribed at the outlets of a hydrocyclone, $\hat{\mathbf{f}} = 0$. Such imposition of boundary condition enables the avoidance any "a priori assumption" concerning velocity field

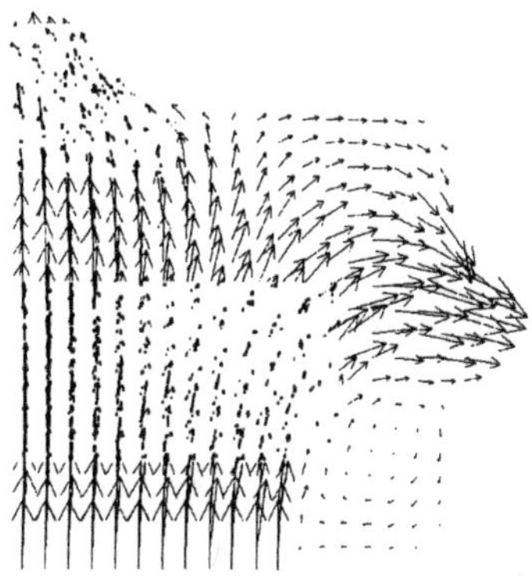

Fig. 2. Velocity vector plots on symmetry plane of 3D container, $Re = 400$

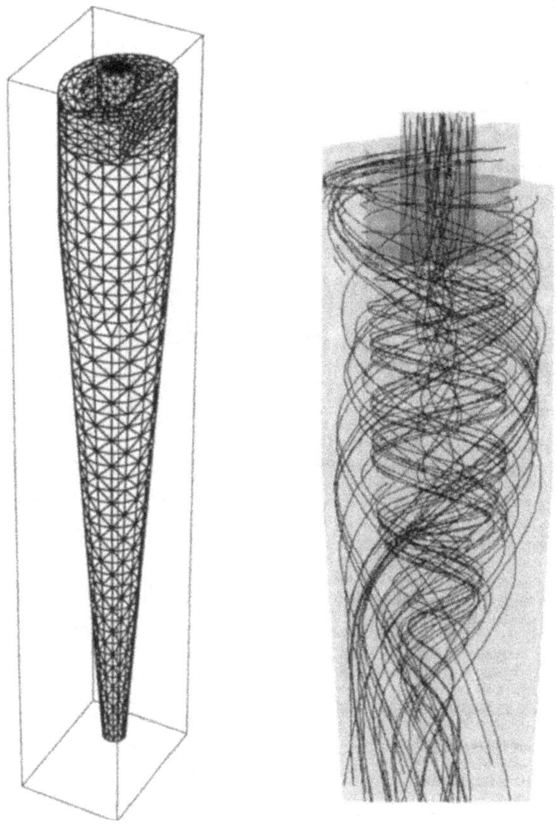

Fig. 3. The unstructured grid consisting of 25816 tetrahedral elements within hydrocylone (left picture) and the simulated three dimensional pathlines showing spiral flow (right picture), $\nu = 5.82 \cdot 10^{-5} m^2/s$ and feed flow rate $= 2.04 \cdot 10^{-4} m^3/s$

and in particular not to define mass split-ratio. The tangential inlet into the hydrocyclone body leads to a high swirl flow-field and particle classification results primarily from centrifugal-drag force balance. Some of the downward flow leaves through the underflow orifice in the apex of the cone while the rest reverses its vertical direction and goes up via the inner helical flow and out through the vortex finder. The P2/P0 element failed to reproduce this flow pattern and P2/P0 gave far more consistency with experimental results. The significant flow features, such as a the double helical flows, short-circuit and recirculation flow, have been demonstrated, Fig. 3, using the P2/P0 element.

4 Concluding Remarks

The numerical results obtained for the class of internal flows having one inlet and more than one outlet showed that the choice of function spaces for the approximation of velocity and pressure fields has a critical effect on physically meaningful solutions. The P2/P1 element generated spurious, non-physical results for the above mentioned class of problems while working properly for popular lid-driven cavity benchmark problems. The lack of element mass balance was considered as a reason for poorer results. This was overcome by changing the pressure approximation field from continuous piecewise linear to piecewise constant.

Acknowledgments

The computations were partly performed at the Institute of Applied Mechanics and Aeronautics of the Warsaw University of Technology. We acknowledge David Silvester and Ron Thatcher from the UMIST Mathematical Department for their helpful suggestions and comments.

References

1. Brezzi F., Fortin M. (1991) Mixed and Hybrid Finite Element Methods. Springer Verlag, Berlin-Heilderberg-New York
2. Gresho P., Sani R. (1998) Incompressible Flow and the Finite Element Method. Wiley, Chichester
3. Haroutunian V, Engelman M.S., Hasbani I. (1993) Segregated finite element algorithms for the numerical solution of large-scale incompressible flow problems. Int. J. Num. Methods Fluids **17**, 323–348
4. Nowakowski A.F., Kraipech W., Dyakowski T., Williams R.A. (2000) The hydrodynamics of a hydrocyclone based on a three-dimensional multi-continuum model. Chemical Engineering Journal **80**, 275–282
5. Nowakowski A.F., Kraipech W., Dyakowski T., Williams R.A. (2002) A 3D numerical simulation of hydrocyclone flow pattern using finite element approach. Proceedings of the 4th World Congress on Particle Technology. Sydney, Australia

6. Shankar P.N., Deshpande M.D. (2000) Fluid mechanics in the driven cavity. Ann. Rev. Fluid Mech. **32**, 93–136
7. Thatcher R.W. (1990) Locally mass-conserving Taylor-Hood elements for two- and three-dimensional flow. Int. J. Num. Methods Fluids **11**, 341–353
8. Thatcher R.W. (1993) The finite element method for three dimensional incompressible flow. In: Gunzburger M.D, Nicolaides R.A. (Ed) Incompressible Computational Fluid Dynamics. Cambridge University Press, Cambridge, 427–445
9. Quarteroni A., Valli A. (1994) Numerical Approximation of Partial Differential Equations. Springer Verlag, Berlin-Heilderberg-New York

A fast multipole boundary element method for a modified hypersingular boundary integral equation

Günther Of[1] and Olaf Steinbach[1]

Institut für Angewandte Analysis und Numerische Simulation,
Universität Stuttgart, Pfaffenwaldring 57, 70569 Stuttgart, Germany

Abstract. For the solution of a Neumann boundary value problem we consider a modified hypersingular boundary integral equation which reflects the solvability and scaling conditions of the non–unique problem. For the Galerkin discretization of the elliptic variational problem we describe an efficient boundary element approach which is based on a fast multipole method for the single layer potential. In the same way, an efficient preconditioner for the resulting linear system is constructed.

1 Boundary integral equations

For a bounded domain $\Omega \subset I\!R^3$ with Lipschitz boundary $\Gamma = \partial\Omega$ we consider a Neumann boundary value problem for the Laplace equation,

$$-\Delta u(x) = 0 \quad \text{for } x \in \Omega, \qquad \frac{\partial}{\partial n_x} u(x) = g(x) \quad \text{for } x \in \Gamma, \tag{1}$$

where we have to assume the compatibility condition

$$\int_\Gamma g(x) ds_x \;=\; 0. \tag{2}$$

Note that the solution of (1) is unique up to a constant only. The boundary value problem (1) is equivalent to the solution of the boundary integral variational formulation to find $u \in H^{1/2}(\Gamma)$ such that

$$\langle Du, v\rangle_{L_2(\Gamma)} \;=\; \langle (\frac{1}{2} I - K')g, v\rangle_{L_2(\Gamma)} \quad \text{for all } v \in H^{1/2}(\Gamma), \tag{3}$$

using the hypersingular boundary integral operator

$$(Du)(x) \;=\; -\frac{1}{4\pi}\frac{\partial}{\partial n_x}\int_\Gamma \frac{\partial}{\partial n_y}\frac{1}{|x-y|} u(y) ds_y \quad \text{for } x \in \Gamma \tag{4}$$

and the adjoint double layer potential operator

$$(K'w)(x) \;=\; \frac{1}{4\pi}\int_\Gamma \frac{\partial}{\partial n_x}\frac{1}{|x-y|} w(y) ds_y \quad \text{for } x \in \Gamma. \tag{5}$$

Since $u_0 = 1$ is an eigensolution of the hypersingular boundary integral operator D with zero eigenvalue, the variational problem (3) has to be considered in appropriate subspaces. Note that (2) implies $\langle(\frac{1}{2}I - K')g, u_0\rangle_{L_2(\Gamma)} = 0$ and therefore the solvability condition for (3).

Let us introduce the natural density $w_{\text{eq}} \in H^{-1/2}(\Gamma)$ as the unique solution of the boundary integral equation

$$(Vw_{\text{eq}})(x) = \frac{1}{4\pi} \int_\Gamma \frac{1}{|x-y|} w_{\text{eq}}(y) ds_y = 1 \quad \text{for } x \in \Gamma, \tag{6}$$

where $\text{cap}_\Gamma := \langle 1, w_{\text{eq}}\rangle_{L_2(\Gamma)} = \langle Vw_{\text{eq}}, w_{\text{eq}}\rangle_{L_2(\Gamma)} > 0$ defines the capacity of the boundary Γ [4]. Now we can define the factor space

$$H_*^{1/2}(\Gamma) := \left\{ v \in H^{1/2}(\Gamma) : \langle v, w_{\text{eq}}\rangle_{L_2(\Gamma)} = 0 \right\}. \tag{7}$$

Note that the hypersingular boundary integral operator $D : H^{1/2}(\Gamma) \to H^{-1/2}(\Gamma)$ is $H_*^{1/2}(\Gamma)$–elliptic. Instead of (3) we now consider a modified variational problem: find $u \in H^{1/2}(\Gamma)$ such that

$$\langle Du, v\rangle_{L_2(\Gamma)} + \alpha\langle u, w_{\text{eq}}\rangle_{L_2(\Gamma)}\langle v, w_{\text{eq}}\rangle_{L_2(\Gamma)} = \langle(\frac{1}{2}I - K')g, v\rangle_{L_2(\Gamma)} \tag{8}$$

for all $v \in H^{1/2}(\Gamma)$ where α is a positive constant to be chosen in an appropriate way. The bilinear form in (8) defines via

$$\langle \widetilde{D}u, v\rangle_{L_2(\Gamma)} := \langle Du, v\rangle_{L_2(\Gamma)} + \alpha\langle u, w_{\text{eq}}\rangle_{L_2(\Gamma)}\langle v, w_{\text{eq}}\rangle_{L_2(\Gamma)} \tag{9}$$

for all $u, v \in H^{1/2}(\Gamma)$ a boundary integral operator $\widetilde{D} : H^{1/2}(\Gamma) \to H^{-1/2}(\Gamma)$.

Theorem 1. [6] *For the single layer potential V and the modified hypersingular boundary integral operator $\widetilde{D}$ there hold the spectral equivalence inequalities*

$$\begin{aligned} \min\{c_1^V c_1^D, \alpha\langle 1, w_{eq}\rangle_{L_2(\Gamma)}\} \cdot \langle V^{-1}v, v\rangle_{L_2(\Gamma)} \leqslant \langle \widetilde{D}v, v\rangle_{L_2(\Gamma)} \leqslant \\ \leqslant \max\{\frac{1}{4}, \alpha\langle 1, w_{eq}\rangle_{L_2(\Gamma)}\} \cdot \langle V^{-1}v, v\rangle_{L_2(\Gamma)} \end{aligned} \tag{10}$$

for all $v \in H^{1/2}(\Gamma)$ where c_1^V and c_1^D denote the ellipticity constants of V in $H^{-1/2}(\Gamma)$ and of D in $H_*^{1/2}(\Gamma)$, respectively.

Hence we obtain unique solvability of the modified variational problem (8) for any given $g \in H^{-1/2}(\Gamma)$. Moreover, if g satisfies the compatibility condition (2) we get $u \in H_*^{1/2}(\Gamma)$. Finally, (10) implies the choice

$$\alpha := \frac{1}{4\langle 1, w_{\text{eq}}\rangle_{L_2(\Gamma)}}$$

to obtain the optimal spectral equivalence inequalities

$$c_1^V c_1^D \cdot \langle V^{-1} v, v \rangle_{L_2(\Gamma)} \leqslant \langle \widetilde{D} v, v, \rangle_{L_2(\Gamma)} \leqslant \frac{1}{4} \cdot \langle V^{-1} v, v \rangle_{L_2(\Gamma)}$$

for all $v \in H^{1/2}(\Gamma)$.

2 A fast multipole boundary element method

Let $W_h = \text{span}\{\varphi_i\}_{i=1}^M \subset H^{1/2}(\Gamma)$ be the boundary element space of piecewise linear continuous basis function φ_i with respect to a decomposition of Γ into plane triangular boundary elements. The Galerkin formulation of (8) is then equivalent to a linear system,

$$\left[D_h + \alpha \underline{a}\, \underline{a}^\top\right] \underline{u} = \underline{f} \tag{11}$$

where

$$D_h[j,i] = \langle D\varphi_i, \varphi_j \rangle_{L_2(\Gamma)}, a_i = \langle \varphi_i, w_{\text{eq}} \rangle_{L_2(\Gamma)}, f_j = \langle (\frac{1}{2} I - K') g, \varphi_j \rangle_{L_2(\Gamma)}$$

for all $i, j = 1, \ldots, M$. The stiffness matrix D_h of the hypersingular integral operator D is dense, hence we want to use some fast boundary element method to derive a more efficient approach. To this end we first consider an equivalent representation of the bilinear form induced by the hypersingular boundary integral operator D, [1]:

$$\langle Du, v \rangle_{L_2(\Gamma)} = \frac{1}{4\pi} \int_\Gamma \int_\Gamma \frac{\text{curl}_\Gamma u(y) \cdot \text{curl}_\Gamma v(x)}{|x - y|} ds_y ds_x \tag{12}$$

where $\text{curl}_\Gamma u(x) := n_x \times \nabla_x u^*(x)$ for $x \in \Gamma$ is the surface curl operator and u^* is any extension of u into the neighbourhood of Γ. Using (12) the Galerkin matrix D_h can be written as

$$D_h = T^\top \begin{pmatrix} V_h & & \\ & V_h & \\ & & V_h \end{pmatrix} T$$

where V_h is the Galerkin matrix of the single layer potential with respect to piecewise constant basis functions, $Z_h = \text{span}\{\psi_k\}_{k=1}^N \subset H^{-1/2}(\Gamma)$. T is an appropriate $3N \times M$ matrix which describes the transformation of the coordinates $\underline{u} \in \mathbb{R}^M$ of $u_h \in W_h$ to the piecewise constant vector–valued result in $\mathbb{R}^{3N}$ of $\text{curl}_\Gamma u_h$. Hence the application of the discrete hypersingular operator D_h is reduced to three applications of the discrete single layer potential avoiding the integration of hypersingular integrals. Therefore we need a fast boundary element method for the discrete single layer potential

V_h only. Here we use the fast multipole method as introduced in [2]. The matrix–vector product $\underline{v} = V_h \underline{w}$ can be written for $\ell = 1, \ldots, N$ as

$$v_\ell = \sum_{k=1}^{N} V_h[\ell,k] w_k \approx \sum_{\text{nearfield}} V_h[\ell,k] w_k + \sum_{\text{farfield}} \sum_{i=1}^{N_k} \sum_{j=1}^{N_\ell} \frac{\omega_{k,i}\omega_{\ell,j}}{|x_{k,i} - x_{\ell,j}|} w_k$$

using some appropriate numerical integration scheme in the farfield. The evaluation of

$$v_{\ell,j} = \sum_{\text{farfield}} w_k \sum_{i=1}^{N_k} \frac{\omega_{k,i}}{|x_{k,i} - x_{\ell,j}|}$$

corresponds exactly to the fast multiple particle simulation as described in [2] and can be implemented efficiently [5]. Choosing both the numerical integration scheme in the farfield and the multipole parameters in an appropriate way, one can show corresponding error estimates for the perturbed single layer potential, see [7] for details. Hence we obtain a fast multipole boundary element method for an efficient solution of (11) preserving the a priori error estimates of the standard boundary element method [8].

To bound the number of conjugate gradient iterations for solving the symmetric and positive definite system (11) an appropriate preconditioner is needed. Due to the spectral equivalence inequalities (10) the Galerkin discretization of the inverse single layer potential yields an almost optimal preconditioner. Following [10], the resulting preconditioner is given by

$$C_{\tilde{D}} := \bar{M}_h \bar{V}_h^{-1} \bar{M}_h \tag{13}$$

where

$$\bar{V}_h[j,i] = \langle V\varphi_i, \varphi_j \rangle_{L_2(\Gamma)}, \quad \bar{M}_h[j,i] = \langle \varphi_i, \varphi_j \rangle_{L_2(\Gamma)}$$

for all $i, j = 1, \ldots, M$. Therefore the application $C_{\tilde{D}}^{-1}$ requires two inversions of the sparse mass matrix $\bar{M}_h$ and one multiplication with the discrete single layer potential $\bar{V}_h$. The latter can be done by applying the multipole algorithm as described before, but using piecewise linear form functions instead of piecewise constants. Since the mass matrix $\bar{M}_h$ is spectrally equivalent to some diagonal matrix, the application of $\bar{M}_h^{-1}$ corresponds to some simple diagonal scaling.

3 Numerical example

We consider a Neumann boundary value problem for the Laplace equation where Ω is a given three–dimensional L shaped domain. The coarse mesh of the boundary $\Gamma = \partial\Omega$ consists of 28 plane triangles, see Fig. 1. Afterwards, a uniform refinement up to 6 levels with finally 114688 boundary elements was performed. The fast multipole Galerkin boundary element method was implemented as described above. For the computation of the right–hand side in (11) a corresponding multipole algorithm for the adjoint double layer potential was used. The natural density $w_{\rm eq}$ was approximated a priori by a Galerkin variational scheme for (6) using piecewise constant basis functions. As preconditioner for the linear system (13) we compare our proposed preconditioner (13) with some diagonal scaling $C_{\widetilde{D}} = \operatorname{diag} \widetilde{D}_h$. The numerical results are shown in Table 1. As stopping criterion in the preconditioned conjugate gradient scheme, a relative error reduction of $\varepsilon = 10^{-8}$ was used. L denotes the refinement level while N is the number of boundary elements and M is the number of boundary nodes, respectively. By matrix we describe the computational costs to generate the nearfield matrices and to set up the multipole algorithm while iter includes the computational costs for the iterative solver. We note that the use of the proposed preconditioning technique is more time consuming but reduces the number of iterations significantly. In Table 2 we present the corresponding extremal eigenvalues and the resulting spectral condition numbers for the preconditioned systems.

4 Conclusions

The described approach for an efficient solution of Neumann boundary value problems can be generalized directly to applications in linear elastostatics. Then the required modifications are chosen such that all rigid body motions are incorporated. Moreover, the modified hypersingular boundary integral operator can be used as a preconditioner when solving Dirichlet boundary value problems. These components can then be used in domain decomposition methods [9], in particular in boundary element tearing and interconnecting methods [3] where local Neumann boundary value problems are to be solved.

References

1. Dautray R., Lions J.L. (1990) Mathematical Analysis and Numerical Methods for Science and Technology. Volume 4: Integral Equations and Numerical Methods. Springer, Berlin.
2. Greengard L., Rokhlin V. (1987) A fast algorithm for particle simulations. J. Comput. Physics **73**, 325–348.
3. Langer U., Steinbach, O. (2002) Boundary element tearing and interconnecting methods. In preparation.
4. McLean, W. (2000) Strongly elliptic systems and boundary integral equations. Cambridge University Press.
5. Of G. (2001) Die Multipolmethode für Randintegralgleichungen. Diploma Thesis, Universität Stuttgart.
6. Of G., Steinbach O. (2003) A preconditioned fast multipole boundary element method for Neumann boundary value problems. In preparation.

Fig. 1. L shaped domain Ω with initial boundary element mesh

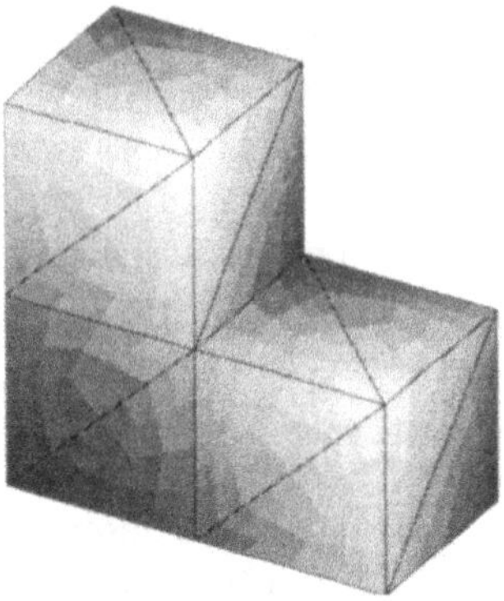

Table 1. Number of iterations and computational costs

			$C_{\widetilde{D}} = \operatorname{diag} \widetilde{D}_h$			$C_{\widetilde{D}} = \bar{M}_h \bar{V}_h^{-1} \bar{M}_h$		
L	N	M	Iter	matrix	solver	Iter	matrix	solver
0	28	16	13	0 sec	0 sec	9	0 sec	0 sec
1	112	58	17	1 sec	0 sec	10	1 sec	0 sec
2	448	226	23	7 sec	1 sec	11	16 sec	1 sec
3	1792	898	28	23 sec	18 sec	11	53 sec	10 sec
4	7168	3586	40	74 sec	166 sec	11	181 sec	74 sec
5	28672	14338	67	6 min	31 min	11	14 min	8 min
6	114688	57346	96	18 min	189 min	11	51 min	41 min

7. Of G., Steinbach, O., Wendland W.L. (2003) The fast multipole method for the symmetric boundary integral formulation. In preparation.
8. Schatz, A., Thomée V., Wendland W.L. (1990) Mathematical Theory of Finite and Boundary Element Methods. Birkhäuser, Basel.
9. Steinbach O. (2003) Stability estimates in hybrid coupled domain decomposition methods. Springer Lecture Notes in Mathematics **1809**.
10. Steinbach O., Wendland W.L. (1998) The construction of some efficient preconditioners in the boundary element method. Adv. Comput. Math. **9**, 191–216.

Table 2. Estimated eigenvalues and spectral condition number

	$C_{\widetilde{D}} = \operatorname{diag} \widetilde{D}_h$			$C_{\widetilde{D}} = \bar{M}_h \bar{V}_h^{-1} \bar{M}_h$		
L	$\lambda_{\min}$	$\lambda_{\max}$	$\kappa(C_{\widetilde{D}}^{-1} \widetilde{D}_h)$	$\lambda_{\min}$	$\lambda_{\max}$	$\kappa(C_{\widetilde{D}}^{-1} \widetilde{D}_h)$
0	9.05 −3	2.88 −2	3.18	1.02 −1	2.56 −1	2.50
1	4.07 −3	2.82 −2	6.94	9.24 −2	2.66 −1	2.88
2	1.98 −3	2.87 −2	14.47	8.96 −2	2.82 −1	3.14
3	9.84 −4	2.90 −2	29.52	8.86 −2	2.89 −1	3.26
4	4.91 −4	2.91 −2	59.35	8.80 −2	2.92 −1	3.31
5	2.46 −4	2.92 −2	118.72	8.79 −2	2.92 −1	3.32
6	1.23 −4	2.92 −2	237.66	8.78 −2	2.92 −1	3.33

Analysis of a Discretization Method for the Richards' Equation

Iuliu Sorin Pop

Department of Mathematics and Computer Science, Technical University Eindhoven, P. O. Box 513, 5600 MB Eindhoven, The Netherlands (I.Pop@tue.nl)

Abstract. A time discretization method for the saturation-based formulation of Richards' equation is analysed numerically. Depending on the flow regime, diffusion may vanish or blow up, leading to degeneracy in the original nonlinear parabolic problem. The numerical approach is based on an Euler implicit time discretization scheme and includes a regularization step, combined with the Kirchhoff transform. Convergence is shown by obtaining error estimates in terms of the time step and of the regularization parameter.

1 Introduction

Richards' equation (see, for example, [4]) is a popular model for porous media flows. Assuming the air pressure in the medium is constant, we are primarily interested in determining the saturation and pressure of the fluid phase (water). Depending on the flow regime (unsaturated, partially saturated, or completely saturated), the Richards' equation admits three forms: saturation based, pressure based, or mixed.

1.1 Physical background

Here we are interested in the unsaturated to partially saturated regime, so saturation becomes the primary unknown. The flow is governed by *Darcy law*,

$$\mathbf{q} = -K\nabla(\Psi + z), \tag{1}$$

where $\mathbf{q}$ $[m/s]$ denotes the *flow rate*, K $[m^2]$ is the *medium permeability* and Ψ $[Pa]$ stands for the *pressure* inside the fluid. Here z $[m]$ denotes the vertical coordinate in the medium. Combining the *continuity condition*

$$\partial_t(\rho\theta) + \nabla\cdot(\rho\mathbf{q}) = 0$$

with Darcy law (1) leads to *Richards' equation*

$$\partial_t(\rho\theta) - \nabla\cdot(\rho K\nabla(\Psi + z)) = 0. \tag{2}$$

Here the *water density* ρ $[kg/m^3]$ is assumed constant, while the *water saturation* θ $[-]$ is reduced (scaled to the $[0, 1]$ interval).

Based on experimental results, different retention curves relating the permeability K and the pressure Ψ to the saturation θ have been proposed in the literature. In this way equation (2) can be rewritten in terms of saturation. Here we tacitly adopt the retention curves proposed by van Genuchten ([8])

$$\theta = (1 + (c|\Psi|)^n)^{-m}, \tag{3}$$

where $c > 0$, $m \in (0,1)$ and $n > 1$ are parameters of the porous medium. As suggested in [8], we assume here the relation $n = \frac{1}{1-m}$ holds. Moreover, due to theoretical restrictions we assume here that $m \geqslant 2/3$ (see [15] for details).

One possibility to predict the hydraulic conductivity knowing the saturation and the pressure inside the fluid is given by Mualem ([13])

$$K = K_s K_r(\theta) = K_s \theta^{\frac{1}{2}} \left[\int_0^\theta \frac{1}{\Psi(v)} dv \Big/ \int_0^1 \frac{1}{\Psi(v)} dv \right]^2,$$

where K_s $[m^2]$ stands for the *absolute permeability* of the medium, possibly depending also on on the position inside the medium, but not on the fluid saturation. K_r $[-]$ denotes the *relative permeability*, which depends on the saturation. Recalling (3), the dependence of the permeability on the reduced saturation can be given explicitly

$$K(\theta) = K_s \theta^{\frac{1}{2}} \left[1 - \left(1 - \theta^{\frac{1}{m}} \right)^m \right]^2. \tag{4}$$

In the saturation based formulation, equation (2) becomes

$$\partial_t \theta - \nabla \cdot \left(D(\theta) \nabla \theta + \frac{K(\theta)}{\theta_s - \theta_r} \nabla z \right) = 0, \tag{5}$$

with θ_s $[-]$ and θ_r $[-]$ denoting the saturated water content, respectively the residual one. The *moisture diffusivity* $D(\theta)$ $[m^2/s]$ is defined by

$$D(\theta) = -K(\theta) \frac{\partial \Psi}{\partial \theta} = \frac{(1-m)K_s}{\alpha m(\theta_s - \theta_r)} \frac{\theta^{\frac{1}{2} - \frac{1}{m}}}{(1 - \theta^{\frac{1}{m}})^m} \left[1 - \left(1 - \theta^{\frac{1}{m}} \right)^m \right]^2.$$

The equation above is nonlinear parabolic. Moreover, for $\theta = 0$ (fluid at residual water content) the diffusivity D vanishes, while for θ going to 1 (completely saturated water content) D goes to infinity, so the equation degenerates at these points. Specifically, for $s \searrow 0$ we have

$$D(s) = O(s^{1/2 + 1/m}), \quad D(1-s) = O((1-s)^{-m}). \tag{6}$$

1.2 A brief survey

Numerous papers are devoted to numerical schemes for the Richards' equation. Time discretization based on the method of lines is studied, for example,

[11] (see also the references therein); adaptive time stepping is considered in [10], or [18]. In case of an implicit discretization, the emerging time discrete elliptic problems are nonlinear. A first class of iterative procedures is proposed in [5] and [9]. Starting from a commonly used procedure for constructing sub- and super-solutions, a linearization scheme is considered in [22] and [14], where convergence is shown through a fixed point argument. The same scheme is also the main theme of [19], where similar arguments are used to demonstrate its efficiency in case of fast diffusion.

Mixed finite elements or finite volumes are considered for the spatial discretization (see, for example, [2],[7], [12], [16], or [21]). Nevertheless, most of the papers listed above (excepting [9]) consider either the slow, or the fast diffusion case, while the van Genuchten model exhibits both types of degeneracy simultaneous.

Our aim here is to present error estimates for an Euler implicit discretization method for the Richards' equation (5). Details, which are omitted here, can be found in [15].

2 Numerical scheme

Our approach is based on a regularization step, combined with the Euler implicit time discretization. Convergence is shown for the semi-discrete case. After chosing an appropriate spatial discretization, the analysis can be extended to the fully discrete situation.

2.1 Approach

Following [1], applying a Kirchhoff transform to the saturation θ leads to an unknown $\Phi := \mathcal{D}(\theta)$ having better regularity properties. Here $\mathcal{D}$ is a primitive of D defined as

$$\mathcal{D} : [0,1] \to (0,\infty), \qquad \mathcal{D}(\theta) = \int_0^\theta D(s)ds. \tag{7}$$

It is easy to see that $\mathcal{D}$ is continuous, bounded and strictly increasing, so after determining Φ, the saturation can be obtained by inverting $\mathcal{D}$. Endowed with boundary and initial conditions, the Richards' equation (5) transforms into

$$\begin{cases} \partial_t \mathcal{D}^{-1}(\Phi) - \nabla \cdot (\nabla\Phi + \mathcal{K}(\Phi)) = 0, & \text{in } Q_T \equiv \Omega \times (0,T), \\ \Phi(x,0) = \mathcal{D}(\theta^0(x)), & \text{in } \Omega, \\ \Phi = 0, & \text{on } \partial\Omega \times (0,T), \end{cases} \tag{8}$$

where Ω (the domain occupied by the porous medium) is open and bounded in $\mathbb{R}^d$ $(d \geqslant 1)$ with Lipschitz-continuous boundary and $T > 0$ an upper time limit, while $\mathcal{K}(\Phi) = K(\mathcal{D}^{-1}(\Phi))/(\theta_s - \theta_r)\nabla z$. For simplicity we consider here homogeneous Dirichlet boundary data, but nonhomogeneous or natural ones can be easily taken into account. We also assume the initial saturation satisfies the physically relevant assumption $0 \leqslant \theta^0 \leqslant 1$ everywhere in Ω.

Nevertheless, even after reformulating (5) in terms of the more regular unknown Φ, it still remains degenerate. To overcome this, a widely used technique is perturbing the nonlinearity $\mathcal{D}$ (or its derivative) so that the resulting parabolic problem becomes regular. This idea has been applied in the analysis of degenerate problems, and also generates efficient numerical algorithms (see, for example, [9] and the references therein).

In what follows we make use of both ideas mentioned above. To do so, let $\varepsilon > 0$ being a small perturbation parameter. $\mathcal{D}$ is approximated by $\mathcal{D}_\varepsilon$ defined as

$$\mathcal{D}_\varepsilon(\theta) = \int_0^\theta \max\{\varepsilon, \min\{1/\varepsilon^r,\ D(s)\}\}\, ds, \tag{9}$$

where $r \geqslant 0$ is a parameter to be chosen conveniently. We have

$$\varepsilon \leqslant \mathcal{D}_\varepsilon'(\theta) \leqslant 1/\varepsilon^r, \qquad |\mathcal{D}_\varepsilon(\theta) - \mathcal{D}(\theta)| \leqslant C\varepsilon^\gamma, \tag{10}$$

for all $\theta \in [0,1]$, with $\gamma = \min\{(3m+2)/(m+2), r(1-m)/m\}$.

In this way we approximate the degenerate problem (8) by

$$\begin{cases} \partial_t \mathcal{D}_\varepsilon^{-1}(\Phi) - \nabla \cdot (\nabla \Phi + \mathcal{K}(\Phi)) = 0, & \text{in } Q_T \equiv \Omega \times (0,T), \\ \Phi(x,0) = \mathcal{D}_\varepsilon(\theta^0(x)), & \text{in } \Omega, \\ \Phi = 0, & \text{on } \partial\Omega \times (0,T), \end{cases} \tag{11}$$

which is regular parabolic. Let now n being an integer and $\tau = T/n$. The Euler implicit discretization for Problem $\mathcal{P}_\varepsilon$ can be written formally as

$$\begin{cases} \mathcal{D}_\varepsilon^{-1}(\Phi^k) - \mathcal{D}_\varepsilon^{-1}(\Phi^{k-1}) = \tau \nabla \cdot \left(\nabla \Phi^k + \mathcal{K}(\Phi^k)\right), \\ \Phi^k|_{\partial\Omega} = 0, \end{cases} \tag{12}$$

for $k = \overline{1,n}$. Here Φ^k approximates the solution Φ of problem (8) at $t_k = k\tau$. Initially we take $\Phi^0 = \mathcal{D}_\varepsilon(\theta^0)$.

2.2 Results

We first consider the nonlinear problems in (12) and give some apriori estimates. For each $k \in \mathbb{N}$, these problems are elliptic. Solution are sought in a weak sense, namely $\Phi^k \in H_0^1(\Omega)$ (the usual *Sobolev space* with vanishing traces on the boundary). Here $(\cdot,\cdot)$ stands for the inner product on $L^2(\Omega)$ or the duality pairing between $H_0^1(\Omega)$ and $H^{-1}(\Omega)$, $\|\cdot\|$ for the norm in $L^2(\Omega)$, $\|\cdot\|_1$ and $\|\cdot\|_{-1}$ are the norms in $H^1(\Omega)$, respectively $H^{-1}(\Omega)$. The inner product and the corresponding norm on $L^2(0,T;V)$ (with V being either $L^2(\Omega)$, $H^1(\Omega)$, or $H^{-1}(\Omega)$) are defined analogously. The following lemma states existence, uniqueness and uniform boundedness for the weak solution of the nonlinear elliptic problems.

Lemma 1. *Given $k > 0$, and Φ^{k-1} satisfying $0 \leqslant \Phi^{k-1} \leqslant M$ almost everywhere in Ω, then the problem in (12) has a unique solution in $\Phi^k \in H_0^1(\Omega)$, which is also bounded by 0 and M.*

Since θ^0 is bounded, the same holds for Φ^0, and by mathematical induction we get existence, uniqueness and boundedness for all k.

Now we can formulate a stability result,

Lemma 2. *Let $p \leqslant n$. If Φ^k is the weak solution of problem (12), we have*

$$\varepsilon^r \|\Phi^p\|^2 + \tau \sum_{k=1}^{p} \|\nabla \Phi^k\|^2 \leqslant C. \tag{13}$$

To state the error estimates we consider the following notations:

$$e_\theta^k := \frac{1}{\tau} \int_{t_{k-1}}^{t_k} \mathcal{D}^{-1}(\Phi(s))ds - \mathcal{D}_\varepsilon^{-1}(\Phi^k), \quad e_\Phi^k := \frac{1}{\tau} \int_{t_{k-1}}^{t_k} \Phi(s)ds - \Phi^k, \tag{14}$$

where $k > 0$ and Φ is the weak solution of problem (8). Then we obtain

Theorem 1. *In the notations above, with γ defined in (10) we have*

$$\sup_{k=\overline{1,n}} \left\|e_\theta^k\right\|_{-1}^2 + \sum_{k=1}^n \int_{t_{k-1}}^{t_k} \left|\left(\mathcal{D}^{-1}(\Phi(s)) - \mathcal{D}_\varepsilon^{-1}(\Phi^k), \Phi(s) - \Phi^k\right)\right| ds$$
$$+\tau^2 \left\|\sum_{k=1}^n \nabla e_\Phi^k\right\|^2 \leqslant C\left\{\tau + \varepsilon^\gamma\right\}.$$

Remark 1. Estimates for saturation are obtained in $L^\infty(0,T;H^{-1}(\Omega))$ The second term on the left is a stronger estimate, since it also involves the more regular unknown. Nevertheless, since $\mathcal{D}'$ can both vanish and blow up, this term does not imply an $L^2(Q_T)$ estimate for Φ or θ.

Remark 2. The last term on the left is similar to an H^1 estimate for the flux after being integrated in time. However, this result is not as good as an estimate in $L^2(0,T;H^1(\Omega))$.

3 An example

In this section we present a numerical experiment for the Richards' equation (5). The retention curves given in (3) and (4) are used, with $m = 2/3$. In applying the Kirchhoff transform we approximate the primitive function $\mathcal{D}$ by an adaptive quadrature method. For solving the nonlinear problems in (12) we have used the linearization procedure suggested in [5] and [9]. Spatial discretization is achieved by an lumped mass upwind box scheme, applied on a regular decomposition of the domain into tetrahedras.

Heterogeneity of the medium is simulated through a variable absolute permeability K_s in (4), this being also the context considered in [20]. However, this construction is quite artificial, since heterogeneity may appear also in other parameters of the retention curves. As proposed in [17], K_s is generated randomly and has a log-normal distribution of mean mean $\overline{log_{10}K_s} = 0$, with a standard correlation $\sigma_{log_{10}K_s} = 0.4$, and a correlation length $\Lambda = 0.02$.

In the simulation we take a unit cube. Initially the medium is assumed to be completely unsaturated ($\theta^0 = 0$) and no flow conditions are imposed at all boundaries excepting the bottom of the domain, which is assumed to be always in contact with a completely saturated region ($\theta = 1$). The algorithm is implemented in *UG* ([3]), while results are visualized with *vrend* ([6]).

Figure 1 displays on the left the decadic logarithm of K_s, which ranges from -2 to 2. Low permeable zones are set to transparency, while elementary volumes where permeability exceeds a certain value reflect the rendering ray, so these become opaque. The picture on the right shows the wetting front after 20 time steps, where now regions where saturation is close to 0 or 1 are set to transparency, so the images display only the wetting front. As before, low saturated regions are blue while highly saturated ones are red. A movie regarding this simulation can be downloaded from *http://www.win.tue.nl/~pop/richards/richards.htm*

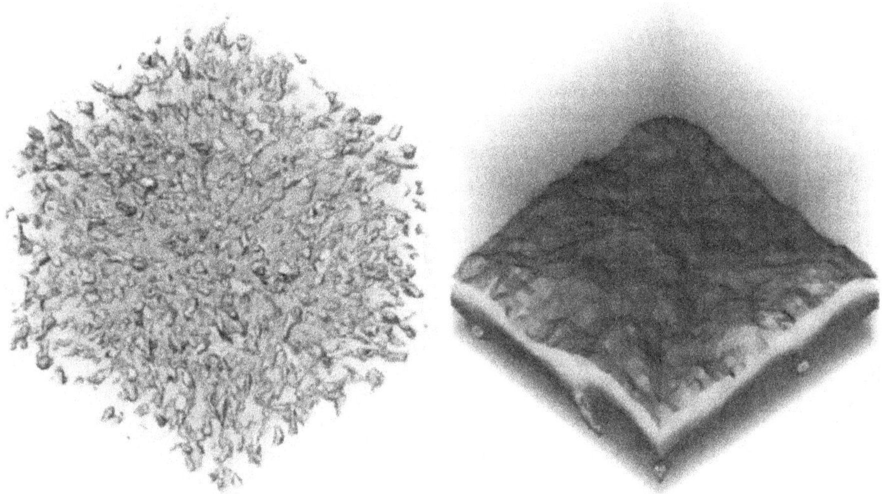

Fig. 1. Absolute permeability (left) and the wetting front (right); colored, see page 374

Acknowledgements. This work was supported by DFG (SFB 359, subproject D1) and by NWO (project 809.62.010 of ALW).

References

1. Alt H. W, Luckhaus S. (1983) Quasilinear elliptic-parabolic differential equations, Math. Z. **183**, 311–341
2. Arbogast T., Wheeler M. F., Zhang N. Y. (1996) A nonlinear mixed finite element method for a degenerate parabolic equation arising in flow in porous media, SIAM J. Numer. Anal. **33**, 1669–1687
3. Bastian P. et al. (1997) UG–A flexible software toolbox for solving partial differential equations, Comput. Vis. Sci. **1**, 27–40
4. Bear J., Bachmat Y. (1991) Introduction to Modelling of Transport Phenomena in Porous Media, Kluwer Academic, Dordrecht
5. Celia M. A., Bouloutas E. T., Zarba R. L. (1990) A general mass-conservative numerical solution for the unsaturated flow equation, Water Resour. Res. **26**, 1483–1496
6. Dărţu C. (2001) Visualization of Volumetric Datasets, Ph.D. Thesis, University of Heidelberg
7. Eymard R., Gutnic M., Hillhorst D. (1999) The finite volume method for Richards equation, Comput. Geosci. **3**, 259–294
8. van Genuchten M. Th. (1980) A closed-form equation for predicting the hydraulic conductivity of unsaturated soils, Soil Sci. Soc. Am. J. **44**, 892–898
9. Jäger W., Kačur J. (1995) Solution of doubly nonlinear and degenerate parabolic problems by relaxation schemes, M2AN Math. Model. Numer. Anal. **29**, 605–627
10. Kavetski D., Binning P., Sloan S. W. (2001) Adaptive time stepping and error control in a mass conservative numerical solution of the mixed form of Richards equation, Adv. Water Res. **24**, 595–605
11. Kelley C. T., Miller C. T., Tocci M. D. (1998) Termination of Newton/chord iterations and the method of lines, SIAM J. Sci. Comput. **19**, 280–290
12. Knabner P., Schneid E. (2002) Adaptive hybrid mixed finite element discretization of instationary variably saturated flow in porous media. In: Breuer M. et al. (Eds.) High Performance Scientific and Engineering Computing, Springer, Berlin Heidelberg, 37–44
13. Mualem Y. (1976) A new model for predicting the hydraulic conductivity of unsaturated porous media, Water Resour. Res. **12**, 513–522
14. Pop I. S., Yong W. A. (2000) On the existence and uniqueness of a solution for an elliptic problem, Studia Univ. Babes-Bolyai Math. **45**, 97–107
15. Pop I. S. (2002) Error estimates for a time discretization method for the Richards' equation, Comput Geosci. **6** (in print)
16. Radu F. A., Pop I. S., Knabner P. (2002) Error estimates for an Euler implicit, mixed finite element discretization of Richards' equation: equivalence between mixed and conformal approaches, RANA Preprint **02-06**, Eindhoven University of Technology
17. Robin M. J. L. et al. (1993) Cross-corelated random field generation with the direct Fourier transform method, Water Resour. Res. **29**, 2385–2397
18. Schneid E. (2000) Hybrid-Gemischte Finite-Elemente-Diskretisierung der Richards-Gleichung. Ph.D. Thesis, University of Erlangen–Nürnberg
19. Slodicka M. (2002) A robust and efficient linearization scheme for doubly nonlinear and degenerate parabolic problems arising in flow in porous media, SIAM J. Sci. Comput. **23**, 1593–1614

20. C. Wagner (1998) Numerical methods for diffusion-reaction-transport processes in unsaturated porous media, Comput. Vis. Sci. **1**, 97–105
21. Woodward C., Dawson C. (2000) Analysis of expanded mixed finite element methods for a nonlinear parabolic equation modeling flow into variably saturated porous media, SIAM J. Numer. Anal. **37**, 701–724
22. Yong W. A., Pop I. S. (1996) A numerical approach to porous medium equations, Preprint **96-50** (SFB 359), IWR, University of Heidelberg

A scheme for construction of algorithms for numerical solution of boundary-value problems on compound domains based on the difference potentials method

Victor S. Ryaben'kii

Keldysh Institute of Applied Mathematics, RAS, Moscow

Abstract. We consider a model problem on two simple subdomains with the unknown function satisfying elliptic equations as well as transition and boundary conditions on common and external boundaries, respectively. A general solution for each subdomain is expressed in terms of the difference potentials with unknown densities. Equations of the difference potential method together with transition and boundary conditions yield the governing equations for the densities.

1 Model problem

We explain the scheme using the following simple example. Let Q be a domain made up of subdomains D_1 and D_2, $\bar{Q} = \bar{D}_1 \cup \bar{D}_2, \quad D_1 \cap D_2 = \emptyset$. We assume that the function $u_Q(x)$

$$u(x) = u_Q(x) = \begin{cases} u_{D_1}(x), & x \in D_1 \\ u_{D_2}(x), & x \in D_2, \end{cases} \quad x = (x_1, x_2)$$

is the solution of the following problem:

$$L_1 u_{D_1} |_x = \frac{\partial^2 u(x)}{\partial x_1^2} + \frac{\partial^2 u}{\partial x_2^2} = 0, \quad x \in D_1$$

$$L_2 u_{D_2} |_x = \frac{\partial^2 u}{\partial x_1^2} + \frac{\partial^2 u}{\partial x_1 \partial x_2} + \frac{\partial^2 u}{\partial x_2} = 0, \quad x \in D_2$$

with the boundary condition

$$u_Q |_{\partial Q} = \alpha(s)$$

and transmission conditions

$$[u]_{\Gamma_{1,2}} = \psi^{(0)}(t)$$

$$[\frac{\partial u}{\partial n}]_{\Gamma_{1,2}} = \psi^{(1)}(t)$$

on $\Gamma_{1,2} = \bar{D}_1 \cap \bar{D}_2$.

The functions $\alpha(s), \psi^{(0)}(t), \psi^{(1)}(t)$ are given; s and t are natural parameters along ∂Q and $\Gamma_{1,2}$.

We denote by $[u]_{\Gamma_{1,2}}$ and $[\frac{\partial u}{\partial n}]_{\Gamma_{1,2}}$ the jumps u_Q and $\frac{\partial u}{\partial n}$ on $\Gamma_{1,2}$.

The scheme of construction is based on *superelements* described below.

2 Difference potentials and superelements

We have to approximate general solutions v_{D_j} to equations $L_j v_{D_j} = 0,\ j = 1, 2$, on each subdomain D_1 and D_2 by means of the expression

$$P_{D_j}(x, c_{1j}, c_{2j}, \ldots, c_{Nj})$$

which depends linearly on arbitrary constants $c_{1j}, c_{2j}, \ldots, c_{Nj}$. We can write

$$v\mid_{\partial D_j} = c_{1j}\psi_1^{(j)}(s) + \ldots + c_{Nj}\psi_N^{(j)}(s)$$

where $\psi_k^{(j)}$ are some basis functions from function spaces on ∂D_j. The approximate equality

$$\frac{\partial v}{\partial n}\mid_{\partial D_j} \approx c_1'\psi_1^{(j)}(s) + c_2'\psi_2^{(j)}(s) + \ldots + c_N'\psi_N^{(j)}(s)$$

is satisfied by the choice of $\bar{c}_j' = (c_{1j}', c_{2j}', \ldots, c_{Nj}')$ depending on $\bar{c}_j = (c_{1j}, c_{2j}, \ldots, c_{Nj})$.

We have to find the matrix $A_j : \bar{c}_j \to \bar{c}_j'$, satisfying the equality $\bar{c}_j' = A_j \bar{c}_j$.

The matrix A_j together with the expression $P_{D_j}(x, c_{1j}, \ldots, c_{Nj})$ such that

$$P_{D_j}(x, c_{1j}, \ldots, c_{Nj}) \approx v_{D_j}(x)$$

will be called *the superelement* for the subdomain $D_j,\ j = 1, 2$.

In order to calculate the superelements, one needs to construct the auxiliary difference problem and the difference potential for each D_j. We consider only the case $j = 1$ (the case $j = 2$ is similar). Recall that

$$L_1 v\mid_x = \frac{\partial^2 v}{\partial x_1^2} + \frac{\partial^2 v}{\partial x_2^2}.$$

We will write D^+ instead D_1. Let D^0 be a square containing D^+, $D^+ \subset D^0$, $D^0 = \{0 \leqslant x_1, x_2 \leqslant L\}$.

Auxiliary difference problem. Let $m = (m_1 h, m_2 h)$, $m_1, m_2 = 0, \pm 1, \ldots$, be a grid with mesh step h, $\frac{L}{h}$ is an integer, and let M^0 be the set of points m which lie inside of D^0. We construct the five-point difference counterpart to the Poisson equation on $\bar{D}^0$ using the stencil

$$N_m = \begin{cases} (m_1 h, m_2 h) \\ ((m_1 \pm 1)h, m_2 h) \\ (m_1 h, (m_2 \pm 1)h). \end{cases}$$

and consider the corresponding difference problem:

$$\sum_{n \in N_m} a_{mn} v_n = f_m, \quad m \in M^0$$

$$v_n = 0, \quad n \in \partial \bar{D}^0.$$

We put $v_n = 0$ if a point n lies on boundary of $\bar{D}^0$. The solution of this problem can be calculated by the Fourier method. This difference boundary-value problem will be called *the auxiliary difference problem.*

We now define the sets

$$M^+ = \{m \mid m \in D^+\}, \quad M^- = M^0 \backslash M^+$$

$$N^+ = \cup N_m, \quad m \in M^+, \quad N^- = \cup N_m, \quad m \in M^-$$

and the grid boundary

$$\gamma = N^+ \cap N^-$$

of the grid domain N^+. The grid domain N^+ is a grid analog of the subdomain D^+.

Difference potential $v_{N^+} = P_{N^+} w_\gamma$. Let w_γ be an arbitrary function on γ.

Definition. Let v_{N^0} be a solution of the auxiliary difference problem with

$$f_m = \begin{cases} 0, & m \in M^+ \\ \sum a_{mn} w_n, & m \in M^-, \end{cases}$$

where

$$w_n = \begin{cases} w_\gamma \mid_n & \text{on} \quad n \in \gamma, \\ 0 \quad \text{on} & n \notin \gamma. \end{cases}$$

The restriction of the solution v_{N^0} to N^+ will be called the *difference potential* $v_{N^+} = P_{n^+} w_\gamma$ with density w_γ.

Theorem, cf. [1]. *A function* v_γ *is the restriction to* γ *of a solution* $v_{N^+} = \{v_n\}$, $n \in N^+$, *of the homogeneous equation*

$$\sum_{n \in N_m} a_{mn} v_n = 0, \quad m \in M^+$$

if and only if

$$P_{N^+} v_\gamma \mid_\gamma = v_\gamma.$$

If this condition is satisfied, then the following equality holds:

$$v_{N^+} = P_{N^+} v_\gamma.$$

We can now construct the superelement, i.e. the matrix A and the expression

$$P_{D^+}(x, \bar{c}).$$

Using the expressions

$$v \mid_{\partial D^+} = \sum_{k=1}^{N} c_k \psi_k(s), \qquad \frac{\partial v}{\partial n} \mid_{\partial D^+} = \sum_{k=1}^{N} c'_k \psi_k(s)$$

we define the function v_γ by means of the Taylor formula. Namely, for an arbitrary point $\nu \in \gamma$ we define v_γ by the following expression:

$$v_\nu = \sum c_k \psi_k(s_\nu) + \rho_\nu \sum c'_k \psi_k(s_\nu).$$

Here s_ν is a point on ∂D such that the normal at this point passes through ν, and ρ_ν is the distance between ν and s_ν with sign plus if ν lies inside D^+ and sign minus if ν lies outside D^+.

The function $v_\gamma = v_\gamma(\bar{c}, \bar{c}')$ is constructed.

We now substitute the function $v_\gamma(\bar{c}, \bar{c}')$ into the condition

$$P_{N^+} v_\gamma \mid_\gamma = v_\gamma.$$

Solving this equation for $\bar{c}'$ by the least square method N times for each elementary $c = \{1, 0, ..., 0\}, \{0, 1, ..., 0\}, ..., \{0, ..., 1\}$, we obtain the matrix A :

$$\bar{c}' = A\bar{c}.$$

Returning to our model problem, we parameterize function u_D using the approximate expression

$$u_D \approx u_D(x, \bar{c}) \equiv u_{N^+} := P_{N^+} u_\gamma$$

where $u_\gamma = u_\gamma(\bar{c}, \bar{c}') = u_\gamma(\bar{c}, A\bar{c})$.

3 Calculation of the solution

Using $\bar{c}'_1 = A_1 \bar{c}_1$ and $\bar{c}'_2 = A_2 \bar{c}_2$ we look for $\bar{c}_1$ and $\bar{c}_2$ so that boundary condition and the transmission condition are satisfied in the sense of the least square method.

With these obtained parameters $\bar{c}_1$ and $\bar{c}_2$, we now have the approximate solution

$$u_Q(x) \approx \begin{cases} u_{D_1}(x, \bar{c}_1), & x \in D_1 \\ u_{D_2}(x, \bar{c}_2), & x \in D_2. \end{cases}$$

Acknowledgement. The work is supported by the RFBR grant No. 02–01–00801

References

1. Ryaben'kii V.S., Difference potentials and its applications, Springer, 2002, p. 538.

The Identification Procedure of the *Cosserat* Material Constants as an Inverse Problem

Bernd Scholz and Wolfgang Ehlers

Institute of Applied Mechanics (CE), Chair II, University of Stuttgart, Germany

Abstract. The *Cosserat* theory is used for the description of granular materials, where – in consequence of the microstructure of the material – additional rotational degrees of freedom come into play. This leads to unsymmetric stress and strain tensors as well as to a couple stress tensor, thus requiring extended constitutive relations with additional material parameters. The goal of the present contribution is the identification of these additional parameters on the basis of the solution of an inverse problem of a non-homogeneous boundary-value problem such as the biaxial test.

1 Physical Model

Kinematics of the Micropolar Theory

In the framework of the *Cosserat* theory, it is assumed that a material point is represented by a rigid microparticle. This applies, in addition to a defined position of the microparticle described by an initial position vector $\mathbf{X}$ at the time t_0, a defined orientation described by the director $\boldsymbol{\Xi}$. Due to the deformation process during time t, the position of the microparticle changes toward $\mathbf{x}$ and the director toward $\boldsymbol{\xi}$ (see Fig. 1).

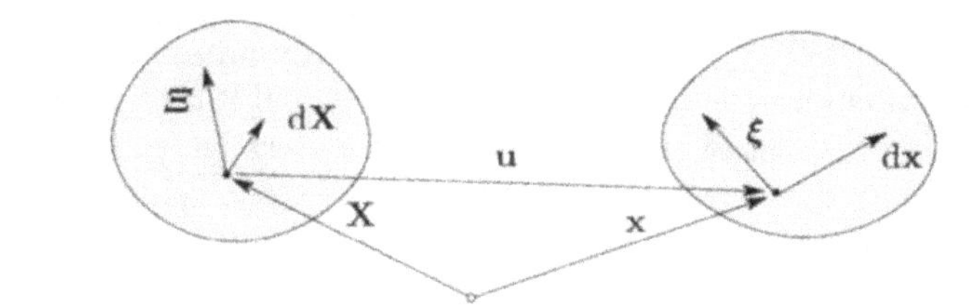

Fig. 1. Micropolar kinematics of a continuum.

In the following description, only the geometrically linear case is considered. The rotation angle $\bar{\varphi}$, which denotes the total rotation of $\boldsymbol{\Xi}$ towards $\boldsymbol{\xi}$, can be split into the continuum rotation $\varphi = \frac{1}{2}\overset{3}{\mathbf{E}}\,(\mathrm{Grad}^T\mathbf{u})$ and the so-called free rotation $\overset{*}{\varphi}$, which is independent of the displacement $\mathbf{u}$ [2,4]. For the construction of *Cosserat* deformation tensors, the difference of the scalar products between line elements and directors are used. This leads to the micropolar strain and curvature tensors

$$\bar{\boldsymbol{\varepsilon}} = \operatorname{Grad}\mathbf{u} + \overset{3}{\mathbf{E}}\,\bar{\boldsymbol{\varphi}} = \tfrac{1}{2}\left(\operatorname{Grad}\mathbf{u} + \operatorname{Grad}^T\mathbf{u}\right) + \overset{3}{\mathbf{E}}\,\overset{*}{\boldsymbol{\varphi}}\,, \tag{1}$$

$$\bar{\boldsymbol{\kappa}} = \operatorname{Grad}\bar{\boldsymbol{\varphi}}\,. \tag{2}$$

Constitutive Equations

The material behaviour is modelled by use of an elasto-plasticity approach with an additive split of the deformation and curvature tensors $\bar{\boldsymbol{\varepsilon}}$ and $\bar{\boldsymbol{\kappa}}$ in elastic and plastic parts:

$$\bar{\boldsymbol{\varepsilon}} = \bar{\boldsymbol{\varepsilon}}\bar{\boldsymbol{\varepsilon}}_{\mathrm{e}} + \bar{\boldsymbol{\varepsilon}}_{\mathrm{p}}\,, \qquad \bar{\boldsymbol{\kappa}} = \bar{\boldsymbol{\kappa}}_{\mathrm{e}} + \bar{\boldsymbol{\kappa}}_{\mathrm{p}}\,. \tag{3}$$

For the relation between the unsymmetric stress and elastic strain tensors $\bar{\boldsymbol{\sigma}}$ and $\bar{\boldsymbol{\varepsilon}}_{\mathrm{e}}$, the *Hooke*an elasticity law is used for the symmetric part $\bar{\boldsymbol{\varepsilon}}_{\mathrm{e\,sym}}$ with the *Lamé* constants μ and λ and is extended by the parameter μ_c to consider the skew-symmetric part $\bar{\boldsymbol{\varepsilon}}_{\mathrm{e\,skw}}$ [1,4]:

$$\bar{\boldsymbol{\sigma}} = 2\,\mu\,\bar{\boldsymbol{\varepsilon}}_{\mathrm{e\,sym}} + \lambda\,(\bar{\boldsymbol{\varepsilon}}_{\mathrm{e\,sym}}\cdot\mathbf{I})\,\mathbf{I} + 2\,\mu_c\,\bar{\boldsymbol{\varepsilon}}_{\mathrm{e\,skw}}\,. \tag{4}$$

The couple stress tensor $\bar{\mathbf{M}}$ is assumed to be directly proportional to the elastic curvature tensor $\bar{\boldsymbol{\kappa}}_{\mathrm{e}}$ including the internal length l_c:

$$\bar{\mathbf{M}} = 2\,\mu_c\,l_c^2\,\bar{\boldsymbol{\kappa}}_{\mathrm{e}}\,. \tag{5}$$

The plastic behaviour can be described by a yield criterion and, in the case of non-associated plasticity, by an additional plastic potential for the evolution of $\bar{\boldsymbol{\varepsilon}}_{\mathrm{p}}$. In the framework of frictional materials, a yield criterion from [3] is used, which implies a set of seven material parameters $\{\alpha,\,\beta,\,\gamma,\,\delta,\,\epsilon,\,\kappa,\,m\}$, and is extended by the skew-symmetric part of the stress tensor as well as by the couple stress tensor in combination with the parameters k_σ and k_M [4]:

$$\begin{aligned} F &= \Phi^{1/2} + k_M\sqrt{\mathrm{II}_M} + \beta\,\mathrm{I} + \epsilon\,\mathrm{I}^2 - \kappa = 0\,,\\ \Phi &= (1+\gamma\,\vartheta)^{\,m}\,\mathrm{II}^D_{\mathrm{sym}} + k_\sigma\mathrm{II}_{\mathrm{skw}} + \tfrac{1}{2}\,\alpha\,\mathrm{I}^2 + \delta^2\,\mathrm{I}^4\,,\\ \vartheta &= \frac{\mathrm{III}^D_{\mathrm{sym}}}{\left(\mathrm{II}^D_{\mathrm{sym}}\right)^{3/2}}\,. \end{aligned} \tag{6}$$

Thereby, I is the first invariant of the stress tensor and $\mathrm{II}^D_{\mathrm{sym}}$ and $\mathrm{III}^D_{\mathrm{sym}}$ are the (negative) second and third invariants of the deviatoric part of the symmetric part of the stress tensor. The (negative) second invariant of the skew-symmetric part of the stress tensor is denoted by $\mathrm{II}_{\mathrm{skw}}$, whereas II_M is the (negative) second invariant of the couple stress tensor. The parameters ϵ and δ of the yield criterion are evolved from the initial values ϵ_0 and δ_0 up to the final values $\overset{\star}{\epsilon}$ and $\overset{\star}{\delta}$ using the hardening constants C_δ and C_ϵ by the following hardening law:

$$\begin{aligned} \dot{\delta} &= C_\delta \, (\overset{\star}{\delta} - \delta) \, \|\dot{\bar{\varepsilon}}_{\mathrm{p\,sym}}\| \,, \\ \dot{\epsilon} &= C_\epsilon \, (\overset{\star}{\epsilon} - \epsilon) \, \|\dot{\bar{\varepsilon}}_{\mathrm{p\,sym}}\| \,. \end{aligned} \tag{7}$$

In order to describe the yield direction, the following plastic potential is used, where ψ_1 and ψ_2 are additional parameters governing the yield direction [7]:

$$\begin{aligned} G &= \Gamma^{1/2} + \psi_2 \, \beta \, \mathrm{I} + \epsilon \, \mathrm{I}^2 \,, \\ \Gamma &= \psi_1 \, \mathrm{II}^D_{\mathrm{sym}} + k_\sigma \mathrm{II}_{\mathrm{skw}} + \tfrac{1}{2} \, \alpha \, \mathrm{I}^2 + \delta^2 \, \mathrm{I}^4 \,. \end{aligned} \tag{8}$$

2 Parameter Identification Method

The identification of material parameters leads to an inverse problem. Following this, the solution $\tilde{\boldsymbol{\Phi}}$ of a given boundary-value problem, i. e. the strain $\bar{\boldsymbol{\varepsilon}}$ or the stress $\bar{\boldsymbol{\sigma}}$, is assumed to be already known from measurements, whereas the unknowns are the material parameters summarized in a set $\boldsymbol{s}$. To solve the inverse problem, the measured values $\tilde{\boldsymbol{\Phi}}$ are compared to the numerical solution $\boldsymbol{\Phi}(\boldsymbol{s})$ of the boundary-value problem. This comparison is carried out by use of the least-squares method, where the minimum of the least-squares function is the solution of the inverse problem

$$f(\boldsymbol{s}) = \sum_{n=1}^{N} \left(\phi_n(\boldsymbol{s}) - \tilde{\phi}_n \right)^2 \quad \rightarrow \quad \min \tag{9}$$

with N discrete points in time and space. In the general case, there exist upper and lower bounds as additional constraints for the solution of the minimizing problem. For the solution of this problem, the *Sequential Quadratic Programming* (SQP) [8] is applied, which requires the computation of the gradient $\nabla_{\boldsymbol{s}} f(\boldsymbol{s})$, where the components of this gradient are denoted as the sensitivities of f. The sensitivities can be calculated by two methods:

Numerical calculation of the gradient

Here, after a calculation of the boundary-value problem with the set $\boldsymbol{s}$ of material parameters, a material parameter s_i is disturbed and the boundary-value problem has to be solved again. With these two results the differential quotient can be numerically evaluated yielding the i-th component of the gradient $\nabla_{\boldsymbol{s}} f(\boldsymbol{s})$. The method causes a high numerical effort, because the boundary-value problem must be solved for every single material parameter.

The semi-analytical sensitivity analysis

For this method, the discrete system equations are directly differentiated. In case of the Finite Element Method, the discretized weak forms of the balance equations (see [4]) are derivated with respect to $\boldsymbol{s}$. This leads to an equation for the determination of the sensitivities of the primary variables ($\nabla_s \mathbf{u}$, $\nabla_s \bar{\varphi}$). With these sensitivities, in the next step, the sensitivities of the solution Φ are calculated leading finally to the sensitivities of f, cf. [5,6].

3 Numerical Results

The efficiency of the above described method is demonstrated by two examples for the determination of the *Cosserat* material parameters, where material parameters of the standard non-polar approach are adapted previously by homogeneous tests. The first one is an example for the sensitivity analysis for the parameters $\boldsymbol{s} = (\mu_c, l_c, k_\sigma, k_M)^T$. The second one for the identification method applied to the parameter re-identification of the parameters $\boldsymbol{s} = (\mu_c, l_c, k_\sigma)^T$ based on a non-homogeneous biaxial test.

Example 1: Sensitivity Analysis Based on the Biaxial Test

In the biaxial test, a quadratic specimen with fixed out-of-plane translations to simulate plane strain conditions is stabilized by a side load. Then, a load platen is driven down and the specimen fails by evolving a shear-band (see Fig. 2).

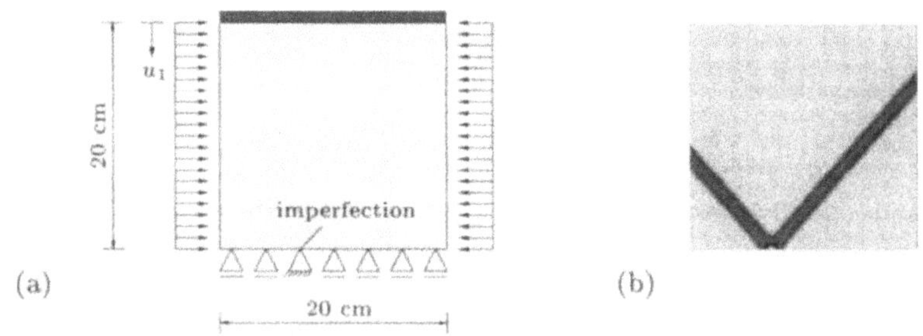

Fig. 2. Biaxial test: (a) set-up, (b) norm of plastic strains in the final state.

For the identification of the material parameters, the measured force on the load platen can be used as well as the thickness and orientation of the shear band. Fig. 3 shows the relative sensitivities of the stress under the load platen in consequence of the parameters of the *Cosserat* theory. Thereby, the sensitivities by use of the semi-analytical sensitivity analysis are compared to the same sensitivities calculated by the numerical method. The semi-analytically calculated curves are shown as solid and the numerically determinated curves as dashed lines. The curves of the sensitivities of the parameters are nearly congruent, so that only a single line for one parameter is visible.

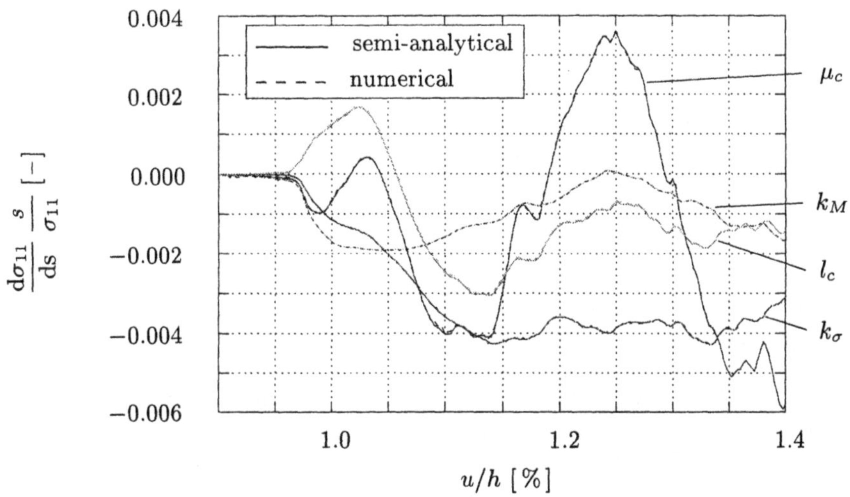

Fig. 3. Comparison between semi-analytical and numerical sensitivities; colored, see page 374

Example 2: Parameter Re-Identification Based on the Biaxial Test

In the parameter re-identification, a calculation with a given set of material parameters is carried out, and the results are taken as virtual test results. Then, the material parameters are varied and, using the method of parameter identification, the original parameters should be recovered. Here, this is done for the parameters μ_c, l_c and k_σ. The parameters are changed in the way

	μ_c	l_c	k_σ
reference parameter	2.0 MN/m^2	0.05 m	12.0 [−]
modified parameter	3.0 MN/m^2	0.01 m	10.0 [−]

Table 1. Modification for the parameter re-identification.

shown in Table 1. After 23 iterations, the original parameters have been re-identified.

4 Conclusions

In the paper, a micropolar model for the description of the behaviour of sand together with a gradient based method for the identification of the material parameters of this model is presented. An example illustrates the application of the proposed method towards a parameter re-identification. The gradient, which is needed for this method, is calculated by the semi-analytical sensitivity analysis, which is also demonstrated by a second numerical example.

In the current research, the method is used for the parameter identification on the basis of real tests, wherefore the gradient based method must be enhanced to find the global minimum of the least-squares function.

References

1. de Borst, R., Sluys, L. (1991) Localisation in a Cosserat continuum under static and dynamic loading conditions. Comput. Methods Appl. Mech. Engrg. **90**, 805–827
2. Diebels, S., Ehlers, W. (1996) On basic equations of multiphase micropolar materials. Technische Mechanik **16**, 77–88
3. Ehlers, W. (1995) A single-surface yield function for geomaterials. Arch. Appl. Mech. **65**, 246–259
4. Ehlers, W., Volk, W. (1998) On theoretical and numerical methods in the theory of porous media based on polar and non-polar elasto-plastic porous solid materials. Int. J. Solids Structures **35**, 4597–4617
5. Kleiber, M. (1997) Parameter Sensitivity in Nonlinear Mechanics. John Wiley & Sons, Chichester
6. Mahnken, R., Stein, E. (1996) A unified approach for parameter identification of inelastic material models in the frame of the finite element method. Comput. Methods Appl. Mech. Engrg. **136**, 225–258
7. Mahnkopf, D. (1999) Lokalisierung fluidgesättigter poröser Medien bei finiten elastoplastischen Deformationen. Dissertation, Bericht aus dem Institut für Mechanik (Bauwesen), Nr. II-5, Universität Stuttgart
8. Spellucci, P. (2001) Program: DONLP2. Technische Universität Darmstadt, www.mathematik.tu-darmstadt.de/ags/ag8/Mitglieder/spellucci_de.html

Taylor-Hood elements in 3D

Christian Wieners

Universität Erlangen-Nürnberg, Institut für Angewandte Mathematik,
Martensstr. 3, D-91058 Erlangen

Abstract. Taylor-Hood elements in 3D yield a stable finite element discretization for saddle point problems on meshes consisting of hexahedra, prisms, pyramids, and tetrahedra. Therefore, they can be used for adaptive computations on locally refined hexahedral meshes with a conforming closure by pyramidal elements.
In this paper we present a suitable construction of conforming pyramidal shape functions, and we establish a corresponding inf-sup condition.

1 Introduction

The construction of hexahedral meshes is a difficult task, especially if one wants to obtain a locally refined mesh for resolving geometrical details or for approximating singularities of the solution. Moreover, hanging nodes are required in a local refinement procedure. Therefore, it is much easier to use tetrahedral meshes. On the other hand, hexahedral meshes provide better approximations than a corresponding tetrahedral mesh with the same number of vertices, in particular in mechanical applications where the approximation with piecewise linear displacements are very poor and severe locking effect can occur.

To achieve optimal flexibility in the choice of meshes, in the refinement procedures, and in the discretization, the software system *UG* [1] supports meshes of mixed types. In particular, a full set of closure rules allows to construct admissible meshes without hanging nodes within one element layer (which is an important feature for minimizing the amount of communication in the parallel version of the mesh adaption). Algorithms for refinement and de-refinement of such meshes are introduced by [5].

Since the conforming connection of hexahedra and tetrahedra require pyramidal elements, it is desirable to have conforming shape functions on pyramids; suitable shape functions are presented in Sect. 2.

Providing a full set of P_1 and P_2 shape functions allows to use Taylor-Hood elements with quadratic velocity and linear pressure for saddle point problems, cf. Fig. 1. For this discretization we establish stability for the Stokes system on mixed type meshes in Sect. 3; this extends the result of VERFÜRTH [6] which is restricted to tetrahedra.

This result makes Taylor-Hood elements now available for a wide range of applications, e. g., for the extension of our numerical simulations of a microscopically incompressible solid skeleton saturated with an incompressible pore-fluid on hexahedra [7] to adaptive meshes. Since this model produces

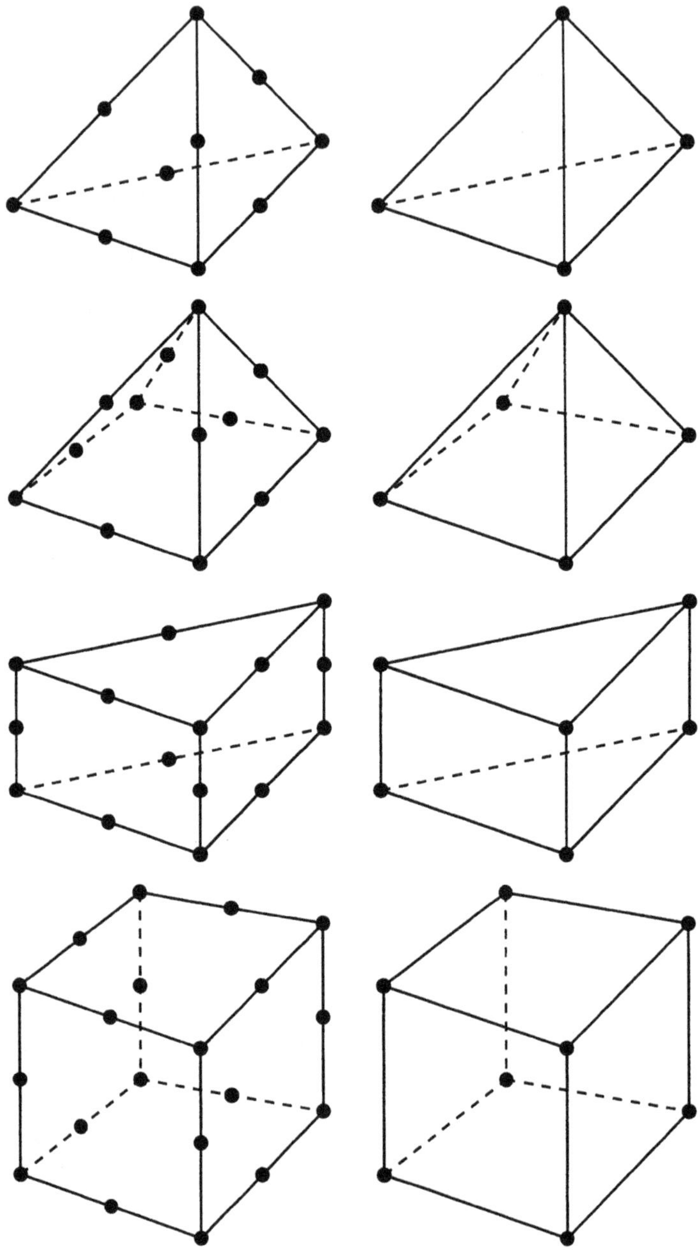

Fig. 1. Taylor-Hood elements in 3D: (multi-)quadratic conforming elements for the displacements u (left) and (multi-)linear elements for the pressure p (right) on a tetrahedron, a pyramid, a prism, and a hexahedron

local effects such as shear bands, we expect a substantial improvement of the accuracy and the efficiency of the computations by local refinement techniques.

2 Conforming shape functions for pyramids

On admissible meshes conforming approximations are desirable. Conforming P_1 and P_2 shape functions for tetrahedra and hexahedra are well known and can be found in many textbooks, e. g. [3]; the extension to prisms is straight forward. The construction of suitable shape functions for pyramids is more involved and not unique. In particular, a conforming family of elements cannot achieved with polynomial shape functions.

Lemma 1. *There exists no continuously differentiable conforming shape functions for the pyramid which are linear resp. bilinear on the faces.*

Proof. Let the reference pyramid be described by the vertices

$$a_0 = (0,0,0),\ a_1 = (1,0,0),\ a_2 = (1,1,0),\ a_3 = (0,1,0),\ a_4 = (0,0,1).$$

We consider a conforming shape function p for the vertex a_0, and we assume that p is continuously differentiable. Then, restricted to the faces we have

$$\begin{array}{lll} \text{(a)} & p(x,y,z) = (1-x)(1-y) & \text{in conv}\{a_0,a_1,a_2,a_3\}, \\ \text{(b)} & p(x,y,z) = 1-x-z & \text{in conv}\{a_0,a_1,a_4\}, \\ \text{(c)} & p(x,y,z) = 1-y-z & \text{in conv}\{a_0,a_3,a_4\}, \\ \text{(d)} & p(x,y,z) = 0 & \text{in conv}\{a_1,a_2,a_4\}, \\ \text{(e)} & p(x,y,z) = 0 & \text{in conv}\{a_2,a_3,a_4\}. \end{array}$$

Using (d) and (e), we obtain for the gradient of p at the vertex a_4

$$(0,-1,1)\cdot\nabla p(a_4) = (-1,0,1)\cdot\nabla p(a_4) = (-1,-1,1)\cdot\nabla p(a_4) = 0.$$

If ∇p is continuous at a_4, this implies $\nabla p(a_4) = 0$, but (b) and (c) gives

$$(0,0,1)\cdot\nabla p(a_4) = -1.$$

This is a contradiction.

The same result holds for quadratic shape functions for pyramids.

Since we have no polynomial shape functions, we describe here one possibility for the construction of appropriate pyramidal shape functions based on a splitting of the pyramid into two tetrahedra. We present P_1 and P_2 shape functions for the reference element in Tables 1 and 2. Restricted to triangular faces the P_2 shape functions are quadratic, and restricted to quadrilateral sides they are elements of the Serendipity class.

The shape functions for the reference element can be transformed easily to arbitrary pyramids. In the isoparametric case, the invertibility of the isoparametric mapping for the given P_1 shape functions are studied by KNABNER-SUMM [4].

$$q_0(x,y,z) = \begin{cases} (1-x)(1-y) + z(y-1) & \text{for } x > y \\ (1-x)(1-y) + z(x-1) & \text{for } x \leqslant y \end{cases}$$
$$q_1(x,y,z) = \begin{cases} x(1-y) - zy & \text{for } x > y \\ x(1-y) - zx & \text{for } x \leqslant y \end{cases}$$
$$q_2(x,y,z) = \begin{cases} xy + zy & \text{for } x > y \\ xy + zx & \text{for } x \leqslant y \end{cases}$$
$$q_3(x,y,z) = \begin{cases} (1-x)y - zy & \text{for } x > y \\ (1-x)y - zx & \text{for } x \leqslant y \end{cases}$$
$$q_4(x,y,z) = z$$

Table 1. P_1 shape functions q_i on pyramids (the polynomials q_i correspond to the nodal points a_i)

$$p_0(x,y,z) = \begin{cases} ((1-x)(1-y) + z(y-1))(1-2x-2y-2z) & \text{for } x > y \\ ((1-x)(1-y) + z(x-1))(1-2x-2y-2z) & \text{for } x \leqslant y \end{cases}$$
$$p_1(x,y,z) = \begin{cases} (x(1-y) - zy)(2x-2y-1) & \text{for } x > y \\ (x(1-y) - zx)(2x-2y-1) & \text{for } x \leqslant y \end{cases}$$
$$p_2(x,y,z) = \begin{cases} (xy + zy)(2x+2y+2z-3) & \text{for } x > y \\ (xy + zx)(2x+2y+2z-3) & \text{for } x \leqslant y \end{cases}$$
$$p_3(x,y,z) = \begin{cases} ((1-x)y - zy)(2y-2x-1) & \text{for } x > y \\ ((1-x)y - zx)(2y-2x-1) & \text{for } x \leqslant y \end{cases}$$
$$p_4(x,y,z) = z(2z-1)$$
$$p_{01}(x,y,z) = \begin{cases} 4x((1-x)(1-y) + z(y-1)) - 2yz(1-x-z) & \text{for } x > y \\ 4x((1-x)(1-y) + z(x-1)) - 2xz(1-y-z) & \text{for } x \leqslant y \end{cases}$$
$$p_{12}(x,y,z) = \begin{cases} 4y(x(1-y) - zy) + 2yz(1-x-z) & \text{for } x > y \\ 4y(x(1-y) - zx) + 2xz(1-y-z) & \text{for } x \leqslant y \end{cases}$$
$$p_{23}(x,y,z) = \begin{cases} 4x((1-x)y - zy) + 2yz(1-x-z) & \text{for } x > y \\ 4x((1-x)y - zx) + 2xz(1-y-z) & \text{for } x \leqslant y \end{cases}$$
$$p_{03}(x,y,z) = \begin{cases} 4y((1-x)(1-y) + z(y-1)) - 2yz(1-x-z) & \text{for } x > y \\ 4y((1-x)(1-y) + z(x-1)) - 2xz(1-y-z) & \text{for } x \leqslant y \end{cases}$$
$$p_{04}(x,y,z) = \begin{cases} 4z((1-x)(1-y) + z(y-1)) & \text{for } x > y \\ 4z((1-x)(1-y) + z(x-1)) & \text{for } x \leqslant y \end{cases}$$
$$p_{14}(x,y,z) = \begin{cases} 4z(x(1-y) - zy) & \text{for } x > y \\ 4z(x(1-y) - zx) & \text{for } x \leqslant y \end{cases}$$
$$p_{24}(x,y,z) = \begin{cases} 4z(xy - zy) & \text{for } x > y \\ 4z(xy - zx) & \text{for } x \leqslant y \end{cases}$$
$$p_{34}(x,y,z) = \begin{cases} 4z((1-x)y + zy) & \text{for } x > y \\ 4z((1-x)y + zx) & \text{for } x \leqslant y \end{cases}$$

Table 2. P_2 shape functions p_i, p_{ij} on pyramids (the polynomials p_i correspond to the nodal points a_i, and the polynomials p_{ij} correspond to the nodal points on the edge mid points $(a_i + a_j)/2$)

3 Stability for Taylor-Hood elements in 3D

Let $\Omega \subset \mathbf{R}^3$ be a polyhedral domain decomposed into tetrahedra, pyramids, prisms and hexahedra $E \in \mathcal{E}_h$ such that $\bar{\Omega} = \bigcup_{E\in\mathcal{E}_h} \bar{E}$. We assume

$$c\, h_E \leqslant |e| \leqslant C\, h_E \qquad \text{for all edges } e \subset \partial E, \tag{1}$$

where h_E denotes the diameter of the element, and $|e|$ denotes the length of e. Here and in the following, all constants will be independent of h_E and only dependent on the angles of the elements. In addition, we assume that every element has only one boundary face, i. e., $\partial\Omega \cap \partial E$ is empty or an element face. Let $V_h \subset H_0^1(\Omega)^3$ be the space of (multi-)quadratic Lagrange elements on Ω, and let $Q_h \subset L^2(\Omega)$ be the space of (multi-)linear Lagrange elements.

We consider the Stokes problem: for a given right hand side $f \in L^2(\Omega)^3$, find $u_h \in V_h$ and $p_h \in Q_h$ such that

$$\begin{array}{llll} (\nabla u_h, \nabla v_h)_{0,\Omega} + (\operatorname{div} v_h, p_h)_{0,\Omega} & = & (f, v_h)_{0,\Omega}, & v_h \in V_h, \\ (\operatorname{div} u_h, q_h)_{0,\Omega} & = & 0 & q_h \in Q_h. \end{array} \tag{2}$$

We want to establish that $V_h \times Q_h$ is a stable discretization, i. e., a constant $\beta_0 > 0$ (independent of the mesh parameter h) exists such that

$$\beta_0 \inf_{c\in\mathbf{R}} \|q_h + c\|_{0,\Omega} \leqslant \sup_{v_h\in V_h} \frac{b(v_h, q_h)}{\|v_h\|_{1,\Omega}} \tag{3}$$

is satisfied.

For the stability proof of Taylor-Hood elements on triangles, quadrilaterals, and tetrahedra we refer to [2]. Here, we describe in detail the modifications of the stability proof which are required for a mesh in 3D with mixed element types.

We follow the approach introduced by [6] for tetrahedra (see also [2, Chap. VI]). Therefore, we introduce the mesh dependent norm

$$\|q\|_{Q_h}^2 = \sum_{E\in\mathcal{E}_h} h_E^2\, \|\nabla q\|_{0,E}^2$$

(note that $\|\cdot\|_{Q_h}$ vanishes for constant functions).

The proof is based on the explicit construction of a bounded linear operator $P_h \colon Q_h \longrightarrow V_h$ satisfying

$$b(P_h(q_h), q_h) \geqslant \beta_1\, \|P_h(q_h)\|_{1,\Omega} \|q_h\|_{Q_h}, \qquad q_h \in Q_h. \tag{4}$$

Then, the inf-sup condition holds with respect to $\|\cdot\|_{Q_h}$, i. e.,

$$\beta_1\, \|q_h\|_{Q_h} \leqslant \sup_{v_h\in V_h} \frac{b(v_h, q_h)}{\|v_h\|_{1,\Omega}}. \tag{5}$$

Together with the inf-sup condition of the continuous problem

$$\beta_2 \inf_{c\in\mathbf{R}} \|q_h + c\|_{0,\Omega} \leqslant \sup_{v\in H_0^1(\Omega)^3} \frac{b(v,q_h)}{\|v\|_{1,\Omega}} \tag{6}$$

we now obtain the stability result for Taylor-Hood elements.

Lemma 2. *Assume that a bounded linear operator $P_h\colon Q_h \longrightarrow V_h$ satisfying* (4) *exists. Then, the discrete inf-sup condition* (3) *is valid.*

The proof follows [2, Lem. VI.6.1].

Proof. Due to (1) a bounded interpolation operator $\Pi_h\colon H_0^1(\Omega)^3 \longrightarrow Q_h$ exists satisfying $\|\Pi_h v\|_{1,\Omega} \leqslant C_1 \|v\|_{1,\Omega}$ and

$$\begin{aligned}(\operatorname{div}(v-\Pi_h v), q)_{0,\Omega} &= -(v-\Pi_h v, \nabla q)_{0,\Omega} \\ &\leqslant C_2 \sum_{E\in\mathcal{E}_h} h_E \|v\|_{1,E}\|\nabla q\|_{0,E} \leqslant C_2 \|v\|_{1,\Omega}\|q\|_{Q_h}\end{aligned} \tag{7}$$

for $v \in H_0^1(\Omega)^3$ with constants $C_1, C_2 > 0$.
We fix $\theta \in (0,1)$. For a given $q_h \in Q_h$ we can find $v \in H_0^1(\Omega)^3$ such that

$$\theta\beta_2 \inf_{c\in\mathbf{R}} \|q_h + c\|_{0,\Omega} \leqslant \frac{b(v,q_h)}{\|v\|_{1,\Omega}} \tag{8}$$

using (6). Inserting Π_h we obtain

$$\begin{aligned}\sup_{v_h\in V_h} \frac{b(v_h,q_h)}{\|v_h\|_{1,\Omega}} &\geqslant \frac{b(\Pi_h v,q_h)}{\|\Pi_h v\|_{1,\Omega}} \\ &\geqslant \frac{1}{C_1}\,\frac{b(\Pi_h v,q_h)}{\|v\|_{1,\Omega}} = \frac{1}{C_1}\,\frac{b(v,q_h)}{\|v\|_{1,\Omega}} + \frac{1}{C_1}\,\frac{b(\Pi_h v - v,q_h)}{\|v\|_{1,\Omega}} \\ &\geqslant \frac{\theta\beta_2}{C_1} \inf_{c\in\mathbf{R}} \|q_h + c\|_{0,\Omega} - \frac{C_2}{C_1}\|q_h\|_{Q_h}.\end{aligned}$$

Combined with the inf-sup condition in the mesh-dependent norm (5) this gives (3) with $\beta_0 = \theta\beta_1\beta_2/(\beta_1 C_1 + C_2)$.

It remains to construct a suitable operator $P_h\colon Q_h \longrightarrow V_h$ on element level. Therefore, let $t_e = a_i - a_j \in \mathbf{R}^3$ be the tangent vector for an edge e with the vertices a_i, a_j of length $|e|$ (with some fixed orientation) and let a_e be the midpoint of the edge. For $E \in \mathcal{E}_h$, $e \subset \partial E \setminus \partial\Omega$ and $q_h \in Q_h$ let $v_e = P_e(q_h) \in V_h$ be defined by the nodal values

$$v_e(a_e) = (q(a_i) - q(a_j))(a_j - a_i)$$

and $v_e(a) = 0$ for all other nodal points of V_h; together, this gives

$$v_h = P_h(q_h) := \sum_{e\not\subset\partial\Omega} P_e(q_h) \in V_h.$$

The construction of P_h is illustrated (for simplicity only on quadrilaterals) in Fig. 2.

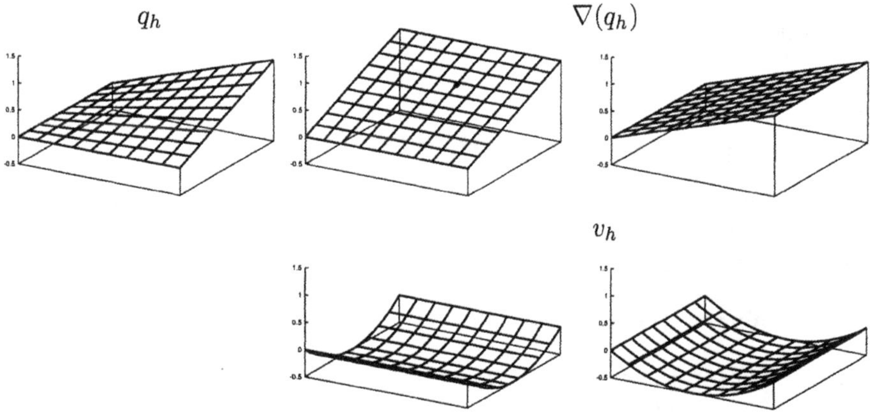

Fig. 2. Construction of $v_h = P_h(q_h)$ on the reference quadrilateral $\hat{E}$: set $v_h(a_i) = 0$ for the element corners a_i, and on the edge e from a_i to a_j define $v_h(a_e) = (q_h(x_i) - q_h(x_j))(x_j - x_i)$

We have $||v_h||_{1,E} \leqslant C\ h_E\ ||\nabla q||_{0,E}$, which yields

$$||v_h||_{1,\Omega} \leqslant C \left(\sum_E h_E^2\ ||\nabla q||_{0,E}^2 \right)^{1/2} . \tag{9}$$

Thus, the operator $P_h\colon Q_h \longrightarrow V_h$ is continuous.

Lemma 3. *For $q_h \in Q_h$, $E \in \mathcal{E}_h$ and $v_h = P_h(q_h)$ we have*

$$-\int_E \nabla q_h \cdot v_h\, dx \geqslant C\ h_E^2\ ||\nabla q_h||_{0,E}^2. \tag{10}$$

Proof. Due to condition (1) the inequality (10) must be checked for the reference elements $\hat{E}$ only; then, the assertion follows directly from the transformation formula for $\hat{E}$ to E giving the factor h_E^{-1} for the transformation of ∇q_h and the factor h_E for the transformation of $P_h(q_h)$.
Let $q_1, ..., q_k$ be the P_1-shape functions of $\hat{E}$. Define the matrices

$$M = (m_{ij}) := \left(-\int_{\hat{E}} \nabla q_i \cdot P_{\hat{E}}(q_j)\, dx \right),$$

where $P_{\hat{E}} = \sum_{e \subset \partial \hat{E}} P_e$, and

$$N = (n_{ij}) := \left(\int_{\hat{E}} \nabla q_i \cdot \nabla q_j\, dx \right).$$

Both matrices are positive semidefinite, and restricted to $Q_h/\mathbf{R}$ they are positive definite. In addition, one has to check the case that one face is a boundary face $F \subset E$, i. e., replace $P_{\hat{E}}$ by

$$P_{\hat{E},F} = \sum_{e \subset \partial\hat{E},\ e \not\subset F} P_e.$$

Direct computation yields

$$\alpha^T M \alpha \geqslant \hat{C}\ \alpha^T N \alpha, \qquad \alpha \in \mathbf{R}^k, \qquad \hat{C} \geqslant 0.0246$$

for the reference tetrahedron, pyramid, prism and hexahedron and all cases of boundary faces. Then, a linear combination gives the assertion for the reference element.

Summing up the elements, (10) gives

$$-(P_h(q_h), \nabla q_h)_{0,\Omega} \geqslant c\ \|q_h\|^2_{Q_h}, \qquad q_h \in Q_h.$$

Inserting (9) yields condition (4), which finally proves stability in Lem. 2.

Remark 1. Only for tetrahedra a simple argument exists providing a positive constant $\hat{C}$, because in this case ∇q_h is piecewise constant. In general, the computation of a lower bound for $\hat{C}$ requires the explicit construction of the matrices M and N and rough eigenvalue bounds for the generalized eigenvalue problem $M\alpha = \lambda N \alpha$ in all 24 cases. This is realized in *UG*.

Remark 2. The restriction to at most one boundary face can be weaken to the condition of at least three inner edges. Moreover, the result extends to the case that Dirichlet boundary conditions are prescribed on $\Gamma \subset \partial\Omega$, where Γ has positive measure.

References

1. P. Bastian, K. Birken, K. Johannsen, S. Lang, N. Neuß, H. Rentz-Reichert, and C. Wieners. UG – a flexible software toolbox for solving partial differential equations. *Computing and Visualization in Science*, 1:27–40, 1997.
2. F. Brezzi and M. Fortin. *Mixed and Hybrid Finite Element Methods.* Springer, 1991.
3. P. G. Ciarlet. *The Finite Element Method for Elliptic Problems.* North Holland, 1978.
4. P. Knabner and G. Summ. The invertibility of the isoparametric mapping for pyramidal and prismatic elements. *Numer. Math.*, 88:661–681, 2001.
5. S. Lang. *Parallele Numerische Simulation instationärer Probleme mit adaptiven Methoden auf unstrukturierten Gittern.* PhD thesis, Universität Stuttgart, 2001.
6. R. Verfürth. Error estimates for a mixed finite element approximation of the Stokes equation. *RAIRO Anal. Numer.*, 18:175–182, 1984.
7. C. Wieners, M. Ammann, S. Diebels, and W. Ehlers. Parallel 3-d simulations for porous media models in soil mechanics. *Comput. Mech.*, 29:75–87, 2002.

Part III

Solid Mechanics

Singularities of surface-breaking cracks in bi-material interfaces

Atanas Dimitrov and Eckart Schnack

Institute of Solid Mechanics, Karlsruhe University,
Kaiserstr. 12, D-76128 Karlsruhe, Germany

Abstract. In the present work we address a numerical method for calculation of corner singularities in three-dimensional elastic domains containing corners, cracks or multi-material interfaces. It is based on a separation of variables and a Galerkin-Petrov finite-element approximation and leads to a quadratic eigenvalue problem from which the singularity exponents and the angular functions in the asymptotical expansion of the solution are obtained as eigenpairs. By this method the practically important problem of a surface-breaking crack in a bi-material interface is studied. The singularity exponents are reported in dependence on different geometrical parameters and the material properties.

Keywords: Bi-material interface, Composite material, Delamination, Surface-breaking crack, Corner singularity.

1 Introduction

The use of composites in various application areas permanently increases during the last years. Especially fiber-reinforced materials are very popular, because some features as high stiffness-to-weight and strength-to-weight ratios make them very attractive for some advanced structures. Nevertheless, the applicability of such materials remains in general limited, due to an essential shortage of design data: As it is well known, elastic structures composed of different materials are often characterized by singular stress fields [1,2]. This undesirable phenomenon usually leads to the appearance and growth of cracks up to destruction.

In order to obtain the fracture behavior of such an structure, the local state near a singular point may be incorporated in a fracture criterion within a generalized Griffith theory [3,4]. However, the investigation of such problems is very difficult if some real 3D-effects and anisotropic material properties are considered, since analytical methods are not available and standard numerical algorithms fail due to lack of convergence.

In Section 2 we address a numerical method for calculation of corner singularities in three-dimensional elastic domains possibly containing corners, cracks or multi-material interfaces. It has been introduced in the previous works [5,6] for isotropic problems and extended in [7] to anisotropic materials. By this method the important problem of a surface-breaking crack in a bi-material interface is studied in Section 3. The singularity exponents are reported in dependence on different geometrical parameters and the material properties.

2 Numerical calculation of corner singularities in 3D

From a theoretical point of view it is well known that the linear elastic solution may contain gradient singularities, if the domain of consideration includes corners or sudden changes in the material properties as in the case of composites, see for instance [1,2] and the references therein for a survey. For the treatment of such situations in $\mathbb{R}^3$ it can be helpful to expand the solution in the neighborhood of the singular point in an asymptotical series

$$\boldsymbol{u} = \sum_{\mathrm{i}} K_{\mathrm{i}}\, r^{\lambda_{\mathrm{i}}}\, \boldsymbol{U}_{\mathrm{i}}(\theta, \varphi), \tag{1}$$

where r, θ, φ are the spherical coordinates, λ_{i} are the singularity exponents, $\boldsymbol{U}_{\mathrm{i}}$ are the so called angular functions and K_{i} are the corner stress intensity factors (CSIFs). The asymptotical series (1) can be explicitly constructed only in very special cases. For general three-dimensional problems, however, some numerical methods are needed. In the following we briefly describe a technique introduced in the previous works [5,6] for isotropic problems and extended in [7] to anisotropic materials.

In order to keep our considerations sufficiently general, we introduce an abstract mixed boundary value problem of linear elasticity in a bounded domain $\Omega \subset \mathbb{R}^3$ which coincides with a cone $\mathcal{K}$ in the ϵ-vicinity U_O^ϵ of the origin O, so that $\Omega_O^\epsilon := \mathcal{K} \cap U_O^\epsilon = \Omega \cap U_O^\epsilon$, see Fig. 1. We denote with S the section of the cone with the unit sphere. By this general model every geometry containing a conical singular point can be described (note that the shape of S is not specified).

Our aim is to find all solutions, often called the *eigenstates*, which satisfy the differential equation of elasticity and the homogeneous boundary conditions in a sufficiently small neighborhood Ω_O^ϵ of O, i.e. on $\Gamma_0 \cup \Gamma_1$, but not the boundary conditions remote from O, i.e. on Γ_{T}. $\Gamma_0, \Gamma_1, \Gamma_T$ denote the Dirichlet, Neumann and transmissional part of the boundary $\partial\Omega_O^\epsilon$, respectively. We obtain the eigenstates as solutions of the following weak formulation:

Problem 1 *Find* $\boldsymbol{u} \in [H^1(\Omega_O^\epsilon)]^3$*, so that*

$$\mathcal{B}(\boldsymbol{u}, \boldsymbol{v}) := \int_{\Omega_O^\epsilon} \boldsymbol{\sigma}^{\mathrm{T}}(\boldsymbol{u})\, \boldsymbol{\varepsilon}(\boldsymbol{v})\, \mathrm{d}\Omega = 0 \quad \forall \boldsymbol{v} \in [H_0^1(\Omega_O^\epsilon)]^3. \tag{2}$$

The r.h.s. in (2) vanishes because of the homogeneous boundary conditions. The stress and strain vectors, which contain all independent components of the stress and strain tensors respectively, are defined by

$$\boldsymbol{\sigma}^{\mathrm{T}} := [\sigma_{\mathrm{x}}, \sigma_{\mathrm{y}}, \sigma_{\mathrm{z}}, \tau_{\mathrm{xy}}, \tau_{\mathrm{xz}}, \tau_{\mathrm{yz}}], \tag{3a}$$

$$\boldsymbol{\varepsilon}^{\mathrm{T}} := [\varepsilon_{\mathrm{x}}, \varepsilon_{\mathrm{y}}, \varepsilon_{\mathrm{z}}, 2\varepsilon_{\mathrm{xy}}, 2\varepsilon_{\mathrm{xz}}, 2\varepsilon_{\mathrm{yz}}]. \tag{3b}$$

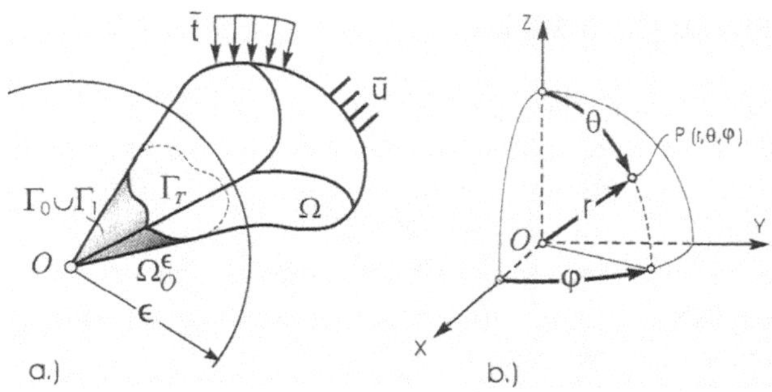

Fig. 1. a.) Boundary value problem with a singular point O. b.) Spherical and cartesian coordinate systems in O.

With $[H_0^1(\Omega_O^\epsilon)]^3$ we denote the space of admissible (with respect to the boundary conditions) displacement fields.

We now consider a finite-element approximation of the weak problem (2). Introducing two different finite-dimensional subspaces $U_h \neq V_h$ as trial and test spaces we obtain

Problem 2 *Find an approximate solution* $\boldsymbol{u}^h \in U_h \subset [H^1(\Omega_O^\epsilon)]^3$, *so that*

$$\mathcal{B}(\boldsymbol{u}^h, \boldsymbol{v}^h) = 0, \quad \forall \boldsymbol{v}^h \in V_h \subset [H_0^1(\Omega_O^\epsilon)]^3. \tag{4}$$

Such a formulation is called *Galerkin-Petrov method.*

The spaces U_h, V_h are obtained by the partitioning of S into triangular finite elements Δ_i. According to (1) the trial and test functions in a typical space sector $(r, \theta, \varphi) \in [0, \epsilon] \times \Delta_i$ can be expressed by

$$\boldsymbol{u}_i^h(r, \theta, \varphi) = r^\lambda \, \boldsymbol{N}(\theta, \varphi) \, \boldsymbol{T}_d^{-1} \, \boldsymbol{d}_i, \tag{5a}$$

$$\boldsymbol{v}_i^h(r, \theta, \varphi) = \Phi(r) \, \boldsymbol{N}(\theta, \varphi) \, \boldsymbol{T}_d^{-1} \, \boldsymbol{b}_i, \tag{5b}$$

where $\boldsymbol{N}$ collects some shape functions and $\boldsymbol{d}_i$ denotes the nodal displacement vector. Equation (5a) comprises the main idea of the method: By a semi-continuous approach, in which the discretization is performed only with respect to θ, φ, a separation of variables can be enforced, so that the final problem is independent of r [5,6].

Introducing (5a) into the expression for the discrete bilinear form (4) and considering the constitutive relation $\boldsymbol{\sigma}(\boldsymbol{u}_i^h) = \boldsymbol{C}\,\boldsymbol{\varepsilon}(\boldsymbol{u}_i^h)$ we obtain after some additional transformations and the substitution $\lambda = \bar{\lambda} - 1/2$ a quadratic eigenvalue problem

$$\left[\boldsymbol{P} + \bar{\lambda}\boldsymbol{Q} + \bar{\lambda}^2\boldsymbol{R}\right] \, \boldsymbol{d} = \boldsymbol{0}, \tag{6}$$

from which the singularity exponents λ_i can be calculated as eigenvalues and the angular functions $\boldsymbol{U}_i$ as eigenvectors. The material matrix $\boldsymbol{C}$ is defined for different material models in [7], the definition of the matrices $\boldsymbol{P}, \boldsymbol{Q}$ and $\boldsymbol{R}$ as well a suitable solution technique for the quadratic eigenvalue problem (6) based on the iterative Arnoldi method can be found in [5,6].

3 Numerical Examples

In this section we study the surface-breaking crack in the perfect interface of two materials. This problem has been analyzed for two equal isotropic materials in [5] and for two different isotropic materials with a fixed Young's moduli ratio E_1/E_2 in [4]. We investigate here the influence of E_1/E_2 on the singular behavior and present also some results for the case of two transversely-isotropic materials, which is a more realistic model of a surface delamination.

The two materials take the space between $(\theta,\varphi)_1 \in (0,\pi/2)\times(0,\pi)$ and $(\theta,\varphi)_2 \in (\pi/2,\pi)\times(0,\pi)$, which is also the domain, which have to be discretized for the numerical solution. The crack is considered as stress free at $(\theta,\varphi) \in (\pi/2)\times(0,\alpha)$, where $\alpha \in (0,\pi)$ denotes a parameter. The x-z plane denotes here the free surface.

Consider first two different, isotropic materials with Poisson's ratio $\nu = 0.3$ and Young's moduli E_1 and E_2. The results, depending on α and E_1/E_2, are shown in Fig. 2. The three exponents, which have been calculated in [5] for $E_1/E_2 = 1$, i.e. in the homogeneous case are also given for comparison. They will be denoted as the symmetrical resp. anti-symmetrical solution, because the corresponding crack-opening modes are either symmetrical (one mode) or anti-symmetrical (two modes) with respect to the crack plane [5].

The exponents are real almost in the whole range $\alpha \in (0,\pi)$, as can be seen from Fig. 2. Only for $\alpha \to \pi$ one real and two complex conjugate exponents appear. The corresponding imaginary parts are shown in the lower right corner of Fig. 2 top and bottom. For small ratios $E_1/E_2 = 2$ only one exponent is slightly affected in comparison with the symmetrical solution, the other two exponents coincide practically with the anti-symmetrical solution, Fig. 2 top. However, for large ratios $E_1/E_2 = 10$ the dominant exponent is significantly smaller and therefore stronger as in the homogeneous case, especially for small angles α, Fig. 2 bottom. The ratio $E_1/E_2 = \infty$ denotes the limit, where material 1 is rigid and the ligament of material 2 is considered as clamped.

Consider next two materials with the following transversely isotropic properties: $E_\mathrm{F} = 1.42\cdot10^5$ MPa, $E_\mathrm{T} = 8.2\cdot10^3$ MPa, $G_\mathrm{FT} = 4.9\cdot10^3$ MPa and $\nu_\mathrm{FT} = \nu_\mathrm{T} = 0.3$, which are typical for some high-modulus graphite-epoxy systems. We denote with F the fiber direction and with T an arbitrary direction in the transversely plane. The fibers lie in the x-y plane and have different orientations ζ_1,ζ_2 in the two layers with respect to the positive x-axis. We consider the so called cross-ply case [±45°] given by $\zeta_1 = -45°$ and $\zeta_2 = +45°$.

The singularity exponents, depending on the angle α are given in Fig. 3. The different parts of the graph are denoted for simplicity with numbers from /1/ to /10/. For small angles α one real /1/ and two complex conjugate exponents /2/ appear. The imaginary part of /2/ decreases and vanishes at a certain angle α, at which the real part of /2/ splits in two real exponents /3/ and /4/. After that, the exponent /4/ joins the exponent /1/ and again

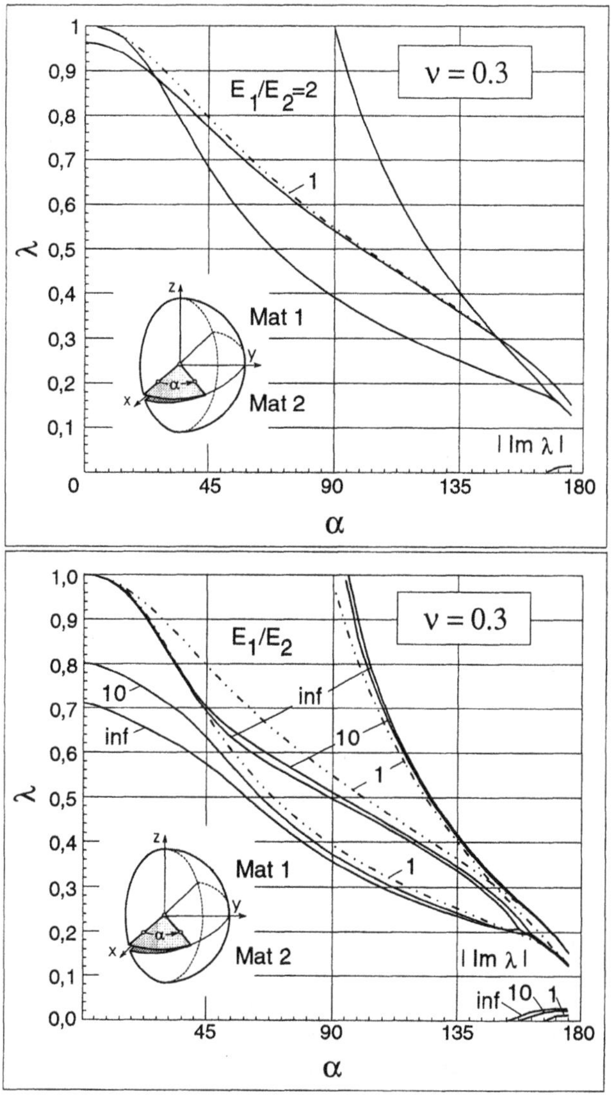

Fig. 2. Singularity exponents λ for the surface-breaking crack in a interface of two isotropic materials.

two complex conjugate exponents /5/ appear. However the imaginary part of /5/ vanishes at another angle α and its real part splits in two real exponents /7/ and /8/. At the same time another real exponent /6/ is coming from the region $\lambda > 1$, joining /7/ and becoming complex /9/. And last but not least, an additional real exponent /10/ is present for large angles α. As can be seen, the resulting singular behavior of this problem is really non-trivial.

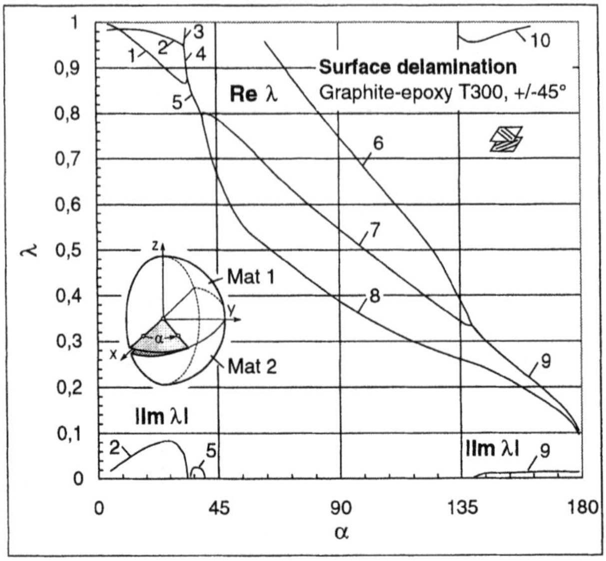

Fig. 3. Singularity exponents λ for the surface-breaking crack in a interface of two transversely-isotropic materials.

By this examples we have demonstrated, that the numerical method, presented in Section 2 is also applicable to multi-material structures with isotropic or anisotropic properties and cracks.

References

1. Leguillon D., Sanchez-Palencia E. (1987) Computation of Singular Solutions in Elliptic Problems and Elasticity. Masson, Wiley, Paris
2. Nicaise S., Sändig A.-M. (1994) General interface problems, Part 1&2. Mathematical Methods in the Applied Sciences, **17**, 395–450
3. Leguillon D., Sanchez-Palencia E. (1992) Fracture in heterogeneous materials. Weak and strong singularities. In P. Ladaveze and O.C. Zienkiewicz (Eds.) New advances in Computational Structural Mechanics. Elsevier, 423–434
4. Leguillon D., Sanchez-Palencia E. (1999) On 3D cracks intersecting a free surface in laminated composites. International Journal of Fracture, **99**, 25–40
5. Dimitrov A., Andrä H., Schnack E. (2001) Efficient computation of order and mode of corner singularities in 3D - elasticity. International Journal for Numerical Methods in Engineering, **52**, 805–827
6. Dimitrov A., Schnack E. (2002) Asymptotical expansion in non-Lipschitzian domains — a numerical approach. Numerical Linear Algebra with Applications, to appear
7. Dimitrov A., Andrä H., Schnack E. (2002) Singularities near three-dimensional corners in composite laminates. International Journal of Fracture, to appear

Multi-time Scale Simulation of Longitudinal Impact Responses

Bin Hu[1], Werner Schiehlen[1], and Peter Eberhard[2]

[1] Institute B of Mechanics, University of Stuttgart, Pfaffenwaldring 9, D-70550 Stuttgart, Germany
[hbi, wos]@mechb.uni-stuttgart.de

[2] Institute of Applied Mechanics, University of Erlangen Egerlandstr. 5, D-91058 Erlangen, Germany
eberhard@ltm.uni-erlangen.de

Abstract. For efficient simulation of impact responses with sufficient accuracy, multibody system methods suitable for large time scales and elastodynamic approaches appropriate for small time scales should be combined. During the contact phase the short time impact response is computed using elastodynamics for accuracy, while during the separation phase the long term motion is obtained using the method of multibody systems for efficiency. This paper deals with a varying time scale simulation for impact responses including both short time elastodynamics and its stepwise transition to long term rigid body dynamics for longitudinal impact responses. The application of this varying time scale simulation is experimentally validated for the impact of a sphere colliding with an elastic rod.

1 Introduction

In the history of mechanics, longitudinal wave propagation in rods has been intensively investigated, see Szabó [8]. Many theoretical results about longitudinal impacts were reviewed by Goldsmith [3]. Recently, computer algebra systems were used to solve longitudinal impact systems and closed-form solutions for the longitudinal wave, based on the St. Venant impact theory developed for perfectly planar contacts, are available, see e.g. Hu and Eberhard [5]. Experimental investigations of longitudinal wave propagations in elastic rods were reviewed by Al-Mousawi [1]. With the advancement of laser techniques, some additional aspects of wave propagation can be observed with Laser-Doppler-Vibrometers nowadays. Recent experimental investigation of a sphere colliding with a rod can be found, for example, in Hu, Eberhard and Schiehlen [6] where some comparisons between the measurements and numerical simulations with different impact theories are also discussed. In this paper, the impact responses of a sphere colliding with an elastic rod during the transition from the wave propagation to the rigid body motion is further simulated by using a multi-time scale simulation. The short time behavior of the wave propagation during contact is modeled using an elastodynamic approach with a very small time scale. During separation the transient behavior of the elastic vibrations for the deformation of the rod is simulated with the method of flexible multibody systems and the long-term rigid body motion is simulated with the method of rigid multibody systems. In the simulation of the impact responses during separation, the amplitudes of elastic vibrations are monitored. If an elastic coordinate after some time remains always small enough, this elastic coordinate will not longer be considered in the simulation. Due to the material damping all elastic vibrations induced by the impact will vanish and the system has then only the rigid body motion. Using this varying time scale method, both efficiency and accuracy of simulation are achieved.

2 Mathematical modeling of impact responses

The schematic description of the impact system and its experimental setup are shown in Fig. 1. A steel sphere with radius $R_1 = 1.5\,cm$ and mass $m_2 = 0.11\,kg$ is considered as a rigid body and used to strike a brass rod with density $\rho = 8430\,kg/m^3$ and cross-sectional area $A = \pi 10^{-4}\,m^2$. The homogeneous cylindrical rod has diameter $20\,mm$ and length $L = 1.5\,m$. The brass rod has Young's modulus $E = 100\,GPa$ and Poisson's ratio $\nu = 0.36$. It is assumed that the rod is initially at rest and is struck on the end $x = 0$ at the time $t = 0$ by the moving rigid sphere with the initial velocity v_0. The displacement of the rigid sphere at time t is denoted by $y(t)$ and the displacement of the rod at position x and time t is given by $u(x,t)$.

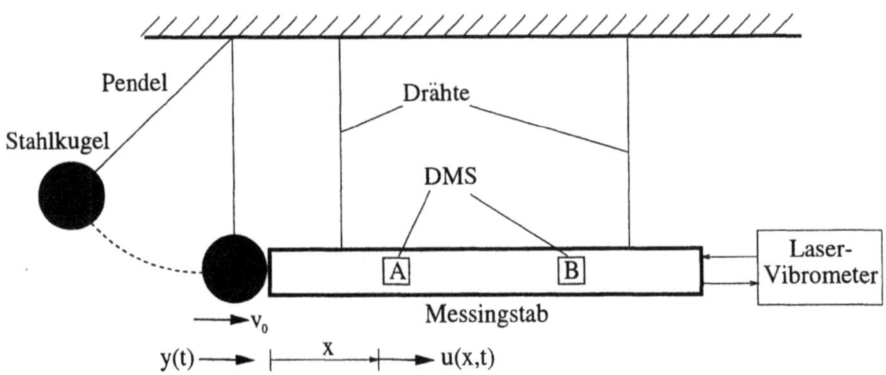

Fig. 1. Schematic description of the impact system and the experimental setup

For the simulation of impact responses, it is advantageous to divide the processes in two phases: contact and separation. The motion states of the colliding bodies can be detected by means of collision detections, see, for example, Eberhard and Jiang [2]. During contact the colliding bodies are coupled by the contact force, while during separation the contact force does not exist any more. In the simulation the boundary approach for the elastodynamics of the rod during contact and the method of multibody systems for the impact responses during separation are combined.

2.1 Model during contact

The governing partial differential equation for the longitudinal wave in the rod reads as

$$\frac{\partial^2 u(x,t)}{\partial t^2} = c^2 \frac{\partial^2 u(x,t)}{\partial x^2} \qquad \text{with} \qquad c = \sqrt{E/\rho}\,. \tag{1}$$

Due to D'Alembert the general solution of (1) is given as

$$u(x,t) = f(ct - x) + g(ct + x) \tag{2}$$

where f and g are two real functions representing a forward traveling wave and a backward traveling wave, respectively. The mathematical model using the boundary approach is the wave equation (1) with the dynamical boundary condition for the struck end of the rod,

$$EA\frac{\partial u(0,t)}{\partial x} = F(t) \tag{3}$$

where the contact force F is assumed to be described by the Hertzian elastostatic contact law. According to this contact law, the contact force F during impact reads as

$$F = -k\delta^{\frac{3}{2}} \quad \text{for} \quad \delta \geqslant 0 \tag{4}$$

where δ is the indentation depth between the striking sphere and the struck end of the rod, i.e.,

$$\delta(t) := y(t) - u(0,t)\,. \tag{5}$$

The coefficient k is a constant which depends on the local geometry and elastic properties of the sphere and the rod in the contact area,

$$k = \frac{4\sqrt{R_1}}{3(\frac{1-\nu_1^2}{E_1} + \frac{1-\nu^2}{E})}\,. \tag{6}$$

The steel sphere has Young's modulus $E_1 = 210\,GPa$ and Poisson's ratio $\nu_1 = 0.30$. For the free rod, the stress at the free end must vanish. The boundary condition for the free end is a Neumann boundary condition,

$$\frac{\partial u(L,t)}{\partial x} = 0\,. \tag{7}$$

The rigid sphere has locally the equation of motion

$$m_1\ddot{y}(t) = -F(t)\,. \tag{8}$$

Combining the boundary condition for the contact end (3) and the boundary condition for the free end (7) with the equation of motion of the striking sphere (8) and using the general solution (2) of the wave equation, one obtains a set of nonlinear ordinary differential equations with time delay for the unknown functions f, g and q which determine the wave propagation in the rod.

2.2 Model during separation

During the separation phase it is assumed that the displacement $u(x,t)$ of the rod at the position x can be described approximately by a modal model using a sum of the rigid body motion $x_1(t)$ and n vibrations with the mode shapes $\phi_i(x)$ and the modal coordinates $q_i(t)$

$$u(x,t) = x_1(t) + \sum_{i=1}^{n} \phi_i(x) q_i(t) . \tag{9}$$

The mode shapes $\phi_i(x)$ are functions of the spatial coordinate x and satisfy the geometric boundary conditions of systems. For the free rod, they can be chosen as

$$\phi_i(x) = \sqrt{2} \cos\left(\frac{i\pi}{L} x\right) \quad \text{for } i = 1, 2, \cdots, n . \tag{10}$$

During contact the material damping is not considered since the duration of contact is very small, while during separation the material damping must be considered so that the decaying of the elastic vibrations can be modeled. It is assumed that the constitutive law of the rod is viscoelastic, i.e. the stress $\sigma(x,t)$ and the strain $\varepsilon(x,t)$ have the relation

$$\sigma(x,t) = E\left[\varepsilon(x,t) + \eta\dot{\varepsilon}(x,t)\right] . \tag{11}$$

In order to get the equations of motion, D'Alembert's virtual work principle

$$\int_0^L \rho\ddot{u}\delta u A dx + \int_0^L \sigma\delta\varepsilon A dx = \delta W^a \tag{12}$$

is used. Since the rod is suspended by thin wires with length $L_h = 20\,cm$ and moves with a small amplitude, its rigid body motion can be modeled as a linear damped pendulum due to the influence of the suspension and the air drag. Therefore, the virtual work of the applied forces is

$$\delta W^a = -mg\frac{x_1}{L_h}\delta x_1 - c_1\dot{x}_1\delta x_1 \tag{13}$$

where g denotes the gravitational constant and c_1 characterizes the linear damping. Using the orthogonality of the mode shapes, one can rewrite Eq.(12)

into

$$m\ddot{x}_1 + c_1 x_1 + m\frac{g}{L_H}x_1 = 0\,, \tag{14}$$

$$m\ddot{q}_i + \frac{\eta EA}{L}i^2\pi^2\dot{q}_i + \frac{EA}{L}i^2\pi^2 q_i = 0 \qquad \text{for} \quad i = 1, 2, \cdots, n\,. \tag{15}$$

It can be seen from (15) that for viscoelastic material higher frequency vibrations are damped more quickly. If after some time, these amplitudes remain small enough, then the corresponding elastic coordinate q_i needs no longer to be considered. From (15) one can compute when the amplitudes of q_i and $\dot{q}_i$ are smaller than the given error bounds for displacement ε_u and for velocity ε_v, respectively. In general, the exact computation is time-consuming. However, one can use the amplitude estimations of linear vibrations available in the literature, for example, Hu and Eberhard [4]. More details about this approach can be found in Hu and Schiehlen [7].

3 Comparisons with experiments

In experiments the strains were measured with strain gauges. The used strain gauges are of type 3/120XY13 made by Hottinger Baldwin Messtechnik GmbH and the used glue is Z70. The two used DC signal conditioners are of type DMS805B made by Rohrer GmbH with a frequency range up to $1\,MHz$. The positions A and B have distances $0.5m$ and $1m$ from the contact end of the rod, respectively. For the measurement of the displacement and velocity at the free end of the rod, a Laser-Doppler-Vibrometer of type OFV-3000/OFV-302 made by Polytec GmbH was used.

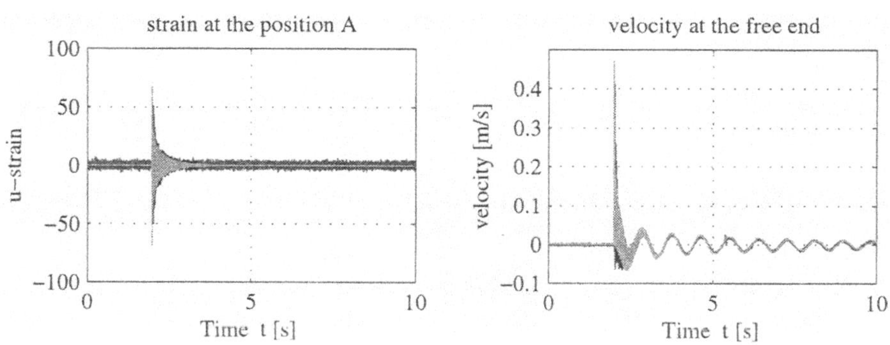

Fig. 2. Comparison between the measured and simulated results; colored, see page 375

In the simulation the specified error bounds ε_u and ε_v are chosen as

$$\varepsilon_u = 10^{-3}\,mm\,, \quad \varepsilon_v = 10^{-3}\,s/m\,. \tag{16}$$

The unknown damping parameters c_1 and η are identified from the measurement to be

$$c_1 = 0.3773 \frac{Ns}{m}, \quad \eta = 8.6 \times 10^{-8}\, s. \tag{17}$$

Though the parameter η is very small, it is an important parameter to describe the decay of the elastic vibrations for the deformation of the rod due to the impact. The comparison between the measured and simulated results is shown in Fig. 2 where the dark lines denote the measured signals and the light lines denote the simulated results. It can be seen that both results agree with each other very well. Using the multi-time scale simulation the required accuracy is satisfied in the simulation and the computing time is greatly reduced since those modes with very small elastic vibrations are longer deleted in the simulation.

4 Conclusions

Experimental and analytical analysis show that the impact behavior comprises of different phenomena: wave propagation, structural vibrations and rigid body motion. For efficient simulation of these different phenomena, different time scales are needed. In this paper, the elastodynamic boundary approach for the wave propagation, the method of flexible multibody systems for the structural vibrations and the method of rigid multibody systems for the rigid body motion are combined and different time scales are used for different phenomena. In the simulation both accuracy and efficiency are considered. The impact responses are simulated as accurate as necessary and as efficient as possible. The experimental results of the steel sphere colliding with the elastic brass rod have validated this varying time scale simulation for longitudinal impact responses.

References

1. Al-Mousawi, M. M. (1986) On experimental studies of longitudinal and flexural wave propagations: An annotated bibliography. Applied Mechanics Reviews **39**, 853–864.
2. Eberhard, P. and Jiang, S. (2000) Collision detection using interpolation schemes. Mathematical and Computer Modelling of Dynamical Systems **6**, 309–322.
3. Goldsmith, W. (1960) Impact: the theory and physical behaviour of colliding solids. Edward Arnold Ltd, London.
4. Hu, B. and Eberhard, P. (1999) Response bounds of linear vibrations. Journal of Applied Mechanics, **66**, 997–1004.
5. Hu, B. and Eberhard, P. (2001) Symbolic computation of longitudinal impact waves. Computer Methods in Applied Mechanics and Engineering **190**, 4805–4815.

6. Hu, B., Eberhard, P. and Schiehlen, W. Comparison of analytical and experimental results for longitudinal impacts on elastic rods. Journal of Vibration and Control, accepted for publication.
7. Hu, B. and Schiehlen, W. (2002) Multi-time scale simulation for impact systems: from wave propagation to rigid body motion. Archive of Applied Mechanics, submitted for publication.
8. Szabó, I. (1977) Geschichte der mechanischen Prinzipien und ihrer wichtigsten Anwendungen, Birkhäuser, Basel.

On admissible coefficients of friction for the solvability of contact problems with friction

Jiří Jarušek[1]* and Christof Eck[2]

[1] Mathematical Institute, Academy of Sciences of the Czech Republic.
[2] Institute of Applied Mathematics, University Erlangen-Nürnberg.

Contact problems with friction have many applications in mechanics and represent an important task of applied mathematics. Despite this fact, there are still many open problems. The main reason for this situation is the non-compact, non-monotone and non-smooth character of the friction term. The aim of this survey is both to overview available theorems about existence of solutions and to outline possible future developments.

We consider a (visco-)elastic body occupying a bounded domain $\Omega \subset \mathbb{R}^N$ being in contact with a rigid undeformable support. With the notation $\boldsymbol{u} = (u_i)$ for the displacement, $\boldsymbol{\varepsilon} \equiv (e_{ij}) := \left(\frac{1}{2}(\partial_i u_j + \partial_j u_i)\right)$ for the linearized strain tensor and $\boldsymbol{\sigma} = (\sigma_{ij})$ for the stress tensor, the constitutive law is assumed in the form

$$\sigma_{ij} \equiv \sigma_{ij}(\boldsymbol{u}) = a^{(0)}_{ijk\ell} e_{k\ell}(\boldsymbol{u}) + a^{(1)}_{ijk\ell} e_{k\ell}(\dot{\boldsymbol{u}}), \ i,j = 1,\dots,N. \tag{1}$$

Here and in the sequel, ∂_i denotes the partial derivative with respect to the i-th space variable while the time derivatives will be denoted by dots and the summation convention is employed. The aim is to find sufficient conditions for the solvability of unilateral contact problems with Coulomb friction with the emphasis on the requirements for the coefficient of friction.

1 The static problem

The static problem can be interpreted as a time discretization of the dynamic problem (resulting in the term with the constant $d \geqslant 0$ in (2)). We look for $\boldsymbol{u}$ such that the relations

$$du_i - \partial_j \sigma_{ij}(\boldsymbol{u}) = f_i, \ i = 1,\dots,N, \ \text{in } \Omega, \tag{2}$$

$$\boldsymbol{u} = U \quad \text{on } \Gamma_U, \tag{3}$$

$$\boldsymbol{\sigma}_\Gamma(\boldsymbol{u}) = T \quad \text{on } \Gamma_T, \tag{4}$$

$$u_n \leqslant g_n, \ \sigma_n(\boldsymbol{u}) \leqslant 0, \ \sigma_n(\boldsymbol{u})(u_n - g_n) = 0 \ \text{on } \Gamma_C, \tag{5}$$

$$\left.\begin{array}{l} \dot{\boldsymbol{u}}_t = 0 \Rightarrow |\boldsymbol{\sigma}_t(\boldsymbol{u})| \leqslant \mathfrak{F}|\sigma_n(\boldsymbol{u})|, \\ \dot{\boldsymbol{u}}_t \neq 0 \Rightarrow \boldsymbol{\sigma}_t(\boldsymbol{u}) = -\mathfrak{F}|\sigma_n(\boldsymbol{u})|\dot{\boldsymbol{u}}_t/|\dot{\boldsymbol{u}}_t| \end{array}\right. \quad \text{on } \Gamma_C \tag{6}$$

are satisfied. Here Γ, the boundary of Ω, is assumed to be Lipschitz and to be composed of three pairwisely disjoint and measurable parts Γ_Y, $Y \in$

* The grant support for the research by Grant Agency of the Academy of Sciences of the Czech Republic, grant Nr. A 1075005 is gratefully acknowledged.

$\{C,T,U\}$, $v_n = \boldsymbol{v}\cdot\boldsymbol{n}$, $\boldsymbol{v}_t - v_n\boldsymbol{n}$ are the normal and tangential components of the respective vector $\boldsymbol{v}$ (for the stress it is $\boldsymbol{\sigma}_\Gamma \equiv \boldsymbol{\sigma n}$) and $\boldsymbol{n}$ is the unit outer normal vector. Since in the static problem the velocity has no sense, we assume purely elastic material, $a^{(0)}_{ijkl} \equiv a_{ijkl}$ and $a^{(1)}_{ijkl} \equiv 0$ in (1)

If the contact of two bodies occupying two domains Ω_κ, $\kappa = 1,2$, is modelled, then $\Omega = \Omega_1 \cup \Omega_2$, Γ_Y, $Y \in \{T,U\}$ is the union of the respective parts of boundaries of the particular domains and Γ_C is the common contact surface. The outer normal vector is chosen with respect to Ω_1 and the relative displacement $\boldsymbol{u}^{\mathrm{rel}} \equiv \boldsymbol{u}_1 - \boldsymbol{u}_2$ is taken in (5) and (6).

With the notation L_p, W^α_p and $H^\alpha \equiv W^\alpha_2$ for Sobolev spaces, $\boldsymbol{L}_p$ and $\boldsymbol{H}^\alpha$ for the space of functions with range in $\mathbb{R}^N$ and $(\cdot,\cdot)_M$ for a (possibly generalized) scalar product in $L_2(M)$ or $\boldsymbol{L}_2(M)$ for a measurable set M, we introduce the cone $\mathfrak{K}_\Omega := \{\boldsymbol{v} \in \boldsymbol{H}^1(\Omega);\ v|_{\Gamma_U} = U \ \&\ v_n - g_n \leqslant 0\}$.
An element $\boldsymbol{u} \in \mathfrak{K}_\Omega$ is a weak solution of Problem (2–6), *if for all $\boldsymbol{v} \in \mathfrak{K}_\Omega$*

$$
\begin{aligned}
(d\boldsymbol{u}, \boldsymbol{v}-\boldsymbol{u})_\Omega + (\sigma_{ij}(\boldsymbol{u}), e_{ij}(\boldsymbol{v}-\boldsymbol{u}))_\Omega + (\mathfrak{F}\sigma_n(\boldsymbol{u}), |\boldsymbol{v}_t| - |\boldsymbol{u}_t|)_{\Gamma_C}& \\
\geqslant (\mathcal{L}, \boldsymbol{v}-\boldsymbol{u}) \quad \text{with } \mathcal{L}: \boldsymbol{v} \mapsto (\boldsymbol{f},\boldsymbol{v})_\Omega + (T,\boldsymbol{v})_{\Gamma_T}.&
\end{aligned}
\tag{7}
$$

The coefficients $a_{ijk\ell}$ are assumed to be symmetric, $a_{ijk\ell} = a_{jik\ell} = a_{k\ell ij}$, Hölder continuous with an exponent $\beta > \frac{1}{2}$ and elliptic and bounded in the sense

$$
|\boldsymbol{\xi}|^{-2} a_{ijk\ell}\xi_{ij}\xi_{k\ell} \in (a, A) \text{ on } \Omega \text{ for all symmetric matrices } \boldsymbol{\xi} \equiv (\xi_{ij}) \tag{8}
$$

with positive real constants a, A. The contact part of the boundary Γ_C is assumed to be $C^{2+\varepsilon}$ smooth for a fixed arbitrarily small $\varepsilon > 0$. The given data satisfy $U \in \boldsymbol{H}^1(\Omega)$ $\operatorname{supp} U \cap \Gamma_C = \emptyset$, $T \in \boldsymbol{H}^{1/2}(\Gamma_T)^*$, $0 \leqslant g_n \in H^1(\Gamma_C)$ and $\boldsymbol{f} \in \boldsymbol{H}^1(\Omega)^* \cap \boldsymbol{H}^{1/2}(\widetilde{\Omega})^*$ for a set $\widetilde{\Omega} \subset \Omega$ such that $\Gamma_C \subset \partial\widetilde{\Omega}$. Moreover, with the space of rigid motions $\mathcal{R} := \{\boldsymbol{v} \in \boldsymbol{H}^1(\Omega); \varepsilon(\boldsymbol{v}) = 0\}$, the set $Z := (\mathfrak{K}_\Omega - g_n) \cap \mathcal{R}$ and its maximal subspace Z_0, we assume

$$
(i)\ \operatorname{mes}\Gamma_U > \dot{0}, \quad \text{or} \quad (ii)\ \mathcal{L}(Z_0) = 0 \ \&\ \mathcal{L}|_{Z\setminus Z_0} < 0 \text{ or } \ (iii)\ d > 0. \tag{9}
$$

For contact problems of two elastic bodies we require appropriate modifications of these conditions. For the admissible coefficient of friction $\mathfrak{F}$ and the set $\Gamma_{\mathfrak{F}} := \operatorname{cl}\{\boldsymbol{x} \in \Gamma_C; \exists \boldsymbol{v} \in \mathbb{R}^N \text{ such that } \mathfrak{F}(\boldsymbol{x},\boldsymbol{v}) \neq 0\}$ the following conditions are required:

$0 \leqslant \mathfrak{F} \equiv \mathfrak{F}(x, \boldsymbol{u}_t)$ *satisfies the Carathéodory conditions,* $\|\mathfrak{F}\|_{L_\infty(\Gamma_C \times \mathbb{R}^N)} < C_{\mathfrak{F}}$ *and one of the conditions* $\operatorname{dist}(\Gamma_{\mathfrak{F}}, \Gamma \setminus \Gamma_C) > 0$, $\operatorname{dist}(\Gamma_C, \Gamma_U \cup \operatorname{supp} T) > 0$.

The upper bound $C_{\mathfrak{F}}$ depends on the material. For isotropic material,

$$
a_{ijk\ell} = \frac{E}{2+2\nu}(\delta_{ik}\delta_{j\ell} + \delta_{i\ell}\delta_{jk}) + \frac{E\nu}{(1+\nu)(1-2\nu)}\delta_{ij}\delta_{k\ell} \tag{10}
$$

with the Kronecker symbol δ, the Posson ratio $\nu \in \left(-1, \frac{1}{2}\right)$ and the Young modulus of elasticity $E > 0$, this bound is

$$
C_{\mathfrak{F}} = \frac{\sqrt{3-4\nu}}{2-2\nu} \text{ for } N = 2, \quad \text{and } C_{\mathfrak{F}} = \sqrt{\frac{3-4\nu}{4-4\nu}} \text{ for } N \geqslant 3,
$$

and for general material it holds $C_{\mathfrak{F}} = \sqrt{3a/(4A)}$. For contact problems of two bodies and isotropic material with the material parameters E_κ, ν_κ of body Ω_κ, the upper bound is given in terms of the quantities $R_\kappa \equiv E_\kappa/(1-\nu_\kappa^2)$, $S_\kappa \equiv (1-\nu_\kappa)^2/(3-4\nu_\kappa)$ for $N=2$ or $R_\kappa \equiv E_\kappa/(1+\nu_\kappa)$, $S_\kappa \equiv (1-\nu_\kappa)/(3-4\nu_\kappa)$ for $N \geqslant 3$, $\kappa = 1,2$, and has the form

$$\begin{aligned} C_{\mathfrak{F}} &= \frac{1}{2\sqrt{S_1S_2}}\frac{R_1S_1+R_2S_2}{R_1\sqrt{S_1}+R_2\sqrt{S_2}} \quad \text{if } 4R_2^2 \geqslant \frac{(R_1S_1-R_2S_2)^2}{S_1S_2} \\ C_{\mathfrak{F}} &= \sqrt{\frac{R_2}{S_1}}\frac{1}{\sqrt{R_1+R_2}+\sqrt{R_2}} \quad \text{if } 4R_2^2 \leqslant \frac{(R_1S_1-R_2S_2)^2}{S_1S_2}. \end{aligned} \tag{11}$$

if $R_2S_2 \geqslant R_1S_1$. For a general material and a general dimension $N \geqslant 2$ the bound $C_{\mathfrak{F}}$ depends on the constants of ellipticity and boundedness a_κ, A_κ of the two bodies Ω_κ via

$$\begin{aligned} C_{\mathfrak{F}} &= \sqrt{\frac{3a_1a_2}{4A_1A_2}}\frac{A_1+A_2}{\sqrt{a_1A_1}+\sqrt{a_2A_2}} \quad \text{if } \frac{a_2}{a_1} \geqslant \frac{(A_1-A_2)^2}{4A_1A_2} \\ C_{\mathfrak{F}} &= \sqrt{\frac{3a_1a_2}{A_1}}\frac{1}{\sqrt{a_1+a_2}+\sqrt{a_2}} \quad \text{if } \frac{a_2}{a_1} \leqslant \frac{(A_1-A_2)^2}{4A_1A_2} \end{aligned} \tag{12}$$

provided $A_2 \geqslant A_1$. If the additional condition to (11) and (12) is reversed, then the corresponing formulae with interchanged indices 1 and 2 hold.

Theorem 1. *Under the above listed assumptions there exists a weak solution to Problem* (2–6).

In fact, due to a localization technique employed in the proof of the theorem, the constant $C_{\mathfrak{F}}$ only depends on the material properties along the contact part Γ_C. Moreover, if $\mathfrak{F}_1$ is an other coefficient of friction such that $\mathfrak{F}_1(\cdot,v) = \mathfrak{F}(\cdot,v)$ for $v \neq 0$ and $\mathfrak{F}_1(\cdot,0) \geqslant \mathfrak{F}(\cdot,0)$ and if $\mathfrak{F}$ satisfies all the above listed requirements, then every weak solution of Problem (2–6) with the coefficient $\mathfrak{F}$ (whose existence is ensured by Theorem 1) solves the same problem with the coefficient $\mathfrak{F}_1$.

The presented result was partially published in [3]. The version presented here is a part of a monograph in preparation [6].

2 The dynamic problem

In this case the viscoelastic constitutive law (1) is employed. The time interval $I = (0,T)$ is assumed to be bounded. We look for $\boldsymbol{u}$ such that

$$\ddot{u}_i - \partial_j\sigma_{ij}(\boldsymbol{u}) = f_i,\ i = 1,\dots,N, \ \text{ in } Q \equiv I \times \Omega, \tag{13}$$

$$\boldsymbol{u} = U \quad \text{on } S_U \equiv I \times \Gamma_U, \tag{14}$$

$$\boldsymbol{\sigma}_\Gamma(\boldsymbol{u}) = T \quad \text{on } S_T \equiv I \times \Gamma_T, \tag{15}$$

$$\dot{u}_n \leqslant 0,\ \sigma_n(\boldsymbol{u}) \leqslant 0,\ \dot{u}_n\sigma_n(\boldsymbol{u}) = 0 \quad \text{on } S_C \equiv I \times \Gamma_C, \tag{16}$$

$$\begin{aligned} \dot{\boldsymbol{u}}_t = 0 &\Rightarrow |\boldsymbol{\sigma}_t(\boldsymbol{u})| \leqslant \mathfrak{F}|\sigma_n(\boldsymbol{u})|, \\ \dot{\boldsymbol{u}}_t \neq 0 &\Rightarrow \boldsymbol{\sigma}_t(\boldsymbol{u}) = -\mathfrak{F}|\sigma_n(\boldsymbol{u})|\dot{\boldsymbol{u}}_t/|\dot{\boldsymbol{u}}_t| \end{aligned} \quad \text{on } S_C, \tag{17}$$

$$u(0,x) = u_0(x),\ \dot{u}(0,x) = u_1(x) \ \text{ for } x \in \Omega. \tag{18}$$

The Signorini condition (16) formulated in velocities represents a first-order approximation of the original condition in displacements (5).

With $\mathfrak{K}_Q := \{v \in H^{1/2,1}(Q);\ v = \dot{U} \text{ on } S_U \ \&\ v_n \leqslant 0 \text{ a.e. on } S_C\}$ we define:

A weak solution of Problem (13–18) *is an element* $u \in B_0(I; H^1(\Omega))$ *satisfying the initial condition (18), the relations* $\dot{u} \in B_0(I; L_2(\Omega)) \cap \mathfrak{K}_Q$, $\ddot{u} \in L_2(I; H^{-1}(\Omega)) \cap H^{1/2}(I; L_2(\Omega))^*$ *and, for all* $v \in \mathfrak{K}_Q$, *the inequality*

$$\begin{aligned}(\ddot{u}, v - \dot{u})_Q + (\sigma_{ij}(u), e_{ij}(v - \dot{u}))_Q + (\mathfrak{F}\sigma_n(u), |v_t| - |\dot{u}_t|)_{S_C} \\ \geqslant (f, v - \dot{u})_Q + (T, v - \dot{u})_{S_T}.\end{aligned} \tag{19}$$

Here, B_0 denotes the space of bounded mappings with the appropriate sup-norm and the first and second index in anisotropic Sobolev spaces represents the time and space regularity of its elements, respectively.

The solvability of this problem is proved under the previous requirements for both Γ_C and the time-independent elasticity tensor $\left(a^{(0)}_{ijk\ell}\right)$. For the viscosity tensor $\left(a^{(1)}_{ijk\ell}\right)$ we assume the same conditions as for the tensor of elasticity, denoting the positive real constants of ellipticity and boundedness by $a_{(\iota)}, A_{(\iota)}$, $\iota = 0, 1$. The given data satisfy $f \in \left(H^{1/4,1/2}(\widetilde{Q})\right)^* \cap \left(H^{1/2-\varepsilon,1}(Q)\right)^*$ for some $\varepsilon > 0$ arbitrarily small and $\widetilde{Q} = I \times \widetilde{\Omega}$ with $\widetilde{\Omega}$ satisfying the condition from the previous section, $T \in \left(H^{1/4+\varepsilon,1/2}(S_T)\right)^*$, $u_0, u_1 \in H^{3/2}(\Omega)$ and $U \in H^{1/2+\varepsilon,1}(Q)$ with some $\varepsilon > 0$. Moreover, U fulfils the compatibility conditions $U = 0$ on S_C, $U(0,\cdot) = u_0$ and $\dot{U}(0,\cdot) = u_1$ on Ω. For the admissible coefficient of friction we have the conditions:

$0 \leqslant \mathfrak{F} \equiv \mathfrak{F}(x, \dot{u}_t)$ *satisfies the Carathéodory conditions, one of the conditions* $\operatorname{dist}(\Gamma_{\mathfrak{F}}, \Gamma \setminus \Gamma_C) > 0$, $\operatorname{dist}(\Gamma_C, \Gamma_U \cup \pi_\Gamma(\operatorname{supp} T)) > 0$ *and, moreover,* $\|\mathfrak{F}\|_{L_\infty(\Gamma_C \times \mathbb{R}^N)} < C_{\mathfrak{F}}$.

Here $\Gamma_{\mathfrak{F}}$ is defined as earlier and π_Γ is the appropriate projection. For a general material the bound $C_{\mathfrak{F}}$ is

$$C_{\mathfrak{F}} = k_{\max}\sqrt{\frac{a_{(1)}}{A_{(1)}}} \ \text{ with } k_{\max} = \sqrt{\frac{z_{\max}}{2(z_{\max}+1)\left(1 + 2\sqrt{2}z_{\max}\right)}}. \tag{20}$$

Here $z_{\max} = \left(\sqrt[3]{3\sqrt{3}+\sqrt{19}} + \sqrt[3]{3\sqrt{3}-\sqrt{19}}\right)\Big/\sqrt{6} \approx 1.2510786$, hence $k_{\max} \approx 0.258342$. For *viscously* isotropic material, i.e. if the relation (10) holds for $a^{(1)}_{ijkl}$ with the modulus of viscosity E_1 and the viscous Poisson ratio ν_1, then

$$C_{\mathfrak{F}} = \sup_{E_1 \geqslant 0} 1\Big/\left(D_0(E_1)\sqrt{D_1^2 + D_2^2\sqrt{E_1}}\right) \ \text{ with} \tag{21}$$

$$D_0^2 \equiv D_0^2(E_1) = \sup_{|\xi|>0} \frac{(\mathsf{a}_2 + 4|\xi|^2\mathsf{a}_1) - 4|\xi|^2(|\mathsf{a}|^2\mathsf{b}_2 + |\xi|^2\mathsf{a}_2)}{C_{\nu_1,|\xi|}(E_1)(\mathsf{a}_2\mathsf{b}_2 + 4|\xi|^2\mathrm{Im}(\mathsf{ab}) - 4|\xi|^2|\mathsf{ab} - |\xi|^2|^2)},$$

$$D_1^2 = \max\left\{ d_{N-1}\left(\nu_1, \tfrac{1}{2}\right),\ \sup_{|\xi|>0} \frac{c_{N-1}\left(\frac{1}{2}\right)(2+2\nu_1)|\xi|(|\mathsf{a}|^2\mathsf{b}_2 + |\xi|^2\mathsf{a}_2)}{\mathsf{a}_2\mathsf{b}_2 + 4|\xi|^2\mathrm{Im}(\mathsf{ab}) - 4|\xi|^2|\mathsf{ab} - |\xi|^2|^2} \right\},$$

$$D_2^2 = \sup_{|\xi|>0} \frac{c_1\left(\frac{1}{4}\right)\sqrt{2+2\nu_1}(|\mathsf{a}|^2\mathsf{b}_2 + |\xi|^2\mathsf{a}_2)}{\mathsf{a}_2\mathsf{b}_2 + 4|\xi|^2\mathrm{Im}(\mathsf{ab}) - 4|\xi|^2|\mathsf{ab} - |\xi|^2|^2}.$$

The constants arising here are defined by $\mathsf{a} = \mathsf{a}_1 + i\mathsf{a}_2$, $\mathsf{b} = \mathsf{b}_1 + i\mathsf{b}_2$ with $\mathsf{a}_{1,2} = \sqrt{\frac{1}{2}\left(\sqrt{|\xi|^4+1} \pm |\xi|^2\right)}$, $\mathsf{b}_{1,2} = \sqrt{\frac{1}{2}\left(\sqrt{|\xi|^4+k^2} \pm |\xi|^2\right)}$ and $k = (1-2\nu_1)/(2-2\nu_1)$. Moreover, $C_{\nu_1,|\xi|}(E_1) = \sqrt{(2+2\nu_1)E_1}c_1\left(\frac{1}{4}\right) + c_{N-1}\left(\frac{1}{2}\right)(2+2\nu_1)|\xi|$, $c_1(\alpha) \equiv 2^{2-2\alpha}\int_{\mathbb{R}} |t|^{-1-2\alpha}\sin^2(t)\,dt$, $c_k(\alpha) \equiv c_1(\alpha)\int_{\mathbb{R}^{k-1}}\left(1+|s|^2\right)^{-k/2-\alpha}ds$, $k \geqslant 2$, $d_1(\alpha) = 0$ and $d_k(\nu_1,\alpha) = (2+2\nu_1)c_k(\alpha)$ for $k \geqslant 2$ and $\alpha \in (0,1)$. The values of D_κ, $\kappa = 0,1,2$, have been computed numerically. The dependence of $C_{\mathfrak{F}}$ on E_1 disappears, because a change of the time unit does not influence the ratio $|\boldsymbol{\sigma}_t(\boldsymbol{u})|/\sigma_n(\boldsymbol{u})$, while it changes the value of $C_{\mathfrak{F}}$. Hence an optimization with respect to the time unit can be done, and the maximal $C_{\mathfrak{F}}$ is then independent of E_1. The resulting $C_{\mathfrak{F}} \equiv C_{\mathfrak{F}}(\nu_1)$ from (21) is depicted in Figure 1. The three-dimensional result is valid also for all higher dimensions. With some modifications of the scheme used in deriving the previous results, the following theorem has been proved:

Theorem 2. *Under the above listed assumptions there exists a weak solution of Problem* (13–18).

The remarks to Theorem 1 remain valid here. The result is contained in [6], its partial cases have been published in [7], [4] and [5].

3 The quasistatic problem

The result to be presented here is due to L.-E. Andersson, [1]. For sets from the preceding section the respective relations (13–15) and (17–18) are again required with the cancellation of $\ddot{\boldsymbol{u}}$ in (13) and of the condition for the velocity $\dot{\boldsymbol{u}}$ in (18). Moreover, the contact condition (16) on S_C is changed to the original contact condition for displacements like in (5). As in the static case a purely elastic material is considered. We introduce the cone $\widetilde{\mathfrak{K}}_Q := \left\{ v \in H^1\left(I; \boldsymbol{H}^1(\Omega)\right);\ v|_{S_U} = U \ \&\ v_n - g_n \leqslant 0 \right\}$.
A weak solution of the quasistatic contact problem with Coulomb friction is an element $\boldsymbol{u} \in \widetilde{\mathfrak{K}}_Q$ *such that for all* $\boldsymbol{v} \in \widetilde{\mathfrak{K}}_Q$

$$\begin{aligned}(\sigma_{ij}(\boldsymbol{u}), e_{ij}(\boldsymbol{v}-\boldsymbol{u}))_Q + (\mathfrak{F}\sigma_n(\boldsymbol{u}), |\boldsymbol{v}_t - \boldsymbol{u}_t + \dot{\boldsymbol{u}}_t| - |\dot{\boldsymbol{u}}_t|)_{S_C} \\ \geqslant (\boldsymbol{f}, \boldsymbol{v}-\boldsymbol{u})_Q + (T, \boldsymbol{v}-\boldsymbol{u})_{S_T}.\end{aligned} \tag{22}$$

For the solvability of this problem the requirements to Γ_C and to the time-independent elasticity tensor $(a_{ijk\ell})$ from the previous sections are

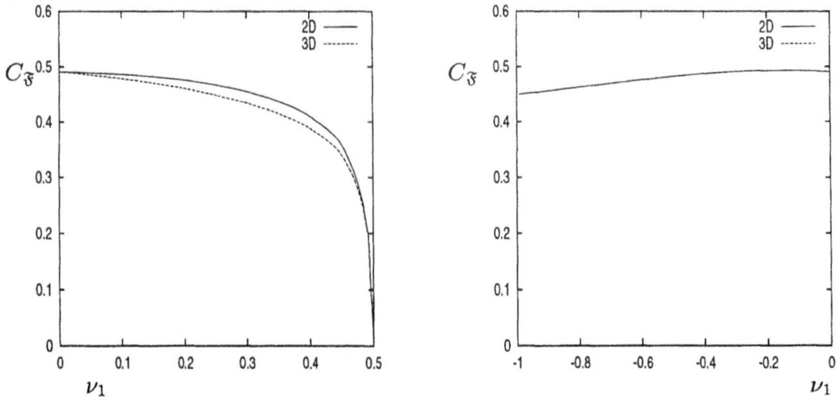

Fig. 1. Optimal upper bound for the coefficient of friction in dependence on the viscous Poisson ratio ν_1

preserved. The given data satisfy $U \in H^1(I; \boldsymbol{H}^1(\Omega))$, $\operatorname{supp} U \cap \Gamma_C = \emptyset$, $T \in H^1(I; \boldsymbol{H}^{1/2}(\Gamma_T)^*)$, $\boldsymbol{f} \in H^1(I; \boldsymbol{H}^1(\Omega)^*) \cap H^1(I; \boldsymbol{H}^{1/2}(\widetilde{\Omega})^*)$ with the same properties of $\widetilde{\Omega}$ as above and $\boldsymbol{f}(0,\cdot) = 0$, $0 \leqslant g_n \in H^1(\Gamma_C)$ and $u_0 \in \boldsymbol{H}^{3/2}(\Omega)$. Only the coercive case of this problem is solved, i.e. $\operatorname{mes}\Gamma_U > 0$ is required (cf. (9)). For the admissible coefficient of friction it is needed:

$0 \leqslant \mathfrak{F} \equiv \mathfrak{F}(\boldsymbol{x})$ *satisfies* $\|\mathfrak{F}\|_{L_\infty(\Gamma_C)} < C_{\mathfrak{F}}$ *and one of the conditions* $\operatorname{dist}(\operatorname{supp}\mathfrak{F}, \Gamma \setminus \Gamma_C) > 0$, $\operatorname{dist}(\Gamma_C, \Gamma_U \cup \pi_\Gamma(\operatorname{supp} T)) > 0$.

The constant $C_{\mathfrak{F}}$ depends on global trace and inverse trace estimates for the domain Ω, it is therefore not possible to present a formula valid for all domains. The proved existence result is:

Theorem 3. *Under the above listed assumptions there exists a weak solution of the quasistatic contact problem with Coulomb friction.*

The proof is done by an implicit time discretization, leading to a sequence of static problems; by the derivation of appropriate a priori estimates for its solutions and a passage to the limit for the time discretization.

4 Conclusions

We have presented various results about the existence of solutions to contact problems with friction for static, quasistatic and fully dynamic cases. The most important open question is the solvability of dynamic problems with the original unilateral contact condition in the displacements (instead of the displacement velocities), but this problem is very hard to solve. In the dynamic case for anisotropic material the upper bound for the coefficient of friction is rather low, hence it could be helpful to derive better bounds for special cases of anisotropy. Other open questions concern the uniqueness of solutions. The nonuniqueness of solutions of the quasistatic problem has been shown by P. Balard [2]; for other problems there are no corresponding results available.

References

1. Andersson, L.-E. (2000) Existence result for quasistatic contact problem with Coulomb friction. Appl. Math. Optimiz. **42**, 169–202.
2. Balard, P. (1999) A counterexample to uniqueness of quasistatic elastic contact problems with small friction. Int. J. Engrg. Sci. **37**, 163–178.
3. Eck, C., Jarušek, J. (1998) Existence results for the static contact problem with Coulomb-friction. Math. Models Meth. Appl. Sci. **8**, 445–468.
4. Eck, C., Jarušek, J. (1998) Existence of solutions for the dynamic frictional contact problem of isotropic viscoelastic bodies. Preprint Nr. 242, Inst. Angew. Math, Univ. Erlangen-Nürnberg. To appear in Nonlin. Anal., Theory Meth. Appl.
5. Eck, C., Jarušek, J. (2002) On the solvability of unilateral dynamic frictional contact problems of isotropic viscoelastic bodies in three and more space dimensions. Preprint Nr. 287, Inst. Angew. Math, Univ. Erlangen-Nürnberg.
6. Eck, C., Jarušek, J. and Krbec M. Unilateral Contact Problems in Mechanics. Variational Methods and Existence Theorems. Monograph in preparation.
7. Jarušek, J., Eck, C. (1999) Dynamic contact problems with small Coulomb friction for viscoelastic bodies. Existence of solutions. Math. Models Meth. Appl. Sci. **9**, 11–34.

Regularity results for transmission problems of linear elasticity on polyhedral domains

Dorothee Knees

Mathematisches Institut A, Universität Stuttgart, Germany
kneesde@mathematik.uni-stuttgart.de

Abstract. Boundary value problems for the Lamé operator with piecewise constant material coefficients are investigated on polyhedral domains. Because of geometric peculiarities and non-smooth material constants, the displacement fields and especially the stress fields have a singular behavior in the neighborhood of corners, edges and those points where the material constants jump. For 3D problems it is not clear if the displacement fields are bounded. In this article we describe sufficient conditions on the distribution of the material parameters and the geometry which guarantee that weak solutions of the BVP are bounded and piecewise continuous.

1 Formulation of the problem

1.1 Domains and spaces

We consider polyhedral domains $\Omega \subset \mathbb{R}^3$ which are divided into polyhedral subdomains Ω_i, Fig. 1: $\overline{\Omega} = \bigcup_{i=1}^{M} \overline{\Omega_i}$, $\Omega_i \cap \Omega_j = \emptyset$ for $i \neq j$. For simplicity we assume that Ω and Ω_i have Lipschitz boundaries. We divide the exterior parts of $\partial\Omega_i$ into pairwise disjoint two dimensional polygons $\gamma_{i,l}$ with Lipschitz boundaries such that $\gamma_{i,l} \subset \partial\Omega_i$ and $\partial\Omega = \bigcup_{i,l} \overline{\gamma_{i,l}}$. The interface $\bigcup_{i\neq j}(\partial\overline{\Omega_i} \cap \partial\overline{\Omega_j})$ is also divided into pairwise disjoint two dimensional polygons $\gamma_{ij,l}$ with Lipschitz boundaries, $\gamma_{ij,l} \subset \partial\Omega_i \cap \partial\Omega_j$. We set $\mathcal{F} = \{\gamma_{i,l};\, i,l\} =: \mathcal{D} \cap \mathcal{N}$, $\mathcal{D}$ and $\mathcal{N}$ disjoint. $\mathcal{D}$ describes the Dirichlet boundary, $\mathcal{N}$ the Neumann boundary. Finally $\mathcal{G} := \{\gamma_{ij,l};\, i,j,l\}$ describes the interface.
For $l \in \mathbb{N}_0$ and $\Omega_i \subset \mathbb{R}^3$ we denote by $H^l(\Omega_i)$ the standard Sobolev spaces. Further, for $l = 1, 2, \ldots$ and $\gamma \in \mathcal{G} \cup \mathcal{F}$ we define the trace spaces $H^{l-\frac{1}{2}}(\gamma)$ by $H^{l-\frac{1}{2}}(\gamma) = \overline{\mathcal{D}(\overline{\gamma})}^{\|\cdot\|_{H^{l-\frac{1}{2}}(\gamma)}}$, where $\mathcal{D}(\overline{\gamma}) = \{u\big|_{\overline{\gamma}} : u \in \mathcal{C}^\infty(\mathbb{R}^3),\ \operatorname{supp} u \subset \overline{\gamma}\}$ and the norm is a Sobolev-Slobodetskij norm $\|\cdot\|_{H^{l-\frac{1}{2}}(\gamma)}$, see [9].
Finally $\tilde{H}^{l-\frac{1}{2}}(\gamma) = \{u \in H^{l-\frac{1}{2}}(h(\gamma)) : \operatorname{supp} u \subset \overline{\gamma}\}$, where $h(\gamma)$ is that two dimensional plane which contains γ. For $l = 1$, $\gamma \in \mathcal{G} \cup \mathcal{F}$, we define the dual spaces: $H^{-\frac{1}{2}}(\gamma) := \left(\tilde{H}^{\frac{1}{2}}(\gamma)\right)'$, $\tilde{H}^{-\frac{1}{2}}(\gamma) := \left(H^{\frac{1}{2}}(\gamma)\right)'$.

1.2 The boundary transmission problem

Before we formulate the transmission problem for linear, homogeneous, isotropic elasticity with piecewise constant material parameters (transmission

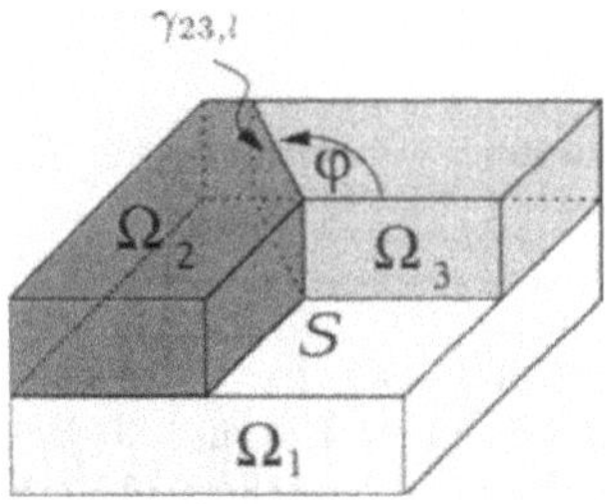

Fig. 1. Fichera's corner

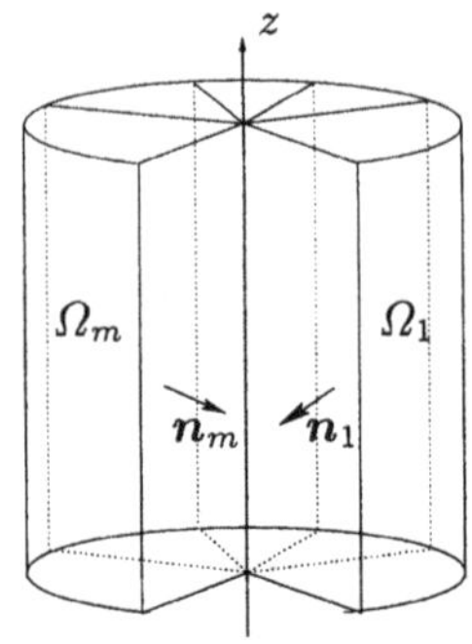

Fig. 2. Exterior edge

problem for the Lamé operator) we introduce some notation:
By $u : \Omega \to \mathbb{R}^3$ we denote the displacement field, $u_i = u\big|_{\Omega_i}$; $\lambda_i, \mu_i \in \mathbb{R}$ are the Lamé constants on Ω_i. The stress tensor is given via Hooke's law by $\sigma_i(u) = \lambda_i \operatorname{tr} \varepsilon(u) + 2\mu_i \varepsilon(u)$, where $\varepsilon(u) = \frac{1}{2}(\nabla u + (\nabla u)^T)$ is the linearized strain tensor. For simplicity we assume homogeneous Dirichlet data in the following. Analogous to [3] we define

$$E(\Omega) := \{u \in L^2(\Omega) : u\big|_{\Omega_i} \in H^1(\Omega_i),\ \operatorname{div}(u\big|_{\Omega_i}) \in L^2(\Omega_i)\}.$$

Definition 1 (Weak solution). Let $\overline{\Omega} = \cup_{i=1}^M \overline{\Omega_i} \subset \mathbb{R}^3$ be a polyhedral domain as defined above. Let further $f_i \in L^2(\Omega_i)$ for $1 \leqslant i \leqslant M$ and $g_{i,k} \in H^{-\frac{1}{2}}(\gamma_{i,k})$ for $\gamma_{i,k} \in \mathcal{N}$.
$u \in E(\Omega)$ is a weak solution of the boundary transmission problem for the Lamé operator if

$$-\operatorname{div}(\sigma_i(u_i)) = -\big(\mu_i \Delta u_i + (\lambda_i + \mu_i) \operatorname{grad} \operatorname{div} u_i\big) = f_i \qquad \text{in } \Omega_i, \qquad (1)$$

$$u_i\big|_{\gamma_{i,k}} = 0, \qquad \gamma_{i,k} \in \mathcal{D}, \quad (2)$$

$$\sigma_i(u_i)\boldsymbol{n}_{i,k} = g_{i,k}, \qquad \gamma_{i,k} \in \mathcal{N}, \quad (3)$$

$$u_i\big|_{\gamma_{ij,k}} - u_j\big|_{\gamma_{ij,k}} = 0, \qquad \gamma_{ij,k} \in \mathcal{G}, \quad (4)$$

$$\sigma_i(u_i)\boldsymbol{n}_{ij,k} + \sigma_j(u_j)\boldsymbol{n}_{ji,k} = 0, \qquad \gamma_{ij,k} \in \mathcal{G}. \quad (5)$$

Here, $\boldsymbol{n}_{i,k}$ is the exterior normal vector of Ω_i with respect to $\gamma_{i,k}$, $\boldsymbol{n}_{st,k}$ is the exterior normal vector of Ω_s with respect to $\gamma_{st,k} \in \mathcal{G}$, $\boldsymbol{n}_{st,k} = -\boldsymbol{n}_{ts,k}$.

Remark 1. Due to [3] the expression $\sigma_i(u_i)\boldsymbol{n}_{i,k} \in H^{-\frac{1}{2}}(\gamma_{i,k})$ is meaningful for $u \in E(\Omega)$.

In the following we assume $\mu_i > 0$, $\lambda_i + \mu_i > 0$ for all i. The transmission problem is then an *elliptic boundary transmission problem.* Existence and uniqueness of weak solutions can be proved with the Lax/Milgram Lemma and Korn's inequality.
The question now is: What about the regularity of weak solutions if $f_i \in L^2(\Omega_i)$ and $g_{i,k} \in H^{\frac{1}{2}}(\gamma_{i,k})$?

2 Asymptotic expansion of weak solutions

The regularity of weak solutions is mainly influenced by the corners and edges of the subdomains Ω_i. The behavior of a solution in the neighborhood of these peculiarities can be described by an asymptotic expansion which is expressed in terms of eigenvalues (the singular exponents) and eigenvectors of eigenvalue problems which are related to the corners and edges.

2.1 Corner singularities

The eigenvalue problem for corners can be deduced in the following way:

Let S be a corner, $\Omega_1, \ldots, \Omega_m$ the subdomains of Ω which contain S. Let $\mathcal{K}_i$ be the infinite cone with tip in S which coincides with Ω_i in a neighborhood of S; $\mathcal{K}^S$ is the cone which coincides with Ω in a neighborhood of S. The model transmission problem for the Lamé transmission operator on the cone $\mathcal{K}^S$ is given by (1)-(5) with zero right hand sides. Shortly: $L(\partial_x, \partial_y, \partial_z)u = 0$ in $\mathcal{K}^S$.
Next we introduce spherical coordinates (ρ, θ, φ) with respect to S. In these coordinates we have $L(\partial_x, \partial_y, \partial_z) = \rho^{-2} L_c(\theta, \varphi, \rho\partial_\rho, \partial_\varphi, \partial_\theta)$. Applying the Mellin transform with respect to ρ, $(\rho\partial_\rho) \to \alpha \in \mathbb{C}$, results in $\mathcal{A}_c(\alpha) := L_c(\varphi, \theta, \alpha, \partial_\varphi, \partial_\theta)$. We set $G := \mathcal{K}^S \cap S^2$, where S^2 is the unit sphere.

Definition 2. The eigenvalue problem which is related to a corner S is the following problem: Find $\alpha \in \mathbb{C}$ and $v \not\equiv 0$ such that $\mathcal{A}_c(\alpha)v(\omega) = 0, \;\; \omega \in G$.

Remark 2. 1. Explicit formulae for $\mathcal{A}_c(\alpha)$ are given in [7]. Properties of the operator $\mathcal{A}_c(\alpha)$ were derived in [1,6].
2. Formally this eigenvalue problem can be obtained by inserting the ansatz $u = \rho^\alpha v(\theta, \varphi)$ in the model problem $L(\partial_x, \partial_y, \partial_z)u = 0$.

Let α be an eigenvalue of $\mathcal{A}_c$, $I(\alpha)$ the geometric multiplicity of α and $\{\Psi_{\alpha,\mu;0},\, 1 \leqslant \mu \leqslant I(\alpha)\}$ a set of eigenfunctions. $M_{\alpha,\mu}$ denotes the range of the eigenfunction $\Psi_{\alpha,\mu;0}$, $\{\Psi_{\alpha,\mu;\kappa}, 0 \leqslant \kappa \leqslant M_{\alpha,\mu}-1\}$ are the associated eigenfunctions of $\Psi_{\alpha,\mu;0}$. We now assume that $\{\Psi_{\alpha,\mu;\kappa},\, 1 \leqslant \mu \leqslant I(\alpha), 0 \leqslant \kappa \leqslant M_{\alpha,\mu}-1\}$ is a canonical system of Jordan chains (Def. in [6]). For $\gamma = (\alpha, \mu, \kappa)$ we set

$$w_{\gamma,c}(\rho, \theta, \varphi) := \sum_{q=0}^{\kappa} \frac{(\ln \rho)^p}{q!} \Psi_{\alpha,\mu;\kappa-q}(\theta, \varphi). \tag{6}$$

2.2 Edge singularities

Let e be an edge. We introduce a Cartesian coordinate system (x, y, z) such that the edge is part of the z-axis, see Fig. 2. In this coordinate system we shortly write $L(\partial_x, \partial_y, \partial_z)$ for the Lamé transmission operator. Since edge singularities are generated by the non-tangential (to the edge) part of L we

introduce the following model problem: $L(\partial_x, \partial_y, 0)u(x,y) = 0,\ (x,y) \in \mathcal{C}^S$. Here $\mathcal{C}^S = \{(x,y) \in \mathbb{R}^2 : (x,y,0) \in \mathcal{D}^S\}$ and $\mathcal{D}^S$ is that infinite dihedron which coincides with Ω in a neighborhood of S.
Next we introduce polar coordinates (r,φ) in the (x,y)-plane: $L(\partial_x, \partial_y, 0) = r^{-2} L_e(\varphi, r\partial_r, \partial_\varphi)$. Applying the Mellin transform with respect to r, $r\partial_r \to \beta \in \mathbb{C}$, we get $\mathcal{A}_e(\beta) := L_e(\varphi, \beta, \partial_\varphi)$. The explicit formulae for $\mathcal{A}_e(\beta)$ are given in [6].

Remark 3. Formally the eigenvalue problem can be deduced by inserting the ansatz $u(x,y) = (u_x(x,y), u_y(x,y), u_z(x,y)) = r^\beta v(\varphi)$ in $L(\partial_x, \partial_y, 0)u = 0$.

Definition 3. The eigenvalue problem which is related to a point S on an edge e is the following problem: Find $\beta \in \mathbb{C}$, $v \not\equiv 0$ such that $\mathcal{A}_e(\beta)v = 0$ for $0 < \varphi < \phi$. Here, ϕ denotes the total opening of the edge at S.

Let β be an eigenvalue of $\mathcal{A}_e$ and $\{\Phi_{\beta,\mu;\kappa}, 1 \leqslant \mu \leqslant I(\beta), 0 \leqslant \kappa \leqslant M_{\beta,\mu}\}$ a canonical set of Jordan chains. For $\delta := (\beta, \mu, \kappa)$ we set

$$v_{\delta,e}(r,\varphi) := \sum_{q=0}^{\kappa} \frac{(\ln r)^q}{q!} \Phi_{\beta,\mu;\kappa-q}(\varphi). \tag{7}$$

2.3 Asymptotic expansion of weak solutions

Theorem 1 (Asymptotic expansion). *[1] Let $\Omega \subset \mathbb{R}^3$ be as described above, $f_i \in L^2(\Omega_i)$, $g_{i,k} \in H^{\frac{1}{2}}(\gamma_{i,k})$ for $\gamma_{i,k} \in \mathcal{N}$ and $u \in E(\Omega)$ a solution of (1)-(5). For a corner point S we denote by η^S a cut-off function with $\eta \equiv 1$ in a neighborhood of S. $\mathcal{E}_S$ is the set of all edges crossing at S.*
Let $0 < s \leqslant 1$ such that there are no eigenvalues of $\mathcal{A}_c(\alpha)$ on the line $\mathrm{Re}\ \alpha = -\frac{1}{2} + s$ and no eigenvalues of $\mathcal{A}_e(\beta)$, $e \in \mathcal{E}_S$, on the line $\mathrm{Re}\ \beta = s$.
Then the weak solution u has the following asymptotic expansion:

$$\eta^S u = u_{\text{reg}} + \eta^S \sum_{\gamma \in \Lambda_S} c_{\alpha,\mu} \rho^\alpha w_{\gamma,c} + \eta^S \sum_{e \in \mathcal{E}_S} \sum_{\delta \in \Lambda_e} \mathcal{R}_e(c_e^{\beta,\mu}) r_e^\beta v_{\delta,e}. \tag{8}$$

Here, $u_{\text{reg}}\big|_{\Omega_i} \in H^{1+s}(\Omega_i)$. Further
$\Lambda_S = \{\gamma = (\alpha,\mu,\kappa) : \alpha =$ eigenvalue of $\mathcal{A}_c(\alpha)$, $\mathrm{Re}\ \alpha \in]-\frac{1}{2}, -\frac{1}{2}+s[\}$,
$\Lambda_e = \{\delta = (\beta,\mu,\kappa) : \beta =$ eigenvalue of $\mathcal{A}_e(\beta)$, $\mathrm{Re}\ \beta \in]0,s[\}$, $e \in \mathcal{E}_S$.
$c_{\beta,\mu}$ are constants, $c_e^{\beta,\mu} \in H^{s-\mathrm{Re}\ \beta}(\mathbb{R})$ and $\mathcal{R}_e$ is a smoothing operator, see [1]. Finally ρ denotes the distance to corner S, r_e the distance to edge e. $v_{\delta,e}$ are the singular functions in (7) with r replaced by r_e, $w_{\gamma,c}$ the singular functions in (6) .

Corollary 1. *If there exists $\varepsilon > 0$ such that for all corners and edges of Ω there are no eigenvalues of the corresponding eigenvalue problems in the strip $\mathrm{Re}\ \alpha \in]-\frac{1}{2}, \varepsilon]$ for corners and $\mathrm{Re}\ \beta \in]0, \frac{1}{2}+\varepsilon]$ for edges, then a weak solution with right hand sides as in Theorem 1 satisfies: $u\big|_{\Omega_i} \in H^{\frac{3}{2}+\varepsilon}(\Omega_i)$. The Sobolev embedding theorem then shows $u\big|_{\Omega_i} \in C(\overline{\Omega_i})$.*

Note that in Theorem 1 negative singular exponents α are not excluded. Thus in principal unbounded deformation fields are possible. Up to now it is an open question whether or not there exist examples where the corner eigenvalue problem has eigenvalues α with Re $\alpha \in]-\frac{1}{2}, 0[$.
In the next section we will present sufficient conditions on the geometry and material parameters which guarantee that the strips in Corollary 1 are free of eigenvalues.

3 Quasi-monotonicity and regularity results

If the edge e or the corner point S is included in only one subdomain and if the boundary conditions do not change, then it can be proved that the strips in Corollary 1 do not contain eigenvalues of the corresponding eigenvalue problems, see the overview in [6]. Unfortunately this need not be the case for mixed boundary conditions or when the edge e or the corner point are included in several subdomains. For edge exponents β there exist examples for which Re β is arbitrarily close to 0, [4]. In contrast to this it is an open question whether or not there are examples for corner exponents with negative real parts. A sufficient condition for nonnegative corner exponents is a quasi-monotone distribution of the material parameters.

3.1 Estimates for edge exponents

Let e be an exterior edge with Dirichlet-conditions on all faces which contain e. Let (r, φ, z) be cylindrical coordinates such that $e = \{(0,0,z),\ z \in \mathbb{R}\}$. Let further $\Omega_1, \dots, \Omega_m$ be those subdomains which contain e. We assume that the numbering of the subdomains is such that $\Omega_i = \{(r, \varphi, z) : \phi_{i-1} < \varphi < \phi_i\}$ where $\phi_0 < \phi_1 < \dots < \phi_m < \phi_0 + 2\pi$, see Fig. 2. By $\boldsymbol{n}_1, \boldsymbol{n}_m$ we denote the exterior normal vectors on Ω_1 and Ω_m, respectively.

Theorem 2. *[5] Let e be an edge with Dirichlet conditions on both faces. Let further the parameters λ_i, μ_i be quasi-monotone with respect to e, i.e. $\exists! j \in \{1, \dots, m\} : \mu_1 \geqslant \dots \geqslant \mu_j \leqslant \dots \leqslant \mu_m$ and $\lambda_1 \geqslant \dots \geqslant \lambda_j \leqslant \dots \leqslant \lambda_m$. If there exists $\boldsymbol{t}_j \in \mathbb{R}^3$ which points into Ω_j and for which $\boldsymbol{n}_1 \boldsymbol{t}_j < 0$ and $\boldsymbol{n}_m \boldsymbol{t}_j < 0$, then there are no eigenvalues of the edge eigenvalue problem in the strip* Re $\beta \in [0, \frac{1}{2}]$.

Remark 4. Quasi-monotonicity was originally defined in [2] for transmission problems for the Laplace-operator.

3.2 Estimates for corner exponents

Let S be an exterior corner point which is contained in an arbitrary number of subdomains with Dirichlet conditions on the exterior faces. For a face $\gamma_{i,l} \in \mathcal{D}$ with $S \in \overline{\gamma}_{i,l}$ we denote by $\boldsymbol{n}_{i,l}$ the exterior normal vector of $\gamma_{i,l}$.

For an interface $\gamma_{ij,l} \in \mathcal{G}$ with $S \in \overline{\gamma}_{ij,l}$, $\boldsymbol{n}_{ij,l}$ is the exterior normal vector of Ω_i with respect to $\gamma_{ij,l}$. Note that $\boldsymbol{n}_{ij,l} = -\boldsymbol{n}_{ji,l}$.

Theorem 3. *[5] If there exists* $\boldsymbol{t} \in \mathbb{R}^3$ *such that*

$$\forall \gamma_{i,l} \in \mathcal{D} \text{ with } S \in \overline{\gamma}_{i,l} : \boldsymbol{n}_{i,l}\boldsymbol{t} < 0, \tag{9}$$

$$\forall \gamma_{ij,l} \in \mathcal{G} \text{ with } S \in \overline{\gamma}_{ij,l} : \boldsymbol{n}_{ij,l}\boldsymbol{t}(\lambda_i - \lambda_j) > 0,\ \boldsymbol{n}_{ij,l}\boldsymbol{t}(\mu_i - \mu_j) > 0, \tag{10}$$

then there are no eigenvalues α *of the corner eigenvalue problem in the strip* $\operatorname{Re} \alpha \in [-\frac{1}{2}, 0]$.

The conditions in Theorem 3 can be regarded as a generalized quasi-monotonicity condition. If for a corner S these conditions are satisfied, then for all edges which contain S, the conditions in Theorem 2 are satisfied as well. For further results (Neumann conditions, mixed conditions, interior points) see [5,8].

Example 1 (Fichera's corner). Consider the domain in Fig. 1 with Dirichlet conditions. The conditions in Theorem 3 are satisfied if:

$$\begin{aligned} \varphi \in (0, \tfrac{\pi}{2}] &: \lambda_1 < \lambda_2 < \lambda_3,\ \mu_1 < \mu_2 < \mu_3, \\ \varphi \in (\tfrac{\pi}{2}, \pi) &: \lambda_1 < \lambda_2 < \lambda_3,\ \mu_1 < \mu_2 < \mu_3 \text{ or } \lambda_1 < \lambda_3 < \lambda_2,\ \mu_1 < \mu_3 < \mu_2, \\ \varphi \in [\pi, \tfrac{3\pi}{2}) &: \lambda_1 < \lambda_3 < \lambda_2,\ \mu_1 < \mu_3 < \mu_2. \end{aligned}$$

References

1. Dauge, M. (1988) Elliptic Boundary Value Problems on Corner Domains. Smoothness and Asymptotic Expansion, Springer Verlag, Lecture Notes in Mathematics, vol. 1341.
2. Dryja, M., Sarkis, M.V., Widlund, O.B. (1996) Multilevel Schwarz Methods for Elliptic Problems with Discontinuous Coefficients in Three Dimensions, Numerische Mathematik, vol.72, pp. 313-348.
3. Grisvard, P. (1985) Elliptic Problems in Nonsmooth Domains, Pitman Press, Bath, Avon.
4. Kellogg, R.B. (1975) On the Poisson Equation with Intersecting Interfaces, Applicable Analysis 4, pp. 101-129.
5. Knees, D. (2002) Regularity results for transmission problems for the Laplace and Lamé operators on polygonal or polyhedral domains, Bericht SFB404, no. 2002/10, Universität Stuttgart.
6. Kozlov, V.A., Maz'ya, V.G., Rossmann,J. (2001) Spectral Problems Associated with Corner Singularities of Solutions to Elliptic Equations, AMS, Mathematical Surveys and Monographs, vol.85.
7. Leguillon, D. (1994) Computations of 3-D singularities in elasticity, In *Boundary Value Problems and Integral Equations in Nonsmooth Domains*, Costabel, Dauge, Nicaise, Eds., vol 167 of Lecture Notes in Pure and Applied Mathematics, Marcel Dekker Inc., pp. 161-170.
8. Nicaise,S., Sändig, A.-M. (1999) Transmission Problems for the Laplace and Elasticity Operators: Regularity and Boundary Integral Formulation, Mathematical Methods in the Applied Sciences, vol. 9, pp. 855-898.
9. Wloka, J. (1982) Partielle Differentialgleichungen, Teubner Verlag, Stuttgart.

Quasistatic Propagation of Cracks

Victor A. Kovtunenko

Lavrent'ev Institute of Hydrodynamics, 630090 Novosibirsk, Russia

Abstract. The problem of shape perturbation for the variational inequality describing solids with cracks under the nonpenetration condition is considered. There are used the iteration-penalty and finite-element methods to calculate numerically an approximate solution of the variational inequality. From the shape sensitivity analysis we deduce energetic characteristics of a solution in the general form. Applying analytical formulas, we describe the quasistatic propagation of a crack under the linear loading applied. The numerical example is presented, when a penetration between the crack faces occur.

1 Perturbation of the crack problem

Variational inequalities for models of solids with cracks under the nonpenetration condition were analysed in [1]. In [2]–[4] the shape sensitivity technique was developed to obtain the first derivative of the potential energy functional. Considering linear crack problems [5], in [6], [7] the asymptotic terms of the second order were applied to describe numerically the quasistatic propagation of rectilinear cracks.

Let Ω be a bounded domain in $\mathbf{R}^N$, $N = 2, 3$, with the boundary, which includes two parts Γ_D and Γ_N, $\Gamma_D \cap \Gamma_N = \emptyset$. Consider the crack $\Gamma(0)$ inside Ω as the $(N-1)$-manifold in $\mathbf{R}^N$. We denote the domain with crack as $\Omega(0) = \Omega \setminus \overline{\Gamma}(0)$. For small enough parameter ε, we consider the function of perturbation $\Phi \in [C^2(-\varepsilon_0, \varepsilon_0; W^{1,\infty}(\mathbf{R}^N))]^N$ with $\Phi(0)(x) = x$, $x \in \mathbf{R}^N$. Let $\Phi(\varepsilon)(\Omega) = \Omega$, then the coordinate transformation $y = \Phi(\varepsilon)(x)$ transforms the initial domain $\Omega(0)$ to the perturbed domain $\Omega(\varepsilon) = \Omega \setminus \overline{\Gamma}(\varepsilon)$ with $\Gamma(\varepsilon) = \Phi(\varepsilon)(\Gamma(0))$. We assume that for each ε the domain Ω can be splitted into two subdomains with Lipschitz-continuous boundaries and common interface, which intersects Γ_D, such that $\Gamma(\varepsilon)$ is a part of this interface.

The Jacobian of transformation has the asymptotic $J(\varepsilon) = 1 + \varepsilon \operatorname{div}(V) + o(\varepsilon)$, where $V \in [W^{1,\infty}(\mathbf{R}^N)]^N$ denotes the velocity field $\partial\Phi/\partial\varepsilon(0)$. Therefore, $\Phi(\varepsilon)$ yields the one-to-one correspondence between domains $\Omega(0)$ and $\Omega(\varepsilon)$, also between spaces $\widetilde{H}^1(\Omega(0))$ and $\widetilde{H}^1(\Omega(\varepsilon))$, when we introduce the space

$$\widetilde{H}^1(\Omega(\varepsilon)) = \{u \in [H^1(\Omega(\varepsilon))]^N, \quad u = 0 \quad \text{on } \Gamma_D\}.$$

For the displacement vector $u = (u_1, ..., u_N)$ we consider the linear elasticity model of nonhomogeneous anisotropic solid

$$\sigma_{ij}(u) = c_{ijkl}\epsilon_{kl}(u), \quad \epsilon_{ij}(u) = 0.5(u_{i,j} + u_{j,i}), \quad i, j = 1, ..., N,$$

with elasticity coefficients $c_{ijkl} \in C^2(\mathbf{R}^N)$, $i,j,k,l = 1,...,N$, which are symmetric and elliptic as it is usually supposed. Let us choose the unit normal vector $\nu(\varepsilon)$ to $\Gamma(\varepsilon)$ to distinguish its two opposite faces $\Gamma^{\pm}(\varepsilon)$, which correspond to $\pm\nu(\varepsilon)$, respectively. Then we introduce the closed convex set of admissible displacements

$$K(\varepsilon) = \{u \in \widetilde{H}^1(\Omega(\varepsilon)), \quad [\![u]\!]\nu(\varepsilon) \geqslant 0 \quad \text{on } \Gamma(\varepsilon)\}, \quad [\![u]\!] = u|_{\Gamma^+(\varepsilon)} - u|_{\Gamma^-(\varepsilon)},$$

which yields the condition of nonpenetration between the crack faces. For the given traction force $g = (g_1,...,g_N) \in [L^2(\Gamma_N)]^N$, there exists the unique solution $u(\varepsilon) \in K(\varepsilon)$ of the variational inequality

$$\int_{\Omega(\varepsilon)} \sigma_{ij}(u(\varepsilon))\epsilon_{ij}(v - u(\varepsilon)) \geqslant \int_{\Gamma_N} g_i(v - u(\varepsilon))_i \quad \forall v \in K(\varepsilon) \tag{1}$$

describing static equilibrium of the solid with crack.

2 Asymptotics of the solution and energy

When $\nu(\varepsilon) \circ \Phi(\varepsilon) = \nu(0)$, then $\Phi(\varepsilon)$ provides the one-to-one correspondence between $K(0)$ and $K(\varepsilon)$, too. In this case, the transformed solution of (1) as the function $u(\varepsilon) \circ \Phi(\varepsilon) \in K(0)$ solves the following variational inequality

$$\begin{aligned} &\textstyle\int_{\Omega(0)} J(\varepsilon)(c_{ijkl} \circ \Phi(\varepsilon))E_{kl}(\Psi(\varepsilon); u(\varepsilon) \circ \Phi(\varepsilon))E_{ij}(\Psi(\varepsilon); v - u(\varepsilon) \circ \Phi(\varepsilon)) \\ &\geqslant \textstyle\int_{\Gamma_N} g_i(v - u(\varepsilon) \circ \Phi(\varepsilon))_i \quad \forall v \in K(0), \end{aligned} \tag{2}$$

where $E_{ij}(\Psi; u) = 0.5(u_{i,k}\Psi_{kj} + u_{j,k}\Psi_{ki})$, $\Psi = (\partial\Phi/\partial x)^{-1}$, $i,j = 1,...,N$. From (2) and (1) as $\varepsilon = 0$ we prove the estimates

$$\|u(\varepsilon) \circ \Phi(\varepsilon)\|_{\widetilde{H}^1(\Omega(0))} \leqslant c_0, \quad \|u(\varepsilon) \circ \Phi(\varepsilon) - u(0)\|_{\widetilde{H}^1(\Omega(0))} \leqslant c_1\varepsilon. \tag{3}$$

This provides existence of the weak material derivative $\dot{u}(\Phi) \in Z(u(0))$ in the sense

$$\varepsilon_n^{-1}(u(\varepsilon_n) \circ \Phi(\varepsilon_n) - u(0)) \;\to\; \dot{u}(\Phi) \quad \text{weakly in } \widetilde{H}^1(\Omega(0)) \text{ as } \varepsilon_n \to 0. \tag{4}$$

Here $Z(u(0))$ is a tangential hyperplane to $K(0)$ at $u(0)$, i.e.

$$\begin{aligned} Z(u(0)) = \{\ & u \in \widetilde{H}^1(\Omega(0)), \quad [\![u]\!]\nu(0) \geqslant 0 \quad \text{on } C(u(0)), \\ & \textstyle\int_{\Omega(0)} \sigma_{ij}(u(0))\epsilon_{ij}(u) = \int_{\Gamma_N} g_i u_i\}, \end{aligned}$$

for $C(u(0)) \subseteq \Gamma(0)$ being the coincidence set where $[\![u(0)]\!]\nu(0) = 0$. Moreover, with the help of (3) and (4) we prove two orthogonality conditions

$$\int_{\Omega(0)} \sigma_{ij}(u(0))\epsilon_{ij}(\dot{u}(\Phi)) = \int_{\Gamma_N} g_i\dot{u}(\Phi)_i = -\int_{\Omega(0)} A_1(V; u(0), u(0)) \tag{5}$$

with the symmetric bilinear form as follows:

$$A_1(V;u,v) = \mathrm{div}(Vc_{ijkl})\epsilon_{kl}(u)\epsilon_{ij}(v) - \sigma_{ij}(u)E_{ij}(\tfrac{\partial V}{\partial x};v) - \sigma_{ij}(v)E_{ij}(\tfrac{\partial V}{\partial x};u).$$

Let us introduce the function of potential energy $P(\Phi) : (-\varepsilon_0, \varepsilon_0) \to \mathbf{R}$ by the relation

$$P(\Phi)(\varepsilon) = 0.5\int_{\Omega(\varepsilon)} \sigma_{ij}(u(\varepsilon))\epsilon_{ij}(u(\varepsilon)) - \int_{\Gamma_{\mathcal{N}}} g_i u(\varepsilon)_i. \tag{6}$$

Applying the coordinate transformation $y = \Phi(\varepsilon)(x)$ to the integrals in (6), due to (3)–(5) the following expansion holds

$$P(\Phi)(\varepsilon) = P(\Phi)(0) + \varepsilon P'(\Phi)(0) + o(\varepsilon),$$

with the first derivative of potential energy as follows:

$$P'(\Phi)(0) = 0.5\int_{\Omega(0)} A_1(V;u(0),u(0)). \tag{7}$$

It is independent on the choice of perturbation function Φ in the sense that, if $\Omega(\varepsilon) = \Phi^1(\varepsilon)(\Omega(0)) = \Phi^2(\varepsilon)(\Omega(0))$, then $P'(\Phi^1)(0) = P'(\Phi^2)(0)$. Similar to (7) formula for constant elasticity coefficients and V of the $W^{2,\infty}$-class was firstly obtained in [3], and for the partial case of perturbation in [2]. When there exists $D \subset \Omega(0)$ such that $u(0) \in H^2(\overline{D})$ and

$$V = 0 \quad \text{or} \quad \{\nabla c_{ijkl}\} = 0 \quad \text{in } \overline{\Omega}(0) \setminus D,$$

$$c_{ijkl}(x)\xi_{kl}[0.5\mathrm{div}(V(x))\xi_{ij} - \xi_{im}V_{m,j}(x)] = 0 \quad \forall\{\xi_{ij}\} \quad \text{a.e. } x \in \Omega(0) \setminus \overline{D},$$

then, integrating by parts, the domain integral in (7) can be represented in the equivalent form of integral over boundary ∂D,

$$I(V) = \int_{\partial D} \sigma_{ij}(u(0))[0.5(V\cdot n)\epsilon_{ij}(u(0)) - n_j(V\cdot\nabla u(0)_i)],$$

where n is the unit outward normal vector to ∂D. This implies the general form of invariant integrals of energy.

3 Numerical example on the crack propagation

Let $\Omega \in \mathbf{R}^2$ be the unit square $\{0 < x_1 < 1,\ |x_2| < 0.5\}$ with the boundary $\overline{\Gamma}_{\mathcal{N}} \cup \overline{\Gamma}_{\mathcal{D}}$, $\Gamma_{\mathcal{D}} = \{x_1 = 1,\ |x_2| < 0.5\}$. Let us take the crack

$$\Gamma(0) = \{0 < x_1 < l,\ x_2 = 0\}$$

of the length $0 < l < 1$. In this case $\nu(0) = (0,1)$. The Lamé model $\sigma_{ij}(u) = 2\mu\epsilon_{ij}(u) + \lambda(\epsilon_{11}(u) + \epsilon_{22}(u))\delta_{ij}$, $i,j = 1,2$, with

$$\mu = \frac{E}{2(1+\nu)}, \quad \lambda = \frac{E\nu}{(1+\nu)(1-2\nu)}, \quad \nu = 0.34, \quad E = 7.3\cdot 10^4\,\mathrm{mPa}$$

is considered. We take the traction force g given on $\Gamma_N = \Gamma_N^1 \cup \Gamma_N^2$ by the following functions

$$\begin{cases} g_1(x) = -6.5t_0\mu|2x_2|, \quad g_2(x) = 0 & \text{on } \Gamma_N^1 = \{x_1 = 0,\ 0 < \pm x_2 < 0.5\} \\ g_1(x) = 0,\ g_2(x) = \pm t_0\mu(1 - |2x_1 - 1|) & \text{on } \Gamma_N^2 = \{0 < x_1 < 1,\ x_2 = \pm 0.5\} \end{cases}$$

with the multiplier $t_0 = 0.9 \cdot 10^{-3}$.

The inequality (1) as $\varepsilon = 0$ is approximated by the penalty equation

$$\int_{\Omega(0)} \sigma_{ij}(u^\delta(0))\epsilon_{ij}(v) - \tfrac{1}{\delta}\int_{\Gamma(0)} [\![u_2^\delta(0)]\!]^- [\![v_2]\!] = \int_{\Gamma_N} g_i v_i, \quad u^- = \max\{0, -u\},$$

with small parameter $\delta > 0$, and then linearized by iterations as follows:

$$\begin{aligned} &\int_{\Omega(0)} \sigma_{ij}(u^{\delta,n}(0))\epsilon_{ij}(v) + \delta^{-1}\int_{\Gamma(0)} [\![u_2^{\delta,n}(0) - u_2^{\delta,n-1}(0)]\!][\![v_2]\!] \\ &= \int_{\Gamma_N} g_i v_i + \delta^{-1}\int_{\Gamma(0)} [\![u_2^{\delta,n-1}(0)]\!]^- [\![v_2]\!], \quad n = 1, 2, \dots . \end{aligned} \tag{8}$$

We have the convergences

$$\|u^\delta(0) - u(0)\|_{\tilde{H}^1(\Omega(0))} \to 0 \quad \text{as } \delta \to 0 \quad (\text{of the order } \sqrt{\delta}),$$

$$\|u^{\delta,n}(0) - u^\delta(0)\|_{\tilde{H}^1(\Omega(0))} \leqslant (1 + C\delta)^{-0.5n}\|u^{\delta,0}(0) - u^\delta(0)\|_{\tilde{H}^1(\Omega(0))}.$$

To solve numerically the linear equation (8), we apply the finite-element method with piecewise-linear functions on uniform triangle mesh in $\Omega(0)$ of the size $h = 0.025$. It supposed the approximation of order $\sqrt{h}$ for $u(0)$ and order h for the energy. Approximate values of the potential energy $P(\Phi)(0)$ from (6), calculated by this scheme, are presented in Fig. 1 in dependence on the parameter l of the crack length.

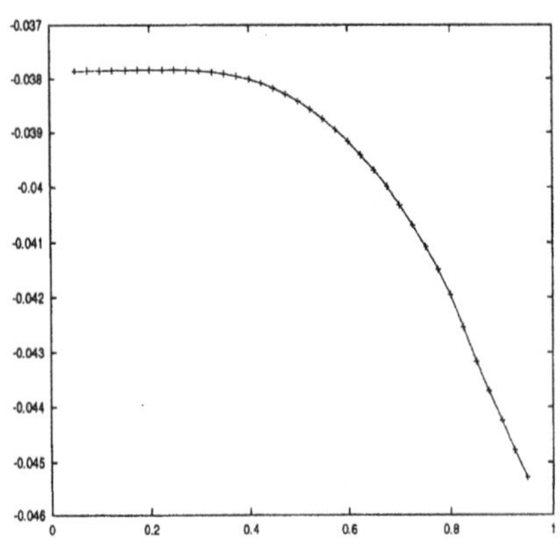

Fig. 1. Potential energy $P(\Phi)(0)$

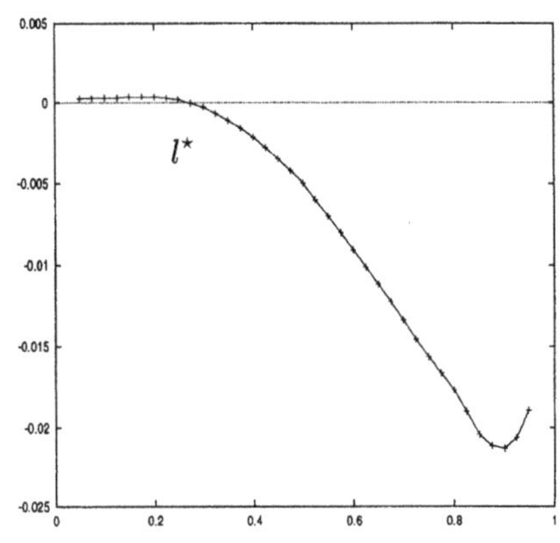

Fig. 2. First derivative $P'(\Phi)(0)$

Let us take the piecewise-linear cut-off function $\chi \in W^{1,\infty}(\mathbf{R}^2)$, which has $\operatorname{supp}(\chi) \subset \Omega$ and $\chi = 1$ in neighborhood of the crack tip $x = (l, 0)$. Applying the perturbation $\Phi(\varepsilon)(x) = (x_1 + \varepsilon\chi(x), x_2)$, it follows from formula (7) the first derivative $P'(\Phi)(0)$ with respect to perturbation of the crack length l (energy release rate), which is numerically calculated as presented in Fig. 2. Here $P'(\Phi)(0) \approx 0$ for $0 < l \leqslant l^\star$, $l^\star \approx 0.25$. Moreover, from the well-known formula in classical fracture mechanics, $-P'(\Phi)(0) = 0.5\mu(1-\nu)|K|^2$, due to $P'(\Phi)(0) \leqslant 0$ we can find the modulus $|K|$ of stress intensity factors as presented in Fig. 3.

For loading parameter $t \geqslant 0$, let us consider the linear loading $g(x)(t) = t\,g(x)$. By the cone property of $K(0)$ we obtain that $u(0)(t) = t\,u(0) \in K(0)$ is a solution of the corresponding quasistatic problem, and, therefore, $P'(\Phi)(0)(t) = t^2 P'(\Phi)(0)$. Let us fix the initial crack length l. From the Griffith fracture criterion $\gamma + P'(\Phi)(0)(t) = 0$, we deduce then the critical value of the maximal loading as

$$g_{cr}(l) = 6.5 t_0 \mu\, t_{cr}(l), \quad t_{cr}(l) = \sqrt{\gamma(-P'(\Phi)(0))^{-1}},$$

and the inverse relation $l = l(g_{cr})$ presented in Fig. 4 as the curve L. Here constant $\gamma > 0$ is the doubled density of surface energy distributed at the crack. The Griffith fracture hypothesis implies that there is no crack growth for $0 \leqslant t < t_{cr}(l)$, and the crack starts to grow when $t = t_{cr}(l)$. In Fig. 4 we see three possible cases of the crack quasistatic propagation in dependence on the initial crack length l. The crack does not grow at all for $l \leqslant l^\star$ with $l^\star$ from Fig. 2. For $l > l^\star$ the crack propagation can be stable (with finite initial velocity), as presented in Fig. 4 for $l = 0.95$, or unstable with a jump (infinite initial velocity), as presented in Fig. 4 for $l = 0.825$.

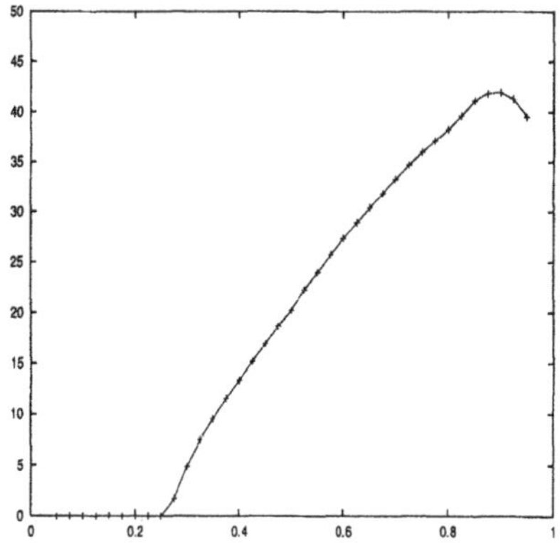

Fig. 3. Modulus $|K|$ of SIF

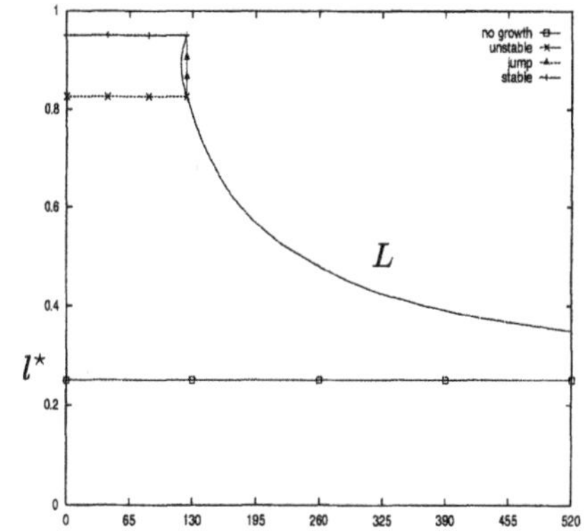

Fig. 4. Critical loading $g_{cr}(l)$

Acknowledgments. *The research results were obtained with support of the INTAS Foundation in framework of the research grant YSF 01/1-33, and visiting Professor W.L. Wendland at the Mathematical Institute A of the University of Stuttgart. The participation at the International Conference on Multifield Problems on 8-10 April, 2002, in Stuttgart was supported by the German Research Foundation (DFG) in framework of the Collaborative Research Programme (SFB) 404.*

References

1. Khludnev A.M., Kovtunenko V.A. (2000) Analysis of Cracks in Solids. WIT-Press, Southampton Boston
2. Khludnev A.M., Sokolowski J. (1999) The Griffith formula and the Cherepanov–Rice integral for crack problems with unilateral conditions in nonsmooth domains. Euro. J. Appl. Math. **10**, 379–394
3. Khludnev A.M., Ohtsuka K., Sokolowski J. (2002) On derivative of energy functional for elastic bodies with cracks and unilateral conditions. Quart. Appl. Math. **60**, 99–109
4. Bach M., Khludnev A.M., Kovtunenko V.A. (2000) Derivatives of the energy functional for 2D-problems with a crack under Signorini and friction conditions. Math. Meth. Appl. Sci. **23**, 515–534
5. Kovtunenko V.A. (2001) Sensitivity of cracks in 2D-Lamé problem via material derivatives. Z. angew. Math. Phys. **52**, 1071–1087
6. Bach M., Kovtunenko V.A., Sukhorukov I.V. (2000) Numerical validation of the shape optimization approach to quasi-static crack propagation. SFB404 Bericht **29**. Universität Stuttgart
7. Bach M., Kovtunenko V.A. (2001) Numerics of quasistatic crack propagation and delamination of interface cracks I&II. SFB404 Bericht **16&17**. Universität Stuttgart

New generalizations of the Cahn-Hilliard equation based on a microforce balance

Alain Miranville[1] and Messoud Efendiev[2]

[1] Université de Poitiers, Laboratoire d'Applications des Mathématiques, SP2MI, Téléport 2, Avenue Marie et Pierre Curie, 86962 Chasseneuil Futuroscope Cedex, France

[2] Universität Stuttgart, Fakultät Mathematik, Pfaffenwaldring 57, D-70569 Stuttgart, Germany

Abstract. In this article, we further generalize the models of Cahn-Hilliard equations proposed by M. Gurtin and based on a balance law for microforces. These models are consistent with the second law of thermodynamics. We also discuss the existence and uniqueness of solutions.

1 Introduction

The Cahn-Hilliard equation is very important in materials science. It is a conservation law which describes important qualitative features of two-phase systems, namely the transport of atoms between unit cells. This phenomenon can be observed, for instance, when a binary alloy is cooled down sufficiently. One then observes a partial nucleation (i.e. the apparition of nucleides in the material) or a total nucleation (the so-called spinodal decomposition : the material quickly becomes inhomogeneous, forming a fine-grained structure ; see e.g. [1] for more details).

Noting that the classical derivation of the Cahn-Hilliard equation is not entirely satisfactory, M. Gurtin proposed in [6] a new approach, based on a new balance law for internal microforces (the microforces describe forces associated with microscopic configurations of atoms, whereas standard forces are associated with macroscopic length scales) and consistent with the second law of thermodynamics (more precisely, M. Gurtin considers a purely mechanical version of the second law which states that the rate at which the free energy increases cannot exceed the sum of the working and of the energy inflow due to mass transport).

The starting point in the Cahn-Hilliard theory is a free energy ψ. The classical choice of free energy (also called Ginzburg-Landau free energy) is derived, at first approximation, in [2] by assuming that the local free energy per molecule in a region of nonuniform composition depends both on the local composition and on that of the immediate environment. It reads

$$\psi = \psi(\rho, \nabla\rho) = \frac{\alpha}{2}|\nabla\rho|^2 + f(\rho), \tag{1}$$

where ρ is the order parameter (corresponding to a density of atoms here), $\alpha > 0$ is related to the surface tension and f is a double-well potential whose wells correspond to the phases of the material. Now, G. Giacomin and J. L. Lebowitz derived rigorously in [5] (based on stochastic arguments) a model for phase separation with long range interactions (see also [4] for the mathematical study of such models) which leads to a Cahn-Hilliard type equation with a (nonlocal) free energy of the form

$$\psi = \psi(\rho) = f(\rho) + \kappa_1 \rho(1-\rho) + \frac{1}{2}\int_\Omega \mathcal{K}(|x-y|)|\rho(x) - \rho(y)|^2 dy, \tag{2}$$

where Ω is the domain occupied by the material and $\kappa_1 = \kappa_1(x) = \int_\Omega \mathcal{K}(|x-y|)dy$, $\mathcal{K}$ being a smooth function.

The models proposed by M. Gurtin are based on the following balance laws (neglecting the external actions) :

$$\frac{\partial \rho}{\partial t} = -\operatorname{div} h \text{ (mass balance)}, \tag{3}$$

$$\operatorname{div} \zeta + \pi = 0 \text{ (microforce balance)}, \tag{4}$$

where h is the mass flux, ζ (a vector) is the microstress and π (a scalar) corresponds to the internal microforces. Then, one obtains constitutive equations by considering the restrictions imposed by the second law of thermodynamics. In particular, in the models proposed by M. Gurtin, this yields that

$$\zeta = \partial_{\nabla\rho}\psi. \tag{5}$$

Therefore, if we take a free energy as in (2), we deduce from (4) and (5) that $\zeta = 0$ and $\pi = 0$, i.e. there are no internal microforces.

Actually, the constitutive equations obtained by M. Gurtin are derived by assuming that the independent constitutive variables are ρ, $\nabla\rho$ (these are the independent constitutive variables for the classical Cahn-Hilliard theory), μ, $\nabla\mu$ (this is reasonable since we want to consider systems that are sufficiently far from equilibrium ; in the classical theory, μ is given, constitutively, in terms of ρ, assuming that the system is close to equilibrium) and $\frac{\partial\rho}{\partial t}$ (the kinetics) ; we refer the reader to [7], [8] and [9] for the mathematical study of such models. So, in order to further generalize these models, but also to have a theory which is compatible with a free energy of the form (2) (in the sense that the microforces do not vanish), we propose to also add $\nabla\frac{\partial\rho}{\partial t}$ to the list of independent constitutive variables.

2 Derivation of the equations

We call $Z = (\rho, \nabla\rho, \frac{\partial\rho}{\partial t}, \nabla\frac{\partial\rho}{\partial t}, \mu, \nabla\mu)$ (independent constitutive variables) and we assume that the constitutive functions h, π, ζ and also the free energy ψ depend on Z.

Then, the second law of thermodynamics yields a dissipation inequality of the form $[\partial_\rho\psi(Z)+\pi(Z)-\mu]\frac{\partial\rho}{\partial t}+[\partial_{\nabla\rho}\psi(Z)-\zeta(Z)].\nabla\frac{\partial\rho}{\partial t}+[\partial_{\frac{\partial\rho}{\partial t}}\psi(Z)]\frac{\partial^2\rho}{\partial t^2}+[\partial_{\nabla\frac{\partial\rho}{\partial t}}\psi(Z)].\nabla\frac{\partial^2\rho}{\partial t^2}+[\partial_\mu\psi(Z)]\frac{\partial\mu}{\partial t}+[\partial_{\nabla\mu}\psi(Z)].\nabla\frac{\partial\mu}{\partial t}+h(Z).\nabla\mu\leqslant 0$, for every fields ρ and μ (see [6] for more details). This yields that, necessarily

$$\psi = \psi(\rho, \nabla\rho), \tag{6}$$

and there remains the dissipation inequality

$$[\partial_\rho\psi(Z)+\pi(Z)-\mu]\frac{\partial\rho}{\partial t}+[\partial_{\nabla\rho}\psi(Z)-\zeta(Z)].\nabla\frac{\partial\rho}{\partial t}+h(Z).\nabla\mu\leqslant 0, \tag{7}$$

for every fields ρ and μ. We can rewrite (7) in the form

$$F(X,Y).Y\leqslant 0,\ \forall X,\ Y, \tag{8}$$

where $X = (\rho, \nabla\rho, \frac{\partial\rho}{\partial t}, \nabla\frac{\partial\rho}{\partial t}, \mu, \nabla\mu)$, $Y = (\frac{\partial\rho}{\partial t}, \nabla\frac{\partial\rho}{\partial t}, \nabla\mu)$ and $F(X,Y) = (\partial_\rho\psi+\pi-\mu, \partial_{\nabla\rho}\psi-\zeta, h)$. It then follows from [6], Appendix B, that there exists a matrix $\mathcal{B}=\mathcal{B}(X,Y)$ such that

$$F(X,Y) = -\mathcal{B}(X,Y)Y, \tag{9}$$

$$\mathcal{B}(X,Y)Y.Y\geqslant 0, \tag{10}$$

for every X and Y. Writing $\mathcal{B}$ in the form $\mathcal{B}=\begin{pmatrix}\beta & {}^ta & {}^tb\\ c & A & B\\ d & C & D\end{pmatrix}$, we finally obtain the constitutive equations

$$\partial_\rho\psi+\pi-\mu = -\beta\frac{\partial\rho}{\partial t}-a.\nabla\frac{\partial\rho}{\partial t}-b.\nabla\mu, \tag{11}$$

$$\partial_{\nabla\rho}\psi-\zeta = -c\frac{\partial\rho}{\partial t}-A\nabla\frac{\partial\rho}{\partial t}-B\nabla\mu, \tag{12}$$

$$h = -d\frac{\partial \rho}{\partial t} - C\nabla\frac{\partial \rho}{\partial t} - D\nabla\mu, \tag{13}$$

such that (7) is satisfied and where the constitutive moduli β (a scalar), a, b, c, d (4 vectors) and A, B, C, D (4 matrices) depend on Z.

We finally deduce from (3)-(4) and (11)-(13) the following equations :

$$\frac{\partial \rho}{\partial t} - \operatorname{div}(d(Z)\frac{\partial \rho}{\partial t}) - \operatorname{div}(C(Z)\nabla\frac{\partial \rho}{\partial t}) = \operatorname{div}(D(Z)\nabla\mu), \tag{14}$$

$$\begin{aligned}&\mu - b(Z).\nabla\mu + \operatorname{div}(B(Z)\nabla\mu) = \beta(Z)\tfrac{\partial \rho}{\partial t} + a(Z).\nabla\tfrac{\partial \rho}{\partial t} - \\ &\operatorname{div}(c(Z)\tfrac{\partial \rho}{\partial t}) - \operatorname{div}(A(Z)\nabla\tfrac{\partial \rho}{\partial t}) + \partial_\rho\psi(\rho,\nabla\rho) - \operatorname{div}(\partial_{\nabla\rho}\psi(\rho,\nabla\rho)).\end{aligned} \tag{15}$$

Remark 1. In [2], the free energy (1) is obtained, at first approximation, by neglecting higher derivatives of ρ. We could further generalize the models of M. Gurtin by adding such derivatives to the list of independent constitutive variables. It is interesting however to note that, if we wish that the free energy does depend on such a derivative of ρ (via the dissipation inequality, consequence of the second law), it is necessary to add also the same derivative of $\frac{\partial \rho}{\partial t}$ to the list of independent constitutive variables.

3 Remarks on the mathematical study of the equations

The mathematical study of the general equations (14)-(15) would be very difficult. However, we have, keeping the same notations as in Section 1

$$F(X,Y) = -\mathcal{B}Y + o(|X| + |Y|), \tag{16}$$

as X and Y tend to 0, where $\mathcal{B}$ is a constant and positive semi-definite matrix (see [6], Appendix B). Thus, we will assume, at first approximation, that the constitutive moduli are constant. Furthermore, we will consider here the classical Ginzburg-Landau free energy (1) only ; the case of nonlocal free energies is much more delicate (see [4]) and will be considered elsewhere, at least for simplified models. In that case, we can rewrite the equations in the form

$$\frac{\partial \rho}{\partial t} - d.\nabla\frac{\partial \rho}{\partial t} - \operatorname{div}(C\nabla\frac{\partial \rho}{\partial t}) = \operatorname{div}(D\nabla\mu), \tag{17}$$

$$\begin{aligned}&\mu - b.\nabla\mu + \operatorname{div}(B\nabla\mu) = \beta\tfrac{\partial \rho}{\partial t} + (a-c).\nabla\tfrac{\partial \rho}{\partial t} - \\ &\operatorname{div}(A\nabla\tfrac{\partial \rho}{\partial t}) - \alpha\Delta\rho + f'(\rho),\end{aligned} \tag{18}$$

where

$$\begin{aligned}&\beta x^2 + Ay.y + Dz.z + (a+c).yx + (b+d).zx + ({}^tB + C)y.z \geqslant 0,\\ &\forall x \in R,\ \forall y,\ z \in R^n,\end{aligned} \tag{19}$$

n being the spatial dimension, which corresponds to the dissipation inequality (7). We associate with (17)-(18) periodic boundary conditions (thus, $\Omega = \Pi_{i=1}^n (0, L_i)$, $L_i > 0$, $i = 1, \ldots, n$). Finally, we assume that $f(s) = \sum_{i=1}^n a_i s^i$, $a_{2p+2} > 0$, $p \in N^\star$ (logarithmic potentials are also very important in the Cahn-Hilliard theory, but are more difficult to treat ; in particular, we then need to prove that ρ remains between 0 and 1. This will be also treated elsewhere).

We set $\mathcal{E}(t) = \int_\Omega (\frac{\alpha}{2}|\nabla\rho|^2 + f(\rho))dx$ (the total (free) energy) and we easily prove, assuming that ρ and μ are regular enough, that

$$\begin{aligned}&\tfrac{d\mathcal{E}}{dt} = -\beta|\tfrac{\partial\rho}{\partial t}|^2 - (A\nabla\tfrac{\partial\rho}{\partial t}, \nabla\tfrac{\partial\rho}{\partial t}) - (D\nabla\mu, \nabla\mu) - ((a+c).\nabla\tfrac{\partial\rho}{\partial t}, \tfrac{\partial\rho}{\partial t}) -\\ &((b+d).\nabla\mu, \tfrac{\partial\rho}{\partial t}) - (({}^tB + C)\nabla\tfrac{\partial\rho}{\partial t}, \nabla\mu),\end{aligned} \tag{20}$$

and it follows from the dissipation inequality (19) that the energy is decreasing, as expected in the Cahn-Hilliard theory.

Now, in order to study the existence and uniqueness of solutions, (19) is not sufficient (in particular, (20) is not sufficient to obtain the necessary a priori estimates). However, if we make the following stronger coercivity condition :

$$\begin{aligned}&\beta x^2 + Ay.y + Dz.z + (a+c).yx + (b+d).zx + ({}^tB + C)y.z \geqslant\\ &c_0(x^2 + |y|^2 + |z|^2), \forall x \in R,\ \forall y,\ z \in R^n,\ c_0 > 0,\end{aligned} \tag{21}$$

we can prove the following result (see [10] for the details) :

Theorem 1. *We assume that (21) holds and that $p = 1$ or 2 when $n = 3$. Then, (17)-(18) possesses a unique weak solution (ρ, μ) such that $\rho \in L^\infty(0, T; H^1_{per}(\Omega)) \cap \mathcal{C}([0, T]; L^2(\Omega))$ and $\frac{\partial\rho}{\partial t}$ and $\mu \in L^2(0, T; H^1_{per}(\Omega))$, $\forall T > 0$ and for an initial data in $H^1_{per}(\Omega)$.*

Remark 2. Actually, the equation that is called the Cahn-Hilliard equation is that obtained by eliminating μ in the system for ρ and μ. Here, taking the $\operatorname{div} D\nabla$ of (18), in which we inject the expression of $\operatorname{div}(D\nabla\mu)$ given by (17), we obtain the following generalized Cahn-Hilliard equation :

$$\begin{aligned}&\tfrac{\partial\rho}{\partial t} - (b+d).\nabla\tfrac{\partial\rho}{\partial t} - \operatorname{div}[(\beta D + C - B + \tfrac{1}{2}(b^t d + d^t b))\nabla\tfrac{\partial\rho}{\partial t}] +\\ &b.\nabla\operatorname{div}(C\nabla\tfrac{\partial\rho}{\partial t}) - d.\nabla\operatorname{div}(B\nabla\tfrac{\partial\rho}{\partial t}) - \operatorname{div}[B\nabla\operatorname{div}(C\nabla\tfrac{\partial\rho}{\partial t})] +\\ &\operatorname{div}[D\nabla\operatorname{div}(A\nabla\tfrac{\partial\rho}{\partial t})] + \alpha\operatorname{div}(D\nabla\Delta\rho) - \operatorname{div}[D\nabla f'(\rho)] = 0.\end{aligned} \tag{22}$$

We shall study this second formulation of the problem, which is much more complicated than (17)-(18) (in particular, in the sense that it is not clear how the dissipation inequality can be used), elsewhere.

Acknowledgements : This work was performed while the second author was visiting the University of Stuttgart within the guest program of the Collaborative Research Center "SFB 404 Mehrfeldprobleme in der Continuummechanik". He wishes to thank Professors M. Efendiev and W. Wendland for having invited him and for their warm hospitality. This work was also partially supported by INTAS project 00-899.

References

1. Cahn, J. W. (1961) On spinodal decomposition, Acta Metall. **9**, 795–801.
2. Cahn, J. W., Hilliard, J. E. (1958) Free energy of a nonuniform system I. Interfacial free energy, J. Chem. Phys. **2**, 258–267.
3. Fried, E., Gurtin, M. (1994) Dynamic solid-solid phase transitions with phase characterized by an order parameter, Physica D **72**, 287–308.
4. Gajewski, H., Zacharias, K. (2001) On a nonlocal phase separation model, Preprint.
5. Giacomin, G., Lebowitz, J. L. (1997) Phase segregation dynamics in particle systems with long range interaction I. Macroscopic limits, J. Statist. Phys. **87**, 37–61.
6. Gurtin, M. (1996) Generalized Ginzburg-Landau and Cahn-Hilliard equations based on a microforce balance, Physica D **92**, 178–192.
7. Miranville, A. (2000) Some generalizations of the Cahn-Hilliard equation, Asymptotic Anal. **22**, 235–259.
8. Miranville, A. (2001) Long-time behavior of some models of Cahn-Hilliard equations in deformable continua, Nonlinear Anal. Ser. B **2**, 273–304.
9. Miranville, A. (2001) Consistent models of Cahn-Hilliard-Gurtin equations with Neumann boundary conditions, Physica D **158**, 233–257.
10. In preparation.

Contact of Elastic Bodies with Negative Poisson's Ratio

G. Szefer and D. Kedzior

Cracow University of Technology, Kraków, Poland

Abstract. A class of large deformation contact problems for linear elastic, isotropic bodies exhibiting negative Poisson's ratio will be considered, Bilateral as well as unilateral contact with friction is analyzed. Essential differences in deformation as well in stress distribution in comparison with conventional materials have been corroborated.

1 Introduction

Linear elastic bodies with negative Poisson's ratio ν exhibits non-typical behaviour: they expands laterally when stretched and contract by compression. This unusual property of solids rather seldom observed in natural materials may appear in manufactured elements owing to special microstructure of the material. Already LOVE [5] was the first who mentioned that some kind of single crystals like pyrid attain negative value of the Poisson's ratio. The same effect was observed in some directions of anisotropic crystal cadmium as reported by LI [4]. Negative values of ν were measured by WILLIAMS and LEWIS [10] in human bones. Microstructures exhibiting negative ν (also closed to -1) was considered by ALMGREN [1], EVANS [2], MILTON [6] and SIGMUND [7]. A polyester foam structure with reentrant cells exhibiting linear compressive stress-strain curve up to about 5% strain and Poisson's ratio about of -0.7 was manufactured by LAKES [3]. Structured composites with negative Poisson's ratio have been described by THEOCARIS, STAVROULAKIS and PANAGIOTOPOULOS in [9]. They conclude, that cellular microstructures with non-convex shape possesing reentrant corners leads to negative values of ν. Computational simulations and experimental verifications performed on star-shaped beam-like structures support this hypothesis. On the other hand, the existence condition for linear elasticity reads (in tensor notation): $c_{ijk\ell}\varepsilon_{ij}\varepsilon_{k\ell} = \sigma_{ij}\varepsilon_{ij} \geqslant \alpha\varepsilon_{ij}\varepsilon_{ij}$, $\alpha > 0$, where $c_{ijk\ell}$ stands for the elasticity matrix and $\varepsilon_{ij}, \sigma_{ij}$ means the components of the small strain and stress tensor, respectively. In case of isotropy it is $c_{ijk\ell} = G(\delta_{ik}\delta_{i\ell} + \delta_{i\ell}\delta_{jk}) + \lambda\delta_{ij}\delta_{k\ell}$, where $G = \frac{E}{2(1-\nu)}$ and $\lambda = \frac{E}{(1+\nu)(1-2\nu)}$ are Lamé coefficients and E – is the Young modulus. By using the deviator and axiator decomposition one obtains

$$\begin{aligned}\sigma_{ij}\varepsilon_{ij} &= \left(s_{ij} + \frac{\sigma_{kk}}{3}\delta_{ij}\right)\left(e_{ij} + \frac{\varepsilon_{kk}}{3}\delta_{ij}\right) \\ &= 2Ge_{ij}e_{ij} + \frac{2G+3\lambda}{3}\varepsilon_{kk}^2 \geqslant \alpha\left(e_{ij}e_{ij} + \frac{\varepsilon_{kk}^2}{3}\right)\end{aligned}$$

where s_{ij} , e_{ij} stand for the deviator components of $\boldsymbol{\sigma}$ and $\boldsymbol{\varepsilon}$. Hence for $G > 0$, $2G+3\lambda > 0$ it follows $-1 < \nu < 0.5$. So the negative value of ν do not violate the existence of solution in the linear theory of elasticity. In the present paper an elastic body with negative Poisson's ratio undergoing contact with a flat, rigid foundation is considered. Static and dynamic problems in terms of bilateral and unilateral contact conditions will be analyzed. Friction will be taken into account, too.

2 Statement of the Contact Problem

Consider a linear elastic isotropic body B which due to mass forces with density $\rho_R \mathbf{b}$ and prescribed surface tractions $\mathbf{p}_R$ (or kinematical boundary excitation Δ) come into contact with a rigid and rough obstacle (target body). For generality we assume large deformations and large displacements of the elastic contactor.

Remark 1 *We assume finite strain formulation not only to describe structured materials with large deformation but also therefore, that nonlinear kinematics gives more precise results whenever the strains are practically small. Independent on that, a rubber-like material exhibiting large deformations is considered too, for comparison.*

Using material description in the cartesian coordinate system $\{OX^K\}$ $K = 1, 2, 3$ the following governing system of equations must be fulfilled:

$$\begin{aligned} &\rho_R \ddot{\mathbf{u}} - \operatorname{div} \mathbf{S}\,(\mathbf{1} + \nabla \mathbf{u}) = \rho_R \mathbf{b} \\ &\mathbf{E} = \frac{1}{2}\left(\nabla \mathbf{u} + \nabla \mathbf{u}^\top + \nabla \mathbf{u}^\top \nabla \mathbf{u}\right) \\ &\mathbf{S} = 2G\mathbf{E} + \lambda I^E \mathbf{1} \end{aligned} \tag{1}$$

Here means:

$\mathbf{u}$, $\nabla \mathbf{u}$ – displacement and displacement gradient respectively
$\mathbf{E}$, $\mathbf{S}$ – strains and II Piola-Kirchhoff stress tensor
$I^E = E_{KK}$ – first invariant of the strain tensor

System (1) must be completed by

i) initial conditions:

$$\begin{aligned} &\mathbf{u}\,(t_0, \mathbf{X}) = \mathbf{u}_0(\mathbf{X}) \\ &\dot{\mathbf{u}}(t_0, \mathbf{X}) = \mathbf{v}_0(\mathbf{X})\,, \ X \in B_R \end{aligned} \tag{2}$$

ii) boundary conditions:

$$\begin{aligned} &\mathbf{u} = \Delta(\mathbf{X}, t) \ \text{ on } \ \Gamma_u \\ &\mathbf{S}(\mathbf{1} + \nabla \mathbf{u})\mathbf{N} = \mathbf{p_R}(\mathbf{X}) \ \text{ on } \ \Gamma_\sigma \end{aligned} \tag{3}$$

where
$\mathbf{N}$ – outward unit vector normal to the boundary
$\Gamma = \Gamma_u \cup \Gamma_\sigma \cup \Gamma_c$ – boundary of B_R

iii) contact conditions:

$$d_n \geqslant 0\,,\; t_n \geqslant 0\,,\; t_n d_n = 0 \tag{4}$$

iv) friction conditions:

$$\begin{aligned} &|\mathbf{t_T}| \leqslant \mu t_n \\ &|\mathbf{t_T}| < \mu t_n \quad \Rightarrow \quad \dot{\mathbf{u}}_\mathbf{T} = 0 \\ &|\mathbf{t_T}| = \mu t_n \quad \Rightarrow \quad \dot{\mathbf{u}}_\mathbf{T} = -\lambda \mathbf{t_T} \quad \Rightarrow \quad \mathbf{t_T} = -\mu t_n \frac{\dot{\mathbf{u}}_\mathbf{T}}{|\dot{\mathbf{u}}_\mathbf{T}|} \end{aligned} \tag{5}$$

Above means:
d_n – the distance between the particles which may come into contact
$t_n=(\mathbf{Tn})\mathbf{n}$ – the normal contact stress
$\mathbf{t_T}$ – the tangential contact stress vector
μ – the friction coefficient
$\dot{\mathbf{u}}_\mathbf{T}$ – the sliding velocity.

Remark 2 *The proof of existence of the solution for the dynamic frictional contact problem is still open. For small deformations the solution exists if the following formula are satisfied*

$$\mu \leqslant \frac{\sqrt{3-4\nu}}{2-2\nu}\,,\ (D=2) \qquad\qquad \mu \leqslant \frac{\sqrt{3-4\nu}}{4-4\nu}\,,\ (D=3)$$

This is valid also for $\nu < 0$.

To solve the strong nonlinear initial-boundary-value problem (1)–(5) the incremental approach combined with the Finite Element technique and Newmark's time integration procedure have been used [8]. A series of numerical calculations were performed.

3 Numerical Examples

As a contactor (master body) an elastic block bonded with a rigid plate on the upper edge is considered. The block is deformed due to prescribed vertical translation Δ of the plate (both compresion and tension are considered). Following data has been used: $E = 170000$ (also 5000 for "soft rubber") $\mathrm{kN/m^2}$, $\rho = 1700\,\mathrm{kg/m^3}$, $\nu = -0.48$, $\mu = 0.5 - 1.0$ (for unilateral contact). The results of deformations and contact stress distributions are shown on Fig. 1–4. The static solution was assumed to be the initial state for dynamics. Thus the block has been pushed by horizontal forces applied to the plate. The results illustrates Fig. 5 and 6. For comparison we show the deformation (Fig. 7) in the neighbourhood of the corner for a block with positive Poisson's ratio $\nu = 0.48$ too. The qualitative as well as the quantitative difference is evident. We discuss also a rubber block with a layer of negative Poisson's ratio on the bottom (Fig. 8). Various thicknesses of the layer were taken into account. Some of these results are given on Fig. 9–10.

4 Conclusions

The performed analysis enable to formulate following corollaries: (i) when the block is clamped on the bottom (bilateral contact) then tension occurs at the corner by compression; (ii) this rather unexpected phenomenon corresponds with the case of unilateral contact, where loss of contact at the corner takes place. This behaviour is visible the more the larger is the friction coefficient; (iii) in the case of the layered block one can observe, that the thiner is the layer with negative Poisson's ratio, the larger is the stress concentration; (iv) in terms of dynamics, materials with positive Poisson's ratio exhibit large deformations at the corner (particularly for larger μ) with separation zone and possible self-contact (see Fig. 7) whereas materials with $\nu < 0$ demonstrate a stiffening property (the material is predicted to become tough evoking permanent contact with a larger stick zone).

References

1. Almgren, R. (1985) An isotropic three-dimensional structure with Poisson's ratio $= -1$. J. Elasticity **15**
2. Evans, L. (1990) Tailoring the negative Poisson's ratio. Chemistry & Industry **15**
3. Lakes, R. (1987) Foam structures with negative Poisson's Ratio. Science **235**
4. Li, Y. (1976) The anisotropic behaviour of Poisson's ratio. Young modulus and shear modulus in hexagonal materials. Phys. Satus Solidi **38**
5. Love, A. (1944) A treatise on the mathematical theory of elasticity. N. York
6. Milton, C. (1992) Composite Materials with Poisson's ratio close to -1. J. Mech. Phys. Solids **40**
7. Sigmund, O. (1994) Tailoring Materials with Prescribed Elastic Properties. Danish Center Appl. Math. Mech. Rep. **480**
8. Szefer, G. (1998) Contact Problems in Terms of Large Deformations. ZAMM **78**
9. Theocaris, P., Stavroulakis, G., Panagiotopoulos, P. (1997) Negative Poisson's ratios in composities with star-shaped inclusions: a numerical homogenization approach. Arch, Appl. Mech. **67**
10. Williams, J., Lewis, J. (1982) Properties and an anisotropic model of cancellous bone from the proximal tibial epiphyris. J. Biom. Eng. **104**

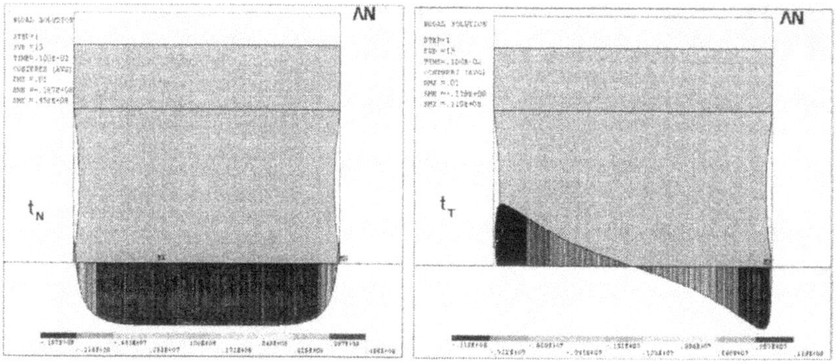

Fig. 1. Bilateral contact under compression

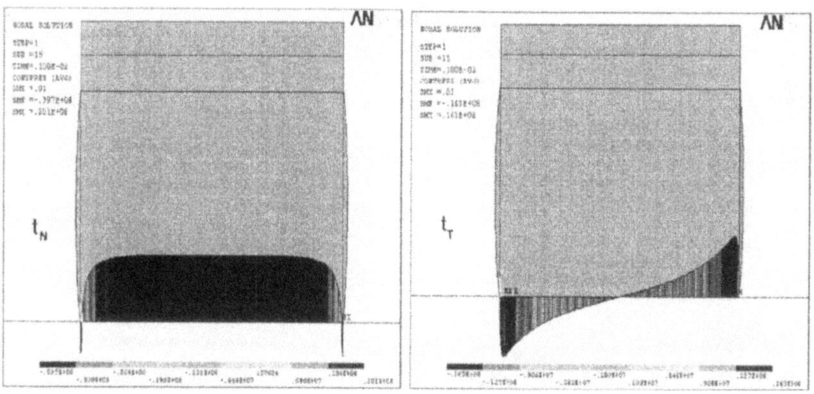

Fig. 2. Bilateral contact under tension

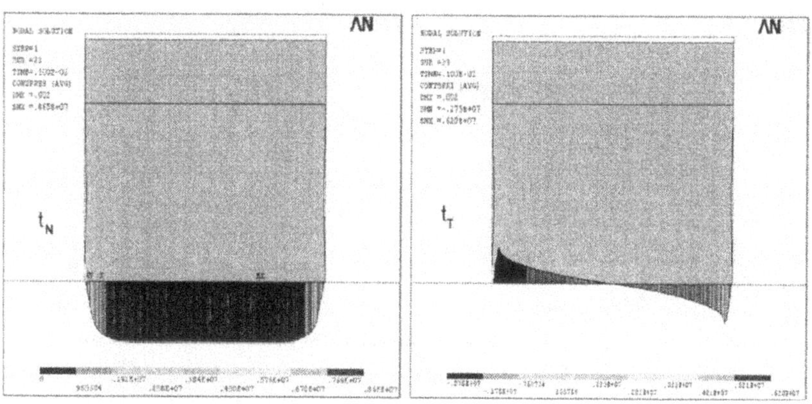

Fig. 3. Unilateral contact with $\mu = 1.0$

Fig. 4. Enlargement at the corner

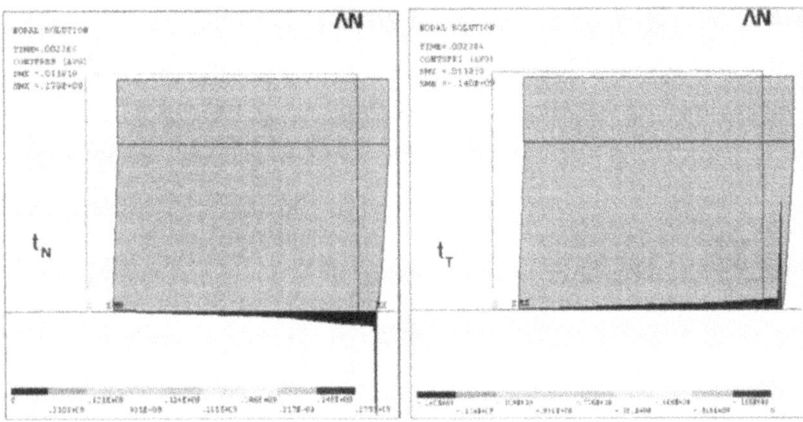

Fig. 5. Dynamic deformation for $\mu = 0.5$

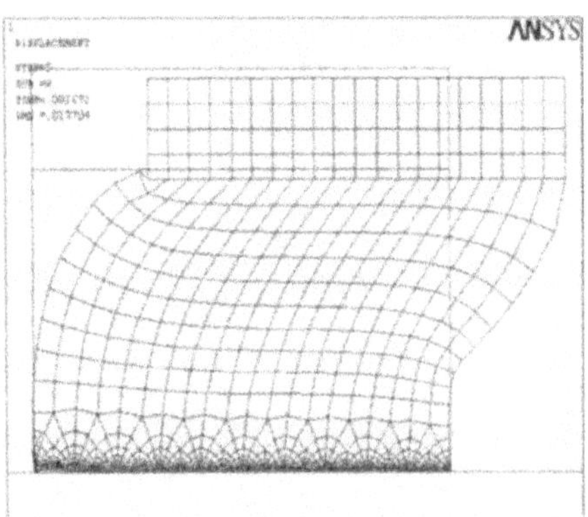

Fig. 6. Deformation for $\mu = 1.0$ ("soft rubber")

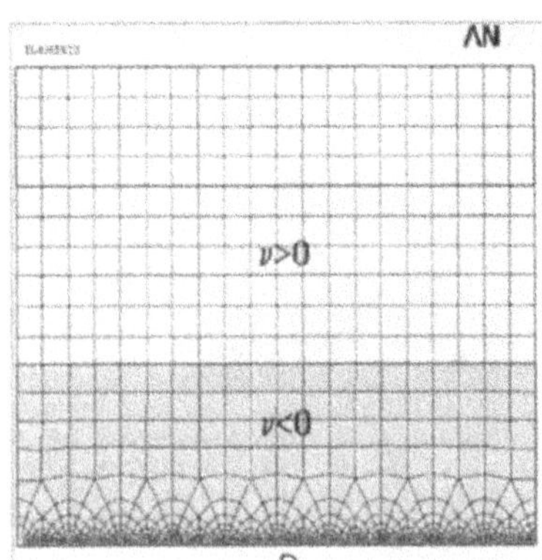

Fig. 8. Layer of thickness d with $\nu < 0$

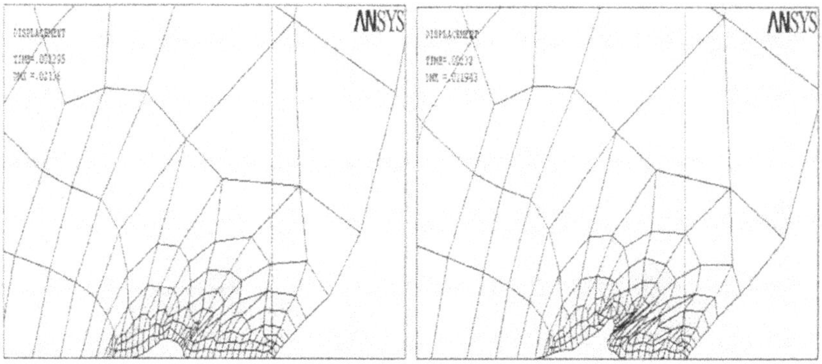

Fig. 7. Deformation at the corner for $\mu = 1.0$ and $\nu = +0.48$

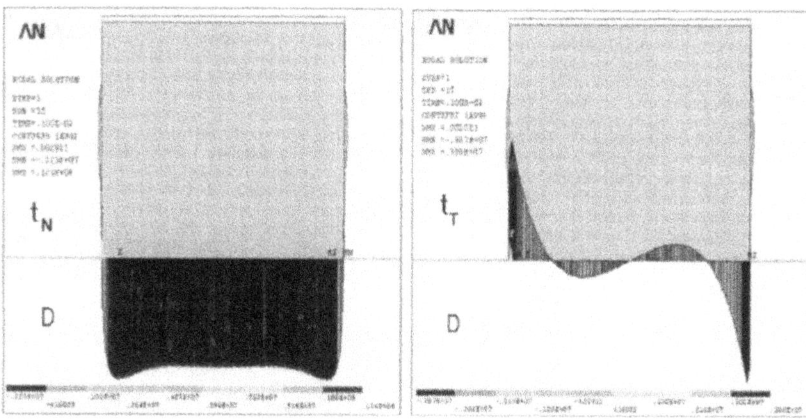

Fig. 9. Layer of thickness $d = \frac{1}{2}h$

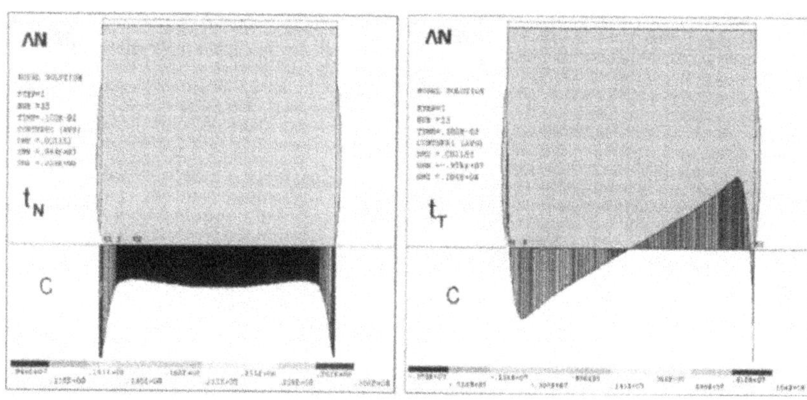

Fig. 10. Layer of thickness $d = \frac{1}{12}h$

Transient Dynamic Response of a Cracked Piezoelectric Solid Under Impact Loading

Ch. Zhang

Hochschule Zittau/Görlitz, University of Applied Sciences, 02763 Zittau, Germany

Abstract. In this paper, a time-domain boundary integral equation method is presented for transient dynamic crack analysis in linear piezoelectric solids. A finite crack in an infinite linear piezoelectric solid subjected to an impact crack-face loading is considered. The corresponding initial-boundary value problem is formulated as a set of hypersingular time-domain boundary integral equations, with the crack-opening-displacement and the electric-potential-discontinuity as unknown boundary quantities. A numerical solution procedure is developed for solving the hypersingular time-domain BIEs, which uses a convolution quadrature formula for approximating the temporal convolution and a Galerkin method for the spatial discretization of the time-domain BIEs.

1 Introduction

The application of time-domain boundary element method (BEM) or boundary integral equation method (BIEM) to linear piezoelectric solids with cracks is yet very limited, due to the fact that the corresponding time-domain dynamic Green's functions have no-closed or simple mathematical forms ([1]-[3]). This inconvenience affects the efficiency and the numerical implementation of the time-domain BEM/BIEM for cracked piezoelectric solids considerably. In this paper, a time-domain BIEM for transient dynamic crack analysis in piezoelectric solids is presented. A finite crack in an infinite linear piezoelectric solid subjected to an antiplane mechanical and inplane electric impact loading is investigated. Hypersingular time-domain boundary integral equations (BIEs) are applied to solve the corresponding initial-boundary value problem, where the crack-opening-displacement (COD) and the electric-potential-discontinuity (EPD) are the unknown boundary quantities. To solve the hypersingular time-domain BIEs, a numerical solution procedure is developed, which uses the convolution quadrature formula of Lubich [4] for approximating the temporal convolution and a Galerkin method for the spatial discretization of the time-domain BIEs. The present time-domain BIEM requires no explicit expressions of the time-dependent Green's functions. In lieu of this, Laplace-domain Green's functions are needed, which can be expressed as Fourier integrals. The spatial variations of the unknown COD and EPD are approximated by Chebyshev polynomials of second kind, which are capable to describe the local behavior of the COD and EPD at crack-tips properly. The application of the convolution quadrature formula of Lubich [4] for temporal and a Galerkin method for spatial discretizations converts the hypersingular time-domain BIEs into to a system of linear algebraic equations, which are solved numerically step by step. Transient dynamic stress and electric displacement intensity factors are obtained numerically from the expansion coefficients of the Galerkin-ansatz. To show the accuracy and the efficiency of the present time-domain BIEM, numerical results are given.

2 Problem Statement and Time-Domain BIEs

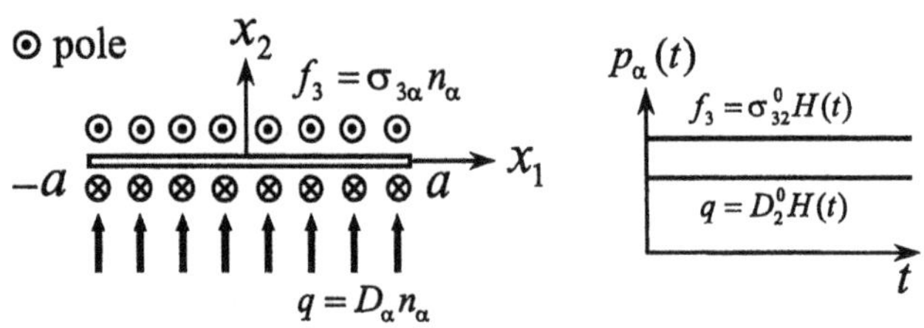

Fig. 1. A finite crack in an infinite piezoelectric solid

Let us consider a finite crack of a length $2a$ in an infinite linear piezoelectric solid as shown in in Fig. 1. The piezoelectric solid under consideration is transversely isotropic and the poling direction is in the x_3-direction. An antiplane mechanical and inplane electric impact loading is applied on the crack-faces. In the absence of body forces and volume electric charges, the equation of motion and the Gauss's law are given by

$$\sigma_{3\alpha,\alpha} = \rho \ddot{u}_3 \, , \tag{1}$$

$$D_{\alpha,\alpha} = 0 \, , \tag{2}$$

where $\sigma_{3\alpha}$ represents the shear stress components, u_3 denotes the displacement component in the x_3-direction, ρ is the mass density, and D_α is the inplane electric displacements, respectively. The cracked piezoelectric solid satisfies the following constitutive equations

$$\sigma_{3\alpha} = C_{44} u_{3,\alpha} - e_{15} E_\alpha \, , \tag{3}$$

$$D_\alpha = e_{15} u_{3,\alpha} + \epsilon_{11} E_\alpha \, , \tag{4}$$

the initial conditions

$$u_3(\mathbf{x}, t) = \dot{u}_3(\mathbf{x}, t) = 0 \, , \qquad \phi(\mathbf{x}, t) = 0 \, , \qquad \text{at} \quad t = 0 \, , \tag{5}$$

and the boundary conditions on the crack-faces

$$f_3 = \sigma_{3\alpha} n_\alpha = \sigma_{32}^0 \cdot H(t) \, , \tag{6}$$

$$q = D_\alpha n_\alpha = D_2^0 \cdot H(t) \,, \tag{7}$$

where C_{44} is the elastic constant, e_{15} is the piezoelectric constant, ϵ_{11} is the dielectric constant, E_α is the electric field components, ϕ is the electric potential, f_3 is the traction component, q is the surface electric charge density, n_α is the unit normal vector, and $H(t)$ is the Heaviside function. Note here that Eq. (7) corresponds to the impermeable electric boundary condition on the crack-faces. Under the quasi-electrostatic assumption, the electric field components E_α and the electric potential ϕ are related by

$$E_\alpha = -\phi_{,\alpha} \,. \tag{8}$$

The initial-boundary value problem governed by Eqs. (1)-(7) can be described by the following hypersingular time-domain BIEs

$$\int_{-a}^{+a} F_{\gamma\alpha}^G(\mathbf{x}, \mathbf{y}; t, \tau) * \Delta d_\gamma(y_1, \tau) dy_1 = \bar{p}_\alpha(x_1, t) \,, \qquad x_1 \in [-a, +a] \,, \tag{9}$$

where an asterisk $*$ denotes Riemann convolution, $F_{\gamma\alpha}^G$ is the time-domain traction Green's function, Δd_γ represents the crack-opening-displacement (COD) and the electric-potential-discontinuity (EPD) which are defined by

$$\Delta d_\gamma = \begin{Bmatrix} \Delta u_3 \\ \Delta\phi \end{Bmatrix} = \begin{Bmatrix} u_3(y_1, y_2 = 0^+; \tau) - u_3(y_1, y_2 = 0^-; \tau) \\ \phi(y_1, y_2 = 0^+; \tau) - \phi(y_1, y_2 = 0^-; \tau) \end{Bmatrix} \,, \tag{10}$$

and the right-hand side $\bar{p}_\alpha(x_1, t)$ is given by

$$\bar{p}_\alpha = \begin{Bmatrix} f_3(x_1, t) \\ q(x_1, t) \end{Bmatrix} = \begin{Bmatrix} \sigma_{32}^0 \cdot H(t) \\ D_2^0 \cdot H(t) \end{Bmatrix} \,. \tag{11}$$

The hypersingular integral in (9) is regarded as Hadamard finite-part integral. The present time-domain BIEM applies a spatial Galerkin method for the spatial discretization of the hypersingular BIEs (9) and does not require any special regularization or integration techniques for computing the hypersingular integral. Furthermore, the present time-domain BIEM does not require an explicit expression of the time-domain Green's function $F_{\gamma\alpha}^G$, but needs its Laplace-transform $\hat{F}_{\gamma\alpha}^G$, which can be expressed as Fourier integrals ([5], [6]).

3 Numerical Solution Procedure

To solve the hypersingular time-domain BIEs (9), the convolution quadrature formula of Lubich [4] for approximating the temporal convolution and a Galerkin method for the spatial discretization of the time-domain BIEs (9) are applied. The unknown vector $\Delta d_\gamma(y_1,\tau)$ is approximated by the following Galerkin-ansatz

$$\Delta d_\gamma(y_1,\tau) = \sqrt{a^2 - y_1^2}\sum_{k=1}^{K} c_{\gamma;k}(\tau) U_{k-1}\left(\frac{y_1}{a}\right), \tag{12}$$

where $c_{\gamma;k}(\tau)$ are the unknown time-dependent expansion coefficients, and $U_{k-1}(y_1/a)$ are Chebyshev polynomials of second kind. The function $\sqrt{a^2-y_1^2}$ in Eq. (12) is introduced to describe the local "square-root" behavior of the unknown COD and EPD at the crack-tips properly. By substituting Eq. (12) into Eq. (9), multiplying both sides of Eq. (9) by $\sqrt{a^2-x_1^2}U_{l-1}(x_1/a)$ and integrating it with respect to x_1 from $-a$ to $+a$, the following time-domain Galerkin-BIEs are obtained

$$\sum_{k=1}^{K}\int_{-a}^{+a}\sqrt{a^2-x_1^2}U_{l-1}\left(\frac{x_1}{a}\right)\int_{-a}^{+a}\sqrt{a^2-y_1^2}U_{k-1}\left(\frac{y_1}{a}\right)F^G_{\gamma\alpha} * c_{\gamma;k}(\tau)dy_1dx_1$$

$$= \int_{-a}^{+a}\bar{p}_\alpha(x_1,t)\sqrt{a^2-x_1^2}U_{l-1}\left(\frac{x_1}{a}\right)dx_1, \qquad l = 1,2,...,K. \tag{13}$$

By applying the following convolution quadrature formula of Lubich [4]

$$f(t) = g * h = \int_0^t g(t-\tau)h(\tau)d\tau \Longrightarrow f(n\Delta t) = \sum_{j=0}^{n}\omega_{n-j}(\Delta t)h(j\Delta t) \tag{14}$$

to Eq. (13), a system of linear algebraic equations for the expansion coefficients $c_{\gamma;k}$ is obtained as

$$\sum_{j=0}^{n}\sum_{k=1}^{K} A^{n-j}_{\gamma\alpha;kl}c^j_{\gamma;k} = p^n_{\alpha;l}, \qquad n = 0,1,2,...,N. \tag{15}$$

where the time-variable t is divided into N equal time-steps Δt and the system matrix in Eq. (15) corresponds to the integration weights $\omega_{n-j}(\Delta t)$ of the convolution quadrature formula (14). The time-domain system matrix and the right-hand side of Eq. (15) are given by

$$A^{n-j}_{\gamma\alpha;kl} = \frac{r^{-(n-j)}}{M}\sum_{m=0}^{M-1}\hat{A}_{\gamma\alpha;kl}(p_m)e^{-2\pi i\cdot(n-j)m/M}, \tag{16}$$

$$p^n_{\alpha;l} = (-1)^l\int_{-a}^{+a}\bar{p}_\alpha(x_1,n\Delta t)\sqrt{a^2-x_1^2}U_{l-1}\left(\frac{x_1}{a}\right)dx_1, \tag{17}$$

where

$$p_m = \delta(\zeta_m)/\Delta t\ , \quad \delta(\zeta_m) = \sum_{j=1}^{2}(1-\zeta_m)^j/j\ , \quad \zeta_m = re^{2\pi i \cdot m/M}\ . \tag{18}$$

When M=N and r^N=$\sqrt{\epsilon}$ are chosen, the error in computing $A^{n-j}_{\gamma\alpha;kl}$ by using the convolution quadrature formula of Lubich [4] is of the order $O(\sqrt{\epsilon})$, with ϵ being the numerical error in computing the Laplace-domain system matrix $\hat{A}_{\gamma\alpha;kl}(p_m)$. The Laplace-domain system matrix can be written as ([5], [6])

$$\hat{A}_{\gamma\alpha;kl} = \begin{cases} 0, & k+l = \text{odd}, \\ \frac{\pi}{2}a^2 i^{k+l}(kl)\begin{bmatrix} \mu I_{kl} + C_{44}D_{kl} & e_{15}D_{kl} \\ e_{15}D_{kl} & -\epsilon_{11}D_{kl} \end{bmatrix}, & k+l = \text{even}, \end{cases} \tag{19}$$

where $J_k(\cdot)$ is the Bessel function of first kind and k-th order, and

$$I_{kl} = \int_{-\infty}^{\infty}\left(\frac{\gamma}{\xi^2} - \frac{1}{\xi}\right) J_k(\xi a)J_l(\xi a)d\xi\ , \qquad D_{kl} = \frac{\delta_{kl}}{k+l}\ , \tag{20}$$

$$\gamma = \sqrt{(\rho/\mu)p^2 + \xi^2}\ , \qquad \mu = C_{44} + e_{15}^2/\epsilon_{11}\ . \tag{21}$$

The time-domain system matrix $A^{n-j}_{\gamma\alpha;kl}$ is symmetric and real-valued. Equation (19) reveals that only a single integral has to be computed numerically in the present time-domain BIEM, though the application of a spatial Galerkin method. The integrand in Eq. (20) behaves as $1/\xi^4$ for $\xi \to \infty$, which converges very fast and can be computed numerically by using an adaptive Romberg quadrature formula in conjunction wit the truncation method. Once the Laplace-domain system matrix has been determined numerically, the time-domain system matrix can be subsequently computed very efficiently by using Eq. (16) and the Fast Fourier Transform (FFT). Equation (15) can be solved numerically to obtain the expansion coefficients $c_{\gamma;k}$ step by step.

4 Numerical Results and Conclusions

Numerical calculations have been performed for PZT-5H with the following material constants: C_{44}=$3.53 \times 10^{10} N/m^2$, e_{15}=$17.0 C/m^2$, ϵ_{11}=$1.51 \times 10^{-8} C/(Vm)$ and $\rho = 7500 kg/m^3$. The parameters used for computing the time-domain system matrix $A^{n-j}_{\gamma\alpha;kl}$ defined by Eq. (16) are: M=N, ε=10^{-12} and r^N=$\sqrt{\varepsilon}$. To describe the combined mechanical and electric impact crack-face loading, a loading parameter λ=$e_{15}D_2^0/\epsilon_{11}\sigma_{32}^0$ is introduced. In what follows, numerical results for the transient dynamic stress intensity factor $K^{\pm}_{III}(t)$ and the transient dynamic electric displacement intensity factor $K^{\pm}_{D}(t)$ are presented, which can be obtained from the expansion coefficients

$c_{\gamma;k}(t)$ in a simple manner. For convenience, the following normalized dynamic intensity factors are introduced

$$\bar{K}^{\pm}_{III}(t) = K^{\pm}_{III}(t)/K^{st}_{III} , \qquad K^{st}_{III} = \sigma^0_{32}\sqrt{\pi a} , \tag{22}$$

$$\bar{K}^{\pm}_{D}(t) = K^{\pm}_{D}(t)/K^{st}_{D} , \qquad K^{st}_{D} = D^0_2\sqrt{\pi a} , \tag{23}$$

where $\pm$ stands for the crack-tips at $\pm a$, while σ^0_{32} and D^0_2 represent the amplitudes of the mechanical and the electric crack-face impact loading.

In the case of a pure mechanical impact (λ=0), the normalized dynamic $\bar{K}^{\pm}_{III}(t)$-factor is shown in Figs. 2 and 3 versus the dimensionless time $c_T t/a$, where $c_T = \sqrt{\mu/\rho}$. Figure 2 shows the influence of the used terms in the Galerkin-ansatz (12) on the normalized dynamic $\bar{K}^{\pm}_{III}(t)$-factor, where a time-step $c_T\Delta t/a$=0.05 is chosen. It can be seen on Fig. 2 that 20 terms in the Galerkin-ansatz (12) are sufficient to obtain accurate results for the normalized dynamic $\bar{K}^{\pm}_{III}(t)$-factor. Figure 3 shows the sensitivity of the present time-domain BIEM to the selected time-steps, where 20 terms in the Galerkin-ansatz (12) have been applied. Fig. 3 reveals that the present time-domain BIEM is pretty insensitive to the selected time-steps. In the time-range considered here, the present time-domain BIEM provides stable numerical results. The normalized dynamic $\bar{K}^{\pm}_{D}$-factor vanishes for a pure mechanical impact, i.e. $\bar{K}^{\pm}_{D}(t)$=0.

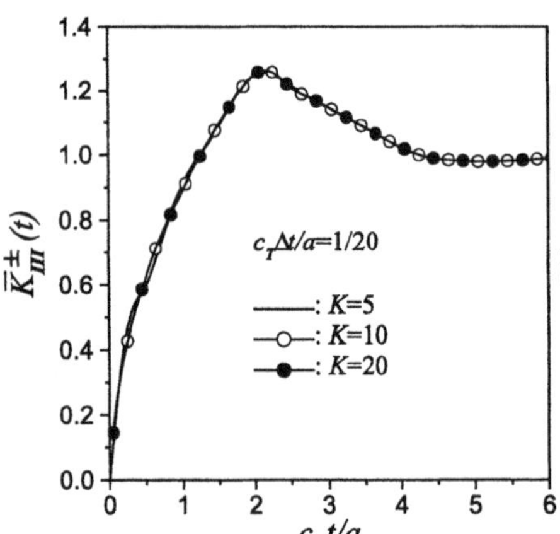

Fig. 2. Normalized $\bar{K}^{\pm}_{III}$-factor (λ=0)

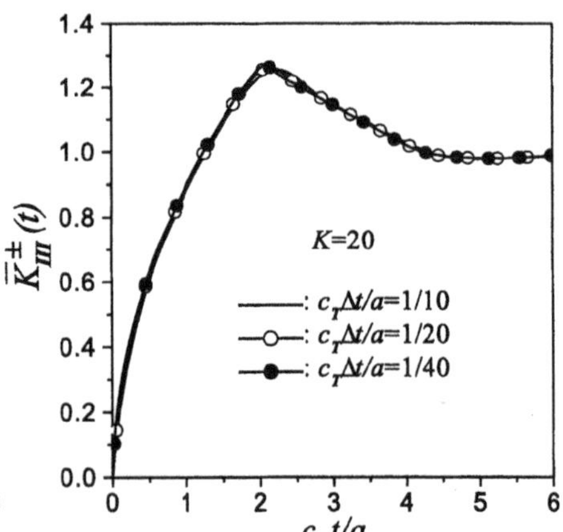

Fig. 3. Normalized $\bar{K}^{\pm}_{III}$-factor (λ=0)

For combined antiplane mechanical and inplane electric impact loading, the corresponding numerical results for the normalized dynamic $\bar{K}^{\pm}_{III}(t)$-factor are given in Figs. 4 and 5. Here, K=20 in the Galerkin-ansatz (12) and a time-step $c_T\Delta t/a$=1/20 are used. Figure 4 is for positive λ-values while Fig. 5 applies for negative λ-values. Figures 4 and 5 show that an inplane electric impact may give rise to an increase or a decrease of the normalized dynamic

$\bar{K}^{\pm}_{III}(t)$-factor, depending on the direction of the applied electric impact (i.e., positive or negative λ-values) and the moment under consideration. The normalized dynamic $\bar{K}^{\pm}_{D}$-factor is independent on the dimensionless time $c_T t/a$ and it takes always the value unity, i.e. $\bar{K}^{\pm}_{D}(t)$=1.

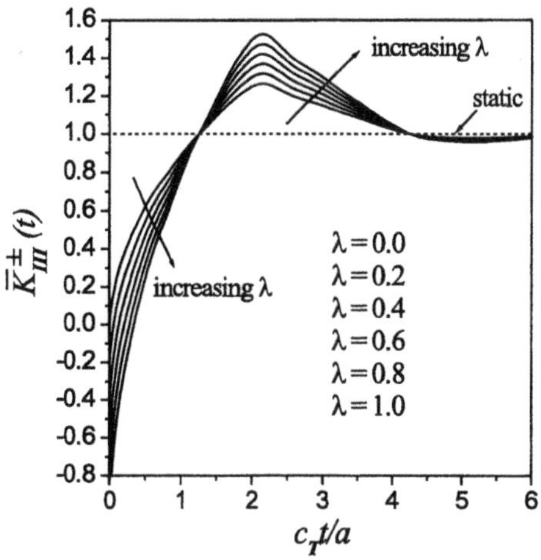

Fig. 4. Normalized $\bar{K}^{\pm}_{III}$-factor ($\lambda \geqslant 0$)

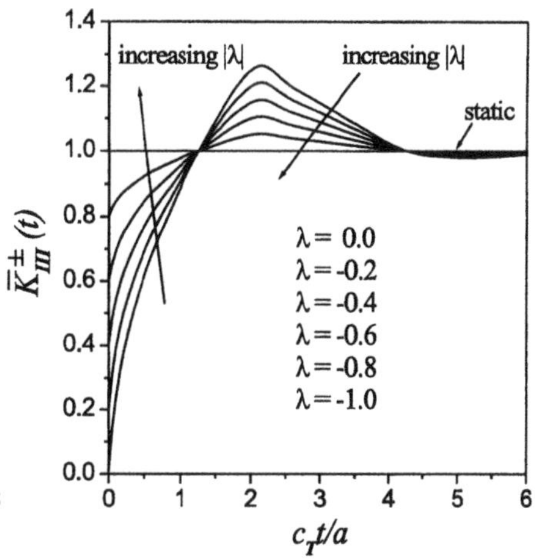

Fig. 5. Normalized $\bar{K}^{\pm}_{III}$-factor ($\lambda \leqslant 0$)

From this analysis it can be concluded that the present time-domain BIEM is highly accurate, efficient and stable for transient dynamic crack analysis in linear piezoelectric solids subjected to antiplane mechanical and inplane electric impact loading.

References

1. Norris, N. (1994) Dynamic Green's functions in anisotropic piezoelectric, thermoelastic and poroelastic solids. Proc. Royal Society London A **447**, 175–188
2. Khutoryansky, N., Sosa, H. (1995) Dynamic representation formulas and fundamental solutions for piezoelectricity. Int. J. Solids Struct. **32**, 3307–3325
3. Daros, H., Antes, H. (2000) Dynamic fundamental solutions for transversely isotropic piezoelectric materials of crystal class 6mm. Int. J. Solids Struct. **37**, 1639–1658
4. Lubich, C. (1988) Convolution quadrature and discretized operational calculus. I. Numerische Mathematik **52**, 129–145
5. Zhang, Ch. (2002) Transient dynamic response of a cracked piezoelectric solid under impact loading. WCCM V - Fifth World Congress on Computational Mechanics, 7–12 July, 2002, Vienna, Austria
6. Zhang, Ch., Wang, C.-Y., Hirose, S. (2002) A time-domain BIEM for dynamic crack analysis of a piezoelectric solid. Boundary Element Techniques III, 10–12 September, 2002, Tsinghua University, Beijing, PR China

Part IV

Porous Media and Fluid – Structure Interaction

Mixed-type systems of convection-diffusion equations modeling polydisperse sedimentation

Stefan Berres[1], Raimund Bürger[1] and Elmer M. Tory[2]

[1] Institute of Mathematics A, University of Stuttgart, Stuttgart, Germany, E-mail: berres|buerger@mathematik.uni-stuttgart.de

[2] Professor Emeritus, Department of Mathematics and Computer Science, Mount Allison University, Sackville, NB, E4L 1E8, Canada, E-mail: sherpa@nbnet.nb.ca

Abstract. Models for the sedimentation of polydisperse suspensions of particles differing in size or density include strictly hyperbolic or mixed hyperbolic-elliptic systems of first-order conservation laws and strongly degenerate parabolic-hyperbolic systems of second-order convection-diffusion equations. The type depends on the properties of the solid particles. We present a summary of recent analyses of such systems and a numerical simulation of the settling of a bidisperse suspension.

1 Introduction

Mathematical models for the sedimentation of suspensions are important to many applications such as solid-liquid separation, mineral processing and wastewater treatment, classification, fluidization, cast formation in the ceramic industry, blood sedimentation, and volcanology. Models for *monodisperse* suspensions of small rigid spheres, all having the same size and density and giving rise to *scalar* partial differential equations, are now fairly well understood [10]. The analysis of models for *polydisperse* suspensions of small particles (of a finite number N of species with diameters $d_1, \dots, d_N$ and densities $\varrho_1, \dots, \varrho_N$, where $d_i \neq d_j$ or $\varrho_i \neq \varrho_j$ for $i \neq j$) involves *systems* of partial differential equations and is much more difficult. We here distinguish polydisperse suspensions of rigid spheres and of solid particles forming compressible sediments. For the former, the effective solid stress σ_e is assumed to vanish, and the resulting mathematical framework is first-order systems of conservation laws, while the latter are characterized by $\sigma_e \not\equiv 0$, and give rise to strongly degenerate systems of second-order convection-diffusion equations.

2 Mathematical model

Starting from the local mass and linear momentum balances for each particle species and the fluid, introducing constitutive assumptions and performing a dimensional analysis, we have derived the following model equations for polydisperse suspensions [4,8]:

$$\frac{\partial \phi_i}{\partial t} + \nabla \cdot \big(\phi_i \mathbf{q} + f_i(\Phi)\mathbf{k}\big) = \nabla \cdot \mathbf{a}_i(\Phi, \nabla\Phi), \quad i = 1, \dots, N; \tag{1}$$

$$\nabla \cdot \mathbf{q} = 0, \tag{2}$$

$$\nabla p = -\varrho(\Phi) g \mathbf{k} + (1-\phi)^{-1} \nabla \cdot \mathbf{T}_{\mathrm{f}}^{\mathrm{E}}. \tag{3}$$

Here ϕ_i denotes the volumetric concentration of solids species i (having diameter d_i and density ϱ_i), t is time, $\mathbf{q}$ is the volume-average velocity of the mixture, $\mathbf{k}$ is the upwards-pointing unit vector, p is the pore pressure, $\varrho(\Phi) := \varrho_1\phi_1 + \dots + \varrho_N\phi_N + (1-\phi)\varrho_{\mathrm{f}}$ is the local density of the mixture where $\Phi := (\phi_1, \dots, \phi_N)^{\mathrm{T}}$, $\phi := \phi_1 + \dots + \phi_N$ is the total solids concentration, g is the acceleration of gravity and ϱ_{f} and $\mathbf{T}_{\mathrm{f}}^{\mathrm{E}}$ are the density and the viscous stress tensor of the fluid, respectively. The key ingredients of the model are the flux density vector $\mathbf{f} := (f_1, \dots, f_N)^{\mathrm{T}}$, where $f_1, \dots, f_N$ are functions of Φ, and the diffusion vectors $\mathbf{a}_i$, $i = 1, \dots, N$. For rigid spheres, we have $\mathbf{a}_i \equiv 0$.

Several authors proposed flux vectors $\mathbf{f}(\Phi)$ as generalizations of the hindered settling function known from sedimentation models for monodisperse suspensions; see [6] for an overview. Some of them extend Batchelor's settling velocity formulas [1,3] for dilute suspensions to the full range of concentration values [11,12]. For analyses of these approaches and numerical examples we refer to [7,8]. In this contribution, we limit ourselves to the Masliyah-Lockett-Bassoon (MLB) model [16,15], which is derived from both the mass *and* linear momentum balance equations for each particle species, in contrast to the models advanced in [11,12]. For spheres differing in size and density, the components $f_i^{\mathrm{M}}(\Phi)$ of the flux vector $\mathbf{f}(\Phi)$ of the MLB model are

$$f_i^{\mathrm{M}}(\Phi) = \mu V(\phi)\phi_i \left[\delta_i(\bar{\varrho}_i - \bar{\boldsymbol{\varrho}}^{\mathrm{T}}\Phi) - \sum_{k=1}^{N} \delta_k \phi_k (\bar{\varrho}_k - \bar{\boldsymbol{\varrho}}^{\mathrm{T}}\Phi)\right] \tag{4}$$

for $i = 1, \dots N$, where $\mu = -g d_1^2/(18\mu_{\mathrm{f}})$ (where μ_{f} is the viscosity of the fluid), $V(\phi)$ is a hindered settling factor that can e.g. be chosen as $V(\phi) = (1-\phi)^{n-2}$ with $n > 2$, $\delta_i = d_i^2/d_1^2$, $\bar{\varrho}_i = \varrho_i - \varrho_{\mathrm{f}}$, $i = 1, \dots, N$, and $\bar{\boldsymbol{\varrho}} := (\bar{\varrho}_1, \dots, \bar{\varrho}_N)^{\mathrm{T}}$.

3 Polydisperse suspensions of rigid spheres

We first briefly discuss the model for polydisperse suspensions of rigid spheres, for which the right-hand part of (1) vanishes. In one space dimension, only Eqns. (1) need to be solved, with $q \equiv 0$ in a closed column of height L. This leads to the initial-boundary value problem of a system of conservation laws

$$\frac{\partial \phi_i}{\partial t} + \frac{\partial f_i(\Phi)}{\partial z} = 0, \quad i = 1, \dots, N, \tag{5}$$

$$\Phi(z, 0) = \Phi_0(z),\ 0 \leqslant z \leqslant L; \quad \mathbf{f}|_{z=0} = \mathbf{f}|_{z=L} = 0, \quad t > 0. \tag{6}$$

It is well known that solutions of (5) are discontinuous in general, and that the propagation speed $\sigma(\Phi^+, \Phi^-)$ of a discontinuity in the concentration field ϕ_i separating the states Φ^+ and Φ^- is given by the Rankine-Hugoniot condition $\sigma = (f_i(\Phi^+) - f_i(\Phi^-))/(\phi_i^+ - \phi_i^-)$.

We recall that the system (5) is called *hyperbolic* if the eigenvalues of the Jacobian $\mathcal{J}_{\mathbf{f}}(\Phi) := (\partial f_i/\partial \phi_k)_{1\leqslant i,k\leqslant N}$ are real, and *strictly hyperbolic* if these are moreover pairwise distinct. For $N = 2$, a system with a pair of complex conjugate eigenvalues is *elliptic*. Some of the flux-density vectors $\mathbf{f}(\Phi)$ proposed in the literature cause the system (5) to be non-hyperbolic, or to be of mixed hyperbolic-elliptic type in the case $N = 2$, depending on the size and density parameters, where the type is *mixed* if for given sizes and densities, the system is non-hyperbolic (or elliptic) for $\Phi \in E$, $E \subset \mathcal{D} := \{\Phi \in \mathbf{R}^N : \Phi \geqslant 0, \phi \leqslant 1\}$, and $\mathcal{D}\backslash E$ is nonempty.

The ellipticity criterion is equivalent to Batchelor and Janse van Rensburg's [2] criterion for the occurrence of instabilities like blobs and viscous fingering in bidisperse sedimentation; see [18] for experimental findings. By a perturbation analysis, we showed [8] that loss of hyperbolicity, that is the occurrence of complex eigenvalues of $\mathcal{J}_{\mathbf{f}}(\Phi)$, provides an instability criterion for polydisperse suspensions of *arbitrary* numbers of species N. For $N = 3$, this criterion can be evaluated by a convenient calculation of the discriminant

$$I_3(\Phi) := 4s^3 - s^2r^2 + 27t^2 + 4r^3t - 18rst, \; r := -\operatorname{tr}\mathcal{J}_{\mathbf{f}}, \; t := -\det\mathcal{J}_{\mathbf{f}},$$

$$s := -\frac{\partial f_1}{\partial \phi_3}\frac{\partial f_3}{\partial \phi_1} - \frac{\partial f_1}{\partial \phi_2}\frac{\partial f_2}{\partial \phi_1} - \frac{\partial f_2}{\partial \phi_3}\frac{\partial f_3}{\partial \phi_2} + \frac{\partial f_1}{\partial \phi_1}\frac{\partial f_2}{\partial \phi_2} + \frac{\partial f_1}{\partial \phi_1}\frac{\partial f_3}{\partial \phi_3} + \frac{\partial f_2}{\partial \phi_2}\frac{\partial f_3}{\partial \phi_3}$$

of the characteristic polynomial of $\mathcal{J}_{\mathbf{f}}(\Phi)$. The model equations are unstable if and only if $I_3 > 0$. This also handles the bidisperse case; we set $I_2 := -I_3(\phi_1, \phi_2, 0)$. In [8] we determine instability regions (evaluating numerically I_3) for three different choices of $\mathbf{f}(\Phi)$. More recently [4] it was proved that for equal-density particles ($\bar{\varrho} = \cdots = \bar{\varrho}_N = \varrho_{\mathrm{s}} - \varrho_{\mathrm{f}}$), arbitrary N and particle size distributions, the system (5) is strictly hyperbolic for all $\Phi \in \mathcal{D}$ with $\phi_1 > 0, \ldots, \phi_N > 0$ and $\phi < 1$ if the flux vector (4) is chosen.

4 Compressible sediments

Mathematical models for sedimentation with compression of monodisperse suspensions lead to scalar strongly degenerate parabolic-hyperbolic partial differential equations [9], which have been investigated intensively. In a recent paper [4], the assumptions of both the monodisperse-compressive and the polydisperse-rigid sphere models were combined to yield a theory of sedimentation of polydisperse suspensions forming compressible sediments. The decisive constitutive assumption for the mathematical analysis is that the effective stress σ_{e} in the networked porous bed formed by the solid particles is a function of the porosity or equivalently, of ϕ only, and that the phase pressure p_i associated with particle species i is given by $p_i = (\phi_i/\phi)(\phi p + \sigma_{\mathrm{e}}(\phi))$, where p is the pore pressure. Since the assumption $\sigma_\mathrm{e} = \sigma_\mathrm{e}(\phi)$ is reasonable for solid particles of the same material and thus of the same density only, the analysis can be limited to equal-density particle species differing in size. Under these assumptions and assuming one space dimension, we can write the components of the diffusion vector $\mathbf{a}$ introduced in (1) as

$$a_i\left(\Phi, \frac{\partial \Phi}{\partial z}\right) = \sum_{j=1}^{N} \eta_{ij}(\Phi)\frac{\partial \phi_j}{\partial z}, \quad i = 1, \dots, N, \tag{7}$$

$$\begin{aligned} \eta_{ij}(\Phi) := -\frac{\mu V(\phi)}{g\phi}\Big\{(1-\phi)\phi_i(\delta_i - \boldsymbol{\delta}^\mathrm{T}\Phi)\sigma_\mathrm{e}'(\phi) + \Big[\delta_i\delta_{ij} - \delta_j\phi_i \\ - \frac{\phi_i}{\phi}(\delta_i - \boldsymbol{\delta}^\mathrm{T}\Phi)\Big]\sigma_\mathrm{e}(\phi)\Big\}, \; i = 1, \dots, N; \; \boldsymbol{\delta} = (\delta_1, \dots, \delta_N)^\mathrm{T}, \end{aligned} \tag{8}$$

where $\delta_{ij} = 1$ if $i = j$ and $\delta_{ij} = 0$ otherwise. Defining the diffusion matrix $\mathbf{A}(\Phi) := (\eta_{ij}(\Phi))_{1 \leqslant i,j \leqslant N}$, we can rewrite the resulting system (Eq. (1) in one dimension and setting $q = 0$) as

$$\frac{\partial \Phi}{\partial t} + \frac{\mathbf{f}^\mathrm{M}(\Phi)}{\partial z} = \frac{\partial}{\partial z}\left(\mathbf{A}(\Phi)\frac{\partial \Phi}{\partial z}\right). \tag{9}$$

The generic assumption on $\sigma_\mathrm{e}(\phi)$ is [9]

$$\sigma_\mathrm{e}(\phi) = \begin{cases} = 0 & \text{for } \phi \leqslant \phi_\mathrm{c}, \\ > 0 & \text{for } \phi > \phi_\mathrm{c}, \end{cases} \quad \sigma_\mathrm{e}'(\phi) = \begin{cases} = 0 & \text{for } \phi \leqslant \phi_\mathrm{c}, \\ > 0 & \text{for } \phi > \phi_\mathrm{c}, \end{cases}$$

where ϕ_c is a critical concentration at which the solid particles are assumed to touch each other. It can then immediately be read off from (8) that the system (9) is degenerate since $\mathbf{A}(\Phi) = 0$ wherever $\phi \leqslant \phi_\mathrm{c}$, and the analysis of the first-order system shows that for $\phi < \phi_\mathrm{c}$ the system (9) is strictly hyperbolic. Moreover, the system (9) has the non-obvious property of being parabolic for $\phi_\mathrm{c} < \phi < 1$, which means that $\mathbf{A}(\Phi)$ has N real, positive eigenvalues. This is proved in [4] by exploiting the similar properties of the characteristic polynomials of $\mathcal{J}_\mathbf{f}^\mathrm{M}(\Phi)$ and $\mathbf{A}(\Phi)$. Furthermore, it turns out that the eigenvalues of $\mathbf{A}(\Phi)$ are pairwise distinct and bounded.

As demonstrated in [4], the properties of both matrices can be utilized to control the numerical diffusion of the Kurganov-Tadmor scheme [13] applied to (9). Although no existence or uniqueness result is currently available for (9) in the non-scalar case, such equations recently gained attention in the context of BGK numerical approximations [14], for which the model of polydisperse sedimentation with compression should provide an interesting application.

5 Numerical simulations

When (5) is hyperbolic on $\mathcal{D}$, modern shock-capturing numerical schemes can be applied to solve the initial-boundary value problem (5), (6), and thereby

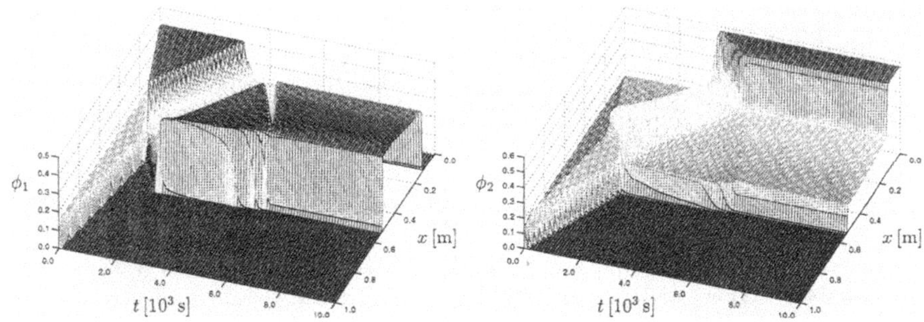

Fig. 1. Simulation of the settling of a bidisperse suspension: concentrations of the larger (left) and the smaller (right) particles; colored, see page 375

to simulate the sedimentation process. These schemes approximate smooth portions of concentration profiles with second-order accuracy, sharply resolve discontinuities (kinematic shocks), and reproduce them only in the physically correct location. In the scalar case they satisfy an entropy criterion ensuring that only physically admissible discontinuities are computed. One of these methods is due to Kurganov and Tadmor [13].

For the simulations, we consider a settling column of height $H = 1.0$ m with water as fluid ($\mu_{\mathrm{f}} = 10^{-3}$ Pa·s, $\varrho_{\mathrm{f}} = 1000\,\mathrm{kg/m^3}$) and $g = 9.81\,\mathrm{m/s^2}$. The hindered settling function $V(\Phi) := (1-\phi)^{n-2}$ with $n = 4.7$ is cut off at $\phi_{\max} = 0.58$. The particles have the properties $d_1 = 1.5\times 10^{-4}$ m, $d_2 = 0.3\times 10^{-4}$ m, $\varrho_1 = 1500$ kg/m^3, $\varrho_2 = 2500$ kg/m^3, and the initial concentration is $\Phi^0 = (0.15, 0.15)^{\mathrm{T}}$. Thus $\boldsymbol{\delta} = (1, 0.04)^{\mathrm{T}}$, $\bar{\boldsymbol{\varrho}} = (500, 1500)$ kg/m^3, $\mu = -1.23\times 10^{-5}$. Following [8, Sect. 5.1], we chose the relatively low maximum concentration to ensure that the model equations are hyperbolic [5].

The diagrams of Figure 1 show that, at a small total solids concentration, the larger species settle faster since they have a higher Stokes velocity and dominate the rising sediment bed. At the same time, the smaller species form a concentratted region above the upper interface of the larger species (Smith effect [17]). After the larger particles have completely settled, the smaller ones do not just settle above the existing sediment, as one might expect, building a layer consisting of just the smaller particles, but the following occurs: The fairly equally partitioned solids concentration at the lower bound of the Smith zone causes the local velocity of the larger species to change sign, so that the larger particles at the bound move upwards making room for the smaller particles and thus the nose settles down. The settling Smith zone (consisting of smaller species) causes the re-formation of a new sediment bed which grows from the bottom. This consists entirely of the smaller particles. After the settling nose has completely settled and displaced larger particles, a mixed bed forms above the sediment where the dominance of the smaller species decreases with height.

References

1. Batchelor, G.K., 'Sedimentation in a dilute polydisperse system of interacting spheres. Part 1. General theory', *J. Fluid Mech.* **119** (1982), 379–408.
2. Batchelor, G.K. and Janse van Rensburg, R.W., 'Structure formation in bidisperse sedimentation', *J. Fluid Mech.* **166** (1986), 379–407.
3. Batchelor, G.K. and Wen, C.S., 'Sedimentation in a dilute polydisperse system of interacting spheres. Part 2. Numerical results', *J. Fluid Mech.* **124** (1982), 495–528.
4. Berres, S., Bürger, R., Karlsen, K.H. and Tory, E.M., 'Strongly degenerate parabolic-hyperbolic systems modeling polydisperse sedimentation with compression', Preprint 2002/07, SFB 404, University of Stuttgart, 2002.
5. Berres, S., Bürger, R. and Tory, E.M., 'Mathematical model and numerical simulation of the liquid fluidization of solid particle mixtures', submitted to *Comp. Visual. Sci.* (2002).
6. Bürger, R., Concha, F., Fjelde, K.-K. and Karlsen, K.H., 'Numerical simulation of the settling of polydisperse suspensions of spheres', *Powder Technol.* **113** (2000), 30–54.
7. Bürger, R., Fjelde, K.-K., Höfler, K. and Karlsen, K.H., 'Central difference solutions of the kinematic model of settling of polydisperse suspensions and three-dimensional particle-scale simulations', *J. Eng. Math.* **41** (2001), 167–187.
8. Bürger, R., Karlsen, K.H., Tory, E.M. and Wendland, W.L., 'Model equations and instability regions for the sedimentation of polydisperse suspensions of spheres', *ZAMM* **82** (2002) 699–722.
9. Bürger, R., Wendland, W.L. and Concha, F., 'Model equations for gravitational sedimentation-consolidation processes', *ZAMM* **80** (2000), 79–92.
10. Bustos, M.C., Concha, F., Bürger, R. and Tory, E.M., *Sedimentation and Thickening*, Kluwer Academic Publishers, Dordrecht 1999.
11. Davis, R.H. and Gecol, H., 'Hindered settling function with no empirical parameters for polydisperse suspensions', *AIChE J.* **40** (1994), 570–575.
12. Höfler, K. and Schwarzer, S., 'The structure of bidisperse suspensions at low Reynolds numbers'. In: Sändig, A.-M., Schiehlen, W. and Wendland, W.L. (eds.), *Multifield Problems: State of the Art*, Springer-Verlag, Berlin (2000), 42–49.
13. Kurganov, A. and Tadmor, E., 'New high resolution central schemes for nonlinear conservation laws and convection-diffusion equations', *J. Comp. Phys.* **160** (2000), 241–282.
14. Lattanzio, C. and Natalini, R., 'Convergence of diffusive BGK approximations for nonlinear strongly parabolic systems', *Proc. Royal Soc. Edinburgh Ser. A: Math.* **132** (2002), 341–358.
15. Lockett, M.J. and Bassoon, K.S., 'Sedimentation of binary particle mixtures', *Powder Technol.* **24** (1979), 1–7.
16. Masliyah, J.H., 'Hindered settling in a multiple-species particle system', *Chem. Eng. Sci.* **34** (1979), 1166–1168.
17. Smith, T.N., 'The sedimentation of particles having a dispersion of sizes'. *Trans. Instn. Chem. Engrs.* **44** (1966), T152–T157.
18. Weiland, R.H., Fessas, Y.P. and Ramarao, B.V., 'On instabilities arising during sedimentation of two-component mixtures of solids', *J. Fluid Mech.* **142** (1984), 383–389.

Simultaneous Solution to Viscous Fluid Flow Interacting with Large Amplitude Structural Vibrations

Dieter Dinkler, Björn Hübner, and Elmar Walhorn

Technische Universität Braunschweig, Institut für Statik,
Beethovenstr. 51, 38106 Braunschweig, Germany

Abstract. The paper presents a simultaneous solution procedure for fluid-structure interaction problems with application to building aeroelasticity. The structural motion is described with a geometrically nonlinear theory, and the flow field is modeled with the incompressible Navier-Stokes equations. The space-time finite element method is applied to both continua leading to a consistent discretization of fluid and structure. In order to enforce momentum conservation between both continua, a weighted residual formulation of coupling conditions is introduced. The monolithic approach ensures high convergence and accuracy of the solution and allows coupled stability analyses. A numerical example of membrane flutter demonstrates efficiency and versatility of the presented methodology.

1 Introduction

The development of accurate solvers for complex fluid-structure interaction problems is still a challenging task in research, and commercial software is available only for some special cases. The solution strategies range from a loose coupling [1–3] of different solvers to time-marching solutions of the strongly coupled system [4–6]. If stability characteristics of the coupled system are investigated, strongly coupled formulations should be preferred.

The time-discontinuous stabilized space-time finite element method was introduced in the field of elastodynamics by Hughes and Hulbert [7]. They have applied the formulation for wave propagation problems with a space dimension of one. For fluid dynamics applications Hughes et al. [8] have presented the Galerkin / least-squares stabilization for advective-diffusive equations, while Tezduyar et al. [9] have applied the space-time formulation for solving the incompressible Navier-Stokes equations on a moving mesh.

The present article deals with the application of time-discontinuous stabilized space-time finite elements for solving fluid-structure interaction problems with strong coupling in a single system of equations and variables. A consistent finite element model of the multi-field problem composed of geometrically nonlinear elastic structure, incompressible viscous flow field and coupling conditions is developed. The formulation of the coupling conditions uses boundary tractions as additional variables in order to enforce momentum conservation between fluid and structure.

2 Space-time discretization

In order to develop an accurate and efficient solution procedure for the strongly coupled system, a monolithic approach is chosen for numerical simulation. Therefore, a single discretization method is applied to the entire system, using stabilized space-time finite elements for fluid and structural dynamics. The space-time finite element method provides a consistent discretization of space and time domain, avoiding semi-discrete formulations. For numerical efficiency the space-time domain Q is divided into a sequence of space-time slabs $Q_n = \Omega_t \times (t_n, t_{n+1})$, see figure 1. Since the isoparametric space-time elements are adaptable in time direction, the method leads to a natural description of the time dependent fluid domain and satisfies the geometric conservation law (GCL) without additional equations, see [10].

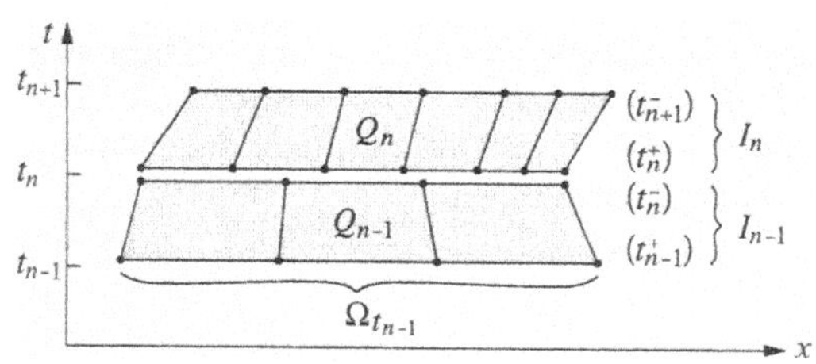

Fig. 1. Moving space-time mesh

The interpolation of physical variables is chosen to be discontinuous in time, see figure 2, with independent degrees of freedom in adjoining time slabs at discrete time levels t_n. The time-discontinuous Galerkin formulation is of higher order accurate, implicit and A-stable for time integration [11]. In case of linear interpolation functions the accuracy is of third order. The continuity of primal variables is achieved in integral form using jump terms in the variational formulation of the model equations.

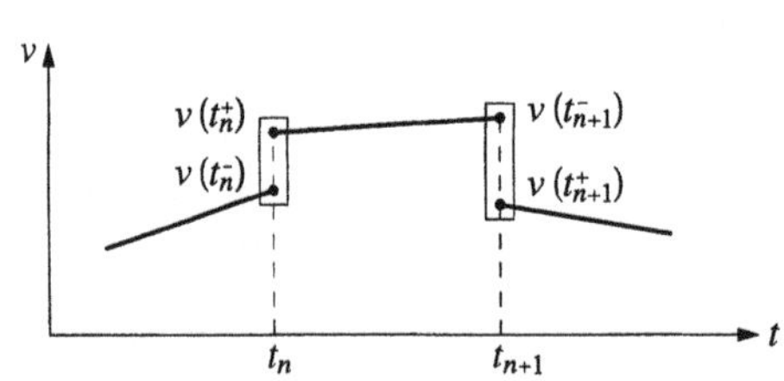

Fig. 2. Time-discontinuous interpolation

The Galerkin / least-squares stabilization of space-time elements suppresses numerical oscillations in solutions to hyperbolic differential equations [8], comparable with second order upwind schemes. The stabilization term contains a least-squares weighted residuum and introduces additional numerical diffusion. The method yields stable solutions to the Navier-Stokes equations for a wide range of Reynolds numbers and allows the application of equal order interpolation functions for velocities and pressure, see [9].

3 Model Equations

The governing equations of the applied model are described in detail by Hübner et al. [6]. In the following, the weak formulation is presented since it is basis for the discretization.

3.1 Fluid dynamics

The stabilized space-time finite element formulation of the fluid model within the time slab Q_n including boundaries P_n reads

$$\int_{Q_n} \delta\boldsymbol{v} \cdot \rho \Big(\frac{\partial \boldsymbol{v}}{\partial t} + \boldsymbol{v} \cdot \operatorname{grad} \boldsymbol{v}\Big) \mathrm{d}Q \; + \; \int_{Q_n} \boldsymbol{D}(\delta\boldsymbol{v}) \cdot\cdot\, 2\mu\, \boldsymbol{D}(\boldsymbol{v})\; \mathrm{d}Q \tag{1a}$$

$$- \int_{Q_n} (\operatorname{div} \delta\boldsymbol{v})\, p \; \mathrm{d}Q \; - \; \int_{Q_n} \delta\boldsymbol{v} \cdot \boldsymbol{f} \; \mathrm{d}Q \tag{1b}$$

$$+ \int_{Q_n} \delta p \operatorname{div} \boldsymbol{v} \; \mathrm{d}Q \tag{1c}$$

$$+ \int_{\Omega_n} \delta\boldsymbol{v}(t_n^+) \cdot \rho \Big(\boldsymbol{v}(t_n^+) - \boldsymbol{v}(t_n^-)\Big) \mathrm{d}\Omega \tag{1d}$$

$$+ \sum_e \int_{Q_n^e} \tau_M^F \, \frac{1}{\rho} \; \mathcal{L}(\delta\boldsymbol{v}, \delta p) \cdot \Big(\mathcal{L}(\boldsymbol{v}, p) - \boldsymbol{f}\Big)\, \mathrm{d}Q \tag{1e}$$

$$+ \sum_e \int_{Q_n^e} \tau_C^F \, \rho \; (\operatorname{div} \delta\boldsymbol{v}) \; \operatorname{div} \boldsymbol{v} \; \mathrm{d}Q \tag{1f}$$

$$- \int_{P_n^g} \delta\boldsymbol{v} \cdot \boldsymbol{t} \; \mathrm{d}P \; + \; \int_{P_n^g} \delta\boldsymbol{t} \cdot (\boldsymbol{v} - \boldsymbol{g}) \mathrm{d}P \tag{1g}$$

$$- \int_{P_n^h} \delta\boldsymbol{v} \cdot \boldsymbol{h} \; \mathrm{d}P \; = \; 0 \tag{1h}$$

for all $\delta\boldsymbol{v}$, δp and $\delta\boldsymbol{t}$. The interpolation functions for all variables are bilinear or linear in space and discontinuous linear in time. In equation (1) lines (a) and (b) represent the weak form of the momentum conservation, while term (c) enforces the incompressibility condition of the continuity equation. The velocity jump term of the time-discontinous formulation is shown in line (d) and the stabilization term of the momentum equation in line (e), where

$$\mathcal{L}(\boldsymbol{v}, p) \; = \; \rho\Big(\frac{\partial \boldsymbol{v}}{\partial t} + \boldsymbol{v} \cdot \operatorname{grad} \boldsymbol{v}\Big) \; - \; \operatorname{div}\Big(2\mu\, \boldsymbol{D}(\boldsymbol{v})\Big) + \operatorname{grad} p \tag{2}$$

is the residuum of the strong form of momentum balance without body force term. Line (f) represents the stabilization term of the continuity equation. For the parameters τ_M^F and τ_C^F the definition by Tezduyar et al. [9] is applied. The weighted form of the Dirichlet boundary condition is shown in line (g) and the last term (h) represents the surface load of the Neumann condition.

3.2 Structural dynamics

The stabilized space-time finite element formulation of the structural part within the time slab $Q_{0,n}$ including boundaries $P_{0,n}$ reads

$$\int_{Q_{0,n}} \delta \boldsymbol{v} \cdot \rho_0 \, \dot{\boldsymbol{v}} \, \mathrm{d}Q_0 + \int_{Q_{0,n}} \dot{\boldsymbol{E}}(\delta \boldsymbol{v}, \boldsymbol{u}) \cdot\cdot \boldsymbol{S} \, \mathrm{d}Q_0 - \int_{Q_{0,n}} \delta \boldsymbol{v} \cdot \boldsymbol{f}_0 \, \mathrm{d}Q_0 \tag{3a}$$

$$+ \sum_e \int_{Q^e_{0,n}} \delta \boldsymbol{S} \cdot\cdot \left(\underline{\boldsymbol{C}}^{-1} \cdot\cdot \dot{\boldsymbol{S}} - \dot{\boldsymbol{E}}(\boldsymbol{v}, \boldsymbol{u}) \right) \mathrm{d}Q_0 \tag{3b}$$

$$+ \int_{\Omega_0} \delta \boldsymbol{v}(t_n^+) \cdot \rho_0 \left(\boldsymbol{v}(t_n^+) - \boldsymbol{v}(t_n^-) \right) \mathrm{d}\Omega_0 \tag{3c}$$

$$+ \sum_e \int_{\Omega^e_0} \delta \boldsymbol{S}(t_n^+) \cdot\cdot \underline{\boldsymbol{C}}^{-1} \cdot\cdot \left(\boldsymbol{S}(t_n^+) - \boldsymbol{S}(t_n^-) \right) \mathrm{d}\Omega_0 \tag{3d}$$

$$+ \sum_e \int_{Q^e_{0,n}} \tau_M^S \, \delta \dot{\boldsymbol{v}} \cdot \left(\rho_0 \, \dot{\boldsymbol{v}} - \mathrm{div}_0(\boldsymbol{F}\boldsymbol{S}) - \boldsymbol{f}_0 \right) \mathrm{d}Q_0 \tag{3e}$$

$$- \int_{P^g_{0,n}} \delta \boldsymbol{v} \cdot \boldsymbol{t}_0 \, \mathrm{d}P_0 + \int_{P^g_{0,n}} \delta \boldsymbol{t}_0 \cdot (\boldsymbol{v} - \boldsymbol{g}) \, \mathrm{d}P_0 \tag{3f}$$

$$- \int_{P^h_{0,n}} \delta \boldsymbol{v} \cdot \boldsymbol{h}_0 \, \mathrm{d}P_0 = 0 \tag{3g}$$

for all $\delta \boldsymbol{v}$, $\delta \boldsymbol{S}$ and $\delta \boldsymbol{t}_0$. In equation (3) line (a) represents the weak form of the momentum conservation and line (b) fulfills the constitutive law on element level, what leads to the mixed-hybrid formulation of Knippers and Harbord [12], where only the velocities are global degrees of freedom. The jump terms for velocities (c) and stresses (d) satisfy the initial conditions of the time slab in integral form. The stabilization term, which is needed for wave propagation problems, is shown in line (e). For the parameter τ_M^S the choice of Hughes and Hulbert [7] is used. The last two lines represent the variational form of the Dirichlet boundary condition (f) and the surface load term (g) of the Neumann condition. Interpolation functions for velocities and boundary tractions are chosen to be bilinear and linear in space and discontinuous linear in time, while the stress interpolation is discontinuous and incompletely linear in space and discontinuous linear in time. All time slabs are solved successively, where initial conditions are given by

$$\begin{aligned} \boldsymbol{v}(t_0^-) &= \boldsymbol{v}_i \,, \\ \boldsymbol{S}(t_0^-) &= \underline{\boldsymbol{C}} \cdot\cdot \tfrac{1}{2} \left(\mathrm{grad}_0 \boldsymbol{u}_i + (\mathrm{grad}_0 \boldsymbol{u}_i)^{\mathrm{T}} + (\mathrm{grad}_0 \boldsymbol{u}_i)^{\mathrm{T}} \, \mathrm{grad}_0 \boldsymbol{u}_i \right) . \end{aligned} \tag{4}$$

Time integration of the velocities leads to the displacement field $\boldsymbol{u}$, which is used for computing the time derivative of the Green-Lagrange strain tensor and for specifying the current position of the fluid-structure interface.

5 Flow over a building with elastic membrane roof

This example deals with the interaction of fluid flow around a building and nonlinear vibrations of the gravity prestressed membrane roof. The fluid density is $\rho = 1.25\,\mathrm{kg/m^3}$ and the viscosity $\mu = 0.1\,\mathrm{Ns/m^2}$, which is much higher than the viscosity of air in the atmosphere. However, the focal point of this contribution is the strongly coupled simulation of the interaction of viscous fluid flow and nonlinear structural vibrations. The inflow velocity profile is defined by $v_x(y) = 35\mathrm{m/s}\,(y/350\,\mathrm{m})^{0.22}$ leading to a velocity value of $v_x(5\,\mathrm{m}) = 13.75\,\mathrm{m/s}$ at roof level. Thus, the Reynolds number becomes $\mathrm{Re} \approx 1700$ using the building length of $l = 10\,\mathrm{m}$ as length scale. The system configuration with dimensions and boundary conditions is given in figure 4.

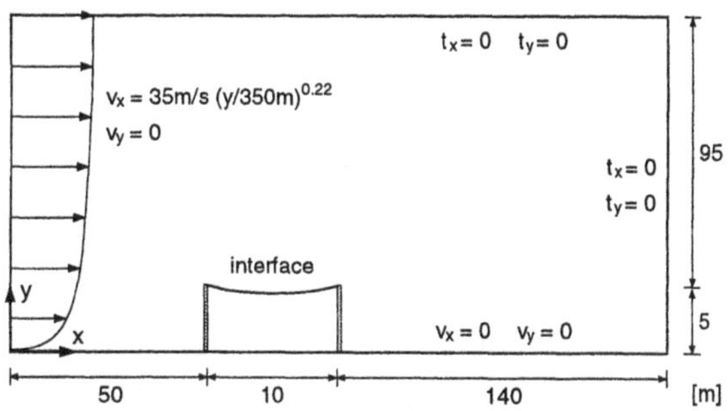

Fig. 4. System configuration

The fluid domain is discretized with 10939 8-noded hexaeder space-time elements, while 100 elements are used for the membrane structure. The spatial finite element mesh and a detailed plot close to the leading edge are shown in figure 5. The bold line in the detail represents the boundary of the fluid domain and the thin elements below the membrane.

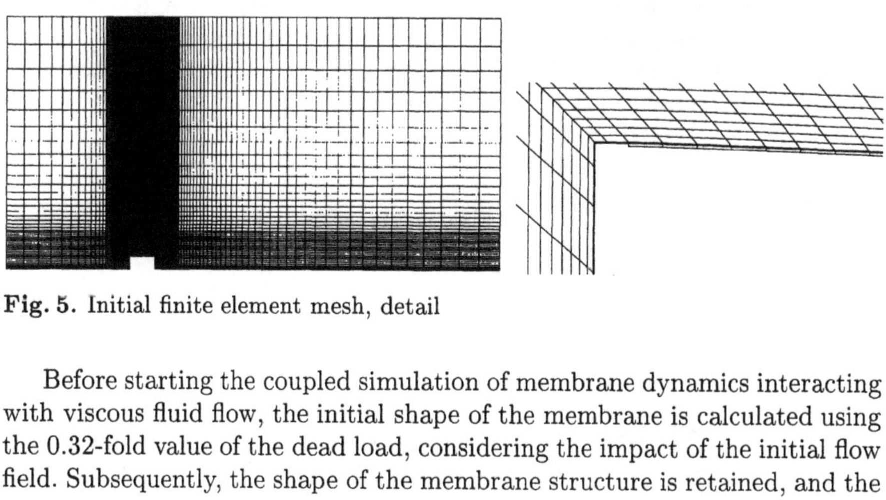

Fig. 5. Initial finite element mesh, detail

Before starting the coupled simulation of membrane dynamics interacting with viscous fluid flow, the initial shape of the membrane is calculated using the 0.32-fold value of the dead load, considering the impact of the initial flow field. Subsequently, the shape of the membrane structure is retained, and the

3.3 Fluid-structure coupling

The space-time finite element formulation of the coupling conditions uses boundary traction variables on the interface in order to fulfill momentum conservation and geometrical continuity between structure and fluid in integral form. In particular, the preset velocity values of the Dirichlet boundary conditions of the fluid are substituted by structural velocities and the fluid boundary tractions act as surface loads onto the structure

$$\int_{P^c} \delta \boldsymbol{t}^F \cdot (\boldsymbol{v}^F - \boldsymbol{v}^S)\, \mathrm{d}P \; - \int_{P^c} \delta \boldsymbol{v}^F \cdot \boldsymbol{t}^F \, \mathrm{d}P \; - \int_{P_0^c} \delta \boldsymbol{v}^S \cdot \Big(-\frac{\mathrm{d}\Gamma_t}{\mathrm{d}\Gamma_0}\, \boldsymbol{t}^F \Big)\, \mathrm{d}P_0 \,. \qquad (5)$$

The tractions are projected onto the reference configuration and change the sign, since the unit outward normal vectors of fluid and structure have opposite directions on the interface.

4 Solution strategy

The deformation of the elastic structure leads to moving boundaries of the fluid, and the fluid mesh must be adapted to the current position of the interface. Since the space-time finite element formulation allows the specification of arbitrary mesh node displacements, the fluid mesh boundary can move with the structure. However, inside of the fluid domain new mesh point positions must be defined in order to avoid overlapping elements. Therefore, the fluid domain is supposed to be an artificial elastic continuum with a stiffness distribution which is correlated with the inverse of the element size. On the interface, the real structural displacements are imprinted as Dirichlet boundary conditions, leading to a displacement field of the pseudo-structure, which determines new coordinates for the fluid mesh nodes.

The complete set of nonlinear equations including fluid mesh dynamics is solved in a single iteration loop, see figure 3, leading to a simultaneous solution of the entire system, which ensures high convergence and accuracy of the solution in conjunction with an efficient numerical approach. The highly nonlinear system due to nonlinear kinematics, convection and unknown mesh motion is solved by a Picard iteration scheme.

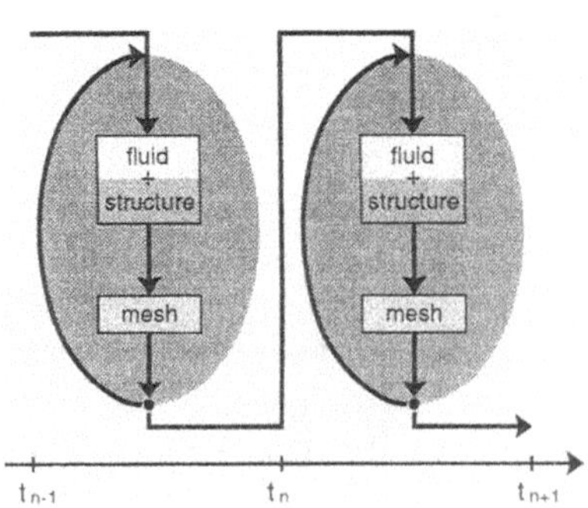

Fig. 3. Simultaneous solution with a single iteration loop

initial flow field is calculated. The first period of the coupled simulation is similar to the uncoupled flow behavior until the membrane starts to oscillate, leading to the development of vortices at the leading edge. The vortices move over the roof structure and induce membrane vibrations, which interact with the fluid flow. After some seconds a more or less periodical system behavior occurs with a coupled frequency of $f \approx 0.80$ Hz and displacement amplitudes in the membrane center of $u_y = 15 - 20$ cm, see figure 6.

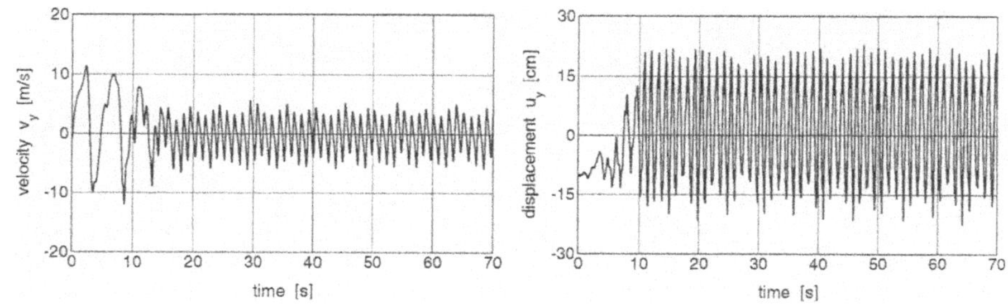

Fig. 6. Time histories of vertical velocity and membrane displacement

Due to strong nonlinearities in membrane kinematics and fluid dynamics, the flow induced membrane vibrations are not fully periodical steady, but similar vibration shapes appear in every period of the coupled motion.

In case of a fixed roof, vortex shedding occurs only at the trailing edge, and vortices do not move over the roof structure. Behind the building, the vortices rest some time and become larger before they move away. This leads to the comparatively low frequency of the fixed case. However, in the coupled simulation, vortices form at the leading edge, move over the roof structure and pass the trailing edge without resting behind the building. Here, the comparatively high frequency is determined mainly by membrane dynamics, see figure 7. The example shows the necessity of a strongly coupled simulation of roof structure and fluid flow in order to predict the coupled system behavior, since the flow field around the fixed building leads to quite different effects on the membrane structure.

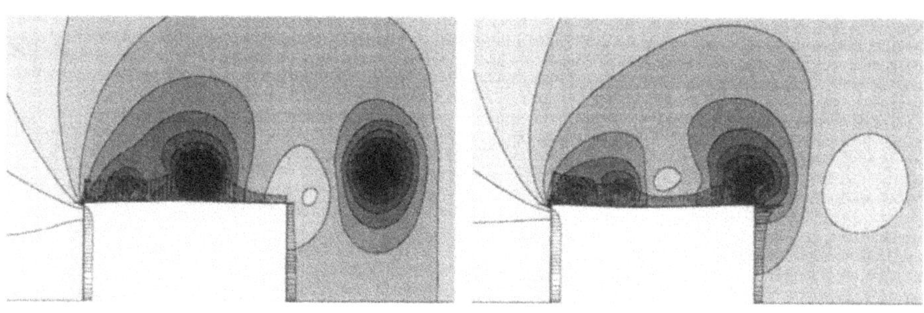

Fig. 7. Pressure fields at $t = 67.0$ s and 67.7 s

References

1. Löhner, R., Yang, C., Cebral, J., Baum, J. D., Luo, H., Pelessone D., Charman, C. (1995) Fluid-structure interaction using a loose coupling algorithm and adaptive unstructured grids. In: Hafez, M., Oshima, K. (eds.) Computational Fluid Dynamics Review 1995. John Wiley & Sons, Chichester
2. Masud, A., Hughes, T. J. R. (1997) A space-time Galerkin / least-squares finite element formulation of the Navier-Stokes equations for moving domain problems. Comput. Methods Appl. Mech. Engrg. **146**, 91–126
3. Wall, W. A., Mok, D. P., Ramm, E. (1999) Partitioned analysis approach for the transient coupled response of viscous fluids and flexible structures. In: Europ. Conf. Comput. Mechanics '99, München
4. Bendiksen, O. O. (1991) A new approach to computational aeroelasticity. AIAA paper 91-0939
5. Grohmann, B. A., Dornberger, R., Dinkler, D. (1997) Time-discontinuous stabilized space-time finite elements for aeroelasticity. Fluid-Structure Interaction, Aeroelasticity, Flow-Induced Vibrations and Noise, Vol. III, ASME, AD-Vol. 53-3
6. Hübner, B., Walhorn, E., Dinkler, D. (2001) Strongly coupled analysis of fluid-structure interaction using space-time finite elements. In: 2nd Europ. Conf. Comput. Mechanics 2001, Cracow
7. Hughes, T. J. R., Hulbert, G. M. (1988) Space-time finite element methods for elastodynamics: Formulations and error estimates. Comput. Methods Appl. Mech. Engrg. **66**, 339–363
8. Hughes, T. J. R., Franca, L. P., Hulbert, G. M. (1989) A new finite element formulation for computational fluid dynamics: VIII. The Galerkin / least-squares method for advective-diffusive equations. Comput. Methods Appl. Mech. Engrg. **73**, 173–189
9. Tezduyar, T. E., Behr, M., Liou, J. (1992) A new strategy for finite element computations involving moving boundaries and interfaces – The deforming-spatial-domain / space-time procedure: I. The concept and the preliminary numerical tests. Comput. Methods Appl. Mech. Engrg. **94**, 339–351
10. Lesoinne, M., Farhat, C. (1996) Geometric conservation laws for flow problems with moving boundaries and deformable meshes, and their impact on aeroelastic computations. Comput. Methods Appl. Mech. Engrg. **134**, 71–90
11. Johnson, C. (1993) Discontinuous Galerkin finite element methods for second order hyperbolic problems. Comput. Methods Appl. Mech. Engrg. **107**, 117–129
12. Knippers, J., Harbord, R. (1994) A mixed hybrid FE formulation for solution of elasto-viscoplastic problems. Computational Mechanics **13**, 231–240

A Continuum Approach for the Swelling of Charged Hydrated Media

Wolfgang Ehlers, Ayhan Acartürk and Bernd Markert

Institute of Applied Mechanics (CE)
University of Stuttgart, Pfaffenwaldring 7, 70569 Stuttgart, Germany
www.mechbau.uni-stuttgart.de/ls2

Abstract. Swelling of charged hydrated tissues is an important topic both in biomechanics and in biomedical surgery. From the mathematical point of view, the problem is based on an electro-chemomechanical multiphase formulation, consisting of a porous solid skeleton saturated by an interstitial fluid which itself is an ideal mixture of a liquid solvent (water) and dissolved electrolytes (ions).

Based on the general framework of the Theory of Porous Media (TPM), the model results in a set of strongly coupled partial differential equations consisting of the overall momentum balance, the overall volume balance and the balance of molar anion concentration.

By use of the finite element method (FEM), the electro-chemomechanical multiphase formulation is incorporated in the numerical tool PANDAS. To set an example for the validity of the presented approach, an initial-boundary-value problem is solved which exhibits a realistic two-dimensional swelling situation.

1 Multiphase Formulation

The TPM is a macroscopic continuum theory which is based on the theory of mixtures extended by the concept of volume fractions. Herein, the constituents are assumed to be in a state of ideal disarrangement [1,2].

1.1 Constituents, Volume Fractions and Densities

Starting from a binary mixture consisting of solid and fluid constituents φ^α denoted by $\alpha = \{S, F\}$, the model is extended by incorporating the fixed charges and the ions as additional components. In particular, the solid skeleton is understood as a molecular meshwork attached by volume free fixed charges φ^{fc}. Furthermore, the interstitial fluid φ^F is composed of three components φ^β, namely the liquid solvent, the cations and the anions in the following indicated by $\beta = \{L, +, -\}$. Following this, by introducing the volume fractions $n^\alpha = \mathrm{d}v^\alpha/\mathrm{d}v$ which relate the partial volumes v^α to the bulk volume v and by assuming fully saturated conditions, the saturation constraint yields

$$\sum_\alpha n^\alpha = n^S + \underbrace{n^L + n^+ + n^-}_{n^F} = 1\,. \tag{1}$$

The introduction of the volume fractions motivates the definition of two density functions, the effective density $\rho^{\alpha R}$ which relates the mass of a constituent to the constituent volume and the partial density ρ^α which relates the mass of φ^α to the bulk volume. The density functions are related by

$$\rho^\alpha = n^\alpha \rho^{\alpha R} . \tag{2}$$

Particularly, the partial density of the components φ^β is given by $\rho^\beta = n^F c^\beta_m M^\beta_m$, where $c^\beta_m = \mathrm{d}n^\beta_m / \mathrm{d}v^F$ is the concentration (molarity) which relates the number of moles n^β_m to the volume v^F of the interstitial fluid and $M^\beta_m = \mathrm{d}m^\beta / \mathrm{d}n^\beta_m$ denotes the molar mass which relates the mass m^β to the number of moles.

1.2 Kinematics

Following the idea of superimposed continua, each constituent follows its individual motion $\chi_\alpha(\mathbf{X}_\alpha, t)$ and has its own velocity field with respect to different reference positions $\mathbf{X}_\alpha$:

$$\mathbf{x} = \chi(\mathbf{X}_\alpha, t) , \quad \overset{\prime}{\mathbf{x}}_\alpha = \frac{\mathrm{d}\chi_\alpha(\mathbf{X}_\alpha, t)}{\mathrm{d}t} . \tag{3}$$

Therein, $(\cdot)'_\alpha$ denotes the material time derivative with respect to φ^α.

The model under consideration incorporates seven independent fields, namely the solid displacement $\mathbf{u}_S = \mathbf{x} - \mathbf{X}_S$, the seepage velocity $\mathbf{w}_F = \overset{\prime}{\mathbf{x}}_F - \overset{\prime}{\mathbf{x}}_S$, the relative ion velocities $\mathbf{w}_\gamma = \overset{\prime}{\mathbf{x}}_\gamma - \overset{\prime}{\mathbf{x}}_S$, the effective fluid pressure p and the ion concentrations c^γ_m. Herein, $\gamma = \{+, -\}$ indicates only the mobile ions. For the liquid solvent, it is assumed that $\overset{\prime}{\mathbf{x}}_L \approx \overset{\prime}{\mathbf{x}}_F$, i. e. $\mathbf{w}_L \approx \mathbf{w}_F$.

According to $(3)_1$ and the solid displacement $\mathbf{u}_S$, the solid deformation gradient is defined by

$$\mathbf{F}_S := \frac{\partial \mathbf{x}}{\partial \mathbf{X}_S} = \mathrm{Grad}_S\, \mathbf{x} = \mathbf{I} + \mathrm{Grad}_S\, \mathbf{u}_S . \tag{4}$$

1.3 Balance relations

Proceeding from materially incompressible constituents and excluding any mass exchanges, volume balances are introduced for the phases φ^α and the components φ^β:

$$(n^\alpha)'_\alpha + n^\alpha \mathrm{div}\, \overset{\prime}{\mathbf{x}}_\alpha = 0 , \quad (n^F c^\beta_m)'_\beta + n^F c^\beta_m \mathrm{div}\, \overset{\prime}{\mathbf{x}}_\beta = 0 . \tag{5}$$

The volume balance of the charged solid and of the fixed charges as a part of the solid, respectively, can be integrated analytically leading to [5]

$$n^S = n^S_{0S} \det \mathbf{F}_S^{-1} , \quad c^{fc}_m = c^{fc}_{m0S} \frac{n^F_{0S}}{n^F} \det \mathbf{F}_S^{-1} , \tag{6}$$

where n^S_{0S}, $n^F_{0S} = 1 - n^S_{0S}$ and c^{fc}_{m0S} are initial values. Following *Truesdell*'s metaphysical principles, the mixture volume balance can be obtained as the sum of the partial volume balances $(5)_1$:

$$(n^S + n^F)'_S = \operatorname{div}\,[(\mathbf{u}_S)'_S + n^F \mathbf{w}_F] = 0\,. \tag{7}$$

For the numerical implementation, the component volume balances are rewritten in terms of the solid motion leading to

$$n^F (c^\beta_m)'_S + c^\beta_m \operatorname{div}(\mathbf{u}_S)'_S + \operatorname{div}(n^F c^\beta_m \mathbf{w}_\beta) = 0\,. \tag{8}$$

Note that under the assumptions c^L_m = const. and n^+, $n^- \ll n^L$, the volume balance of the liquid component φ^L is equivalent to that of the entire fluid.

Moreover, the partial momentum balances are given by

$$\mathbf{0} = \operatorname{div}\mathbf{T}^\alpha + \rho^\alpha\,\mathbf{b} + \hat{\mathbf{p}}^\alpha\,. \tag{9}$$

Herein, $\mathbf{T}^\alpha = (\mathbf{T}^\alpha)^T$ are the symmetric partial *Cauchy* stress tensors, $\rho^\alpha\,\mathbf{b}$ represent the body forces and $\hat{\mathbf{p}}^\alpha$ are the momentum productions, where $\hat{\mathbf{p}}^S + \hat{\mathbf{p}}^F = \mathbf{0}$ must hold due the overall conservation of momentum.

1.4 Constitutive Equations

To close the set of governing equations, additional constitutive relations have to be introduced. Proceeding from the general effective stress principle, cf. [2], the following relations hold:

$$\mathbf{T}^\alpha = -n^\alpha p\,\mathbf{I} + \mathbf{T}^\alpha_E\,, \qquad \hat{\mathbf{p}}^F = p\,\operatorname{grad} n^F + \hat{\mathbf{p}}^F_E\,. \tag{10}$$

Herein, $(\,\cdot\,)_E$ indicates the so-called extra quantities for which further constitutive assumptions have to be formulated. From thermodynamical considerations, it follows that the solid extra stress and the fluid extra stress have the form

$$\mathbf{T}^S_E = -n^S \sum_\beta \mu^\beta_F\,\mathbf{I} + \mathbf{T}^S_{E\,mech.}\,, \quad \mathbf{T}^F_E = -n^F \sum_\beta \mu^\beta_F\,\mathbf{I} + \mathbf{T}^F_{E\,mech.}\,, \tag{11}$$

where μ^β_F are the electrochemical potentials of φ^β per fluid volume $[\mathrm{J}/v^F]$ and $\mathbf{T}^S_{E\,mech.}$ is the purely mechanical part that can be computed from an appropriate viscoelasticity law [3]. The purely mechanical fluid friction stress can be neglected *a priori*, i. e. $\mathbf{T}^F_{E\,mech.} \approx \mathbf{0}$.

Furthermore, the entropy principle leads to the following ansatz for the extra momentum production

$$\hat{\mathbf{p}}^F_E = \sum_\beta \mu^\beta_F\,\operatorname{grad} n^F - \frac{(n^F)^2 \gamma^{FR}}{k^F}\,\mathbf{w}_F\,, \tag{12}$$

wherein γ^{FR} is the effective fluid weight and k^F denotes the *Darcy* permeability which may depend on the deformation state [3].

For the chemical equations the molar electrochemical potentials [J/mol] of the fluid components are needed which are given by

$$\mu_m^L = \mu_{m0}^L + R\,\theta\,\ln\frac{c_m^L}{\sum_\beta c_m^\beta}\,, \qquad \mu_m^\gamma = \mu_{m0}^\gamma + R\,\theta\,\ln c_m^\gamma + z^\gamma F\xi\,. \tag{13}$$

These terms are related to μ_F^β via $\mu_m^\beta = \mu_F^\beta / c_m^\beta$. Moreover, μ_{m0}^β are the reference chemical potentials of the components, R is the universal gas constant, θ is the absolute *Kelvin* temperature, z^γ are the valences of the ions, F is the *Faraday* constant and ξ is the electrical potential.

1.5 Interstitial Fluid Flow, Ion Diffusion and Osmotic Pressure

The interstitial fluid flow can be described by an extended *Darcy* filter law and the ion diffusion is governed by the extended *Nernst-Planck* equation

$$\begin{aligned} n^F \mathbf{w}_F &= -\frac{k^F}{\gamma^{FR}}\Big(\operatorname{grad} p - \rho^{FR}\,\mathbf{b} + \sum_\beta \operatorname{grad}\mu_F^\beta\Big)\,, \\ c_m^\gamma \mathbf{w}_\gamma &= -D^\gamma\Big[\operatorname{grad} c_m^\gamma + z^\gamma c_m^\gamma \frac{F}{R\,\theta}\operatorname{grad}\xi\Big] + c_m^\gamma \mathbf{w}_F\,, \end{aligned} \tag{14}$$

where $n^F\,\mathbf{w}_F$ is the so-called filter velocity. Note that $c_m^\gamma \mathbf{w}_F$ is added to the ion velocities to get the velocities relative to the solid motion.

To describe macromolecular solutions, the *Donnan* theory is used in form of the *Donnan* equation

$$c_m^+\,c_m^- = \bar{c}_m^+\,\bar{c}_m^-\,. \tag{15}$$

Herein, the quantities $\bar{c}_m^+$ and $\bar{c}_m^-$ indicate the ion concentrations of the external solution. The osmotic pressure π is computed from the osmolarity difference of the internal and the external solution:

$$\pi = R\,\theta\,[(c_m^+ + c_m^-) - (\bar{c}_m^+ + \bar{c}_m^-)]\,. \tag{16}$$

2 Numerical Implementation

2.1 Reduction to Monovalent Solutions

For simplicity, the model equations are reduced for the description of monovalent salt solutions. In particular, the model is verified for a Na^+Cl^- solution. Thus, with the valences $z^+ = 1$, $z^- = -1$ and $z^{fc} = -1$, we define an external concentration $\bar{c}_m$ and an internal concentration c_m via the electroneutrality condition of the external and the internal solution, respectively:

$$
\begin{aligned}
z^+\bar{c}_m^+ + z^-\bar{c}_m^- &= 0 && \longrightarrow && \bar{c}_m^+ = \bar{c}_m^- =: \bar{c}_m\,, \\
z^+\bar{c}_m^+ + z^-\bar{c}_m^- + z^{fc}c^{fc} &= 0 && \longrightarrow && c_m := c_m^-\,.
\end{aligned}
\tag{17}
$$

Therefore, the internal concentration of the cations is given by $c_m^+ = c_m + c_m^{fc}$ and is no more an independent field.

Moreover, with the electroneutrality condition $(17)_2$, the equations (15) and (16) can be used to compute the equilibrium concentration and the osmotic pressure of the internal solution [4]:

$$
c_m = \sqrt{\bar{c}_m^2 + \left(\frac{c_m^{fc}}{2}\right)^2} - \frac{c_m^{fc}}{2}\,, \qquad \pi = R\,\theta\,(2\,c_m + c_m^{fc} - 2\,\bar{c}_m)\,. \tag{18}
$$

The assumption that there is no electrical current, i. e.

$$
\boldsymbol{I} = F A\,(c_m^+ \mathbf{w}_+ - c_m^- \mathbf{w}_-) = \mathbf{0}\,, \tag{19}
$$

yields with the ion diffusion equations $(14)_2$ a conditional equation for the gradient of the electrical potential which is needed in (14).

2.2 Weak Formulation

For the numerical treatment within the FEM, weak forms of the governing field equations (7), (8) and (9) are required. Therefore, the balance relations are weighted by independent test functions and are integrated over the spatial domain Ω with the surface $\partial\Omega$ resulting in the respective weak formulations for the mixture volume balance, the anion volume balance and the mixture momentum balance:

$$
-\int_\Omega \delta p \operatorname{div}(\mathbf{u}_S)'_S\,\mathrm{d}v + \int_\Omega \operatorname{grad}\delta p \cdot n^F \mathbf{w}_F\,\mathrm{d}v = \int_{\partial\Omega} \delta p\,\bar{q}\,\mathrm{d}a \tag{20}
$$

$$
-\int_\Omega \delta c_m \left[n^F (c_m)'_S + c_m \operatorname{div}(\mathbf{u}_S)'_S \right] \mathrm{d}v + \tag{21}
$$

$$
+\int_\Omega \operatorname{grad}\delta c_m \cdot n^F c_m\,\mathbf{w}_{\mathrm{d}} v = \int_{\partial\Omega} \delta c_m\,\bar{\jmath}\,\mathrm{d}a \tag{22}
$$

$$
\int_\Omega \operatorname{grad}\delta\mathbf{u}_S \cdot (\mathbf{T}_E^S - p\,\mathbf{I})\,\mathrm{d}v = \int_{\partial\Omega} \delta\mathbf{u}_S \cdot \bar{\mathbf{t}}\,\mathrm{d}a \tag{23}
$$

By eliminating the seepage velocity and the anion diffusion in $(20)_1$ and $(21)_2$ by use of the extended *Darcy* filter law $(14)_1$ and the *Nernst-Planck* equation

$(14)_2$, one finally ends up with a displacement-pressure-concentration formulation in the primary variables $\mathbf{u}_S$, p and c_m. Moreover, $\bar{q}$ denotes the fluid volume efflux and $\bar{\jmath}$ is the anion diffusion over the surface $\partial\Omega$ and $\bar{\mathbf{t}}$ is the external load vector acting on the entire mixture.

2.3 Initial and Boundary Conditions

The initial values of the concentration and the osmotic pressure inside the tissue are obtained from

$$c_{m0} = \sqrt{\bar{c}_{m0}^2 + \left(\frac{c_{m0S}^{fc}}{2}\right)^2} - \frac{c_{m0S}^{fc}}{2}, \quad \pi_0 = R\theta\,(2\,c_{m0} + c_{m0S}^{fc} - 2\,\bar{c}_{m0})\,, \tag{24}$$

where $\bar{c}_{m0}$ is the initial concentration of the external solution. Due to an initial osmotic pressure, the charged solid meshwork is prestressed, i. e. $\mathbf{T}_{E0}^S = \pi_0\,\mathbf{I}$. Assuming that the outer surface is instantly in equilibrium with the external solution, the boundary concentration and osmotic pressure are computed from

$$c_m = \sqrt{\bar{c}_m^2(t) + \left(\frac{c_m^{fc}}{2}\right)^2} - \frac{c_m^{fc}}{2}, \qquad \pi = R\theta\,(2\,c_m + c_m^{fc} - 2\,\bar{c}_m)\,. \tag{25}$$

3 Example

To demonstrate the capability of the presented model, a free swelling experiment on hydrogel is simulated by the FEM tool PANDAS. Therefore, a 2-dimensional block is discretized with quadrilateral extended *Taylor-Hood* elements with a quadratic approximation of $\mathbf{u}_S$ and linear approximations of p and c_m. To initiate swelling, the concentration of the external solution is decreased from 0.45 mol/l to 0.15 mol/l within 5 sec.

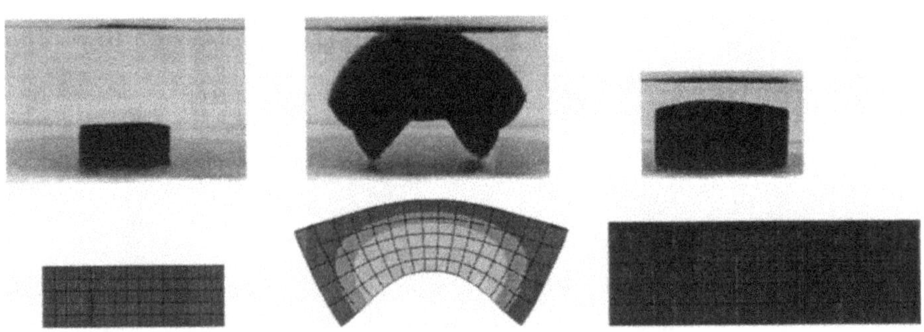

Fig. 1. Qualitative comparison of a free swelling experiment on stoot-colored hydrogel disc (J. M. Huyghe, 1999) with the FEM simulation; colored, see page 376

References

1. Bowen R. M. (1980) Incompressible porous media models by use of the theory of mixtures. Int J Engng Sci 18:1129–1148
2. Ehlers W. (2002) Foundations of multiphasic and porous materials. In Ehlers W and Bluhm J. (Eds.) Porous Media: Theory, Experiments and Numerical Applications. Springer-Verlag, Berlin, 3–86
3. Ehlers W., Markert B. (2001) A linear viscoelastic biphasic model for soft tissues based on the Theory of Porous Media. ASME J Biomech Eng 123:418–424
4. Huyghe J. M., Janssen J. D. (1997) Quadriphasic mechanics of swelling incompressible porous media. Int J Engng Sci 35:793–802
5. Lai W. M., Hou J. S., Mow V. C. (1991) A triphasic theory for the swelling and deformation behaviours of articular cartilage. ASME J Biomech Eng 113:245–258

Simulation of Acoustics-Structure-Interaction on Non-Conforming Discretizations

Matthias Fischer and Lothar Gaul

Institut A für Mechanik, Universität Stuttgart, Germany

Abstract. A mortar coupling algorithm is proposed for acoustic-structure-interaction. Finite plate elements are coupled to a boundary element discretization of the acoustic domain. The interface pressure is interpolated as a Lagrange multiplier, thus, allowing the coupling of non-matching grids. The method is demonstrated on the example of a cavity backed elastic panel.

1 Introduction

A low noise level is a major demand on technical products. Thus, the reduction of noise and vibration has high priority in the design process. Numerical simulations of the structural acoustic behaviour give essential information for optimising the design of the structure. Very often one can perform those simulations neglecting the influence of the acoustic field on the vibration behaviour of the structure. However, this is not acceptable in situations where thin and flexible structures are excited by the acoustic pressure. The interaction effect is enhanced when the structure is submerged in a heavy fluid or the system is operating close to an acoustic resonance. For these applications the acoustic field has to be fully coupled to the vibrating structure.

The numerical cost for the simulation of coupled structural-acoustic systems is tremendous and the choice of suitable discretization methods crucial. The finite element method (FEM) is the preferred tool for the simulation of structural vibrations. For the acoustic part the boundary element method (BEM) offers major advantages: only the boundary of the domain must be discretized and unbounded media can be computed very easily. In the paper at hand, a FEM – BEM coupling algorithm is develped that allows non-conforming discretizations of the coupled domains. This permits the use of different interpolation functions in the sub-domains or non-matching grids in the case that a coarser mesh is sufficient in one of the domains.

The non-conforming coupling approach is based on the mortar element method that was originally introduced by Bernardi et al. [1]. The formulation of the method with Lagrange multipliers is proposed by Ben Belgacem [2] leading to a saddle point problem. Stability estimates and definition of the mortar spaces for 3-dimensional problems can be found in Braess and Dahmen [3].

2 Problem Definition

Harmonic vibrations of thin plates can be described by Kirchhoff's plate theory leading to the biharmonic equation

$$\nabla^2\nabla^2 w - \kappa_s^2 w = \frac{f}{K} \ , \tag{1}$$

where for simplicity the harmonic time-dependency $\exp(i\omega t)$ is ommited from the normal displacement w and the forcing f. The structural wavenumber is defined as $\kappa_s = \rho_s a\omega^2/K$ with density ρ_s, thickness a and the bending stiffness of the plate K. The governing equation for the acoustic field is the Helmholtz equation

$$\nabla^2\phi + \kappa_f^2\phi = 0 \ , \tag{2}$$

where ϕ is the velocity potential which relates to the particle velocity $\boldsymbol{v}$ in the fluid, $\nabla\phi = -\boldsymbol{v}$. Balance of momentum gives the relation between velocity potential and the acoustic pressure $p = \rho_f i\omega\phi$. The acoustic wave number is defined as $\kappa_f = \omega/c_f$ with speed of sound c_f.

On the acoustic–structure interface Γ_i, continuity of the normal velocities is required. Using the normal directions as defined in Fig. 1, one obtains the coupling condition

$$\nabla\phi \cdot \boldsymbol{n}_f = -\boldsymbol{v} \cdot \boldsymbol{n}_f = -i\omega w \ . \tag{3}$$

The forcing of the plate $f = f_f + f_{FEM}$ is split into forces due to the fluid loading and externally applied forces, respectively. Requiring equilibrium at the interface leads to the second coupling condition

$$f_f = p = \rho_f i\omega\phi \ . \tag{4}$$

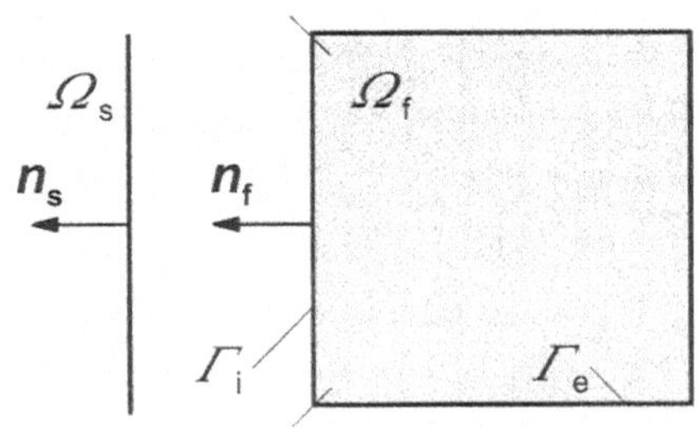

Fig. 1. Normal direction of plate (Ω_s) and acoustic domain (Ω_f) at interface Γ_i

3 Mapping of the Acoustic Domain to the Interface

The direct boudary element method for acoustics is derived from Helmholtz equation (2) by weighting it with the foundamental solution $\phi^* = 1/(4\pi r)$, where r is the Euclidean distance between field point $\boldsymbol{x}$ and load point $\boldsymbol{\xi}$. Integrating by parts two times leads to the Helmholtz integral equation

$$\int_{\Omega} (\phi^* \nabla^2 \phi - \phi \nabla^2 \phi^*) \, \mathrm{d}\Omega = \int_{\Gamma} (\phi^* \nabla \phi \cdot \boldsymbol{n} - \phi \nabla \phi^* \cdot \boldsymbol{n}) \, \mathrm{d}\Gamma \ . \tag{5}$$

The acoustic flux $\psi \equiv \nabla \phi \cdot \boldsymbol{n}$ is defined as the directional derivative of the velocity potential on the boundary. The domain integral in(5) is eliminated by moving the load point $\boldsymbol{\xi}$ to the boundary

$$c(\boldsymbol{\xi}) p(\boldsymbol{\xi}) = \int_{\Gamma} (\phi^* \psi - \phi \psi^*) \, \mathrm{d}\Gamma \tag{6}$$

where the boundary integral has to be evaluated in a Cauchy principle value sense and $c(\boldsymbol{\xi})$ is the so-called integral free term. Discretization of the boundary with standard shape functions $\phi = \mathbf{N}_{\phi}^{\mathrm{T}} \boldsymbol{\phi}$, $\psi = \mathbf{N}_{\psi}^{\mathrm{T}} \boldsymbol{\psi}$ and collocating the load points with the discretization nodes leads to the system of equations

$$\mathbf{H} \boldsymbol{\phi} = \mathbf{G} \boldsymbol{\psi} \ . \tag{7}$$

The boundary of the acoustic domain is split into interface and the remaining external boundary (Fig. 1), $\Gamma = \Gamma_{\mathrm{i}} \cup \Gamma_{\mathrm{e}}$. For simplicity, only Neumann boundary conditions $\bar{\boldsymbol{\psi}}$ are assumed on Γ_{e}. Then (7) can be written as

$$\begin{aligned} \left(\mathbf{H}_{\mathrm{ii}} - \mathbf{H}_{\mathrm{ie}} \mathbf{H}_{\mathrm{ee}}^{-1} \mathbf{H}_{\mathrm{ei}}\right) \boldsymbol{\phi}_{\mathrm{i}} &= \left(\mathbf{G}_{\mathrm{ii}} - \mathbf{H}_{\mathrm{ie}} \mathbf{H}_{\mathrm{ee}}^{-1} \mathbf{G}_{\mathrm{ei}}\right) \boldsymbol{\psi}_{\mathrm{i}} \\ &+ \left(\mathbf{G}_{\mathrm{ie}} - \mathbf{H}_{\mathrm{ie}} \mathbf{H}_{\mathrm{ee}}^{-1} \mathbf{G}_{\mathrm{ee}}\right) \bar{\boldsymbol{\psi}}_{\mathrm{e}} \ , \end{aligned} \tag{8}$$

which defines the mapping $\boldsymbol{\phi}_{\mathrm{i}} = \mathbf{D} \boldsymbol{\psi}_{\mathrm{i}} + \mathbf{f}_{\mathrm{BEM}}$ from the flux on the interface and the given boundary conditions to the velocity potential on the interface.

4 Mortar Coupling Algorithm

The finite element formulation for the harmonic plate vibration [5] is derived from the virtual work equivalence of (1) which writes in operator notation

$$\int_{\Gamma_{\mathrm{i}}} \delta w \mathcal{K} w \, \mathrm{d}\Gamma - \omega^2 \int_{\Gamma_{\mathrm{i}}} \delta w \rho_{\mathrm{S}} w \, \mathrm{d}\Gamma = \int_{\Gamma_{\mathrm{i}}} \delta w (f_{\mathrm{f}} + f_{\mathrm{FEM}}) \, \mathrm{d}\Gamma \ , \tag{9}$$

where the plate is assumed to be completely coupled to the acoustic domain, i. e. $\Omega_{\mathrm{s}} \equiv \Gamma_{\mathrm{i}}$. For the acoustic domain, the discrete form of the mapping $\mathcal{D}$ from Dirichlet to Neumann data on the interface is derived in Sect. 3. It is required in a weak sense weighted with the acoustic flux

$$\int_{\Gamma_{\mathrm{i}}} \delta \psi \phi \, \mathrm{d}\Gamma = \int_{\Gamma_{\mathrm{i}}} \delta \psi \mathcal{D} \psi \, \mathrm{d}\Gamma + \int_{\Gamma_{\mathrm{i}}} \delta \psi f_{\mathrm{BEM}} \, \mathrm{d}\Gamma \ . \tag{10}$$

Enforcing equilibrium on the coupling interface identically and defining the Lagrange multiplier

$$f_{\mathrm{f}} = i\omega\rho_{\mathrm{f}}\phi_{\mathrm{i}} \equiv \lambda \ , \tag{11}$$

leads to the coupled equation

$$\begin{aligned} \int_{\Gamma_{\mathrm{i}}} \delta w \mathcal{K} w \, \mathrm{d}\Gamma \omega^2 \int_{\Gamma_{\mathrm{i}}} \delta w \rho_S w \, \mathrm{d}\Gamma + \rho_{\mathrm{f}} \int_{\Gamma_{\mathrm{i}}} \delta \psi \mathcal{D} \psi \, \mathrm{d}\Gamma \\ - \rho_{\mathrm{f}} \int_{\Gamma_{\mathrm{i}}} (\delta\psi + i\omega\delta w) \, \lambda \, \mathrm{d}\Gamma = \int_{\Gamma_{\mathrm{i}}} \delta w f_{\mathrm{FEM}} \, \mathrm{d}\Gamma + \rho_{\mathrm{f}} \int_{\Gamma_{\mathrm{i}}} \delta\psi f_{\mathrm{BEM}} \, \mathrm{d}\Gamma \ . \end{aligned} \tag{12}$$

The continuity of acoustic flux and normal plate velocity (3) is enforced in a weak sense weighted by the Lagrange multiplier

$$\int_{\Gamma_{\mathrm{i}}} \delta\lambda\psi \, \mathrm{d}\Gamma = -i\omega \int_{\Gamma_{\mathrm{i}}} \delta\lambda w \, \mathrm{d}\Gamma \ . \tag{13}$$

The Lagrange multiplier is interpolated independently on the coupling interface. For the discretisation of the acoustic flux, acoustic pressure and the Lagrange multiplier, bilinear shape functions on quadrilateral elements are chosen. For the Kirchhoff plate, a 12 degree of freedom element is selected as proposed by Zienkiewicz and Taylor [5]. On each corner node of the quadrilateral, the out-of-plane displacement w and the rotations about the x and y axes are degrees of freedom. This element does not fulfil C^1-continuity but is stable for rectangular elements.

The discretisation leads to the following system of equations

$$\begin{bmatrix} \rho_{\mathrm{f}}\mathbf{D} & \mathbf{0} & \mathbf{C}_{\mathrm{BEM}} \\ \mathbf{0} & \mathbf{K} - \omega^2\mathbf{M} & \mathbf{C}_{\mathrm{FEM}} \\ \mathbf{C}_{\mathrm{BEM}}^{\mathrm{T}} & \mathbf{C}_{\mathrm{FEM}}^{\mathrm{T}} & \mathbf{0} \end{bmatrix} \begin{bmatrix} \boldsymbol{\psi} \\ \mathbf{w} \\ \boldsymbol{\lambda} \end{bmatrix} = \begin{bmatrix} \rho_{\mathrm{f}}\mathbf{f}_{\mathrm{BEM}} \\ \mathbf{f}_{\mathrm{FEM}} \\ \mathbf{0} \end{bmatrix} , \tag{14}$$

with the coupling matrices $\mathbf{C}_{\mathrm{BEM}}$ and $\mathbf{C}_{\mathrm{FEM}}$ which are given by

$$\mathbf{C}_{\mathrm{BEM}} = -\rho_{\mathrm{f}} \int_{\Gamma_{\mathrm{i}}} \mathbf{N}_\psi \mathbf{N}_\lambda^{\mathrm{T}} \mathrm{d}\Gamma \quad \mathbf{C}_{\mathrm{FEM}} = -i\omega\rho_{\mathrm{f}} \int_{\Gamma_{\mathrm{i}}} \mathbf{N}_w \mathbf{N}_\lambda^{\mathrm{T}} \mathrm{d}\Gamma \ . \tag{15}$$

The system of equations (14) corresponds to the saddle point problem: find $(u_h, v_h) \in X_h \times M_h$,

$$\begin{aligned} a(u_h, v_h) + b(v_h, \lambda_h) &= f(v_h) \ \forall v_h \in X_h \\ b(u_h, \mu_h) &= 0 \quad \forall \mu_h \in M_h \ . \end{aligned} \tag{16}$$

Stability of the saddle point formulation can be shown by proving the existence of the LBB condition [3].

5 Example: Panel Backed by a Closed Cavity

The coupling algorithm is demonstrated for the example of a flexible panel backed by an acoustic cavity with rigid walls. The panel is excited by a harmonic point force in the center of the plate. For this problem, an analytical series solution was derived by Pretlove [4] which is used as reference for the numerical simulation.

The steel panel has dimensions of $1\,\mathrm{m} \times 1\,\mathrm{m} \times 0.01\,\mathrm{m}$ and is simply supported on all edges. The first eigenfrequency for this panel in vacuum is $f_{11}^0 = 49\,\mathrm{Hz}$ as predicted by Kirchhoff's plate theory. The closed cavity with dimensions $1\,\mathrm{m} \times 1\,\mathrm{m} \times 1\,\mathrm{m}$ is filled with water. The boundaries appart from the interface to the panel are assumed rigid, i. e., homogenous Neumann conditions $\psi = 0$ are applied.

The series solution predicts the first eigenfrequency of the coupled system at $f_{11}^c = 134.3\,\mathrm{Hz}$. One notices an increase from the eigenfrequency in vacuum, i. e. the cavity is not acting as added mass but stiffens the structure. As basis discretization, equal meshes are chosen for the structure and the acoustic domain. The element size is $h_s = h_f = 0.1\,\mathrm{m}$, or 10 elements in each direction. This corresponds to 18 elements per wavelength at the first coupled mode, well above the rule of thumb of at least 6 elements per wavelength for uncoupled acoustic simulations. The computed frequency response is plotted in Fig. 2 and one reads the coupled eigenfrequency at 130.5 Hz. The second frequency response in Fig. 2 is computed with non-conforming meshes. For the discretization of the acoustic cavity, again 10 boundary elements are used in each direction, whereas the plate is now discretized using 14×14 finite elements. The computed igenfrequency is slightly closer to the reference solution than the result using equal meshes.

Considering the fine discretization, the frequency responses show significant deviation from the series solution. Thus, the convergence of the simulation result at 100 Hz is examined in Fig. 3. The error of the normal displacement in the center of the plate is plotted over the inverse of the element size $1/h$. One notices, that for coarse discretizations the error is large and drops rapidly with refinement. For fine discretizations one finds a linear convergence rate. There is no signigicant difference in convergence rate for conforming and non-conforming meshes. However, refining the structure model only does not lead to significant improvement of the result as shown in Fig. 3.

References

1. Bernardi, C., Maday, Y., Patera A. T. (1994) A new nonconforming approach to domain decomposition: the mortar element method. In: Brezis, H., Lions, J.-L. (Eds.) Nonlinear Partial Differential Equations and Their Aplications, Pitman, New York, 13–51
2. Ben Belgacem, F. (1999) The Mortar finite element method with Lagrange multipliers. Numer. Math. 84:173–197

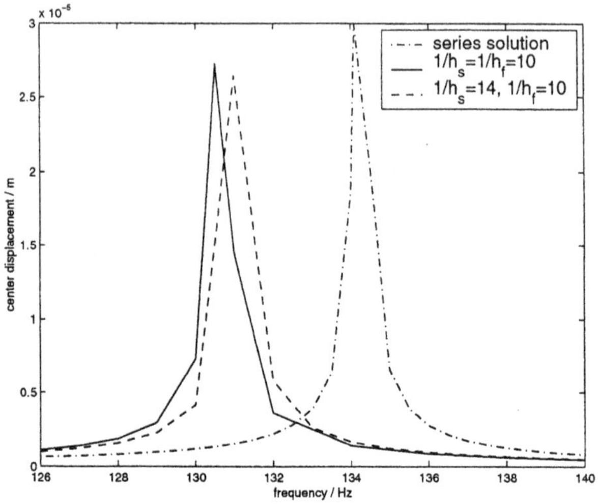

Fig. 2. Frequency response of the coupled system

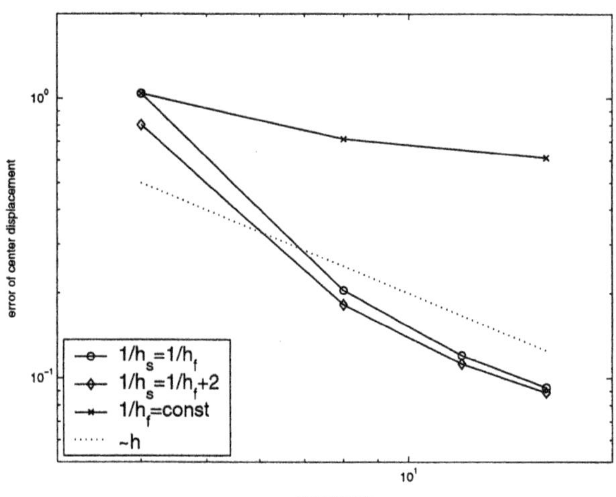

Fig. 3. Error of center displacement at 100 Hz

3. Braess, D., Dahmen, W. (1998) Stabiltiy estimates of the mortar finite element method for 3-dimensional problems. East-West J. Numer. Math. 6(4):249–263
4. Pretlove, A. J. (1966) Forced vibrations of a rectangular panel backed by a closed rectangular cavity. J. Sound Vib. 3(3):252–261
5. Zienkiewicz, O. C., Taylor R. L. (2000) The Finite Element Method Volume 2 Solid Mechanics. Butterworth-Heinemann, Oxford

Injection-ejection fluid influence on velocity and vorticity distributions in laminar-turbulent transition of unsteady incompressible boundary layer on porous surface

Decan J. Ivanovic[1] and Viktor N. Saljnikov[2]

[1] Faculty of mechanical Engineering, University of Montenegro, 81000 Podgorica, Yugoslavia, email: decan@l.cis.cg.ac.yu

[2] Faculty of Mechanical Engineering, University of Belgrade, 11000 Beograd, Yugoslavia

Abstract. Through the porous contour in pependicular direction, the fluid of the same properties as the incompressible fluid in the basic flow, has been injected or ejected. The corresponding equations of unsteady boundary layer, by introducing the appropriate variable transformations, momentum and energy equations and two similarity parameter sets, being transformed into generalized form. These parameters are expressing the influence of the outer flow velocity, the injection or ejection velocity and the flow history in the boundary layer, on the boundary layer characteristics. The numerical integration of the generalized equation with boundary conditions has been performed by means of difference schemes and by using the Tridiagonal Algorithm Method with iterations in the three once localized approximations, where the first unsteady, dynamic and porous parameters will remain, while all others will be let to be equal to zero, and where the derivatives with respect to the first porous parameter will be considered equal to zero. The obtained generalized solutions are used to calculate the shear stress, velocity and vorticity distributions in the laminar-turbulent transition of the unsteady boundary layer on the porous aerofoil when the center velocity changes with time as a degree function and when the potential external velocity on the aerofoil is measured in free flight. As well as for both the accelerating and decelerating flows, the ejection of fluid postpones the boundary layer separation, and vice versa the injection of fluid favours the separation.

1 Introduction

The behaviour of a boundary layer in the presence of a positive or negative pressure gradient along the surface is particularly important for the calculation of the drag on aerofoil as well as for the understanding of the processes which take in a diffuser, because there is transition from laminar to turbulent flow which determines the dividing line between a region with low drag and one where drag is dramatically increased. So one can control laminar-turbulent transition by fluid which has been injected or ejected through the porous contour. Apart from skin friction we are interested in knowing whether the boundary layer will separate under given circumstances and if so, we shall wish to determine the point of separation. As the present time these very complicated phenomena are far from being understood completely but there are in existence several methods of calculation which lead to comparatively satisfactory results, and one of them is a modern parameter method for calculation boundary layers, known as generalized similar method. Similar solutions of the boundary layer equations play an important role in the investigation of the stability of hydrodynamic flows and in developing semi-empirical criteria for the transition to turbulence.

The results obtained by means of boundary layer theory dispose, in comparison with numerical solutions of complete Navier-Stokes equations, significant preference, because they are exact corresponding to the structure of solutions for a great Re numbers, i.e. represent solution which possess the boundary layer character. That is why the numerical method for calculation of Navier-Stokes equations, for a great Re numbers, only is appropriate if algorithm formed for their solving, concerning asymptotic behaviour, gives results which cover up with solutions of boundary layer equations. The generalized similarity equation represents one of the ways for improvement of modern analytical methods for the calculation of boundary layers. As a result of these procedure which consists of introducing a conveniently chosen set of parameters, quantities characterizing any special problem are eliminated from the governing set of equations and the corresponding boundary conditions. A numerical solution of the generalized similarity equation can be found once for all and then it can be used in any special problem of the boundary layer theory.

2 Mathematical model and generalized similarity equation

Generalized similarity method [1–3], is exposed to the problem of unsteady incompressible plane boundary layer on the porous surface [4–7,9], when the fluid of the same properties as fluid in basic flow has been injected or ejected through the surface in perpendicular direction with velocity v_w. The mathematical model of the noticed problem is described by the following equation:

$$\Psi_{ty} + \Psi_y \Psi_{xy} + (v_w - \Psi_x)\Psi_{yy} = U_t + UU_x + \nu\Psi_{yyy} \tag{1}$$

with boundary and initial conditions:

$$\begin{aligned} &y = 0 : \Psi = \Psi_y = 0 ; \quad y \to \infty : \Psi_y \to U(x,t) ; \\ &t = t_0 : \Psi_y = u_1(x,y) ; \quad x = x_0 : \Psi_y = u_0(t,y) , \end{aligned} \tag{2}$$

where is used:

$\Psi(x,y,t)$ – stream function,
$U(x,t)$ – free-stream velocity,
ν – kinematic viscosity,
$u_1(x,y)$ – the streamwise velocity distribution in boundary layer in some determined point of time $t=t_0$,
$u_0(t,y)$ – the streamwise velocity distribution in boundary layer in cross-section $x=x_0$,
x – streamwise coordinate,
y – crosswise coordinate,
t – time.

Introducing new variables in the form [3,6–9]:

$$\begin{aligned} & x=x\,,\ t=t\,,\ \eta=yU^{b_0/2}\left(a_0\nu\int_0^x U^{b_0-1}\,dx\right)^{-1/2}, \\ & \Phi(x,\eta,t)=\Psi U^{b_0/2-1}\left(a_0\nu\int_0^x U^{b_0-1}\,dx\right)^{-1/2} \end{aligned} \tag{3}$$

where is [3] $a_0=0.4408$, $b_0=5.714$, we transform the equation (1) to the new form. Afterwards, we introduce the group of parameters:

$$\begin{aligned} f_{k,n} &= U^{k-1}U^{(k+n)}_{x^{(k)}t^{(n)}}\,z^{**\,k+n} \qquad (k,n=0,1,2,\ldots;k\vee n\neq 0) \\ \lambda_{k,n} &= -\nu^{-1/2}U^{k}(v_w)^{(k+n)}_{x^{(k)}t^{(n)}}\,z^{k+n+1/2} \qquad (k,n=0,1,2,\ldots) \end{aligned} \tag{4}$$

as new independent variables, where is:

$$\begin{aligned} & z^{**}=\delta^{**\,2}/\nu\,,\ \delta^{**}=\left(a_0\nu U^{-b_0}\int_0^x U^{b_0-1}\,dx\right)^{1/2}B\,, \\ & B=\int_0^\infty \Phi_\eta(1-\Phi_\eta)\,d\eta \end{aligned} \tag{5}$$

Now, the already transformed equation (1) being transformed to the new form:

$$\begin{aligned} & B^2\Phi_{\eta\eta\eta}+0.5\left[a_0B^2+(2-b_0)f_{1,0}\right]\Phi\Phi_{\eta\eta}+f_{1,0}\left(1-\Phi_\eta^2\right) \\ & \qquad +f_{0,1}(1-\Phi_\eta)+(0.5\eta T^{**}+B\lambda_{0,0})\,\Phi_{\eta\eta} \\ & =\eta B^{-1}\left(\sum_{\substack{k,n=0\\k\vee n\neq 0}}^{\infty}C_{k,n}B_{f_{k,n}}+\sum_{k,n=0}^{\infty}R_{k,n}B_{\lambda_{k,n}}\right)\Phi_{\eta\eta} \\ & \qquad +\left\{\sum_{\substack{k,n=0\\k\vee n\neq 0}}^{\infty}\left[C_{k,n}\Phi_{\eta f_{k,n}}+A_{k,n}\left(\Phi_\eta\Phi_{\eta f_{k,n}}-\Phi_{f_{k,n}}\Phi_{\eta\eta}\right)\right]\right. \\ & \qquad \left.+\sum_{k,n=0}^{\infty}\left[R_{k,n}\Phi_{\eta\lambda_{k,n}}+E_{k,n}\left(\Phi_\eta\Phi_{\eta\lambda_{k,n}}-\Phi_{\lambda_{k,n}}\Phi_{\eta\eta}\right)\right]\right\} \end{aligned} \tag{6}$$

with corresponding boundary conditions:

$$\begin{aligned} &\eta = 0 \, : \, \Phi = \Phi_\eta \, ; \quad \eta \to \infty \, : \, \Phi_\eta \to 1 \, ; \\ &f_{k,n} = \lambda_{k,n} = 0 \; (k = 0, 1, \ldots ; k \vee n \neq 0) \, : \, \Phi = \Phi_0(\eta) \, , \end{aligned} \tag{7}$$

where $\Phi_0(\eta)$ is Blausius' solution for the problem of flat plate. In the equation (6) the following notations have been used:

$$\begin{aligned} A_{k,n} &= (k-1) f_{1,0} f_{k,n} + f_{k+1,n} + (k+n) f_{k,n} \overset{**}{F} ; \\ E_{k,n} &= k f_{1,0} \lambda_{k,n} + \lambda_{k+1,n} + (k+n+0.5) \lambda_{k,n} \overset{**}{F} ; \\ C_{k,n} &= (k-1) f_{0,1} f_{k,n} + f_{k,n+1} + (k+n) f_{k,n} \overset{**}{T} ; \\ R_{k,n} &= k f_{0,1} \lambda_{k,n} + \lambda_{k,n+1} + (k+n+0.5) \lambda_{k,n} \overset{**}{T} ; \\ \overset{**}{T} &= z_t^{**} \, ; \; \overset{**}{F} = U z_x^{**} \, . \end{aligned} \tag{8}$$

In order to take the equation (6) universal, the multipliers $\overset{**}{F}$ and $\overset{**}{T}$ have to be expressed by means of quantities which are explicit functions only of parameters (4). In the determination of this functions, one can use the momentum and energy equations of the considered problem:

$$\begin{aligned} &(U\delta^*)_t + \left(U^2 \delta^{**}\right)_x U U_x \delta^* - U v_w - \tau_w / \rho = 0 \, ; \\ &\left(U^2 \delta^{**}\right)_t + \left(U^3 \delta_1^{**}\right)_x + U^2 \left(\delta_t^* + 3 \delta_1^{**} U_x - 2 \nu e\right) = 0 \end{aligned} \tag{9}$$

where is:

$$\begin{aligned} &\delta^* = L^{1/2} \int_0^\infty (1 - \Phi_\eta) \, d\eta \, ; \; \tau_w = \rho \nu U^{b_0/2+1} L^{-1/2} \left(\Phi_{\eta\eta}\right)_{\eta=0} \, ; \\ &\delta_1^{**} = L^{1/2} \int_0^\infty \Phi_\eta \left(1 - \Phi_\eta^2\right) \, d\eta \, ; \; e = L^{-1/2} \int_0^\infty \Phi_{\eta\eta}^2 \, d\eta \, ; \\ &L = a_0 \nu U^{-b_0} \int_0^x U^{b_0 - 1} \, dx \, . \end{aligned} \tag{10}$$

After certain transformations, the expressions for $\overset{**}{F}$ and $\overset{**}{T}$ have been obtained as universal, i.e. they do not depend on outer flow characteristics. In equation (6) are not involved in explicit form the velocity at outer border of boundary layer and its derivatives, as well as ejection or injection fluid velocity, thus this equation can be called the generalized i.e. universal equation. The universal boundary conditions have the form as (7).

3 Approximative generalized similarity equation

The numerical integration of the equation (6), with the corresponding universal boundary conditions (7), can be performed "once and forever" only for its approximative form. It means, that the solution of universal equation in practice needs limitation of the number of the independent variables. It leads to the necessity of application of the "segment" method, in which all variables from someone have to be equal to zero. In such a way, the approximative universal equation is obtained. Having the above procedure in mind, the parameters $f_{1,0}, f_{0,1}, \lambda_{0,0}$ will be remained, while all others will be let to be equal to zero. Also, the derivative with respect to the first porous parameter $\lambda_{0,0}$ will be considered as equal to zero. The equation (6) in these approximation i.e. in three parametric "once localized" environment, has the form:

$$\begin{aligned}
&B^2\Phi_{\eta\eta} + 0.5\left[a_0B^2 + (2-b_0)f_{1,0}\right]\Phi\Phi_{\eta\eta} + f_{1,0}\left(1-\Phi_\eta^2\right)\\
&\quad + f_{0,1}(1-\Phi_\eta) + \left(0.5\eta \overset{**}{T} + B\lambda_{0,0}\right)\Phi_{\eta\eta}\\
&\quad = \eta B^{-1}\left[\overset{**}{T}\left(f_{1,0}B_{f_{1,0}} + f_{0,1}F_{f_{0,1}}\right) - f_{0,1}^2 B_{f_{0,1}}\right]\Phi_{\eta\eta}\\
&\quad + \left[\overset{**}{T}\left(f_{1,0}\Phi_{\eta f_{1,0}} + f_{0,1}\Phi_{\eta f_{0,1}}\right) - f_{0,1}^2\Phi_{\eta f_{0,1}}\right.\\
&\quad + f_{1,0}\overset{**}{F}\left(\Phi_\eta\Phi_{\eta f_{1,0}} - \Phi_{f_{1,0}}\Phi_{\eta\eta}\right)\\
&\quad \left. + f_{0,1}\left(\overset{**}{F} - f_{1,0}\right)\left(\Phi_\eta\Phi_{\eta f_{0,1}} - \Phi_{f_{0,1}}\Phi_{\eta\eta}\right)\right],
\end{aligned} \tag{11}$$

and the corresponding boundary conditions (7) are reduced to the following:

$$\begin{aligned}
&\eta = 0 : \Phi = \Phi_\eta = 0; \quad \eta \to \infty : \Phi_\eta \to 1;\\
&f_{1,0} = f_{0,1} = \lambda_{0,0} = 0 : \Phi = \Phi_0(\eta),
\end{aligned} \tag{12}$$

where the functions $\overset{**}{T}$ and $\overset{**}{F}$, after same approximation have the following forms:

$$\begin{aligned}
\overset{**}{T} =& \left\{2\left[2\left(f_{1,0}\overset{**}{H}_{1f_{1,0}} + f_{0,1}\overset{**}{H}_{1f_{0,1}}\right) + \overset{**}{H}_1\right]\right.\\
&\cdot\left[\zeta - 2f_{1,0} - \overset{**}{H}\left(f_{1,0} + f_{0,1}\right) - \lambda_{0,0} + f_{0,1}^2\overset{**}{H}_{f_{1,0}}\right]\\
&\left. + 2f_{0,1}\left(2 - f_{1,0}\overset{**}{H}_{1f_{0,1}} - f_{0,1}\overset{**}{H}_{f_{0,1}}\right) + 6\overset{**}{H}_1 f_{1,0} - 4\alpha\right\}\\
&\cdot\left\{\left[\overset{**}{H} + 2\left(f_{1,0}\overset{**}{H}_{f_{1,0}} + f_{0,1}\overset{**}{H}_{f_{0,1}}\right)\right]\right.\\
&\left.\cdot\left[2\left(f_{1,0}\overset{**}{H}_{1f_{1,0}} + f_{0,1}\overset{**}{H}_{1f_{0,1}}\right) + \overset{**}{H}_1 - 1\right] - 1\right\}^{-1};\\
\overset{**}{F} =& 2\left\{\zeta - 2f_{1,0} - \overset{**}{H}\left(f_{1,0} + f_{0,1} + 0.5\overset{**}{T}\right) - \lambda_{0,0}\right.\\
&\left. + f_{0,1}\overset{**}{H}_{f_{0,1}}\left(f_{0,1} - \overset{**}{T}\right) - \overset{**}{T} f_{1,0}\overset{**}{H}_{f_{1,0}}\right\},
\end{aligned} \tag{13}$$

where is:

$$\overset{**}{H} = B^{-1} \int_0^{\infty} (1 - \Phi_\eta)\, d\eta = A/B\,;$$

$$\overset{**}{H}_1 = B^{-1} \int_0^{\infty} \Phi_\eta \left(1 - \Phi_\eta^2\right) d\eta\,;$$

$$\zeta = B\,(\Phi_{\eta\eta})_{\eta=0}\ ;\ \alpha = B \int_0^{\infty} \Phi_{\eta\eta}^2\, d\eta\,.$$

The numerical integration of the equation (11) with boundary conditions (12) has been performed by means of the difference schemes and by using Tridiagonal Algorithm method with iterations.The obtained results can be used in the withdrawing of general conclusions of boundary layer development and in calculation of particular problems.

4 Unsteady boundary layer on porous wing aerofoil

Universal solutions of the equation (11) $\Phi''(0)$, A, B are used to calculate the characteristic properties of unsteady boundary layer on wing aerofoil whose center velocity changes with time as a degree function. Substituting nondimensional coordinates: $\tilde{x} = x/\ell$ and $\tilde{t} = \tilde{U}t/1$, where is ℓ - chord and U_∞ - endlessly velocity, nondimensional potential external velocity seems

$$\tilde{U}\left(\tilde{x},\tilde{t}\right) = \tilde{U}_1\left(\tilde{t}\right)\tilde{U}_2\left(\tilde{x}\right) = \left(\tilde{B} + \tilde{A}\tilde{t}^n\right)\tilde{U}_2\left(\tilde{x}\right) \tag{14}$$

with constant values for $\tilde{A}, \tilde{B}, n$. The Figure 5 shows potential external velocity $\tilde{U}_2\left(\tilde{x}\right) = U/U_\infty$ on wing aerofoil measured by J. Stueper in free flight [10], where is lift coefficient $c_1 = 0.4$, Reynolds number $R = 4.10^6$ and chord $\ell = 1800\,\text{mm}$. Substituting (14) in (4),(5) yields the following relations for the universal functions:

$$\begin{aligned} &f_{1,0}/B^2 = a_0\tilde{U}^{-b_0}\tilde{U}_x Q\,;\ f_{0,1}/B^2 = a_0\tilde{U}^{-(b_0-1)}\tilde{U}_t Q\,; \\ &\lambda_{0,0}/B = -v_w\left(a_0 Q/\nu\tilde{U}^{b_0}\right)^{1/2}\ ;\ Q = \int_0^{\tilde{x}} \tilde{U}^{b_0-1}\, d\tilde{x}\,. \end{aligned} \tag{15}$$

Using (3) and (13) the expression for the dimensionless skin friction $\tilde{\tau}_w$ has the form

$$\tilde{\tau}_w = \frac{2\tau_w R^{1/2}}{\rho U_\infty^2} = 2\tilde{U}^{b/2+1}\left(a_0 \int_0^{\tilde{x}} \tilde{U}^{b_0-1}\, d\tilde{x}\right)^{-1/2} \Phi''(0)\,. \tag{16}$$

With obtained universal quantities A and B one can determine on a same previous way the expressions for dimensionless displacement thickness $\tilde{\delta}^*$ and dimensionless momentum thickness $\tilde{\delta}^{**}$. Now, we select a given set of the constants $\tilde{A}, \tilde{B}, n$ and for particular point on contour $\tilde{x}_0$ and time $\tilde{t}_0$ searching by (15) the obtained universal functions $\left(f_{1,0}/B^2\right)_0, \left(f_{0,1}/B^2\right)_0, \left(\lambda_{0,0}/B\right)_0$, concerning $\left[\Phi''(0)\right]_0, A_0, B_0$, for different values of porous parameter $\lambda_{0,0}$. Afterwards, using (16) one can determine $\tilde{\tau}_w$ distribution on contour. Preliminary calculations of expressions (15) and (16) have been made for: $\tilde{A} = \pm 1$; $\tilde{B} = 1$; $n = 1, 2, 3$; $\tilde{t} = 0.0, 0.1$; $\lambda_{0,0} = 0.0, \pm 0.2$. For all n values sufficient universal quantities could be found to cover the contour of wing aerofoil, and it means that $\tilde{\tau}_w, \tilde{\delta}^*, \tilde{\delta}^{**}$ can be calculated for all variations of contour accelerating and decelerating through the fluid. It was not the case in reference [9] in which there were no porous parameter, and where the universal solutions are obtained using the approximative momentum and energy equation, so it was reason why we considered only very slow cylinder accelerating and decelerating.

For determination of the nondimensional streamwise velocity distribution $\tilde{u}\left(\tilde{x}, \tilde{t}\right)$ in some boundary layer cross-section $\tilde{x}_0$ and for time $\tilde{t}_0$, one can use the relation:

$$\frac{\tilde{u}\left(\tilde{x}_0, \tilde{t}_0\right)}{\tilde{U}\left(\tilde{x}_0, \tilde{t}_0\right)} = \frac{\partial}{\partial \eta} \Phi\left[\eta; f_{1,0}\left(\tilde{x}_0, \tilde{t}_0\right); f_{0,1}\left(\tilde{x}_0, \tilde{t}_0\right) \lambda_{0,0}\left(\tilde{x}_0, \tilde{t}_0\right)\right], \tag{17}$$

where $\partial\Phi/\partial\eta$ is universal solution obtained from numerical integration of equation (11). There are the distributions of velocities in different cross-sections of porous contour, i.e. for $\tilde{x}/\tilde{\ell} = 0.68$; 0.83; (see Fig. 5) when porous parameter is $\lambda = -0.2$; 0.0; 0.2. In diffuser region for both the high accelerating $\left(\tilde{A} = 1; n = 1\right)$ and decelerating flows $\left(\tilde{A} = -1; n = 1\right)$ and also for ejection ($\lambda_{0,0} = 0.2$) and injection ($\lambda_{0,0} = -0.2$), velocity distribution receives the ridge point, so it is beginning the laminar-turbulent transition, as one can see on Fig. 5.

For controlling the boundary layer separation point, shear stress and drag, the expression for nondimensional ejection and injection velocity distribution $\tilde{v}_w$ can be applicable in form:

$$\tilde{v}_w = -\lambda_{0,0} B^{-1} \left(\nu \tilde{U}^{b_0}\right)^{1/2} \left(a_0 \int_0^{\tilde{x}} \tilde{U}^{b_0 - 1}\, d\tilde{x}\right)^{-1/2} \tag{18}$$

obtained by (4) for $k = n = 0$, from whom for wanted separation point $\tilde{x}_0$ in time $\tilde{t}_0$, one can determine the needful value for $\tilde{v}_w$. On that way we can control the boundary layer separation.

Vorticity distributions are calculated using known expression: $\omega_z \approx -(1/2)\partial u/\partial y$, also in cross-section $\tilde{x}/\tilde{\ell} = 0.83$ for $\lambda_{0,0} = -0.2$; 0.0; 0.2, and it is shown on Fig. 5.

5 Conclusions

It's found that for both in confuser and in diffuser contour regions the accelerating flow increases the shear stress and postpones the separation of boundary layer i.e. laminar-turbulent transition section, and vice versa the decelerating flow reduces the shear stress and favours the separation of flow, see Fig. 5. It can be noted that the unsteady parameter has a significant influence on a shear stress distribution and especially on the laminar-turbulent transition location obtained by zero skin friction criteria. When this parameter is increasing $(\tilde{t} = 0.1;\ 0.2)$ the shear stress magnitude is increasing on whole contour and the separation point is removing along the surface. It means, that the acceleration leads to the postponing of the boundary layer in diffuser region from 73.8% for steady flow i.e. $\tilde{t} = 0.0$ to the 78.7% of contour for $\tilde{t} = 0.1$, and it is when there are no fluid injection or ejection through the porous contour. It is important fact, because the achievement of laminar flow on 73.8%–78.7% of contour significantly reduce the contour drag. The decelerating flow $(\tilde{A} = -1)$ favours the occurring of the separation and for steady flow the separation is occurring at lower contour values i.e. on 72.2% of contour for $\tilde{t} = 0.1$, also when there is no fluid injection and ejection. As well as for both the accelerating and decelerating flows, the ejection of fluid increases the shear stress, especially in confuser region about stagnation point, where shear stress is dramatically increased in time. It's not good for drag, so one can control this great shear stress with fluid injection, when his value is noticeably reduction. Also, the ejection of fluid postpones the boundary layer separation, and vice versa the injection of fluid reduces the shear stress and favours the flow separation. The great fluid ejection, i.e. when the porous parameter is $\lambda_{0,0} = 0.2$, leads to the postponing of separation to 87.4% of contour for $\tilde{t} = 0.1$. It is important fact for commercial transport aircraft, because the achievement of laminar flow on 70%–87% of aerofoil contour significantly reduce the drag, and hence, the fuel consumption of the aircraft. The ejection of fluid increases the shear stress, especially in confuser region about stagnation point , i.e. 4.13%–11.37% of contour, for a different values of acceleration, where shear stress is dramatically increased. It is not good for a drag, so one can control this great shear stress with fluid injection, when his value is noticeably reduction. And for a great fluid injection $(\lambda_{0,0} = -0.2)$ the separation is occurring at lower contour values, about 67% of contour for a decelerating flow and for $\tilde{t} = 0.1$. An important advantage of the generalized similarity method demonstrated in this paper is that the velocity and vorticity distributions, skin friction and laminar-turbulent transition with separation point are found directly, no further numerical integration of momentum equation being involved as it was done in references [5–7,10].

References

1. Gersten, K. (1989) Die Bedeutung der Prandtlschen Grenzschichttheorie nach 85 Jahren. *Z. Flugwiss. Weltraumforsch.*, **13**, 209–218
2. Lojcjanski, L. (1965) Universal. uravn. paramet. pridbli'. v teor. lam. pogr. sl., AN SSSR, *Pr. mat. meh.*, **29**, 1, 28–32

3. Saljnikov, V.N. (1978) Contr. univer. solut. bound. layer, *Theoreth. Applied Mech.*, **4**, 139–163
4. Busmarin, O., Basin, B. (1972) Paramet. metod v teor. nest. pogr. sl., *In'. Fiz. 'ur.*, **22**, 2, 282–292
5. Busmarin, O., Saraev, Á. (1974) Paramet. metod v teor. nest. pogr. sl., *In'. Fiz. 'ur.*, **27**, 1, 110–118
6. Boricic, Z., Nikodijevic, D., Milenkovic., D. (1995) Unsteady MHD bound. lay. on porous surface, *Facta Universitates*, **1**, 5, 631–645
7. Boricic, Z., Nikodijevic, D. (1997) Gen. sim. solut. in theory of unst. MHD boun. lay. on porous surface, *XXII YUCTAM*, Vrnjacka Banja, 7–12
8. Saljnikov, V., Ivanovic, D. (1988) Grenzsch. an porösen Wänden mit absaug. bzw. ausbl., *Z. angew. Math. Mech.*, **68**, 5, 349–352
9. Saljnikov, V., Ivanovic, D. (1994) Verall. Ahnlich. 2-D lam. instat. inkompr. Grenzsch. an profil. Zylindern, *Z. angew. Math. Mech.*, **74**, 5, 391–394
10. Schlichting, H. (1964) *Grenzschicht-Theorie*, Braun-Verlag, Karlsruhe

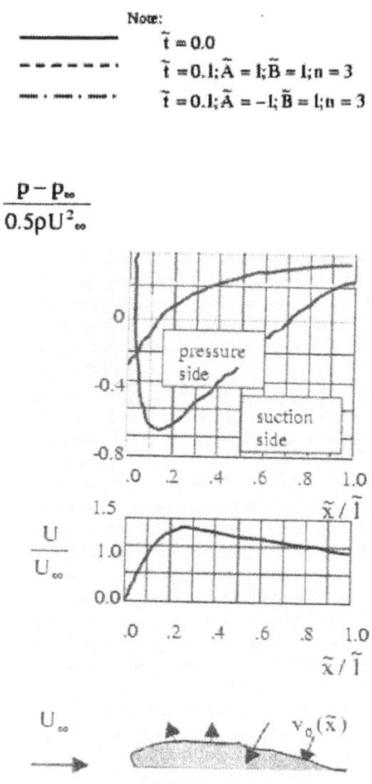

Fig. 1. Potential external velocity on wing aerofoil [10]

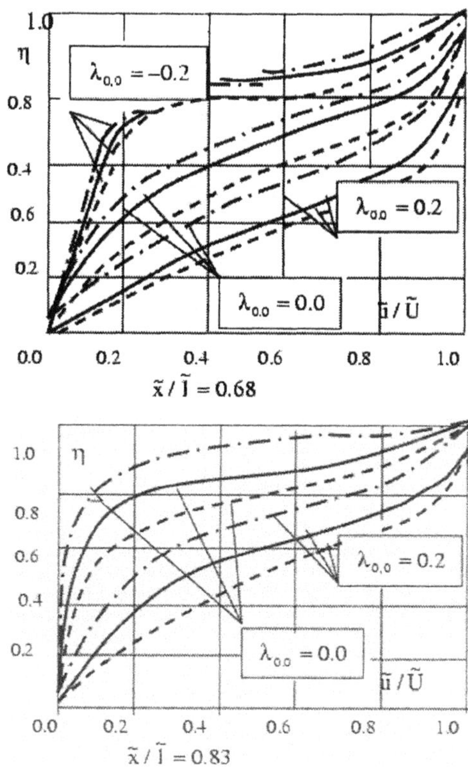

Fig. 2. Velocity distribution

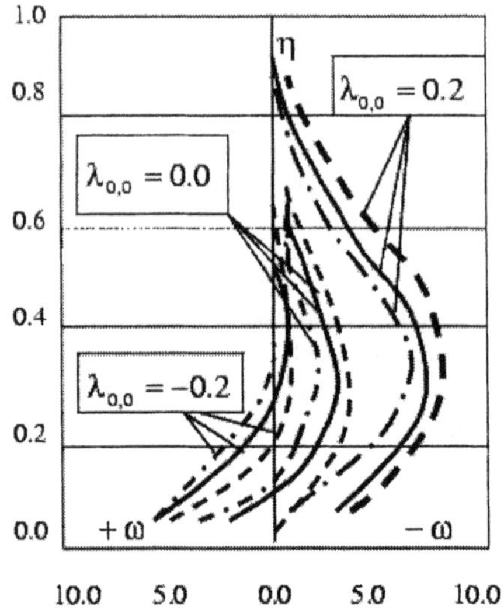

Fig. 3. Vorticity distribution

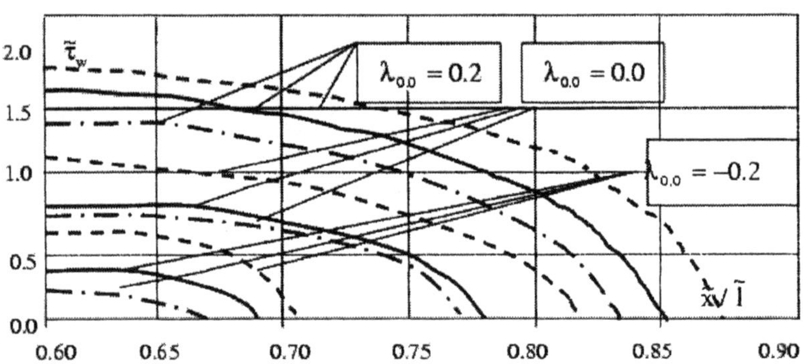

Fig. 4. Shear stress distribution

Modeling of Heterogeneities for Twophase Flow in the Subsurface

Hartmut Jakobs, Andreas Bielinski, Holger Class, and Rainer Helmig

Stuttgart University, Stuttgart, Germany

Abstract. We discuss the influence of heterogeneities on multiphase flow, focussing on the numerical handling of critical effects at the interfaces between two domains with different soil properties and capillary pressures. An example is given, where we show the different results obtained with different numerical methods applied to a 1D heterogeneous two-phase system.

1 Introduction

In the last few decades numerical models for the simulation of multiphase flow and transport processes in porous media became a field of intensive research. These models first came up in the petroleum industry, for example in order to help optimize tertiary oil production by steam flooding, e.g. [1,10,14]. Also an increasing number of problems in environmental engineering takes advantage of sophisticated numerical multiphase and multiphase multicomponent models. A good overview of the mathematical system descriptions, different numerical techniques and codes give for example [5,11,8]. Groundwater contamination as a result of human activity is a major problem in many industrialized countries. The use of hazardous materials like oils, petrochemicals, solvents and many other organic substances makes accidental releases of these materials to the environment unavoidable. The development and the optimization of in-situ remediation techniques for the removal of soil contamination sources is an important task for ensuring high groundwater quality. Within this context, the numerical simulation of the physical processes during such remediations is a very helpful tool, e.g., to get an improved understanding of the coupled processes and to identify significant mechanisms, but also to reduce the number of costly laboratory and field experiments. Furthermore, the modeling of the spreading behavior of contaminants in the subsurface can help in the risk assessment process. For analyzing the spreading of the organic compounds and for the choice of a suitable remediation technique, the correct handling of the influence of heterogeneities plays an important role. Depending on the observed scales this influence must be taken into account to the selection of a relevant model.

2 Governing Equations

We use the multiphase flow differential equation as derived by e.g. [8] with an equation for each phase as follows:

$$\frac{\partial(\phi S_\alpha \varrho_\alpha)}{\partial t} - \nabla \cdot \left(\varrho_\alpha \frac{k_{r\alpha}}{\mu_\alpha} \mathbf{K} \cdot (\nabla p_\alpha - \varrho_\alpha \mathbf{g}) \right) - q_\alpha = 0 \,. \tag{1}$$

where $\mathbf{K}$ is the tensor of absolute permeability, $\mathbf{g}$ the gravitation vector. p as the pressure, μ as the dynamic viscosity, ρ asthe density, and k_r as relative permeability are defined with regard to the phase α with α as the wetting phase we or the non–wetting phase nw.

Additional supplementary constraints are that the saturation must add up to one ($\sum_\alpha S_\alpha = 1$) and that the pressures are calculated according to their order of wettability by saturation-dependent capillary pressure functions.

$$p_c = p_{nw} - p_{we} = f(S) \,, \tag{2}$$

Say, we consider a two-phase system. The fluid phase with the highest affinity to the porous medium is the wetting phase we (normally: water), the second phase is the non-wetting phase nw (e.g., gas or NAPL[1]). Additional state relations for the density $\varrho(p)$, the viscosity $\mu(p)$ and the relative permeability $k_r(S)$ complete the set of equations in case of isothermal systems. The primary variables can be chosen in different ways. One is to use the pressure–saturation formulation, where the two unknowns are the pressure of the wetting phase p_{we} and the saturation of the non-wetting phase S_{nw}, respectively vice versa p_{nw} and S_{we}.

3 Discretization Methods

For the space discretization of the equations, we use a Subdomain Collocation Method (BOX) [8] with fully upwinding of the mobility term, for time discretization a fully implicit Euler method. The nonlinearities in the equations are handled with a Newton–Raphson method, where the elements of the Jacobian matrix are computed by numerical differentiation. The Jacobian system is to be solved by a linear solver. As linear solver we use the Bi-Conjugate Gradient Stabilized solver (Bi–CGSTAB) (e.g. [2]) with a preconditioner based on a multigrid technique [3].

4 Capillary Pressure at Interfaces

Figure 1 shows an example, where DNAPL[2] enters from the top into a fully water-saturated column containing a fine sand lense embedded into ambient

[1] Non-Aqueous Phase Liquid

[2] NAPL with higher density than water

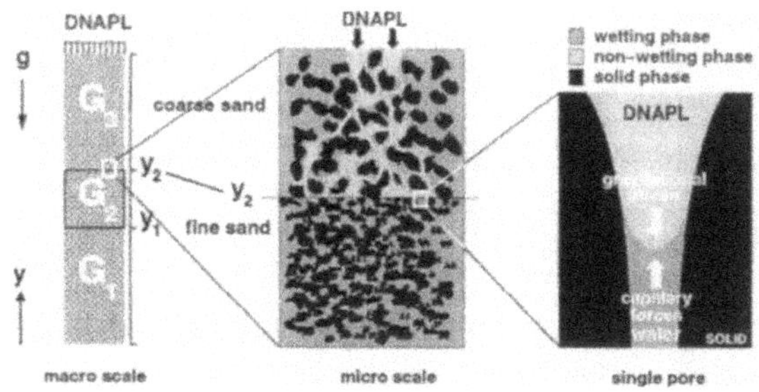

Fig. 1. 1D heterogeneous problem. DNAPL entering from the top into a water-saturated domain.

coarse sand. We devide the domain into three subdomains G_1, G_2, G_3 with the interfaces y_1 and y_2. G_1 and G_3 are coarse sand with identical parameters, G_2 is the fine sand lense. Now consider the saturation S_{we} and the capillary pressure p_c above y_2 $(\lim_{y\to y_2+0} S_{we}, p_c)$, and below y_2 $(\lim_{y\to y_2-0} S_{we}, p_c)$.

We draw our attention to that moment when the DNAPL reaches the fine sand lense, i.e. y_2. On the microscale, the entering of the DNAPL is influenced by two kinds of forces. These are gravitational forces, resulting from the density difference between the fluids, and capillary forces, resulting from different adhesion of the fluids to the solid (see Fig. 1 on the right).

However, the multiphase equations on the macroscale show that the saturation of DNAPL at the interface above y_2 is only defined by the capillary pressure. We first investigate van Genuchten-like types of capillary pressure functions, i.e. we assume $p_c(S_{we} = 1) = 0$ (see Fig. 2 on the left). Looking at the Darcy velocity for the DNAPL (non-wetting phase)

$$\mathbf{v}_{nw} = -\frac{k_{r,nw}}{\mu_{nw}} \mathbf{K} \cdot (\nabla p_{nw} + \nabla p_c - \varrho_{nw} \mathbf{g}) \,, \tag{3}$$

one can see that there can't be any jumps in capillary pressure at any value of y, as otherwise we would have $|\nabla p_c(y)| \to \infty$, thus producing a non-definable flux. So we postulate the capillary equilibrium condition for interfaces:

$$\lim_{y\to y_2+0} pc(y) = \lim_{y\to y_2-0} pc(y). \tag{4}$$

Since we have different capillary pressure curves for G_3 and G_2, we can conclude that there is a jump in saturation at the interface y_2. We can compute $\lim_{y\to y_2-0} S_{we}$ via the inverse capillary pressure–saturation function $Inv_{pc}^{G_2}(pc)$ of domain G_2:

$$\lim_{y\to y_2-0} S_{we} = Inv_{pc}^{G_2}(pc^{G3}(\lim_{y\to y_2+0} S_{we})). \tag{5}$$

Analogous considerations for Brooks-Corey-like capillary pressure–saturation functions (i.e. functions with $p_c(S_{we} = 1) = p_e > 0$, see

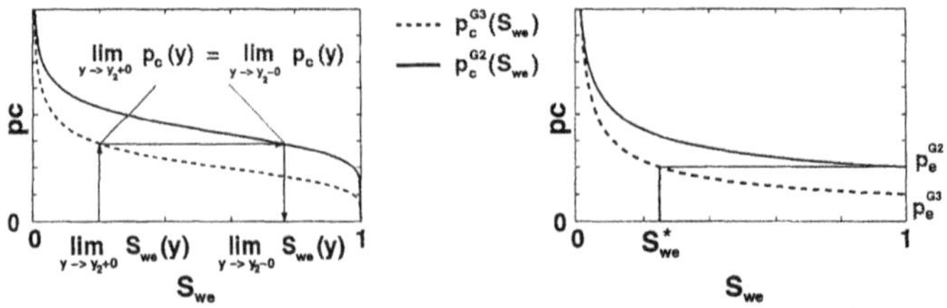

Fig. 2. Capillary pressure equilibrium condition for van Genuchten-like (left) and Brooks-Corey-like (right) type of $p_c(S_{we})$–functions

Fig. 2 on the right) can be carried out. For the entry pressures, we can state that $p_e^{G_3} < p_e^{G_2}$. Having fully water-saturated initial conditions, the capillary pressure gradient at y_2 can not be described, since $\lim_{y\to y_2+0} p_c(y) < \lim_{y\to y_2-0} p_c(y)$. For wetting-phase saturations higher than $\lim_{y\to y_2+0} S_{we}(y) = Inv_{pc}^{G_3}(p_e^{G_2})$, we get for the mobility $\lambda_{nw}(y_2) = \frac{k_{r,nw}}{\mu_{nw}} = 0$ and thus $v_{nw} = 0$. An extended capillary pressure equilibrium condition at heterogeneities as defined by [7] is illustrated in Fig. 2 on the right. Accordingly, we define the threshold saturation S_{we}^* as the saturation which must be reached in G_3 to overcome the entry pressure of G_2 as seen in Fig. 2 on the right: $p_c^{G_3}(S_{we}) = p_e^{G_2}$. So for $\lim_{y\to y_2+0} S_{we} > S_{we}^*$, we have $\lim_{y\to y_2-0} S_{we} = 1$; for $\lim_{y\to y_2+0} S_{we} \leqslant S_{we}^*$, the condition of continous capillary pressure (5) is still valid.

A further possibility to describe discontinuos saturations at the interface is to use the capillary pressure p_c as kind of a quasi-primary variable at the interface. We call this method a Phase Pressure Saturation Formulation with Interface Condition (PPSIC). $S_{we}^{G_2}$ can be recalculated at the interface using the extended capillary pressure condition, i.e.
$\texttt{if}(S_{we}^{G_3} < S_{we}^* \Rightarrow S_{we}^{G_2} = 1)$, $\texttt{else}\ S_{we}^{G_2} = Inv_{pc}^{G_2}(pc^{G_3}(S_{we}^{G_3}))$.

This ensures that the non-wetting phase does not enter into subdomain G_2 until the threshold saturation has been reached. After overcoming S_{we}^* the continuity of capillary pressure at the interface is guaranteed. We refer to [4] for a detailed discussion of this method and to [12] and [13] for its validation on the basis of a theoretical solution for a self-similar problem with a discontinuity. Implementation of the interface condition into the numerical discretization requires an element-wise evaluation of the parameters.

5 Numerical Example

We applied the PPS and the PPSIC schemes to the simple system of Fig. 1. DNAPL spills from the top into the fully water-saturated system with $q_{nw} = 0.05\ [kg/(m^2s)]$. The parameters are those for coarse and medium sand carried out in the VEGAS facility in Stuttgart, Germany, described in detail in [6]. The threshold saturation derived from these relationships is $S_{we}^* \approx 0.13$. The element length for the numerical computation is $\Delta h = 0.25/2^3[m]$.

As mentioned above, we use an fully implicit Euler scheme as time discretization. The time step for the fully implicit Euler scheme is controlled by the numerical convergence of the nonlinear solver, a newton solver combined with a line search. We gave a start and maximum value of $\Delta t_{start} = 80.0$ $[s]$ and a very small minimum time step $\Delta t_{start} = 10.0E - 8$ $[s]$. The simulations run until DNAPL reaches the bottom of the column which occurs slightly after $6000[s]$.

We show the DNAPL front $S_n(y)$ at five time steps in Figs. 3. The five time steps were chosen according to location of the DNAPL front in relation to the block heterogeneities so that we show the DNAPL front 1) in G_3, 2) at y_2, 3) crossing y_2 4) crossing y_1, and 5) in G_1.

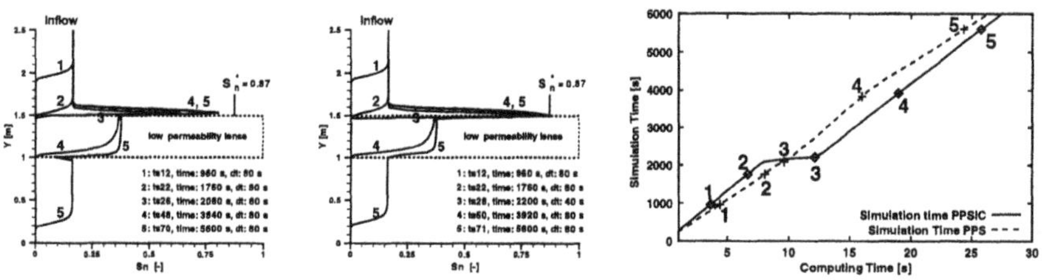

Fig. 3. DNAPL saturation at different times; $\Delta h = 0.25[m]/2^3$; left: without Interface Condition (PPS); middle: with Interface Condition (PPSIC); right: simulated time over computing time, numbers refer to timesteps shown on the left.

In Fig. 3 on the right, we show the simulation time over the computing time. Thus we make sure to eliminate the effects of different sizes within the time steps. The gradient of the curves is a measure for the efficiency of the algorithm. We can see that the infiltration of the DNAPL into G_2 has much more impact on the efficiency while using the PPSIC method in comparison to the PPS method. However, as we can see in Fig. 3 comparing the two plots of the DNAPL front the threshold saturation is fully reproduced by the PPSIC method, but not by the PPS method.

6 Conclusions

The PPSIC-based method, which explicitly enforces the extended capillary equilibrium condition, performs not as good numerically, but more accurately represents the DNAPL saturation discontinuities at block heterogeneities in comparison to the PPS model, which relies on full upwinding. The analysis of more than those runs displayed in this section made it evident that the ratio of simulation time over computing time for PPSIC method worsens compared to the PPS method if 1.) the viscosity μ_n decreases, 2.) the capillary pressure–saturation function is steeper at S^*_{we}, 3.) the relative permeability function for the wetting phase is steeper at $S_w = 1$. Knowing this, we could also construct cases (not based on realistic constitutive relationships) where the PPSIC method was more effective than the PPS method. We attribute some of this behavior to the inexactness of the used Newton method. As a rough rule of thumb we conclude that the ratio of simulation time over computing time is the better for the PPS method the bigger the aberration for the infiltration saturation between PPS method and PPSIC method.

7 Acknowledgement

Our efforts were supported by the Federal Ministry of Economics and Technology (BMWi) under the identification number 02E9370 and the German Research Foundation SFB 404.

References

1. K. Aziz and A. Settari. *Petroleum Reservoir Simulation.* Applied Science Publishers, London, 1979.
2. R. Barrett, M. Berry, T. F. Chan, J. Demmel, J. Donato, J. Dongarra, V. Eijkhout, R. Pozo, C. Romine and H. Van der Vorst. *Templates for the Solution of Linear Systems: Building Blocks for Iterative Methods, 2nd Edition.* SIAM, Philadelphia, PA, 1994.
3. P. Bastian. *Parallele adaptive Mehrgitterverfahren.* Teubner–Verlag, 1996.
4. P. Bastian. *Numerical Computation of Multiphase Flows in Porous Media.* Habilitation thesis, Christian-Albrechts-Universität Kiel, 1999.
5. J. Bear. *Dynamics of Fluids in Porous Media.* Elsevier, New York, 1972.
6. J. Braun. Ausbreitung von NAPL in gesättigten und ungesättigten Böden *VEGAS-Workshop und BMBF/PWAB-Seminar In-Situ Technologien zur Grundwasser- und Altlastensanierung,* University of Stuttgart, 1996.
7. C. van Duijn, J. Molenaar, and M. de Neef, Effects of capillary forces on immiscible two–phase flow in heterogeneous porous media. *Transport in Porous Media,* **21**, 71–93, 1995.
8. R. Helmig. *Multiphase Flow and Transport Processes in the Subsurface - A Contribution to the Modeling of Hydrosystems.* Springer Verlag, 1997.
9. R. Helmig and R. Huber, Comparison of galerkin-type discretization techniques for two–phase flow in heterogeneous porous media. *Advances in Water Resources,* **21**, 697–711, 1998.
10. L.W. Lake. *Enhanced Oil Recovery.* Prentice-Hall, Inc., Englewood Cliffs, New Jersey, 1989.
11. B.B. Looney and R.W. Falta. *Vadose Zone.* Battelle Press, Columbus, OH, 2000.
12. M. de Neef, *Modelling capillary effects in heterogeneous porous media.* Ph.D. thesis, Technische Universiteit Delft, 2000.
13. M. de Neef and J. Molenaar, Analysis of dnapl infiltration in a medium with a low–permeable lens. *Computational Geosciences,* **1**, 191–214, 1997.
14. D.W. Peaceman. *Fundamentals of Numerical Reservoir Simulation.* Elsevier, New York, 1977.

Low-Dimensional Models of Flow Over a Body of Revolution at High Incidence

Yuli Lifshitz and David Degani

Faculty of Mechanical Engineering
Technion - Israel Institute of Technology
Haifa 32000, Israel

Abstract. The asymmetric part of flow over a body of revolution at high angle of attack is considered as coherent structures of the flow and a method of Proper Orthogonal Decomposition (POD) is applied to construct its low-dimensional model in the form of superposition of orthogonal functions. Results of the analysis show that the flow can be represented in the form of linear combination of odd and even functions, and the first two terms of the expansion capture 98.5% of its kinetic energy. Their coefficients are determined by the value of side force and the increment of normal force. The transient behavior of the flow is studied using Fourier transformation.

1 Introduction

The flow around slender body of revolution is often chosen for basic research because of simplicity of the geometry of the body and its similarity to the shape of missiles or aircraft forebodies. However, despite the simplicity of the geometry, the flow developed around a body of revolution at high incidence is highly complex. High–angle–of–attack flow is governed by large region of three–dimensional separated flow where the boundary layers leave the surface of the body along lines of separation and roll up on the leeward side of the body to form strong concentrated vortical flows. Typical sharp–nosed body experiments [1–3] show that, when the angle of attack is kept below $\alpha = 20°$, the mean flow and its wake remains symmetric. When the angle of attack is increased, the mean flow becomes asymmetric. For angle of attack from 30° to 65°, just minute imperfections at the tip (e.g. dust accumulation, surface roughness) can cause large asymmetries in the flow patterns. For angle of attack beyond 65°, the crossflow past the cylindrical part of the body becomes virtually identical to the flow past a two–dimensional cylinder and vortex shedding occurs.

The fact that minute perturbations of body shape can result in finite asymmetries in the flow pattern suggests the existence of inherent instabilities in the symmetric flow. The flow instability for angles of attack between 30° and 65° has been studied numerically, both for laminar [4,5] and turbulent flows [6]. However, these 3D time–accurate computations are very costly and often cannot be used as a design tool for industrial applications. The purpose of this study is to develop for such flows a less expensive tool based on low–dimensional modeling of large systems.

The Karhunen–Loeve expansion also known as the POD or the method of empirical eigenfunctions is widely used to generate low dimensional systems, which have the same behavior as some large scale simulations of partial differential equations (PDEs). The principal idea behind the Karhunen–Loeve method is that, given an ensemble of data, one can find a basis of a given dimension that spans the data optimally in the L_2 sense. After Sirovich's papers with snapshot version of POD [7] in order to analysis large data set with less computational effort, the method became a very useful tool in contemporary fluid dynamics. Holmes et al. [8] provided a thorough review in the important aspects of the POD as applied to the analysis of turbulent flows. The POD identifies structure of the flow using spatial velocity correlation tensor obtained experimentally or simulated by numerical methods. In the last decade a large number of its applications and POD–based low–dimensional models have been reported for several flows when both data of experiments and numerical simulation were taken for calculation of the correlation tensor. There are hundreds of publications on the subject, including boundary layers, free shear layers and jets, separation, delta wing flow, combustion, turbomachines, and flow control. We refer to the paper [9], which is most closely to the flow over a body of revolution at high angle of attack.

2 Proper Orthogonal Decomposition

Here we give a brief introduction to the snapshot method of the Karhunen-Loeve analyses. This is a well known procedure, which has become a frequently used tool to analyze data, generated experimentally or numerically. The main goal of such data analysis is to determine an optimal orthogonal basis spanning the data available. The snapshot method operates with a dataset U, which is generated as a numerical solution to a nonlinear PDE. Assuming that we have ensemble consisting of M datavectors of length N, i.e., $U = \{u^m, m = 1, \ldots, M\}$, (here N reflects the size of the mesh used in the space and M is a number of the time values where the solution is taken) we represent this ensemble as the matrix of column vectors $u^m = [u_1, u_2, \ldots, u_N]^m, m = 1, \ldots, M$ and assume that the average of the ensemble, defined as

$$< \mathbf{u}^m >= \frac{1}{M}\sum_{m=1}^{M} \mathbf{u}^m,$$

is zero. If we denote an inner product between vectors $(\mathbf{v}, \mathbf{u})$ then, the Karhunen–Loeve theory shows that it is possible to find an orthonormal coordinate system $\Phi = \{\varphi^k, k = 1, \ldots, N\}$ that is optimal in the sense that the variance of the dataset in the coordinate direction becomes maximal, i.e.,

$$< (\varphi^k, \mathbf{u}^m)^2 >= \max_{\varphi} < (\varphi, \mathbf{u}^m)^2 >, \quad \| \varphi^k \|= 1. \tag{1}$$

The coordinate vectors φ^k are eigenvectors of the covariance matrix

$$\mathbf{B}_{i,j} = \frac{1}{M} \sum_{m=1}^{M} u_i^m u_j^m. \tag{2}$$

The eigenvectors φ^k are often called the empirical eigenfunctions or the POD modes. They have the form a complete orthonormal set (when properly normalized) and have eigenvalues $\lambda_k =< (\varphi^k, \mathbf{u})^2 >$ that quantify the probability of their occurrence in the flow. The sum of these eigenvalues gives the total "energy" $E = \sum_{k=1}^{N} \lambda_k$ of the system. Determination of the eigenvectors φ^k is the solution an $N \times N$ eigenvalue problem. This is called the direct method. If the data $\mathbf{u}^m$ are linearly independent and if $M \ll N$ then it is convenient to look for a representation of the POD modes as linear combinations of the data

$$\varphi^k = \sum_{m=1}^{M} \mathbf{u}^m v_m^k = \mathbf{U}\mathbf{v}^k. \tag{3}$$

Collecting the set of vectors φ^k in a matrix of column vectors $\mathbf{\Phi} = [\varphi^1, \ldots, \varphi^M]$, we can write the linear combination as $\mathbf{\Phi} = \mathbf{UV}$. In this case the columns of the coefficient matrix $\mathbf{V} = [\mathbf{v}^1, \ldots, \mathbf{v}^M]$ are the eigenvectors of the symmetric $M \times M$ matrix $\mathbf{K}$ with

$$\mathbf{K}_{i,j} = \frac{1}{M} (\mathbf{u}^i, \mathbf{u}^j). \tag{4}$$

The latter method is called the snapshot method or the sample method, and we refer to $\mathbf{K}$ as the sample covariance matrix.

One of applications of the Karhunen–Loeve analysis is to generate a low dimensional systems of ordinary differential equations (ODEs) representing the dynamics of the PDE via Galerkin projection onto the optimal POD modes. For L–dimensional Galerkin projection, the numerical solution of a PDE represented by an N–dimensional vector $\mathbf{u}(t)$ is expanded into $\mathbf{u}(t) = \sum_{k=1}^{L} z_k(t)\varphi^k$. In the event that the eigenvalues of the covariance matrix decay fast, one can decide on a cutoff $L \ll N$ and in that way to generate an approximate system of ODEs that is much smaller than the original numerical simulation for the PDE.

3 Slender Body of Revolution

In this paper we consider the flow over a slender body of revolution of $10D$ length with a $3.5D$ sharp ogive forebody, where D is the diameter of the cylindrical part of the body, at $M_\infty = 0.12$, angle of attack $\alpha = 55°$, and Reynolds number $Re_D = 6000$. The goal is to construct a low–dimensional model of the asymmetric part of the flow, using the numerical simulation, to obtain the detailed database for flow over this body of revolution.

3.1 Governing equation and numerical algorithm

The thin–layer approximation of the conservation equations of mass, momentum, and energy can be represented in a curvilinear coordinates as

$$\frac{\partial \hat{Q}}{\partial \tau} + \frac{\partial \hat{F}}{\partial \xi} + \frac{\partial \hat{G}}{\partial \eta} + \frac{\partial \hat{H}}{\partial \zeta} = \frac{1}{Re}\frac{\partial \hat{S}}{\partial \zeta} \tag{5}$$

where τ is time and the independent spatial variables ξ, η, and ζ are chosen to map a curvilinear body–conforming grid into a uniform computational space. The coordinate ξ is directed along the body, η characterizes the circumferential angle, and ζ is the coordinate leading away from the surface. In Eq. (5) $\hat{Q}$ is the vector of independent flow variables $\hat{F} = \hat{F}(\hat{Q})$, $\hat{G} = \hat{G}(\hat{Q})$ and $\hat{H} = \hat{H}(\hat{Q})$ are the inviscid flux vectors and $\hat{S}$ multiplied by Re^{-1} is the vector of viscous flux terms. The implicit scheme employed for calculations was reported by Steger et al. [10]. The algorithm uses flux-vector splitting technique, upwind spatial differencing for the convection terms in the streamwise direction and central differencing in the other directions. This leads to two–factor implicit approximately factored scheme. The flow at the outer boundary of computational grid is assumed to be the undisturbed freestream. A no–slip condition is set at the body surface and a simple zero axial–gradient extrapolation condition is applied at the downstream end of the computational domain. An implicit periodic condition is imposed in the circumferential edge of the grid.

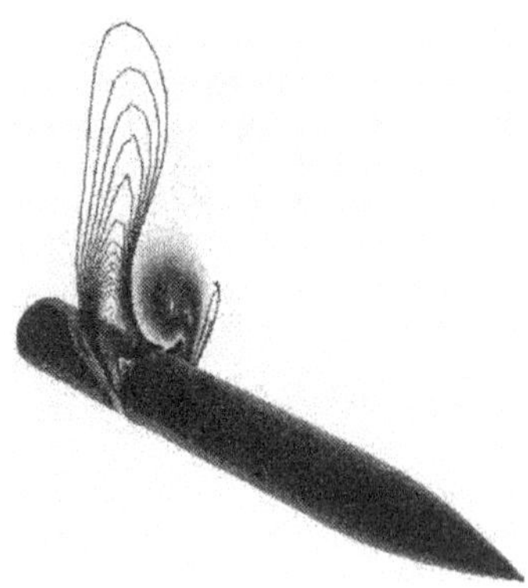

Fig. 1. Body of revolution and helicity density contours in crossflow plane

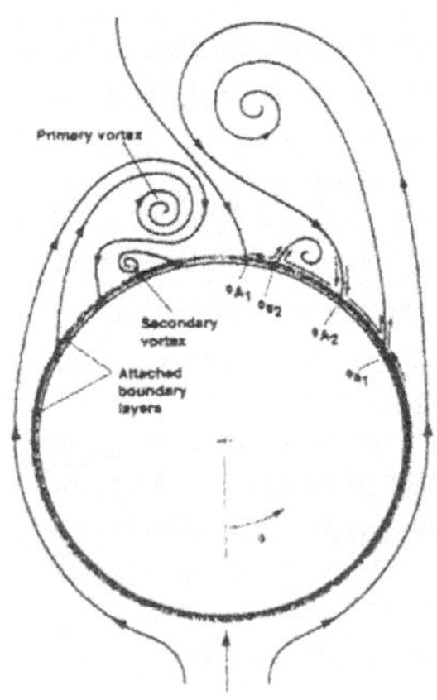

Fig. 2. Schematic of flow structure in crossflow plane

Figure 1 shows typical computations in the form of helicity density contours, H_d, in a crossflow plane ($H_d = \mathbf{V} \cdot \omega$, where $\mathbf{V}$ and ω are the velocity and the vorticity). Figure 2, taken from [11], gives some details of the crossflow pattern. The flow approaching the line of attachment near the windward line of symmetry turns and spreads circumferentially outward and downstream along the body from the windward toward the leeward side under the action of circumferential pressure gradients. Since the main component of the velocity is along the body, a well-defined boundary layer is developed. On the right-hand side, the boundary layer separates from the body at φ_{S1}, the primary separation line, and fluid leaves the body within a shear layer which rolls up to form a primary vortex structure on the leeward side of the body. A similar primary separation occurs on the left-hand side. The primary vortices induce a flow toward the body surface, resulting in a line of attachment at φ_{A1}. The induced downflow turns and runs circumferentially outward and downstream along the body surface toward the region of low pressure generated by the primary vortices. Since the angle of attack of the body is high, and thus, the primary vortices are strong, this induced boundary-layer flow cannot overcome the adverse circumferential pressure gradient it encounters beyond the region of low pressure and it separates at φ_{S2}, the secondary separation line. Once again, the fluid leaves body within a shear layer and rolls up to form the secondary vortex structure. Another line of attachment is located on the body surface at φ_{A2}. Similar behavior occurs on the left-hand side. Under certain conditions a tertiary crossflow separation can occur. The vortex structure itself is essentially inviscid and is controlled by convection of vortical fluid generated within the viscous boundary layer at the body surface rather than by dissipation of vorticity.

3.2 Generation of asymmetric database; POD modes

A time-accurate solution was computed on a grid of $60 \times 100 \times 50$ points, starting from symmetric field. To simulate a symmetry-breaking imperfection, a geometrical disturbance was added near the body apex. A small bumps of the length l = 0.05D were placed at the distance of $0.01D$ from the apex. Five samples of the flow were calculated. In the first one, the bump of the height $h = 0.01D$ was inserted at the circumference angle of 45° at $t = 200$. Here the time t is normalized by a ratio D/a_∞ , where a_∞ is the sonic velocity of the incoming flow. A time variation of side and normal force coefficients is shown on Figure 3. The initially unsteady solution was considered to be converged to a steady state after L_2-norm dropped at least three order of magnitude. The velocity field of such steady state solution can be represented in the form of sum $\mathbf{V} = \mathbf{V}_s + \mathbf{V}_a$, where $\mathbf{V}_s$ is the velocity of symmetric solution while the asymmetric term $\mathbf{V}_a$ is generated by the bump. $\mathbf{V}_a$ was included into the database for further Karhunen-Loeve decomposition. At $t = 1800$ the next bump of height $h = 0.02D$ was set at the circumference angle of 270°, the flow became unsteady again and, after its convergence to a steady state solution, the corresponding field $\mathbf{V}_a$ was included into the database. The procedure was repeated three times for bumps of various height $h = 0.017D$, $0.014D$, and

$0.012D$. Then, the database was doubled according to its symmetry relative to vertical plane.

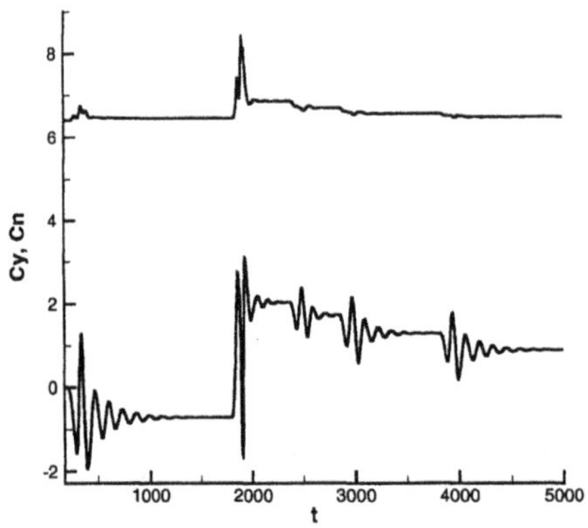

Fig. 3. Time variation of side and normal force coefficients

It is clear from Figure 3, that placement of small bumps near the body apex result in large asymmetry. This confirms our decision to represent the asymmetric part of flow as superposition of coherent structures and the application of the Karhunen–Loeve decomposition for its analysis. The functions in the database are the three Cartesian components vectors on a grid, therefore, the inner product is the integral of the scalar product of two vectors on the whole computational region Ω : $(\mathbf{f}, \mathbf{g}) = \int\limits_{\Omega} (\mathbf{f} \cdot \mathbf{g}) d\Omega$.

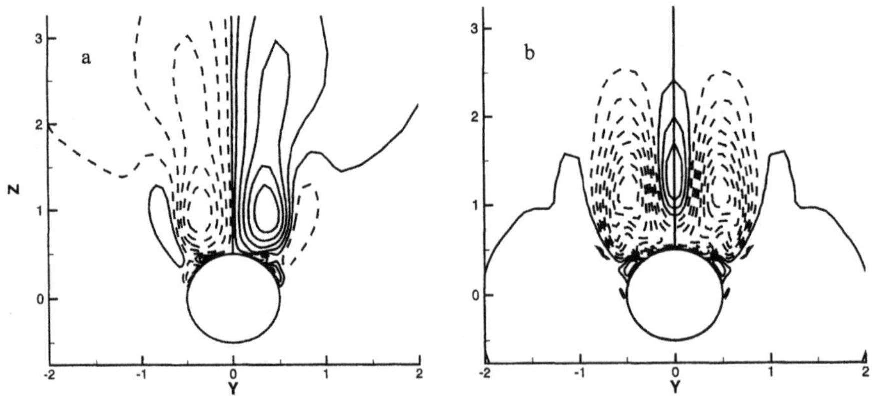

Fig. 4. u–component of additional velocity. (**a**) First POD mode, range of contour lines: -0.065, 0.065. (**b**) Second POD mode, range of contour lines: -0.12, 0.12.

The POD modes resulted from the decomposition are also the vector functions of grid points with three components referred below as u, v, and w as the corresponding components of velocity in the direction of axes of

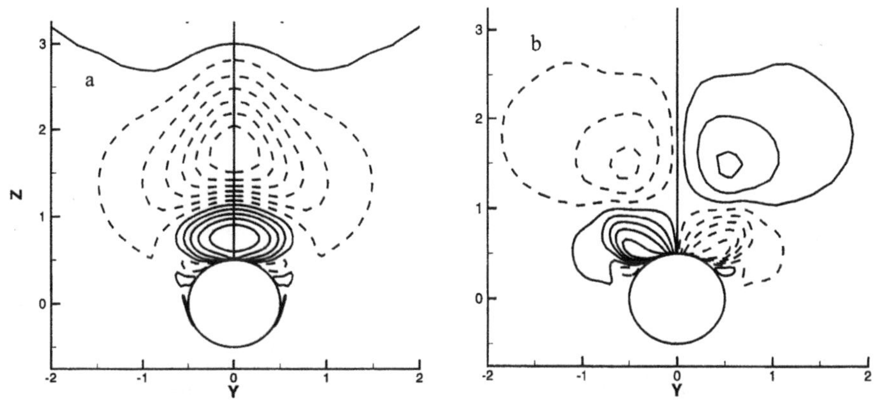

Fig. 5. v–component of additional velocity. (**a**) First POD mode, range of contour lines: −0.1, 0.08. (**b**) Second POD mode, range of contour lines: −0.12, 0.12.

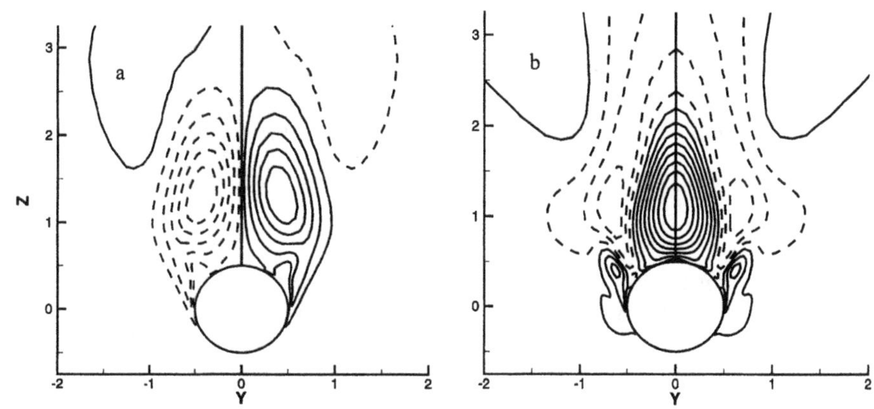

Fig. 6. w–component of additional velocity. (**a**) First POD mode, range of contour lines: −0.12, 0.12. (**b**) Second POD mode, range of contour lines: −0.08, 0.18.

symmetry and the side and vertical direction in the normal cross–section. The odd modes are asymmetric (u and w are the asymmetric functions, and v is the symmetric one) and the even modes are symmetric. The first two modes catch 98.5% of "kinetic energy" $|\mathbf{V}_a|^2$, and the first four modes accumulate 99.95%. We present here the first two POD modes in a cross-section at $8.0D$ from the body apex, the same location as in Figure 1. Figures 4–6 demonstrate the modes in the form of contour lines. The broken lines represent negative values, the solid ones positive values. Range of the contour lines is shown in the figure captions. Figures 4a, 5a, and 6a show u, v, and w components of the first mode. The structure of additional velocity field is well seen in the figures. In particular, we can see the concentration of contour lines in the regions of primary and secondary vortices. The symmetric second mode presented in Figures 4b, 5b, and 6b is the natural supplement to the asymmetric first mode. This allows us to obtain a flow with any degree of asymmetry.

3.3 Transient behavior

If the velocity magnitude of the asymmetric part of the flow is many times smaller than the symmetric one, it is possible to represent the velocity field in the form

$$\mathbf{V} = \mathbf{V}_0(x, y, z) + \epsilon \mathbf{U}(t, x, y, z), \quad \epsilon \ll 1. \tag{6}$$

When substituting equation (6) into the Navier–Stokes equations, we obtain as first approximation a linear equations for the asymmetric flowfield $\partial \mathbf{U}/\partial t = L(\mathbf{U})$ that has a solution in the form of

$$\mathbf{U} = \sum_k \psi_k(x, y, z) e^{(i\omega_k - \beta_k)t}.$$

The frequencies ω_k and damping constants β_k obey the corresponding proper–value problem, however this spectrum can be evaluated by the Fourier analysis of the side force coefficient, which is also represented in the form of

$$c_y = A_0 + \sum_k A_k e^{(i\omega_k - \beta_k)t}.$$

Fourier transformation of the side force in Figure 3 for all five variants shows that only one point of the discrete spectrum can be well picked out from these numerical results. It corresponds to the weakly damped tail of the unsteady response. The values of ω and β are presented in Table 1.

Table 1. Fourier analysis of side force

h/D	β	ω	St_ω
0.010	0.0049	0.0081	0.068
0.020	0.0125	0.0105	0.087
0.017	0.0106	0.0087	0.072
0.014	0.0067	0.0083	0.069
0.012	0.0050	0.0082	0.068

Here the first column contains height of the bumps, the second and the third columns are ω and β (multiplied by D/a_∞), and the fourth column is the Struhal numbers $St_\omega = \omega/M_\infty$. These values are about 0.07 except for the second variant, where variation of the bump is relatively high and the linear representation leads to an appreciable error.

4 Conclusions

The flow past a slender body of revolution has been decomposed into empirical eigenfunctions. The structure of these modes is such that the asymmetric part of the flow can be understood in terms of pairs of modes ordered in their contribution to the additional kinetic energy of the system. Thus, the motion obtained by the model system is at least grossly similar to that of the full one. The low–dimensional description of the flow has been obtained in terms of these modes. A model, which retains 98.5% of the energy of asymmetric motion, is obtained from the first two modes. This can appreciably reduce the computational time needed for numerical study of asymmetric instability of the flow. It is found that Struhal number for all cases is about 0.07 except for the case where the bump is relatively large.

5 Acknowledgements

This research was supported by the Center for Absorbtion in Science Ministry of Emigrant Absorbtion State of Israel. The authors would like to thank Alex Khibnik of UTRC who has drawn our attention to the capabilities of the Karhunen–Loeve decomposition.

References

1. Hunt, B. L., Dexter, P. C. (1978) Pressures on a slender body at high angle of attack in very low turbulence level airstream. AGARD CP 247, Paper 17.
2. Lamont, P. J. (1982) Pressure around an inclined ogive cylinder with laminar, transitional, or turbulent separation. AIAA Journ. **20**, 1492–1499.
3. Zilliac, G. G., Degani, D. et al. (1991) Asymmetric vortices on an slender body of revolution. AIAA Journ. **29**, 667–675.
4. Degani, D. (1990) Numerical investigation of the origin of vortex asymmetry. AIAA Paper 90–0593.
5. Degani, D. (1991) Effect of geometrical disturbances on vortex asymmetry. AIAA Journ. **29**, 560–566.
6. Degani, D., Levy, Y. (1992) Asymmetric turbulent flows over slender bodies. AIAA Journ. **30**, 2267–2273.
7. Sirovich, L. (1987) Turbulence and dynamics of coherent structures: I–III. Quart. Appl. Math. **45**, 561–590.
8. Holmes, P., Lumley, J. L. et al. (1966) Turbulence, Coherent Structures, Dynamical Systems and Symmetry. Cambridge Univ. Press, Great Britain.
9. Deane, A. E., Mavriplis C. (1994) Low–dimensional description of the dynamics in separated flow past thick airfoils. AIAA Journ. **32**, 1222–1227.
10. Steger, J. L., Ying, S. X. et al. (1986) A Partially flux–split algorithm for numerical simulation of unsteady viscous flows. Proc. Workshop on Comp. Fluid Dyn., Univ. of California, Davis, Davis, CA.
11. Degani, D., Schiff, L. B. (1986) Computation of turbulent supersonic flows around pointed bodies having crossflow separation. Journ. Comp. Phys. **66**, 173–196.

Multiphysics Coupling of Three-Phase and Two-Phase Models of Flow in Porous Media

Małgorzata Peszyńska[1] *

University of Texas at Austin, Austin, TX 78712

Abstract. In this paper we show that multiphysics couplings for the simulation of multiphase flow can be realized without loss of accuracy and with decreased computational complexity when compared to single model codes. Specifically, we address the couplings of oil-gas-water models of flow in porous media. An interface is introduced between three-phase and two-phase regions and the problem is solved by combining local simpler solutions coupled with conditions across the interface. Here we discuss issues of matching of variables across the interface and show computational results which demonstrate efficiency and accuracy of the coupling.

1 Introduction

In geosciences, various applications have relied on multiphysics couplings which connect models or algorithms in adjacent or in overlapping computational domains. As examples of the latter, multiphase flow models and the geomechanics, or reactive transport models have been coupled together [5,9].

In this paper we address couplings of oil-gas-water models of flow in adjacent domains in porous media. Our method falls into the category of *heterogeneous domain decomposition* methods, which include, e.g., *fluid-structure interactions*. In contrast to the latter, in our algorithm, the interface across which the coupling of models occurs, is not a physical interface but rather it is one introduced for computational reasons. Specifically, it is fairly common that in a reservoir $\Omega \subset \mathbb{R}^3$ some phases or some components are absent in large parts of Ω over large time intervals $I = (0, T)$. For example, due to gravitational separation, free gas (*gas phase*) typically collects at the top of reservoir. Traditional algorithms simulate the flow by applying a model that can handle a superset of all phases and components. In our multiphysics approach we apply locally simpler models wherever appropriate. For the global solution, coupling conditions are defined. This approach allows the reduction of computational time.

In what follows, we consider an isothermal, three-phase *black-oil* model, and a simple two-phase *oil-water* model [6]. Their coupling was first defined and discussed in [8,3]. Similar issues for the coupling of the oil-water and single-phase (water) models were discussed in [7]. For details on discretization using expanded mixed finite element methods and mortar spaces, fully implicit solver issues, as well as multiscale features see [1,10] and refs. therein.

* Supported by NSF grants: SBR 9873326, ITR EIA-0121523, NPACI 10181410

Below we briefly formulate the models, discuss the coupling, and present computational results which were obtained using the multiphase flow and transport framework IPARS, see [11] and refs. therein. We would like to acknowledge Qin Lu, Mary F. Wheeler and John A. Wheeler for discussions related to this work.

2 Multiphase flow models

Here we consider three components $M = W, O, G$: water, oil (or an ensemble of heavy hydrocarbons), and gas (or an ensemble of light hydrocarbons) [4], respectively, in reservoir of porosity $\phi(x)$ and permeability $K(x)$, with gravity g, and depth $D(x)$, $x \in \Omega$. We use the mass conservation equation

$$\frac{\partial}{\partial t}(\phi N_M) + \nabla \cdot \boldsymbol{U}_M = q_M, \quad x \in \Omega, \tag{1}$$

where N_M denotes the component's concentration, and $\boldsymbol{U}_M, q_M$ denote its flux and source(s), respectively. Equation (1) is a special case of general multicomponent / multiphase equations given in [6,2].

These three components can be present in one or more of the three fluid *phases* denoted by subscript $m = w, o, g$ (water, oleic, gaseous). The distribution between phases is described using η_{Mm} which denotes the mass fraction of component M in phase m. Phase saturations, relative permeabilities, pressures, viscosities, densities, and reference densities, are denoted by $S_m, k_m, , P_m, \mu_m, \varrho_m, \varrho_{mr}$, respectively. We say that a phase m is absent if $S_m = 0$, and we refer to [7] for discussion of residual conditions and phases that are present but immobile.

We use the multiphase extension of Darcy's law for the momentum equation of phase m which defines the volumetric phase velocity $\boldsymbol{V}_m$:

$$\boldsymbol{V}_m = -\frac{Kk_m}{\mu_m}\left(\nabla P_m - \varrho_m g \nabla D\right). \tag{2}$$

The saturations must satisfy the constraint: $S_g + S_o + S_w = 1$, and capillary pressure data is given for the differences $P_o - P_w$ and $P_g - P_o$. The model is complete when concentrations, fluxes and densities are defined. In most cases, $N_M = \sum_m \eta_{mM} S_m \varrho_m / \varrho_{*r}$; $\boldsymbol{U}_M = \sum_m \eta_{mM} \boldsymbol{V}_m \varrho_m / \varrho_{*r}$ where $*$ is a specially chosen phase. However, definitions, units, and constraints on the distribution of components between phases, vary between models. Here we are interested in a black-oil, two-phase and single-phase models.

Black-oil model. Here we assume that $\eta_{Ww} = 1, \eta_{Gg} = 1$ and that $\eta_{Oo} + \eta_{Go} = 1, \eta_{Oo} > 0$. We use formation volume factors $B_m(P_m) = \varrho_{mr}/\varrho_m$ for $m = w, g$. We also define $B_o = \varrho_{or}/\eta_{Oo}\varrho_o$. The amount of G in o at a given pressure is represented using the *gas/oil ratio* $R_o = \eta_{Go}\varrho_{or}/\eta_{Oo}\varrho_{gr}$ which is less than or equal to the *saturated gas/oil ratio* $R_{so}(P_o)$ given as reservoir data. We consider a *phase change* occuring in component G at *bubble pressure* P_{sat}, at which $R_o = R_{so}(P_{sat})$. At $P_o < P_{sat}$, part of the gas component enters gas phase, or vice-versa, if $P_o > P_{sat}$, all gas component dissolves in oil phase. The concentrations and mass fluxes are defined as $N_M = S_m/B_m$, $\boldsymbol{U}_M = \boldsymbol{V}_m/B_m$, for those components $M = W, O$, which are naturally associated with their phases $m = w, o$. For gas component, we have $N_G = S_g/B_g + S_oR_o/B_o$ and $\boldsymbol{U}_G = \boldsymbol{V}_oR_o/B_o + \boldsymbol{V}_g/B_g$.

To complete the formulation, we define the oil phase density which depends on the pressure and on the amount of dissolved gas. Usually, reservoir data is available [2,4] for the formation volume factor of oil i) without any gas dissolved B_{do} and ii) in saturated conditions B_{so}. If $P > P_{sat}$, then $\varrho_o = (\varrho_{or} + R_o\varrho_{gr})/B_o$, where the compressibility of the oil-gas mixture is given as $B_o = B_{do} + (B_{so}(P_{sat}) - B_{do}(P_{sat}))R_o/R_{so}$ and we have here $R_o = N_G/N_O$. If $P_o < P_{sat}$, then $S_g > 0$, $R_o = R_{so}$ and $\varrho_o = (\varrho_{or} + R_{so}\varrho_{gr})/B_{so}$.

Two-phase model. The popular, immiscible, two-phase oil-water model can be seen as a subset of the black-oil model with no gas component present ($\eta_{Go} = 0, \eta_{Oo} = 1$) where obviously $S_g = 0, k_g = 0$. Oil and water are slightly compressible: $\varrho_m = \varrho_{mr} exp(c_m P_m)$ with constants c_m. Concentrations are $N_M = S_m\varrho_m/\varrho_{mr}$ for $m = o, w$ and $M = O, W$, respectively.

Single-phase model. Here additionally we set $S_o = 1$.

3 Interface coupling of oil-water model with black-oil model

Consider $t \in I$ and a decomposition of $\Omega = \tilde{\Omega}^2(t) \cup \tilde{\Omega}^3(t)$ defined as follows: $\tilde{\Omega}^2(t) = \{x : S_g(x;t) = 0\}, \tilde{\Omega}^3(t) = \Omega \setminus \tilde{\Omega}^2(t)$. Note that at a given t, each of $\tilde{\Omega}^2(t), \tilde{\Omega}^3(t)$ (but not both) may be empty if the pressure is low (high) enough. Define the *free boundary* $\tilde{\Gamma}_{23}(t) = \partial\tilde{\Omega}^2(t) \cap \partial\tilde{\Omega}^3(t)$ between the two- and three-phase region. In general, it is difficult to trace $\tilde{\Gamma}_{23}(t)$ in numerical computations.

Let Ω_h be a union of the finite elements covering Ω. For simplicity assume $\Omega_h = \Omega$. Now consider the time-independent decomposition of $\Omega_h = \Omega_h^2 \cup \Omega_h^3$ where $\Omega_h^2 \subseteq \bigcap_{t\in(0,T)} \tilde{\Omega}^2(t)$. The actual definition of Ω_h^2 is allowed to vary, see below. For completeness, $\Omega_h^3 = \Omega_h \setminus \Omega_h^2$. Define the interface $\Gamma_h^{23} = \partial\Omega_h^2 \cap \partial\Omega_h^3$.

The whole idea of multiphysics coupling between the two-phase model and the black-oil model is in applying each of these models, respectively, in Ω_h^2 and Ω_h^3, coupled by interface conditions on Γ_h^{23}. Note that Ω_h^2 can be chosen in any way that is convenient for modeling or computational purposes. In particular, it is desirable for Ω_h^2 to be as large as possible in order to save computational time. Furthermore, we choose Ω_h^2 so that Γ_h^{23} has a convenient geometry. In particular, when using mortar spaces [1] to resolve non-matching grids over Ω_h^2 and Ω_h^3, it is advantageous for Γ_h^{23} to have the most "straight" pieces possible to allow for multiscale and adaptive approximations [10].

In analogy to the above, we can define $\tilde{\Omega}^1(t) =: \{x : S_w(x;t) = 1\}$ and $\Omega_1^h \subseteq \bigcap_{t\in(0,T)} \tilde{\Omega}^1(t)$. These single-phase regions are used in computational examples shown below. Note that $\tilde{\Omega}^2(t) \supseteq \tilde{\Omega}^1(t)$, $\Omega_h^2 \supseteq \Omega_h^1$.

Coupling conditions. Here we use $\Gamma =: \Gamma_h^{23}$. Note that Γ is "artificial", i.e., it does not follow any material boundaries. Therefore, we apply the same conservation principles of momentum and mass on $\Gamma =: \Gamma_h^{23}$, as those applied across boundaries of finite elements. These are realized, respectively, by matching of phase pressures $P_m^2|_\Gamma = P_m^3|_\Gamma, \forall m$, and by matching of component mass fluxes. In consequence, from capillary pressure relationships, phase saturations must match (unless rock types do not match, see [3]), and we have to account properly for the existence of phases and components on each side of Γ. In particular, phase densities computed on Γ from the side Ω_h^2 must match those from the side Ω_h^3.

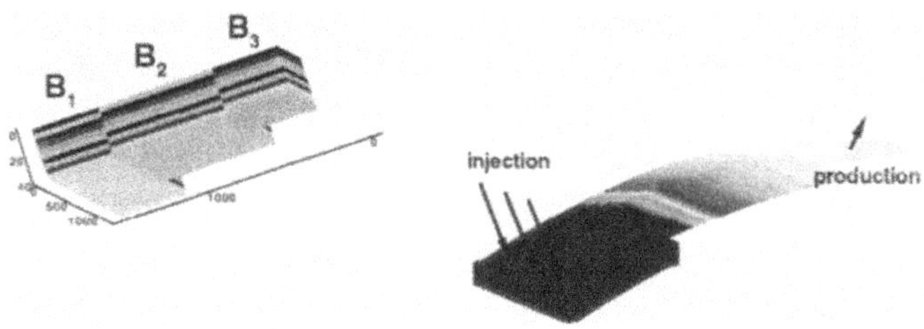

Fig. 1. Left: permeability field and geometry. Right: oil concentration contours at $t = 0$ and well locations indicate that $\Omega^2(t) \supset B_1 \cup B_2$. Gas saturation (not shown) indicates that $\Omega^3(t) \subset B_3$; colored, see page 377

Consider $x \in \Gamma$ and the amount of gas at x. Obviously, $P_o(x) < P_{sat}(x)$. During simulation, we assume that $R_o(x,t)|_\Gamma \approx R_o(t)$. This is reasonable if the reservoir is not too thick: for a typical reservoir (100' thick), the difference is around 1% and for a thick reservoir (300' thick), it can reach 2%. Furthermore, we assume that this ratio remains approximately constant in time, i.e., $R_o(t)|_I \approx R_o^\Gamma$ with R_o^Γ known at initialization. Finally, we assume that the oil phase in Ω_h^2 contains gas component with a fixed gas/oil ratio $\approx R_o^\Gamma$. Consequently, we need to find ϱ_{or}^2 and c_o^2 on Ω_h^2 such that over a certain range of pressures (P_{min}, P_{max}), $x \in \Gamma$,

$$\varrho_{or}^2 exp(P_o c_o^2) = \varrho_o^2|_{\Omega_h^2} \approx \varrho_o^3|_{\Omega_h^3} = \frac{(\varrho_{or}^3 + R_o^\Gamma \varrho_{gr}^3)}{B_o(P_o)}, P_o \in (P_{min}, P_{max}). \quad (3)$$

Given $R_o^\Gamma, B_o, \varrho_{mr}^3$, the parameters ϱ_{or}, c_o can be found, for example, by a least squares procedure. Obviously, for $R_o^\Gamma > 0$, the quality of such a match depends, in particular, on the size of (P_{min}, P_{max}). From (3) it follows that the concentrations (and fluxes) computed on each side of the interface match only if a conversion factor is applied. To see that, multiply (3) on Γ by the saturations $S_o^2|_\Gamma = S_o^3|_\Gamma$ to get

$$\hat{\varrho}_{or}^2 N_o^2 = S_o^2 \hat{\varrho}_{or}^2 exp(P_o \hat{c}_o) \approx S_o^3 \frac{(\varrho_{or}^3 + R_o^\Gamma \varrho_{gr}^3)}{B_o(P_o)} = (\varrho_{or}^3 + R_o^\Gamma \varrho_{gr}^3) N_o^3.$$

Similar matching is obtained for fluxes, using the fact that in two-phase conditions, $k_o^2(S_o^2)|_\Gamma = k_0^3(S_o^3)|_\Gamma$.

4 Examples

Here we present simulation results illustrating the concepts discussed above. The computational domain $\Omega_h = B_1 \cup B_2 \cup B_3$ is presented in Fig. 1. We use the mortar approach to resolve nonmatching grids and an iterative solver on Γ. Γ is a subset of the union of faces between B_1, B_2, B_3. Simulations can be performed either using a traditional approach $\Omega_h^3 \equiv \Omega_h$ denoted by **b**, or using the domain decomposition approach in several variants. The most natural choice (and one which proves to be most efficient) is one in which we assign $\Omega_h^1 \equiv B_1$, $\Omega_h^2 \equiv B_2$ and $\Omega_h^3 \equiv B_3$ (we shall call it **shb**).

Now we discuss results in Fig, 2. First, we compare the solutions obtained **b** with those obtained with **shb**. It turns out that their agreement is excellent, regardless of how much gas is present in the reservoir. Next, we consider

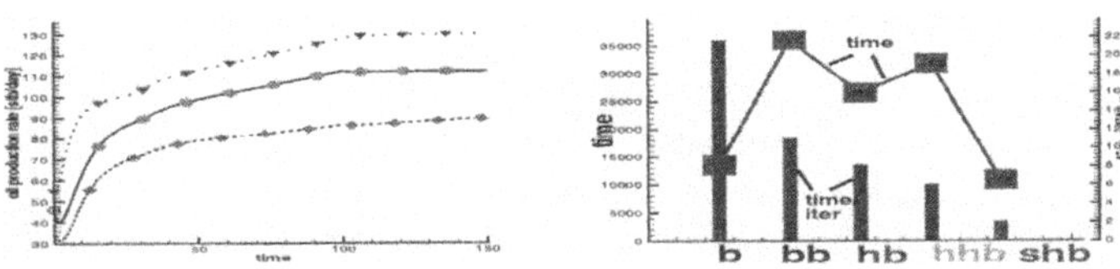

Fig. 2. Left: comparison of the oil production rates obtained with multiphysics (symbols only) and with single model (dotted, solid and dashed lines, respectively). From top to bottom: in block B1: unsaturated oil; saturated oil but no gas cap; gas cap. Right: efficiency of multiphysics couplings; colored, see page 377

the computational efficiency of the multiphysics procedure. In addition to scenarios discussed above, we consider scenario **hb** in which $\Omega_h^2 \equiv B_1 \cup B_2$ with $\Omega_h^3 \equiv B_3$, as well as two additional scenarios: i) **bb** which is like **hb** with black-oil model functionally replacing two-phase model, and ii) **hhb** which is like **shb** in which two-phase model functionally replaces the single-phase model over B_1.

As expected, the time per iteration which corresponds to total subdomain time decreases, as more simple models are used. However, the number of interface iterations increases with the number of subdomains. Therefore, the total time does not decrease monotonically, but multiphysics **shb** is faster than traditional **b** simulation. In general, the efficiency of multiphysics depends mainly on the size of subdomains associated with individual models and on the number of iterations necessary to achieve a desired accuracy of the coupling.

Finally, we study sensitivity to the matching of densities: Fig. 3 contains plots of $\varrho_o^3|_\Gamma$ and of $\varrho_o^2|_\Gamma$ for two different intervals (P_{min}, P_{max}) used in the matching procedure. When this interval is small, the multiphysics results match very well those obtained by traditional simulation. When the interval is large, a large discrepancy due to the density modeling error may occur.

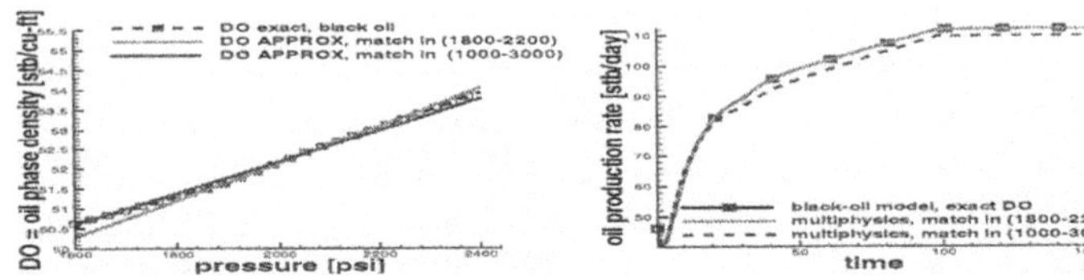

Fig. 3. Sensitivity to density matching. Left: density of oil (exact and approximated using (3)). Right: corresponding well rates; colored, see page 377

References

1. T. Arbogast, L. C. Cowsar, M. F. Wheeler, and I. Yotov. Mixed finite element methods on non-matching multiblock grids. *SIAM J. Numer. Anal.*, 37:1295–1315, 2000.
2. L. W. Lake. *Enhanced oil recovery.* Prentice Hall, 1989.
3. Q. Lu, M. Peszynska, and M. F. Wheeler. A parallel multi-block black-oil model in multi-model implementation. *SPE Journal*, 2002. to appear.
4. W. D. Jr. McCain. *The properties of Petroleum Fluids.* PennWell Books, Tulsa, Oklahoma, second edition, 1990.
5. S. Minkoff, C. M. Stone, S. Bryant, M. Peszynska, and M. F. Wheeler. A loose coupling algorithm for fluid flow and geomechanical deformation modeling. submitted.
6. D. W. Peaceman. *Fundamentals of numerical reservoir simulation.* Elsevier Scientfic Publishing Company, Amsterdam-Oxford-New York, first edition, 1977.
7. M. Peszynska. Advanced techniques and algorithms for reservoir simulation III. Multiphysics coupling for two phase flow in degenerate conditions. In J. Chadam, A. Cunningham, R. E. Ewing, P. Ortoleva, and M. F. Wheeler, editors, *IMA Volume 131: Resource Recovery, Confinement, and Remediation of Environmental Hazards*, pages 21–40. Springer, 2002.
8. M. Peszynska, Q. Lu, and M. F. Wheeler. Multiphysics coupling of codes. In L. R. Bentley, J. F. Sykes, C. A. Brebbia, W. G. Gray, and G. F. Pinder, editors, *Computational Methods in Water Resources*, pages 175–182. A. A. Balkema, 2000.
9. M. Peszynska and S. Sun. Reactive transport model coupled to multiphase flow models. to appear.
10. M. Peszynska, M. F. Wheeler, and I. Yotov. Mortar upscaling for multiphase flow in porous media. Computational Geosciences, to appear.
11. M. F. Wheeler, J. A. Wheeler, and M. Peszynska. A distributed computing portal for coupling multi-physics and multiple domains in porous media. In L. R. Bentley, J. F. Sykes, C. A. Brebbia, W. G. Gray, and G. F. Pinder, editors, *Computational Methods in Water Resources*, pages 167–174. A. A. Balkema, 2000.

On solving boundary value problem in fluid mechanics by Fourier's method: Wall Interference of Transonic Wind Tunnels

Boško Rašuo

Aeronautical Department, Faculty of Mechanical Engineering, University of Belgrade, 27 Marta 80, 11 000 Belgrade, Yugoslavia

Abstract. The influence of perforated walls of transonic wind tunnels at two-dimensional investigations by employing the advanced numerical method for solving Dirichlet's problem formulated for a rectangle of wind tunnel's work section is given in this paper. To demonstrate the appropriateness of the presented advanced algorithm for calculation of transonic wind tunnel wall interference at two-dimensional investigations, the algorithm has been applied to the aerodynamic experimental results from investigations of BGK 1 airfoil obtained in NAE (Canada). The comparative results of calculations for the same test which were performed by finite difference method (obtained in USA, Canada, Japan, France and Holland) and by the presented advanced numerical method (obtained at the University of Belgrade) are given as well.

1 Introduction

One of the fundamental questions imposed upon us when determining transonic wind tunnel work section wall interference is that one related to its correct experimental and mathematical simulation. Since it is impossible to avoid completely its undesirable influence on measured aerodynamic characteristics (lift, drag, lift-curve slope, pitching moment, flow-field quality etc.), the only practical possibility at our disposal is the combination of experimental data from the characteristics investigations and correct mathematical simulations in the course of calculation of their influence on measured values.

In order to preserve the realistic features of flow at the work section's boundaries the boundary conditions, which are to be known to solve this type of boundary problem, are experimentally determined by measuring static pressure distribution at the vicinity of work section walls.

In the paper the boundary value problem has been analyzed by adapting the solution for application of Fourier's transformation. Fourier coefficients have been computed with fast Fourier transformation.

The corrections have been interpreted in the way that the pressure distribution measured over airfoil's surface at wind tunnel investigation at undisturbed flow Mach number M_∞, corresponds to the distribution which could be obtained if the airfoil were investigated in a free air (unconstrained by wind tunnel walls) at a Mach number $M_\infty + \Delta M$.

2 Problem Formulation

The basic idea of the method was given in the works of Mokry, Ohman Chapelier, Bouniol, Chevalier and Paquet [1] with a modification of the part related to modeling of boundary conditions on walls. Instead of using the solution obtained for a infinite segment, the problem is formulated for a rectangle $x_1 < x < x_2$, $y_1 < y < y_2$ (Figure 1). This approach is considered to be more appropriate for investigations of actual finite-length work section.

For transonic inviscid flow governed by the small disturbance equation:

$$\beta^2 \frac{\partial^2 \phi}{\partial x^2} + \frac{\partial^2 \phi}{\partial y^2} = \frac{\kappa+1}{2} \frac{\partial}{\partial x} (\frac{\partial \phi}{\partial x})^2 , \quad \beta = \sqrt{1 - {M_\infty}^2} \tag{1}$$

where ϕ is potential of the disturbance velocity and κ is the ratio of specific heats. In the region of linearized flow (the flow domain between walls and depicted rectangle, Figure 1) we can use decomposition of potential

$$\phi(x,y) = \phi_w(x,y) + \phi_f(x,y) \tag{2}$$

where ϕ_w satisfies (1) in $R \cup G$, and ϕ_f is in infinite region external to G. For a thin airfoil in transonic flow, we can expand potential ϕ_f in form which are given S. Goldstein and M. Mokry [1]:

$$\phi_f(x,y) = {\phi_f}^{\sigma}(x,y) + {\phi_f}^{\gamma}(x,y) + {\phi_f}^{\mu}(x,y) + {\phi_f}^{\omega}(x,y) \tag{3}$$

where the right hand terms denote potentials due to a *source* of strength σ, a *vortex* of strength γ, a *doublet* in the x direction of strength μ and a *doublet* in the y direction of strength ω.

By expanding the complex distributed velocity in the domain of linearized flow in Laurent series, from the main part of the series we obtain the following approximation for $\phi_f(x,y)$

$$\phi_f(x,y) = \frac{\sigma}{2\pi\beta} \log \sqrt{x^2 + (\beta y)^2} - \frac{\gamma}{2\pi} atan \frac{\beta y}{x} + \frac{(\mu/2\pi\beta)x}{x^2 + (\beta y)^2} + \frac{(\omega/2\pi)\beta y}{x^2 + (\beta y)^2} \tag{4}$$

where are the strength of singularities for a thin airfoil of a small angle of attack and camber:

$$\sigma = \frac{1}{2} c C_{Dw}, \quad \gamma = \frac{1}{2} c C_L, \quad \mu = c^2 (A + \delta A), \quad \omega = \frac{1}{2} c^2 \, (\frac{1}{4} C_L + C_{M1/4}) \tag{5}$$

where c is the airfoil chord, C_{Dw} the wake drag coefficient, C_L the lift coefficient, $C_{Ml/4}$ is the quarter-chord pitching moment coefficient, A the cross-sectional area of the airfoil (nondimensionalized by c^2), and δA the nonlinear contribution (for transonic regime) to the airfoil area is [2]:

$$\delta A = \frac{\kappa+1}{2\pi\beta} {M_\infty}^{7/4} \left(\frac{C_L}{4} \right)^2 \left\{ \frac{7}{4} + \log \left[{M_\infty}^{1/2} \left(\frac{t}{c} \right)^{1/3} \frac{|y|}{c} \right] \right\} \tag{6}$$

here t is maximal airfoil thickness. This additional term in the transonic velocity regime considerably increases solid blockage interference of an airfoil in 2-D tunnel. This conception of singularities calculation for the transonic regime is given specially good results which are in good agreement with results of other calculation methods. Very far from the airfoil, the nonlinear term becomes negligible, so that the far field is governed by the linearized Equation

$$\beta^2 \frac{\partial^2 \phi}{\partial x^2}(x,y) + \frac{\partial^2 \phi}{\partial y^2}(x,y) = 0, \quad C_p(x,y) = -2\frac{\partial \phi}{\partial x}(x,y) \tag{7}$$

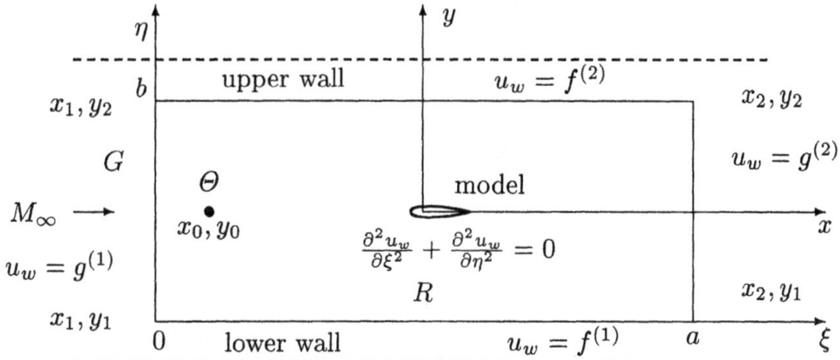

Fig. 1. The co-ordinate system for finite-length working section of wind tunnel

3 Solution of Wall Interference Problem

For the calculation of ϕ_w, we can utilize the transformation: $\xi = \frac{1}{\beta}(x - x_1)$ and $\eta = y - y_1$ which reduce (7) into Laplace equation. As for differentiability of harmonic function velocity, the reduced x - component of interference velocity $u(\xi,\eta) = \frac{\partial \phi_w}{\partial \xi}(x,y) = \beta \frac{\partial \phi_w}{\partial x}(x,y)$ satisfies the equation

$$\frac{\partial^2 u}{\partial \xi^2}(\xi,\eta) + \frac{\partial^2 u}{\partial \eta^2}(\xi,\eta) = 0 \tag{8}$$

in domain of rectangle $0 < \xi < a$, $0 < \eta < b$, where $a = \frac{1}{\beta}(x_2 - x_1)$, $b = y_2 - y_1$ are the sides of transformed rectangle. For transformed x-component of velocity $u(\xi,\eta)$, we can formulate the following Dirichlet's problem:

$$\frac{\partial^2 u_w}{\partial \xi^2}(\xi,\eta) + \frac{\partial^2 u_w}{\partial \eta^2}(\xi,\eta) = 0, \qquad 0 < \xi < a,\ 0 < \eta < b \tag{9}$$

$$\begin{array}{ll} u_w(\xi,0) = f^{(1)}(\xi)\ \ 0 < \xi < a, & u_w(\xi,b) = f^{(2)}(\xi)\ \ 0 < \xi < a \\ u_w(0,\eta) = g^{(1)}(\eta)\ \ 0 < \eta < b, & u_w(a,\eta) = g^{(2)}(\eta)\ \ 0 < \eta < b \end{array}$$

as illustrated in Figure 1. Boundary conditions

$$\begin{array}{ll} f^{(1)}(\xi) = -\beta\Big[\frac{C_p(x,y_1)}{2} + \frac{\partial \phi_f}{\partial x}(x,y_1)\Big], & f^{(2)}(\xi) = -\beta\Big[\frac{C_p(x,y_2)}{2} + \frac{\partial \phi_f}{\partial x}(x,y_2)\Big] \\ g^{(1)}(\eta) = f^{(1)}(0) + \frac{\eta}{b} f^{(2)}(0) - f^{(1)}(0), & g^{(2)}(\eta) = f^{(1)}(a) + \frac{\eta}{b} f^{(2)}(a) - f^{(1)}(a) \end{array}$$

are obtained by measuring the static pressure along sides $y = y_1$ and $y = y_2$ and substituting the potential ϕ_f from (4).

The boundary values problem given by (9) and (10) can be solved by Fourier's method of separation of variables [3]. Solution, adapted for the application of Fourier transform is: $u(\xi,\eta) = f(\xi,\eta) + g(\xi,\eta)$, that is, the reduced component of wall interference

$$u_w(\xi,\eta) = \sum_{k=1}^{\infty} \sin\mu_k\xi \left[A_k^{(1)} \frac{\sinh\mu_k(b-\eta)}{\sinh\mu_k b} + A_k^{(2)} \frac{\sinh\mu_k\eta}{\sinh\mu_k b}\right] + \sum_{k=1}^{\infty} \sin\nu_k\eta \left[B_k^{(1)} \frac{\sinh\nu_k(a-\xi)}{\sinh\nu_k a} + B_k^{(2)} \frac{\sinh\nu_k\xi}{\sinh\nu_k a}\right] \tag{10}$$

where the eigenvalues are: $\mu_k = \frac{k\pi}{a}$ and $\nu_k = \frac{k\pi}{b}$.

The series coefficients are described with integrals [1] and [4]:

$$A_k^{(\ )} = \frac{2}{a}\int_0^a f^{(\ )}(\xi)\sin\mu_k\xi d\xi, \quad B_k^{(\ )} = \frac{2}{b}\int_0^b g^{(\ )}(\eta)\sin\nu_k\eta d\eta \tag{11}$$

where empty space within parentheses should have the index (1) or (2) depending what boundary is involved, lower (1) or upper (2), fore (1) or back (2). The coefficients $A_k^{(\ )}$ can be approximated according to the rule of rectangle [5]

$$A_k^{(1)} = \frac{2}{m}\sum_{j=0}^{m-1} f^{(1)}\left(a\frac{2j+1}{m}\right)\sin\frac{2\pi jk}{m}, \quad A_k^{(2)} = \frac{2}{m}\sum_{j=0}^{m-1} f^{(2)}\left(a\frac{2j+1}{m}\right)\sin\frac{2\pi jk}{m} \tag{12}$$

If m is chosen to be an integer exponent of 2 (2^1,2^2,2^3,...), the above given sum can be solved very efficiently by employing Fast Fourier Transform algorithm for $k = 1, 2, 3, ..., m/2-1$. Derivation of coefficients $B_k^{(\ )}$ is always simpler by using (11) since in that case a closed form of integral is obtained $B_k^{(1)} = \frac{2}{k\pi}[f^{(1)}(0) - (-1)^k f^{(2)}(0)]$ and $B_k^{(2)} = \frac{2}{k\pi}[f^{(1)}(a) - (-1)^k f^{(2)}(a)]$ [6].

4 Wall Interference Corrections

By differentiating the adiabatic relation between velocity and Mach number, we may obtain the correction of Mach number

$$\Delta M = \left(1 + \frac{\kappa - 1}{2}M^2\right) M \frac{\partial\phi_w}{\partial x}(0,0) \tag{13}$$

where, we get velocity correction and correction of angle of attack for model position $x = y = 0$ in accordance with x-component of interference velocity

$$u_w(0,0) = \frac{1}{\beta}u_w\left(-\frac{x_1}{\beta}, -y_1\right) = \frac{\partial\phi_w}{\partial x}(0,0), \ \Delta\alpha = \frac{\partial\phi_w}{\partial y}(0,0) = v_w(0,0) \tag{14}$$

From equations for ξ,η and $u(\xi,\eta)$ we obtain:

$$\frac{\partial\phi_w}{\partial y}(0,0) - \frac{\partial\phi_w}{\partial y}(x_0,y_0) = v_w\left(-\frac{x_1}{\beta},-y_1\right) - v_w\left(\frac{x_0-x_1}{\beta},y_0-y_1\right) \tag{15}$$

where v_w is a function of conjugated velocity

$$v_w(\xi,\eta) = \sum_{k=1}^{\infty}\cos\mu_k\xi\left[A_k^{(1)}\frac{\cosh\mu_k(b-\eta)}{\sinh\mu_k b} - A_k^{(2)}\frac{\cosh\mu_k\eta}{\sinh\mu_k b}\right] + \sum_{k=1}^{\infty}\cos\nu_k\eta\left[-B_k^{(1)}\frac{\cosh\nu_k(a-\xi)}{\sinh\nu_k a} + B_k^{(2)}\frac{\cosh\nu_k\xi}{\sinh\nu_k a}\right] \tag{16}$$

which can be obtained from (12). If we now select the reference point (x_0,y_0), to be inside the linearized flow (Figure 1), then from (2) we have:

$$\frac{\partial\phi_w}{\partial y}(x_0,y_0) = \Theta(x_0,y_0) - \frac{\partial\phi_f}{\partial y}(x_0,y_0),\quad \Theta(x_0,y_0) = \frac{\partial\phi}{\partial y}(x_0,y_0) \tag{17}$$

where $\Theta(x_0,y_0)$ represents the angle of flow at the reference point.

It follows: $\Delta\alpha = v_w(0,0)$ and

$$\Delta\alpha = v_w\left(-\frac{x_1}{\beta},-y_1\right) - v_w\left(\frac{x_0-x_1}{\beta},y_0-y_1\right) + \Theta(x_0,y_0) - \frac{\partial\phi_f}{\partial y}(x_0,y_0) \tag{18}$$

The first two terms on the right-hand side of (20) are derived from (19) and the last term from (4). In this equation, the angle of flow Θ is the only unknown term, which is to be obtained by measurements.

5 Results of Calculations

In order to validate the present method, Mokry's data for BGK 1 airfoil from the NAE 15 × 60-in. transonic wind tunnel were used. The airfoil pressure distribution and the wall pressure distribution along the upper and lower wall are given in [1] and [5]. The principal parameters for that Canadian's investigations were: a reference Mach number $M = 0.784$, the angle of attack $\alpha = 2.56°$, the lift coefficient $C_L = 0.764$, the airfoil area (normalized with c^2) $A = 0.075$ and a Reynolds number of 21.03×10^6.

In numerical calculations, effect of selection of integer exponent n ($m = 2^n$) on the solution convergence has been tested. The complete solution convergence has been realized for Mach number correction, and a satisfactory one for correction of angle of attack with the value of exponent n of only 8 which is illustrated in [5] for selected case Canadian's investigation.

The obtained calculation results of the wind tunnel wall interference performed by the present numerical method show a completely satisfactory agreement together with some calculation results from others authors (Table 1). This is particularly evident to the calculation results of the Mach number corrections and those results are completely identical one another (besides, the Gopinath's calculation results). The calculation results of the attack angle correction vary from -0.58° to -0.89°, which in good size depend from values of an initial angularity of flow in the reference point (in present case $\Theta = 0°$).

Table 1. Corrections for the NAE test case obtained by several calculation methods

Method	References	ΔM_∞	$\Delta\alpha$
Presented Method – *Yugoslavia*	[2,5]	-0.015	-0.70°
NAE (Mokry and Ohman) – *Canada*	[1]	-0.015	-0.67°
ONERA (Capelier et al) – *France*	[1]	-0.015	-0.67°
NASA (Gopinath) – *USA*	[5]	-0.017	-0.67°
NASA (Kemp) – *USA*	[1]	-0.015	-0.64° (-0.89°)
NLR (Smith) – *Holland*	[1]	-0.015	-0.59°
NAL (Sawada) – *Japan*	[5]	—	-0.58°

6 Conclusion

Due to large absolute values of required corrections of angle of attack, it follows that all measurements which are related to the determination of angle of attack, that is, to the evaluation of lift-curve slope, have no significance for practical application without the suggested wind tunnel corrections.

In order to levelage the wind tunnel's role with the ones that the computer analysis and free air investigations have, that is, in order to be supplied with the satisfactory aerodynamic data obtained from wind tunnel investigations for airplanes design applications, it is necessary to supplement the wind tunnel data with calculation of wind tunnel wall corrections as their integral and inseparable part.

References

1. Mokry M. et al. (1983) Two–Dimensional Wind Tunnel Wall Interference, AGARDograph **281**, 86–92
2. Rašuo B. (1998) Advanced Numerical Method for Solving 2–D Transonic Wind Tunnel Wall Interference Problem. 4th WCCM, CIMNE, Barcelona, 1–16
3. Powers, D. (1972) Boundary Value Problems. Academic Press. NY, 113–117
4. Lavrent'ev M., Shabat B. (1987) Metody teorii funktsii kompleksnogo peremennogo. Nauka, Moscow, 200–207
5. Rašuo B. (1999) Transonic Wind Tunnel Wall Interference on Aerodynamic Testing. (in serbian). University of Belgrade, Belgrade, 69–96
6. Rašuo B. (1997) Two–Dimensional Wall Interference in Transonic Wind Tunnels. ZAMM. **77**, 275–276

Acknowledgments: The author is grateful to the International Conference "Multifield Problems" organizers for their support.

Boundary Element Formulations for Linear Poroelastodynamic Continua

Martin Schanz and Dobromil Pryl

Institute of Applied Mechanics, Technical University Braunschweig, D-38023 Braunschweig, P.O. Box 3329, Germany

Abstract. Under the assumption of a linear geometry description and linear constitutive equations, the governing equations are derived for two poroelastic theories, Biot's theory and Theory of Porous Media (TPM), using solid displacements and pore pressure as unknowns. Comparing the sets of differential equations of Biot's theory and of TPM, they show different constant coefficients. Identifying these coefficients with the material data and correlating them for both theories leads to the known problem with Biot's 'apparent mass density' and yet unsolved inconsistencies in finding a counterpart to Biot's stress coefficient.

As both poroelastic theories yield the same mathematical operator the same time-dependent Boundary Element Method based on the 'Convolution Quadrature Method', proposed by LUBICH, is developed. To show the differences in both theories wave propagation phenomena are studied.

1 Introduction

Nowadays, two theories are used to describe the constitutive behaviour of two-phase porous media. Based on the work of von Terzaghi the so-called Biot theory was developed, first, for consolidation problems [1] and, later, also for dynamic applications [2]. Based on the work of Fillunger the Theory of Porous Media (TPM) was developed. This approach utilises the axioms of continuum theories of mixtures [3] extended by the concept of volume fractions, e.g., [4–6].

Remarks on the equivalence of both theories are found in the work of Bowen [5]. In this paper, Biot's theory is shown to be a special case of a linearised theory of mixtures with constant volume fractions, if it is assumed that the parameter Q in Biot's theory vanish. Physically interpreted, the interaction between both constituents is neglected. Further, Ehlers and Kubik [7] compared and discussed the linear versions of both theories claiming that these are equivalent if Biot's apparent mass density is assumed to be zero.

Assuming a linear geometry description and a linear elastic constitutive law for the solid skeleton, in the following, for both theories the sets of governing differential equations are given using solid displacements and pore pressure as unknowns. It turns out that both theories are governed by the same mathematical operator, so that the same boundary element formulation can be used. Aiming on wave propagation problems, the time-dependent poroelastic boundary element formulation developed by Schanz [10] is used to present the differences between both theories.

2 Governing Equations

To introduce the linearised version of TPM, first, the linearised balance of momentum for the mixture as well as for the fluid is given

$$n_0^S \varrho_0^{SR} u_i'' + n_0^F \varrho_0^{FR} \left(u_i'' + w_i'\right) = G u_{i,jj} + \left(K + \frac{1}{3}G\right) u_{j,ji} - n_0^F \left(1 + n_0^S\right) p_{,i} + n_0^S \varrho^{SR} b_i^S + n_0^F \varrho^{FR} b_i^F \tag{1a}$$

$$n_0^F \varrho_0^{FR} \left[u_i'' + w_i'\right] = - n_0^F p_{,i} - \frac{\left(n_0^F\right)^2 \gamma^{FR}}{k^F} w_i + n_0^F \varrho^{FR} b_i^F \tag{1b}$$

using the solid displacement u_i, the seepage velocity w_i, and the pore pressure p as unknowns. Due to the linearisation, constant densities for the solid ϱ_0^{SR} and the fluid ϱ_0^{FR} are used with the exception of the body force terms $n_0^S \varrho^{SR} b_i^S, n_0^F \varrho^{FR} b_i^F$. The constant volume fractions are denoted by n_0^S and n_0^F, respectively ($n_0^F + n_0^S = 1$). The solid skeleton is described by the bulk compression modulus K and the shear modulus G. The second term on the right hand side of (1b) is the interaction force between the solid and the fluid characterised by the permeability k^F and the specific weight γ^{FR} of the fluid. The behaviour of the fluid itself is governed by the continuity equation

$$n_0^F \frac{p'}{R\vartheta} + \varrho_0^{FR} n_0^F \left(1 + n_0^S\right) u_{i,i}' + n_0^F \varrho_0^{FR} w_{i,i} = 0 \tag{2}$$

written in the solid reference coordinate system. In (2), the continuity equation of the solid and the ideal gas equation with the absolute temperature ϑ and the gas constant R is included.

With (1) and (2), there are in 3-d seven equations for the seven unknowns, solid displacement u_i, seepage velocity w_i, and pore pressure p. However, from a physical point of view it is sufficient to describe the problem with four unknowns, e.g., the solid displacement and the pore pressure, because the fluid is modelled without any shear components and, therefore, can be governed by the scalar pressure. Hence, (1b) is used to eliminate the seepage velocity in both other equations. Since w_i is given as time derivative in (1b), this is only possible in Laplace domain. Denoting the Laplace transformed functions with $\hat{()}$ a coupled system of partial differential equations for the unknowns solid displacement $\hat{u}_i$ and pore pressure $\hat{p}$ is established

$$G\hat{u}_{i,jj} + \left(K + \frac{1}{3}G\right) \hat{u}_{j,ji} - s^2 \left(n_0^S \varrho_0^{SR} + n_0^F \varrho_0^{FR} - \beta \varrho_0^{FR}\right) \hat{u}_i - \left(n_0^F \left(1 + n_0^S\right) - \beta\right) \hat{p}_{,i} = \beta \varrho^{FR} \hat{b}_i^F - n_0^S \varrho^{SR} b_i^S - n_0^F \varrho^{FR} b_i^F \tag{3a}$$

$$\hat{p}_{,ii} - n_0^F \frac{s^2}{\beta R\vartheta} \hat{p} - \frac{s^2 \varrho_0^{FR}}{\beta} \left(n_0^F \left(1 + n_0^S\right) - \beta\right) \hat{u}_{i,i} = \varrho^{FR} \hat{b}_{i,i}^F . \tag{3b}$$

In (3), β is an abbreviation dependent on material parameters and the Laplace variable.

Using Biot's theory, the same set of differential equations is achieved if the solid displacement and the pore pressure are also chosen as unknowns. The difference between both theories is the material parameters used, i.e., the constant coefficients in (3). A direct comparison of the differential equations for both theories yields beside the equivalence of obvious material data the identification

$$\frac{\varrho_0^{FR} n_0^F}{R^B} = \frac{1}{R\vartheta} \qquad \text{and} \quad \alpha = n_0^F \left(1 + n_0^S\right) \qquad \text{and} \quad \varrho_a \equiv 0 \, , \tag{4}$$

with Biot's stress coefficient α and R^B. Further, as already reported in [7], it must be required that the apparent mass density ϱ_a introduced by Biot vanishes. More details may be found in [12].

3 Boundary Element Method

As shown in the last section, the set of governing differential equation for both theories differ only in the coefficients, i.e., in different material constants (see (4)), and not in the mathematical operator. Hence, for both theories the same boundary integral equation

$$\int\limits_{\Gamma} \begin{bmatrix} \hat{U}_{ij}^S & -\hat{P}_j^S \\ \hat{U}_i^F & -\hat{P}^F \end{bmatrix} \begin{bmatrix} \hat{t}_i \\ \hat{q} \end{bmatrix} \mathrm{d}\Gamma = \oint\limits_{\Gamma} \begin{bmatrix} \hat{T}_{ij}^S & \hat{Q}_j^S \\ \hat{T}_i^F & \hat{Q}^F \end{bmatrix} \begin{bmatrix} \hat{u}_i \\ \hat{p} \end{bmatrix} \mathrm{d}\Gamma + \begin{bmatrix} c_{ij} & 0 \\ 0 & c \end{bmatrix} \begin{bmatrix} \hat{u}_i \\ \hat{p} \end{bmatrix} \tag{5}$$

can be considered. There, the total stress vector is denoted by $\hat{t}_i$ and the normal flux by $\hat{q}$. Further, the domain is Ω with boundary Γ and several abbreviations $\hat{T}_{ij}^S, \hat{Q}_j^S, \hat{T}_i^F$, and $\hat{Q}^F$ are introduced (for details, see [10]).

The integral free terms c_{ij} and c in (5) result from the limiting behaviour of the fundamental solutions if $\boldsymbol{y}$ approaches $\boldsymbol{x}$. A series expansion of the fundamental solutions with respect to $r = |\boldsymbol{y} - \boldsymbol{x}|$ shows that $\hat{T}_{ij}^S$ and $\hat{Q}^F$ are strongly singular and in the limit $r \to 0$ equal to their elastostatic and acoustic counterparts, respectively. Therefore, these integral free terms are calculated as known from elastostatic and acoustic BE formulations. The other fundamental solutions are either regular $\hat{P}_j^S$ and $\hat{U}_i^F$ or weakly singular $\hat{U}_{ij}^S, \hat{P}^F, \hat{T}_i^F, \hat{Q}_j^S$. A transformation to time domain gives, finally, the time dependent integral equation for poroelasticity

$$\int\limits_{\Gamma} \begin{bmatrix} U_{ij}^S & -P_j^S \\ U_i^F & -P^F \end{bmatrix} * \begin{bmatrix} t_i \\ q \end{bmatrix} \mathrm{d}\Gamma = \oint\limits_{\Gamma} \begin{bmatrix} T_{ij}^S & Q_j^S \\ T_i^F & Q^F \end{bmatrix} * \begin{bmatrix} u_i \\ p \end{bmatrix} \mathrm{d}\Gamma + \begin{bmatrix} c_{ij} & 0 \\ 0 & c \end{bmatrix} \begin{bmatrix} u_i \\ p \end{bmatrix} \tag{6}$$

with the convolution integral $f * g = \int_0^t f(t - \tau)\, g(\tau)\, \mathrm{d}\tau$.

According to the boundary element method the boundary surface Γ is discretized by E iso-parametric elements Γ_e and for the state variables F

polynomial shape functions $N_e^f(\boldsymbol{x})$ are defined. Further, the convolution integrals are approximated by the Convolution Quadrature Method [8]. A backward differential formula of order 2 and the parameter choice $L = N$ and $\mathscr{R}^N = \sqrt{10^{-10}}$ yield the best results [11,9]. Applying these approximations to the integral equation (6) results in the boundary element time stepping formulation for $n = 0, 1, \ldots, N$

$$\begin{bmatrix} c_{ij} u_i(n\Delta t) \\ c\, p(n\Delta t) \end{bmatrix} = \sum_{e,f=1}^{E,F} \sum_{k=0}^{n} \left\{ \begin{bmatrix} \omega_{n-k}^{ef}\left(\hat{U}_{ij}^S, \Delta t\right) & -\omega_{n-k}^{ef}\left(\hat{P}_j^S, \Delta t\right) \\ \omega_{n-k}^{ef}\left(\hat{U}_i^F, \Delta t\right) & -\omega_{n-k}^{ef}\left(\hat{P}^F, \Delta t\right) \end{bmatrix} \begin{bmatrix} t_i^{ef}(k\Delta t) \\ q^{ef}(k\Delta t) \end{bmatrix} \right.$$
$$\left. - \begin{bmatrix} \omega_{n-k}^{ef}\left(\hat{T}_{ij}^S, \Delta t\right) & \omega_{n-k}^{ef}\left(\hat{Q}_j^S, \Delta t\right) \\ \omega_{n-k}^{ef}\left(\hat{T}_i^F, \Delta t\right) & \omega_{n-k}^{ef}\left(\hat{Q}^F, \Delta t\right) \end{bmatrix} \begin{bmatrix} u_i^{ef}(k\Delta t) \\ p^{ef}(k\Delta t) \end{bmatrix} \right\} . \tag{7}$$

The integration weights are calculated corresponding to

$$\omega_n^{ef}\left(\hat{U}_{ij}^S, \Delta t\right) = \frac{\mathscr{R}^{-n}}{L} \sum_{\ell=0}^{L-1} \int_{\Gamma} \hat{U}_{ij}^S \left(\boldsymbol{x}, \boldsymbol{y}, \frac{\gamma\left(\mathscr{R} e^{i\ell \frac{2\pi}{L}}\right)}{\Delta t} \right) N_e^f(\boldsymbol{x})\, \mathrm{d}\Gamma\, e^{-in\ell \frac{2\pi}{L}} , \tag{8}$$

respectively. Note, the calculation of the integration weights is only based on the Laplace transformed fundamental solutions which are available. In order to arrive at a system of algebraic equations, point collocation is used and a direct equation solver is applied.

4 Numerical Example

In order to demonstrate the differences between both poroelastic theories, wave propagation phenomena in a poroelastic half space are considered. For the comparison, a long strip (6 m × 33 m) is discretized with 396 triangular linear elements on 242 nodes (see Fig. 1). Material data used are those of a soil ($K = 2.1 \cdot 10^8\,\mathrm{N/m^2}, G = 9.8 \cdot 10^7\,\mathrm{N/m^2}, \varrho = 1884\,\mathrm{kg/m^3}, \Phi = 0.48, K^S = 1.1 \cdot 10^{10}\,\mathrm{N/m^2}, \varrho_F = 1000\,\mathrm{kg/m^3}, K^F = 3.3 \cdot 10^9\,\mathrm{N/m^2}, \kappa = 3.55 \cdot 10^{-9}\,\mathrm{m^4/Ns}$) and the time step size is $\Delta t = 0.0008\,\mathrm{s}$.

In Fig. 2, the displacements at the surface in point P and in different depths are depicted versus time. The measurement points under the surface are placed on a circle with radius of 15 m, i.e., the point in 15 m depth is exactly 15 m below the excitation (see the small figure in Fig. 2). In principle, the results for both theories show the same behaviour. First, at $t \approx 0.017\,\mathrm{s}$ in Biot's theory and at $t \approx 0.028\,\mathrm{s}$ in TPM the arrival of the compressional wave is observed. Later, for both theories at $t \approx 0.07s$ the shear wave and at nearly the same time the Rayleigh wave arrives. As expected, with increasing depth

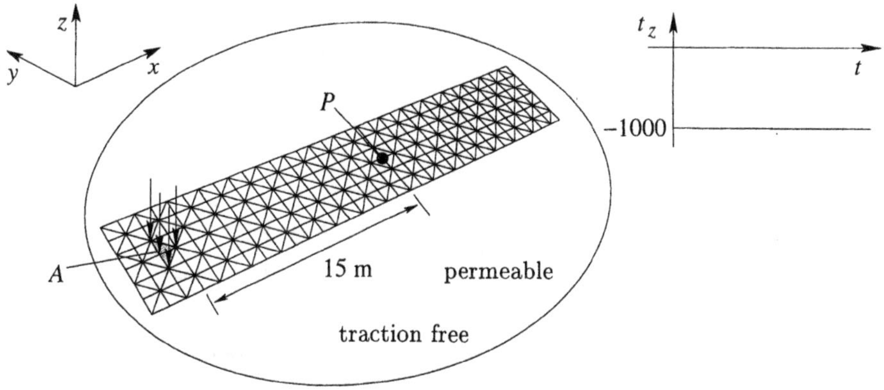

Fig. 1. Half space under vertical load: Discretization and load history

the influence of the Rayleigh wave vanishes noticeably due to the decrease of the Rayleigh pole.

However, also visible is the different modelling of the fast compressional wave by the TPM compared to Biot's theory. Not only the later arrival but also the stronger slope of the results indicate that these theories model something different. Keeping the identification (4) in mind, the results can be interpreted in the way that the influence of the water is modelled differently in both theories. This is in accordance with a more extensive study in [12].

References

1. Biot, M.A.: General Theory of Three-Dimensional Consolidation. *Journal of Applied Physics*, **12**, 155–164, 1941.
2. Biot, M.A.: Theory of Propagation of Elastic Waves in a Fluid-Saturated Porous Solid. I. Low-Frequency Range, II. Higher Frequency Range. *Journal of the Acoustical Society of America*, **28**(2), 168–191, 1956.
3. Bowen, R.M.: Theory of Mixtures. In *Continuum Physics.* (Eringen, A.C., Ed.), Vol. III, Academic Press, New York, 1–127, 1976.
4. Bowen, R.M.: Incompressible Porous Media Models by use of the Theory of Mixtures. *International Journal of Engineering Science*, **18**, 1129–1148, 1980.
5. Bowen, R.M.: Compressible Porous Media Models by use of the Theory of Mixtures. *International Journal of Engineering Science*, **20**(6), 697–735, 1982.
6. Ehlers, W.: Compressible, Incompressible and Hybrid Two-phase Models in Porous Media Theories. *ASME: AMD-Vol.*, **158**, 25–38, 1993.
7. Ehlers, W.; Kubik, J.: On Finite Dynamic Equations for Fluid-Saturated Porous Media. *Acta Mechanica*, **105**, 101–117, 1994.
8. Lubich, C.: Convolution Quadrature and Discretized Operational Calculus. I/II. *Numerische Mathematik*, **52**, 129–145/413–425, 1988.
9. Schanz, M.: A Boundary Element Formulation in Time Domain for Viscoelastic Solids. *Communications in Numerical Methods in Engineering*, **15**, 799–809, 1999.

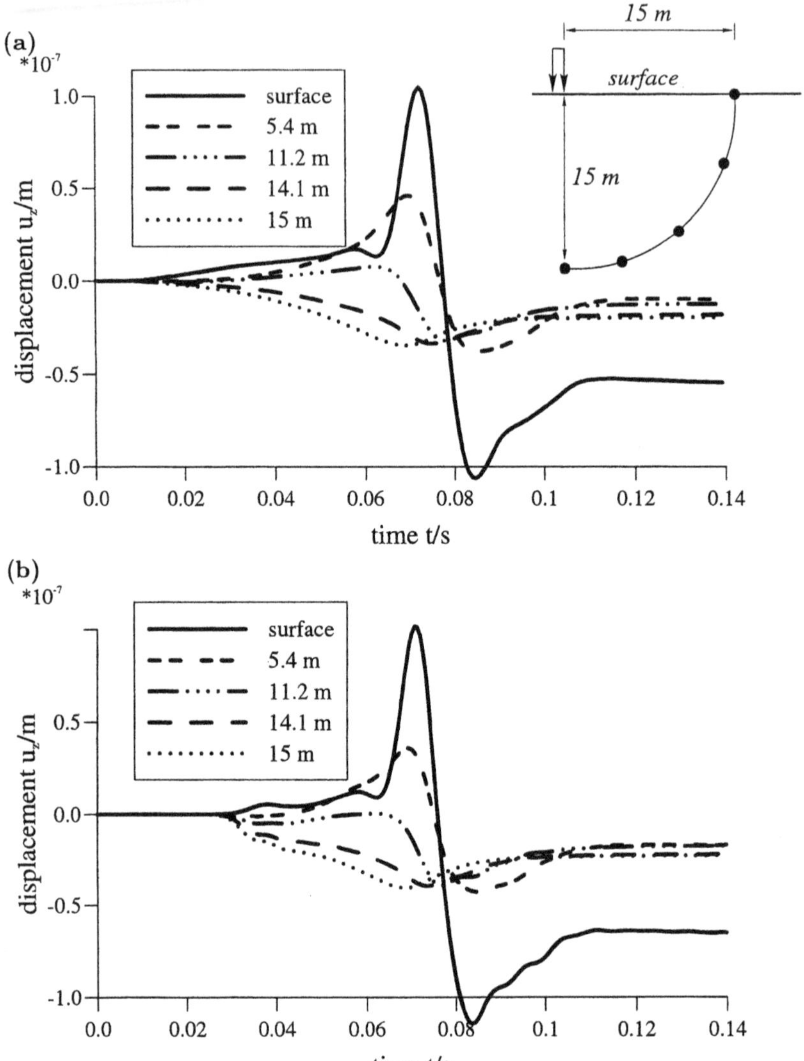

Fig. 2. Displacement at point P on the surface and in different depth versus time **(a)** Biot's theory **(b)** TPM

10. Schanz, M.: *Wave Propagation in Viscoelastic and Poroelastic Continua: A Boundary Element Approach.* Lecture Notes in Applied Mechanics. Springer-Verlag, Berlin, Heidelberg, New York, 2001.
11. Schanz, M.; Antes, H.: Application of 'Operational Quadrature Methods' in Time Domain Boundary Element Methods. *Meccanica*, **32**(3), 179–186, 1997.
12. Schanz, M.; Diebels, S.: A Comparative Study of Biot's Theory and the Linear Theory of Porous Media for Wave Propagation Problems. *Acta Mechanica*, submitted.

Aerodynamic, structural and electromagnetic interaction applied to conformal antennas

Harmen Schippers[1], Guus Vos[1], and Hans van Tongeren[1]

National Aerospace Laboratory NLR, P.O. Box 153, 8300 AD Emmeloord, The Netherlands

Abstract. Conformal phased array antennas on fighter aircraft are subject to unsteady aerodynamic loads. Mechanical forces and these aerodynamic loads will cause vibrations of the antenna aperture. As a consequence, the positions and slopes of the elements of the phased array antenna change, which in turn will change the antenna's radiation pattern: the main beam direction can change and the beam width and/or side lobe level can increase. This paper presents a methodology to compute vibrations of the aperture of the conformal antenna in relation to atmospheric gust and unsteady aerodynamic loads. Computational electromagnetic modelling will assess the subsequent distortion of the antenna radiation pattern.

1 Introduction

Modern fighter aircraft are equipped with an increasing number of antennas in order to fulfil more and more operational avionics functions which are mainly related to radar, Electromagnetic Counter Measures (ECM) and Communication, Navigation and Identification (CNI). To satisfy the increasing requirements on avionics functions, there is a world-wide interest in conformal phased array antennas, which can fulfil multiple functions and which can be flush-mounted on the exterior of the aircraft surface. When phased array antennas are located on the skin of the aircraft, they are subject to unsteady aerodynamic loads. The inertial forces and these aerodynamic loads will cause vibrations of the antenna aperture. The effects of such distortion on the antenna performance will be most significant for highly directional antennas such as Synthetic Aperture Radar (SAR) antennas.

This paper presents a methodology to compute vibrations of the aperture of the conformal antenna due to atmospheric gust loads taking into account unsteady aerodynamic loads. The involved mathematical formulations are related to the following fields:

1. Structural dynamics, to provide knowledge about eigenfrequencies and vibration modes of the unloaded structure supporting the antenna,
2. Aerodynamics, to provide unsteady aerodynamic loads at the antenna,
3. Aero-elastics, to provide knowledge about the magnitude of the vibrations,
4. Electromagnetics, to assess the radiation pattern of the distorted antenna aperture.

The methodology is applied to compute the radiation pattern of a deformed C-band phased array designed on a reconnaissance pod under a fighter aircraft. The C-band antenna consists of a faceted two-dimensional conformal array of patch antennas with 23 elements in across track and 30 elements in along-track. (see Fig. 1).

2 Structural and aerodynamic interaction

The dynamic analysis of the structure containing the phased array antenna is performed by a modal frequency response analysis. This method uses structural vibration mode shapes of the unloaded structure to reduce the size of the aero-elastic equations of motion. Let X_N denote the space of vibration modes belonging to the lowest N eigenfrequencies. The dynamic displacement $\boldsymbol{d}$ is approximated by

$$\boldsymbol{d} = \sum_{\nu=1}^{N} q_\nu \boldsymbol{\phi}_\nu, \boldsymbol{\phi}_\nu \in X_N. \tag{1}$$

The coefficients q_ν are known as the generalised co-ordinates of the vibration modes. The values of q_ν follow from the solution of the generalised modal deflection system which reads

$$M\frac{d^2\boldsymbol{q}}{dt^2} + D\frac{d\boldsymbol{q}}{dt} + K\boldsymbol{q} = \boldsymbol{F}^a + \boldsymbol{F}^e \tag{2}$$

The matrices M, D and K are the generalised mass, damping and stiffness matrices, which result from the free vibration analysis. These matrices are related to the selected set of vibration modes. Furthermore, $\boldsymbol{F}^a$ denotes the generalised aerodynamic force and $\boldsymbol{F}^e$ denotes a prescribed external force, which is typical for representative flight conditions.

The actual definition of the reduced space X_N of vibration modes depends on the relevant part of the aircraft surface to be analysed. For the front part of the reconnaissance pod the structural model is displayed in Fig. 2. This model contains shell elements for the representation of the skin and the frames, beam elements for the stiffeners on the skin. The vibration modes and the eigenfrequencies have been computed by means of NASTRAN. A characteristic vibration mode is shown in Fig. 3. The modes belonging to the lowest ten eigenfrequencies are used to define X_N.

Assume that $\boldsymbol{d}$ varies in time harmonically with radian frequency ω ($\omega = 2\pi F_V$) . Then, equation (2) can be written as:

$$(K + j\omega D - \omega^2 M)\boldsymbol{q} = \boldsymbol{F}^a + \boldsymbol{F}^e \tag{3}$$

The components of the generalised aerodynamic force $\boldsymbol{F}^a$ are given by

$$F_i^a = -\frac{1}{2}\rho_\infty U_\infty^2 \int_S C_p(w)\ (\hat{n} \cdot \boldsymbol{\phi}_i) dS \tag{4}$$

Here ρ_∞ and U_∞ denote the density and velocity at infinite distance from the aircraft. The objective of the unsteady aerodynamic analysis is to determine the value of the pressure coefficient C_p in equation (4) in relation to the normal velocity w of the dynamic displacement of the boundary surface. For relatively low frequency oscillations, C_p is related to the inverse of an integral operator which can be approximated by means of so-called doublet lattice methods. For relatively high frequency oscillations C_p can be computed by means of piston theory, where the pressure difference $p-p_\infty$ is proportional to the normal velocity w. The pressure difference is approximated by $p - p_\infty = \rho_\infty a_\infty w$, with a_∞ the speed of sound at infinity. Then, the pressure coefficient C_p becomes

$$C_p(w) = \frac{2}{M_\infty}\frac{w}{U_\infty}. \tag{5}$$

Here M_∞ denotes the Mach number at infinity. Other expressions for C_p in relation to w can be found in Ref [1]. The normal velocity w follows from the total derivative of the dynamic displacement with respect to time, i.e.,

$$w = \frac{d(\hat{n}\cdot \boldsymbol{d})}{dt} = \frac{\partial(\hat{n}\cdot \boldsymbol{d})}{\partial t} + U_\infty\frac{\partial(\hat{n}\cdot \boldsymbol{d})}{\partial x}, \tag{6}$$

with $\boldsymbol{d}$ given by (1). For harmonic motions, this equation can be rewritten as

$$\frac{w}{U_\infty} = j\frac{\omega}{U_\infty}(\hat{n}\cdot \boldsymbol{d}) + \frac{\partial(\hat{n}\cdot \boldsymbol{d})}{\partial x}. \tag{7}$$

Next substitute (7) into (5). Then, by equations (1) and (4), the ith component of the generalised aerodynamic force $\boldsymbol{F}^a$ can be written as $F_i^a = \sum_{\nu=1}^{N} A_{i,\nu}q_\nu$ with $A_{i,\nu}$ given by

$$A_{i,\nu} = -\frac{1}{2}\rho_\infty U_\infty^2 \int_S (\hat{n}\cdot \boldsymbol{\phi}_i)\frac{2}{M_\infty}(j\frac{\omega}{U_\infty}(\hat{n}\cdot \boldsymbol{\phi}_\nu) + \frac{\partial(\hat{n}\cdot \boldsymbol{\phi}_\nu)}{\partial x})dS \tag{8}$$

Inspection of (8) reveals that the first term of the integral yields a contribution to the damping matrix, while the second term yields a contribution to the stiffness matrix. As a consequence, the unsteady aerodynamic force can be written as $\boldsymbol{F}^a = (K^a + j\omega D^a)\boldsymbol{q}$. Equation (3) becomes

$$(K - K^a + j\omega(D - D^a) - \omega^2 M)\boldsymbol{q} = \boldsymbol{F}^e \tag{9}$$

The external excitation $\boldsymbol{F}^e$ may be due to atmospheric turbulence or due to control surface movements. In this paper, excitation due to gust loading is considered. Then, the ith component element of $\boldsymbol{F}^e$ is given by

$$F_i^e = 2\frac{\rho_\infty U_\infty}{M_\infty} w_g \int_S e^{j\omega\alpha/U_\infty i}(\hat{n}\cdot \hat{e}_z)(\hat{n}\cdot \boldsymbol{\phi}_i)dS, \tag{10}$$

with w_g the amplitude of the gust velocity. In many scenarios the values of w_g follow from the Dryden gust spectrum.

The methodology to compute the deformation of the aperture of a phased array antenna is as follows:

1. Specify a structural finite element model for the structure that supports the phased array antenna.
2. Calculate the vibration modes and eigenfrequencies of this model by means of structural finite element analysis code.
3. Specify the reduced space X_N of vibration modes.
4. Calculate the mass, damping and stiffness matrices M, D and K.
5. Calculate the aerodynamic damping and stiffness matrices D^a and K^a taking into account the above considerations.
6. Calculate the external loads for a range of frequencies ω with the $w_g = 1$.
7. Solve equation (9) for this range of frequencies.
8. Calculate the absolute value of the normal displacement at a certain point of the structural model.

This methodology has been used to compute the frequency response function of the front part of the reconnaissance pod. For the node with maximum deflection of the seventh vibration mode (at 229.6 Hz) the results are displayed in Fig. 4). The calculations reveal a maximum of 0.45 mm at this node for a gust speed amplitude of 1 m/sec. From in-flight measurements fighter aircraft gust speed amplitudes of 3.5 to 16 m/sec have been observed. Hence, for the seventh vibration mode a maximum deflection between about 1 and 7 mm can be expected.

For a deflection of 7 mm the deformation of the aperture of the C-band antenna is shown in Fig. 5.

3 Electromagnetic analysis

The vibration of the surface will effect the electromagnetic radiation of the antenna array. The radiation analysis requires knowledge about the modelling of single patch antenna elements and beam forming for conformal phased array antennas.

A single patch antenna element consists of a ground plane and a rectangular microstrip patch (with length L and width W) on a substrate. It is assumed that a coax cable feeds the patch. This feed generates standing waves in the substrate with frequency f_{EM}. The electric field lines are perpendicular to the ground plane and the patch, as required by boundary conditions for perfectly electric conductors (see figure 6). The substrate between the ground plane and the patch can be considered as a magnetic cavity with leaky magnetic walls. This cavity will support Transverse Magnetic modes (transverse to z, i.e. normal to the patch). The radiating mode of interest is the TM_{10} mode, which is obtained for $L = \lambda_s/2$ with λ_s the wavelength in the substrate, i.e.

$\lambda_s = c/(f_{EM}\sqrt{\epsilon_r})$, c is the speed of light and ϵ_r is the relative permittivity of the substrate. The standing wave with a half-wavelength separation leads to electric fields that are of opposite phase at the radiating sides. The radiation occurs mainly from fringing fields at these radiating sides (see figure 7).
The frequency f_{EM} of the electromagnetic field is much higher than the frequency f_V of the vibrating structure. As a consequence, for the electromagnetic analysis the displacement of the antenna surface (due to the vibration) can be considered as a quasi-steady deformation. The electromagnetic field responds instantaneously to the deformation.
The radiation pattern can be computed by means of an aperture analysis approach (see references [2] and [4]). The patch antenna has two radiating slot apertures in the plane of the patch. The electric field can be resolved into normal and tangential components with respect to this plane (see figure 6). The normal components in the apertures are out of phase and the far field produced by them cancels in the broadside direction. The tangential components of the electric field in the apertures are in phase. The resulting fields combine to give a maximum radiated field normal to the surface of the patch. Therefore, the radiating patch may be represented as two slots spaced at a distance L apart from each other, excited in phase and radiating in space above the ground plane (see figure 8). Each slot radiates the same field as a magnetic dipole with magnetic surface current $\boldsymbol{M} = (2V/h)\hat{m}$, with $\hat{m}$ the unit direction vector of the radiating edge, V the voltage across the slot, and h the height of the substrate. The magnetic surface currents on the surface of the patch are taken piecewise constant at the radiating slots, while they are assumed to be zero elsewhere on the surface. By this aperture approach, the radiated field due to a single patch antenna element can be approximated by (see Ref. [3])

$$\boldsymbol{E}(\boldsymbol{r}) = -jk\frac{e^{-jkR}}{4\pi R}\int_S \boldsymbol{M}(\boldsymbol{r}') \times \hat{R}\; e^{jk\boldsymbol{r}'\cdot\hat{R}}dS', \tag{11}$$

where R is the distance from the antenna to the observer, $\hat{R}$ the unit direction vector from the antenna to the observer, $k = 2\pi/\lambda$. The surface S is given by the two radiating apertures. The patch elements of the antenna array are assumed to be locally planar, both on the undisturbed and the deformed surface. Let p denote the index for patch number p of the antenna array. By the above modelling the radiated electric field of patch p can be approximated by

$$\boldsymbol{E}_p(\boldsymbol{r}) = -jk\frac{e^{-jkR}}{2\pi R}\frac{V_p}{h}(\hat{m}_p \times \hat{R})\int_S e^{jk\boldsymbol{r}'\cdot\hat{R}}dS', \tag{12}$$

where V_p is the voltage excited at patch p. The phase integral can be computed in closed form. The radiation pattern depends on the position of the patch on the surface and the orientation of $\hat{m}_p$. In the case the surface of the antenna array is deformed by vibrations, the electric field has to be computed

according to the new disturbed surface. Inspection of equation (12) reveals that the phase integral and $\hat{m}_p$ have to be computed anew for the deformed antenna surface.
The radiated field of the antenna array follows from the superposition of the fields fromf all elements $\boldsymbol{E}(\boldsymbol{r}) = \sum_p \boldsymbol{E}_p(\boldsymbol{r})$. By an appropriate choice of the values for the voltages V_p the pattern of the array can be directed. For the conformal antenna array of figure 1 the values of V_p have been determined by means of a least square method. A destination field was defined in such a way that the main beam of the array was in the direction $\theta = 135^o$ and $\phi = 90^o$. Furthermore, a side lobe level of -20 dB was specified. The radiation pattern of the synthesis is shown in figure 9. The deterioration of the antenna performance is predicted for a steady deformation that is related to a prescribed amplitude of the vibration mode as displayed in figure 5. As follows from the elasto- mechanical modelling gust loads loads can cause maximum deflections of about 4 to 6 mm. Notice that a deflection 6 mm corresponds with an amplitude 1/8 λ. For this amplitude, the distorted radiation pattern is displayed in Figure 10. A comparison of figures 9 and 10 reveals that the beam-width of the antenna array will increase slightly, while the side lobe levels near the main beam increase significantly. Figure 10 can be interpreted as a snapshot of the radiation pattern of the C-band antenna, at maximum deformation for the specific vibration mode.

References

1. Bisplinghoff R.L., H. Ashley, Principles of Elasticity (1975), Dover Publications, New York.
2. Bahl I.J., Bhartia P., Micro Strip Antennas (1980) , Artech House, Dedham.
3. Silver, S., Microwave Antenna Theory and Design (1984), Peter Peregrinus, London.
4. Stutzman W.L., Thiele G.A., Antenna Theory and Design (1998), John Wiley and Sons Inc., New York, etc.

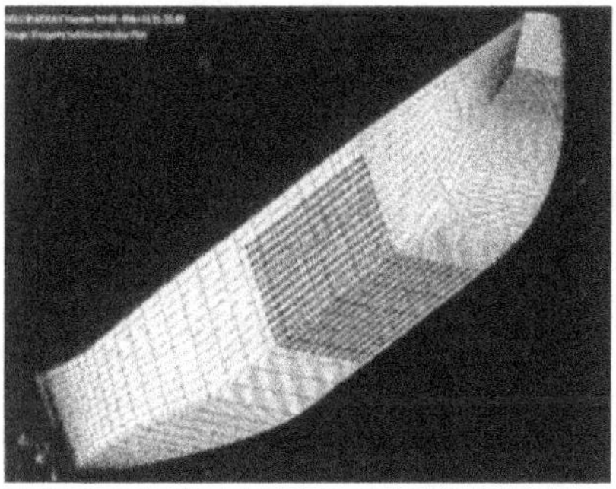

Fig. 1. Artist impression of phased array antenna on reconnaissance pod; col ored, see page 378

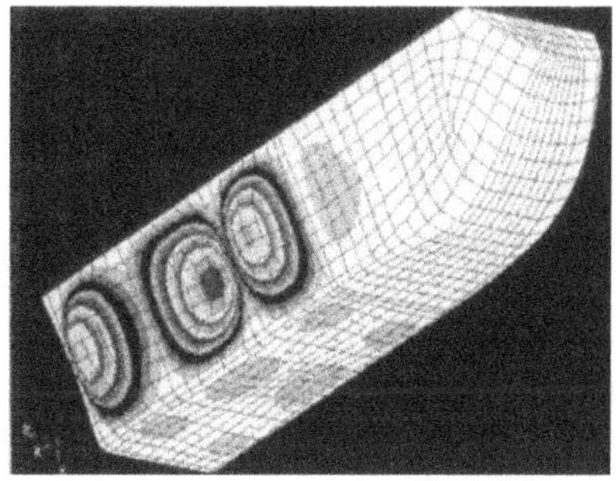

Fig. 3. Vibration mode at 229.6 Hz; col-

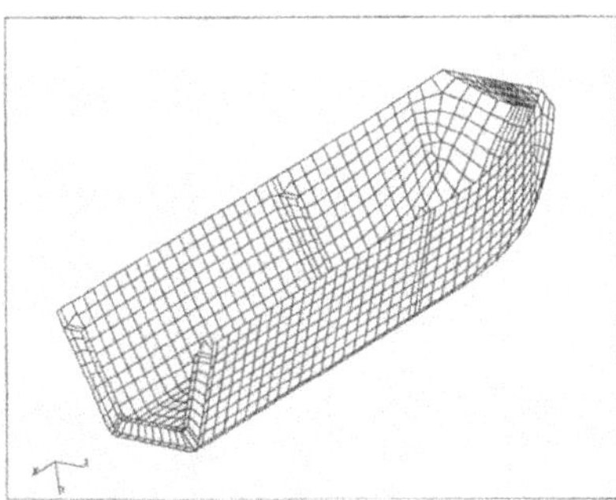

Fig. 2. Structural model of front part of reconnaissance pod

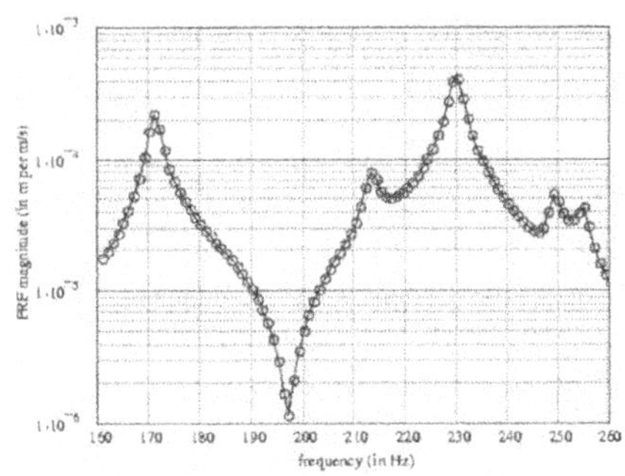

Fig. 4. Frequency response function related to gust load $w_g = 1$; colored, see page 378

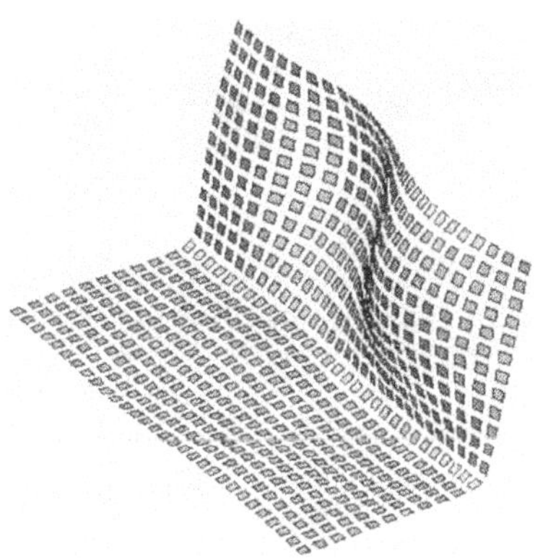

Fig. 5. Deformation of antenna aperture; colored, see page 378

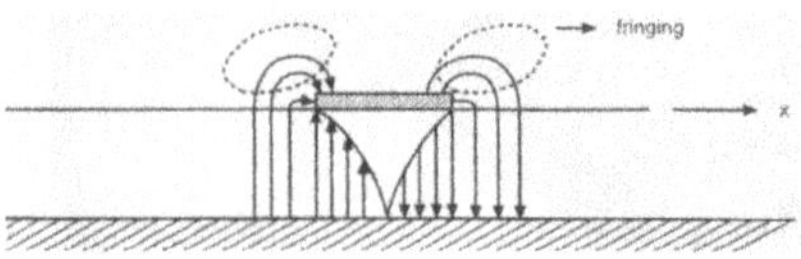

Fig. 6. Electric field in substrate

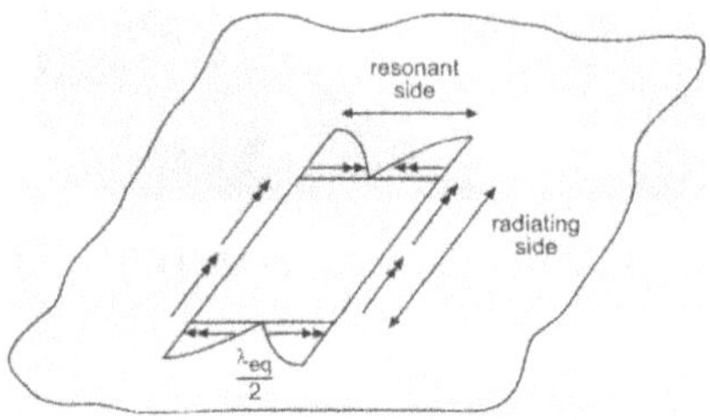

Fig. 7. Resonant and radiating sides of patch

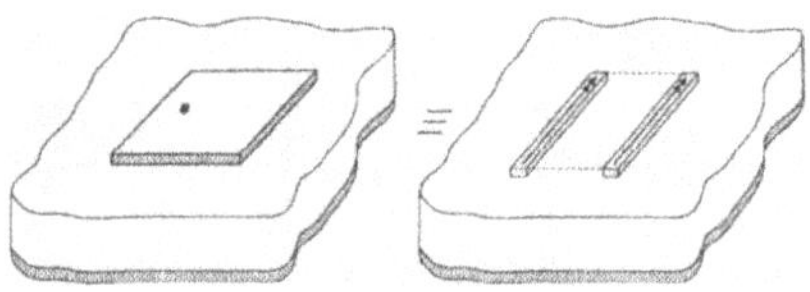

Fig. 8. Equivalent representation of patch by two radiating slots

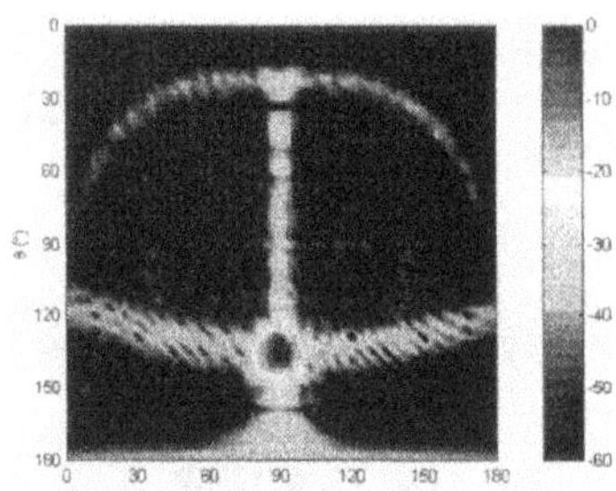

Fig. 9. 3D radiation pattern of the non-deformed SLAR antenna; colored, see page 379

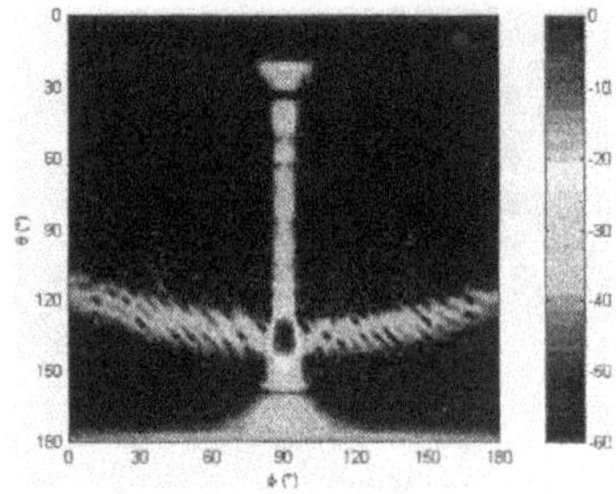

Fig. 10. 3D radiation pattern of the SLAR antenna, surface deformed by vibration mode with resonance frequency 229.6 Hz; colored, see page 379

Direct Simulation of Aeroacoustics

Thomas Schwartzkopff[1] and Claus-Dieter Munz[1]

Stuttgart University, Institute for Aero- and Gasdynamics, Pfaffenwaldring 21, 70550 Stuttgart, Germany

Abstract. Direct simulation of both fluid flow and noise within one numerical calculation offers a valuable means of testing the validity and applicability of aeroacoustic theories and analogies because one does not suffer from uncertainties concerning the empirical modelling of aeroacoustic source terms. To reduce computational costs we propose to apply domain decomposition methods. The linearised Euler equations are used in the regime where nonlinear effects can be neglected. Furthermore different grids and time steps are used in each domain, thus leading to a much more efficient code. The main difficulty here is to avoid non-physical reflections at the artificial interface between different mathematical models, grids and numerical methods. In this paper a new idea is presented, how to reduce these reflections by two orders of magnitude. We restrict ourselves to one space dimension.

1 Introduction

The main difficulty in the calculation of sound generated by fluid flow at low Mach numbers is the occurrence of different scales [5,2,7]. The fluid flow is characterised by small spatial fluid structures containing a large amount of energy that propagate with small convective velocity, such as small vortices in a turbulent flow. The radiated acoustic waves are phenomena with small amplitudes and small energy amount, but with long wavelength due to fast propagation with sound velocity. In addition the propagation of sound over long distances is of interest. These different scales and different physical behaviours of fluid flow and sound propagation lead to a difficult task for the construction of numerical methods to capture sound propagation and generation well.

Classical approaches in aeroacoustics are mainly based on analytical solutions of the linear wave equation in the frequency domain which is valid in the far field. The sound generation is approximated by source terms obtained from a flow simulation. This procedure is called 'acoustic analogy' and was initiated by the classical work of Lighthill [3]. The increasing power of modern computers in connection with the improvement of numerical methods enables to advance into the direction of direct simulation of fluid flow together with acoustics in the time domain. Direct simulations may give better insight into the mechanism of sound generation and propagation in the vicinity of the flow regime. Thus, the gap between computational fluid dynamics (CFD) and classical acoustics may be closed[4]. This discipline is often referred as computational aeroacoustics (CAA).

However the requirements for a flow solution and an acoustic solution are different although they obey the same equations. On the one hand the resolution of shocks or strong local gradients in the fluid flow has to be accurate. Therefore one uses numerical schemes with relatively high, inherent dissipation around discontinuities and introduces artificial viscosity. On the other hand, the acoustic solver should propagate smooth solutions without dissipation. Thus a numerical scheme with low numerical dissipation and viscosity is required in this case[5].
In this paper a non-overlapping domain decomposition is considered[1]. The global domain Ω is decomposed into several subdomains Ω^i which are adapted to the local requirements. The flow domain around an obstacle Ω^E is calculated with the full, non-linear Euler or Navier-Stokes equations on an unstructured grid. In the neighbouring domain Ω^{LE} the linearised Euler equations on a structured grid are used for the acoustic wave propagation if non linear effects can be neglected.
We note, that high order schemes for solving the flow problem have to be used, because the acoustics are inherent in the flow phenomena and have to be resolved without dispersion and damping.

2 Equations

This paper deals with some surveys of the one dimensional coupling problem of an Euler domain (Ω^E) to a linearised Euler domain (Ω^{LE}). The nonlinear Euler equations are described by:

$$\boldsymbol{U}_t^E + \boldsymbol{F}^E(\boldsymbol{U}^E)_x = 0 \tag{1a}$$

$$\boldsymbol{U}^E = \begin{bmatrix} \rho \\ M \\ E \end{bmatrix}, \quad \boldsymbol{F}^E(\boldsymbol{U}^E) = \begin{bmatrix} M \\ \frac{3-\gamma}{2}\frac{M^2}{\rho} + (\gamma-1)E \\ \gamma\frac{ME}{\rho} - \frac{\gamma-1}{2}\frac{M^3}{\rho^2} \end{bmatrix} \tag{1b}$$

and the linearised Euler equations in conservative variables are given by

$$\boldsymbol{U}_t^{LE} + \boldsymbol{F}^{LE}(\boldsymbol{U}^{LE})_x = 0 \tag{2a}$$

$$\boldsymbol{U}^{LE} = \begin{bmatrix} \rho' \\ M' \\ E' \end{bmatrix} = \begin{bmatrix} \rho - \rho_0 \\ M - M_0 \\ E - E_0 \end{bmatrix} \tag{2b}$$

$$\boldsymbol{F}^{LE}(\boldsymbol{U}^{LE}) = \begin{bmatrix} M' \\ \frac{3-\gamma}{2}\frac{2\rho_0 M_0 M' - M_0^2 \rho'}{\rho_0^2} + (\gamma-1)E' \\ \gamma\frac{\rho_0 M_0 E' + \rho_0 E_0 M' - M_0 E_0 \rho'}{\rho_0^2} - \frac{\gamma-1}{2}\frac{3\rho_0 M_0^2 M' - 2M_0^3 \rho'}{\rho_0^3} \end{bmatrix} \tag{2c}$$

The linearised equations are obtained from the nonlinear equations by introducing the perturbation (2b) in (1a) and neglect terms which are of $\mathcal{O}(\boldsymbol{U}'^2)$.

3 Physical Considerations

The problem of coupling two different sets of equations, namely nonlinear Euler (1a) and linearised Euler equations (2a), may be considered as coupling materials with different wave propagation speeds. The wave speeds (eigenvalues) of (1a) are $\lambda_1^E = u$ and $\lambda_{2/3}^E = u \pm c$ whereas the wave speeds of (2a) are given by $\lambda_1^{LE} = u_0$ and $\lambda_{2/3}^{LE} = u_0 \pm c_0$.
From the physics of waves it is known [8] that the reflection index r for a wave passing this interface Γ is characterised by the refraction indices n of the materials. The indices are given by

$$n^{LE} = \frac{1}{u_0 + c_0} \qquad n^E = \frac{1}{u + c} \tag{3}$$

$$r = \left(\frac{n^{LE} - n^E}{n^{LE} + n^E}\right)^2 \approx \left(\frac{\mathcal{O}(U'^2)}{\mathcal{O}(U_0^2)}\right)^2 \tag{4}$$

Furthermore the waves are diffracted at the interface. If the wave front has the angle α^E with respect to the normal at the interface, the diffracted wave in Ω^{LE} will have the angle

$$\alpha^{LE} = \arcsin\left(\frac{n^E}{n^{LE}} \sin(\alpha^E)\right) \tag{5}$$

and even total reflection is possible for $n^E/n^{LE} \cdot \sin(\alpha^E) > 1$.
One can conclude, that if a numerical coupling scheme is applied, which simulates the artificially introduced interface like a material interface without any errors, then there will certainly be reflections at the interface which are of the magnitude $\mathcal{O}(U'^2)$. The aim is now to find a transmission condition which avoids this physical behaviour at the "artificial" interface.

4 Numerical Considerations

The main idea is to combine the two different mathematical models into one, so that they can be handled in an easier way. Therefore a new parameter s is introduced and a new flux function is generated

$$F_s^\star(U) = \frac{1+s}{2} F^E(U) + \frac{1-s}{2} F^{LE}(U) \tag{6}$$

which is now valid in Ω^E $(s = +1)$ and Ω^{LE} $(s = -1)$. At the interface $s \in [-1, +1]$ is used. A discussion about this flux function is given in [6].

4.1 Numerical Tests

A Gaussian pulse in the pressure with amplitude $\hat{p}' = 0.1 \cdot p_0$ is prescribed as initial condition. It divides into a left moving and a right moving part, one leaving the domain and the other one passes the interface between $\Omega^E \leftrightarrow \Omega^{LE}$ (see Fig. 1(**a**)). Calculations were carried out for a pulse passing the interface in both directions and for variational domain discretisations in space and time (see Table 1).

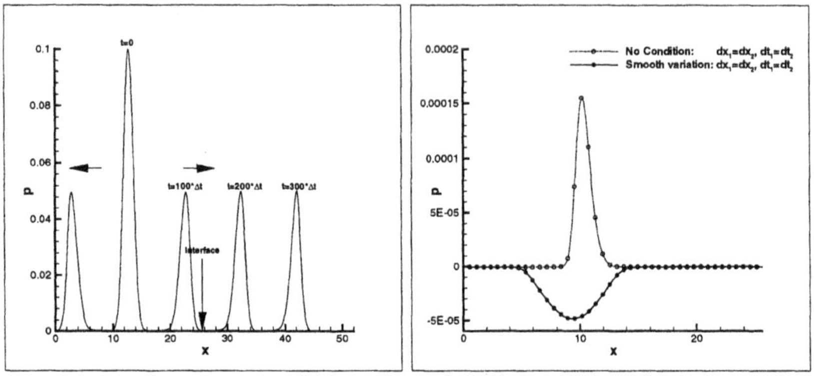

Fig. 1. (**a**) Evolution of an Gaussian pressure pulse as initial condition (**b**) Reflections generated at the interface for a smooth variation of s

4.2 Definition of the Interface

Two ideas how to choose s at the interface are presented. First a smooth variation of s is investigated and secondly an unequal distribution of s for Ω^E and Ω^{LE} is analysed.

Smooth Variation Perhaps the most natural idea is to use a smooth variation of s which corresponds to a smooth variation from nonlinear to linearised Euler equations. The parameter s varies from $s = +1$ to $s = -1$ with a cosine function, spread over 40 cells. Fig. 1(**b**) gives a comparison of the reflections produced by this condition compared to a reflection produced without special treatment of the interface. Note that this reflection is indeed of the order $\mathcal{O}(\boldsymbol{U}'^2)$ as predicted by (4). It can be seen clearly that the amplitude of the reflection decreases by about 1/3. Unfortunately the spatial stretch grows by the same amount. Thus the integral over the reflection is more or less the same. The question still remains: Is it possible to reduce the reflection at the interface?

Discontinuous Interface Flux In this approach different s^E and s^{LE} distributions are used at the interface. The fluxes at the interface Γ are:

$$F_{\partial\Omega^E \cap \Gamma} = F^{HLL}\left(F^\star_{s=1}(\boldsymbol{U}_L), F^\star_{s=0}(\boldsymbol{U}_R)\right) \tag{7a}$$

$$F_{\partial\Omega^{LE} \cap \Gamma} = F^{HLL}\left(F^\star_{s=0}(\boldsymbol{U}_L), F^\star_{s=-1}(\boldsymbol{U}_R)\right) \tag{7b}$$

This definition leads to a flux difference $F_{\partial\Omega^E \cap \Gamma} - F_{\partial\Omega^{LE} \cap \Gamma}$ at the interface and thus this scheme is not conservative at that point.
The numerical results for a pulse passing the interface from Ω^E to Ω^{LE} can be found in Fig. 2(**a**) and for a pulse in the other direction in Fig. 2(**b**). The reflections with this new condition are compared to the reflections which are generated without special treatment at the boundary. Table 1 shows that the

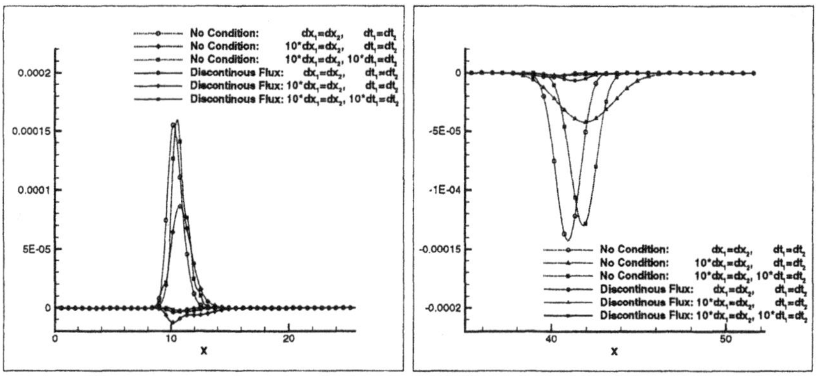

Fig. 2. Reflections generated at the interface for discontinuous interface flux. (**a**) Gaussian pulse passes interface from Ω^E to Ω^{LE}. (**b**) Gaussian pulse passes interface from Ω^{LE} to Ω^E. Test cases are given in Table 1

amplitude of the reflections decreases by about two orders of magnitude with this new interface flux.
Furthermore the reflection predicted by (4) has been validated. No matter

Table 1. Reduction rate of the reflections produced by discontinuous interface flux

No.	$\Delta x^{LE}/\Delta x^E$	$\Delta t_{\Omega^{LE}}/\Delta t_{\Omega^E}$	Reduction rate $\Omega^E \rightarrow \Omega^{LE}$	Reduction rate $\Omega^E \leftarrow \Omega^{LE}$
1	1	1	$2.62 \cdot 10^{-2}$	$2.64 \cdot 10^{-2}$
2	10	1	$14.5 \cdot 10^{-2}$	$4.53 \cdot 10^{-2}$
3	10	10	$2.15 \cdot 10^{-2}$	$5.15 \cdot 10^{-2}$

how the spatial or temporal parameters are chosen, the reflection rate is more or less constant for the case, where no special treatment is done. This is an indicator, that the physical reflections from the "unphysical" interface are much stronger, than the reflections produced by discretisation errors resulting from non–conforming discretisations.

5 Conclusion

The coupling of nonlinear Euler equations with linearised Euler equations has been studied under the aspect of reflections produced by sound waves which pass the interface Γ. If the fluid flow is considered as a nonlinear Euler fluid and as a linearised Euler fluid, each one with different wave speeds (eigenvalues), then some properties of the interface can be shown from considerations with respect to the physics of waves. This leads to a completely new interpretation of this interface.

Numerical tests have shown, that even if a smooth variation from Euler to linear Euler domain is used it is not possible to reduce the reflections. Each Finite–Volume may be regarded to contain a numerical fluid with a different wave speed, leading to smaller reflections at every interface but in the overall sum to the same reflection rate. A new interface condition, using discontinuous fluxes at the interface leads to an reduction rate of two orders of magnitude.

The extension to two dimensions is part of present work. Here an other effect may occur: Diffraction or even total reflection. The interface conditions have to be modified to prevent these two and three dimensional effects.

Acknowledgements

Part of this work was supported by the SFB404 *Mehrfeldprobleme in der Kontinuumsmechanik*, funded by the Deutsche Forschungsgesellschaft DFG.

References

1. J.B. Freund, S.K. Lele, and P. Moin. Matching of near/far-field equation sets for direct computation of aerodynamic sound. In *15th AIAA Aeroacoustics Conference*, October 1993.
2. J.C. Hardin and M.Y. Hussaini. *Computational Aeroacoustics.* ICASE/NASA LaRC Series. Springer Verlag, 1993.
3. M.J. Lighthill. On sound generated aerodynamically - 1. general theory. *Proc. Roy. Soc. London*, 211(A1107):564–587, 1952.
4. B.E. Mitchell, S.K. Lele, and P. Moin. Direct computation of the sound from a compressible co-rotating vortex pair. *J. Fluid Mech.*, 285:181–202, 1995.
5. C.-D. Munz. Computational fluid dynamics and aeroacoustics for low mach number flow. In A. Meister and J. Struckmeier, editors, *Hyperbolic Partial Differential Equations*, pages 269–320. Vieweg, 2002.
6. C.-D. Munz and T. Schwartzkopff. Direct simulation of sound generation and sound propagation using heterogenous domain decomposition. In *Proccedings of WCCM*, July 2002. Preprint.
7. A. Powell. Some aspects of aeroacoustics: From rayleigh until today. *J. of Vibration and Acoustics*, 112:145–159, 1990.
8. P.A. Tipler. *Physik.* Spektrum Akademischer Verlag, 1994.

Simulation of Sedimentation of Monodisperse and Polydisperse Suspensions

E.M. Tory[1], R.A. Ford[2], and M. Bargieł[3]

[1] Professor Emeritus, Mount Allison University, Sackville, NB, Canada
[2] Spielo Gaming International, Moncton, NB, Canada
[3] Institute Of Computer Science, University of Mining and Metallurgy, Cracow, Poland

Abstract. The behaviour of sedimenting monodisperse suspensions is usually deduced from the flux plot, but this approach is not available for polydisperse suspensions. Also, the diversity of velocities, even in monodisperse suspensions, produces a hydrodynamic diffusion that is not taken into account in Kynch's theory. The rapid improvement in computing power has made simulation an attractive method. Sedimentation of suspensions with many species can be handled easily, and stochastic effects can be included, if desired. We show that two sources of difficulty, generation of a concentration gradient and control of fluctuations in concentration, can be overcome by choosing the controlling concentration as that immediately below the test sphere.

1 Introduction

The theoretical framework for the sedimentation of polydisperse suspensions of spheres [3] is the system of ODEs

$$\frac{d\boldsymbol{x}_i}{dt} = \boldsymbol{v}_i = -\frac{4\pi}{3\eta}\sum_{j=1}^{N} a_j^3(\rho_j - \rho_f)g\boldsymbol{A}_{ij}\cdot\boldsymbol{k},\ i = 1, 2, \ldots, N \tag{1}$$

where $\boldsymbol{x}_i$ and $\boldsymbol{v}_i$ are the position and translational velocity of the ith sphere, t is time, a_j is the radius of the jth sphere, g is the gravitational acceleration, $\boldsymbol{k}$ is a unit vector in the vertical direction, and $\boldsymbol{A}_{ij}$ is the translational mobility tensor, which is independent of the density, ρ_j, of the particles and the density, ρ_f, and viscosity, η, of the fluid. $\boldsymbol{A}_{ij}$ depends on the size and position of every sphere and on the nature and position of the boundaries. Noting that the Stokes velocity of the jth sphere is

$$u_{\infty j} = -2a_j^2(\rho_j - \rho_f)g/9\eta\ , \tag{2}$$

we can write (1) as

$$\frac{d\boldsymbol{x}_i}{dt} = \sum_{j=1}^{N} u_{\infty j}\boldsymbol{A}_{ij}^{+}\cdot\boldsymbol{k},\ i = 1, \ldots, N \tag{3}$$

where $\boldsymbol{A}_{ij}^{+} = 6\pi a_j \boldsymbol{A}_{ij}$. Given the initial conditions,

$$\boldsymbol{x}_i(0) = \boldsymbol{x}_{i0},\ i = 1, \ldots, N\ , \tag{4}$$

integration of (1) provides a complete description of the sedimentation path of every sphere.

2 Theory

We restrict our treatment to one dimension. Then, a three-parameter Markov model [6] closely approximates the behaviour of a monodisperse suspension [5,8]. We follow Tory and Ford [8] in generalizing the Markov model to m species with $\rho_1 = \rho_2 = \cdots = \rho_m = \rho_s > \rho_f$. We can distribute $N = \sum_{j=1}^{m} N_j$ sphere centres along the initial height of the suspension and choose a hypothetical cross-sectional area such that the required initial concentration $\boldsymbol{\Phi}_0$ is attained. However, concentrations vary with height and time. We assume that each sphere's parameters are determined by the local solids concentration, $\boldsymbol{\Phi}_i = (\phi_{i1}, \phi_{i2}, \ldots, \phi_{iN})^T$, where Φ_i is the solids concentration vector in the ith bin [8]. Then

$$dX_{jk}(t) = V_{jk}(t)dt \tag{5}$$

$$dV_{jk}(t) = -\beta_j(\Phi_i)[V_{jk}(t) - \mu_j(\Phi_i)]dt + [2\beta_j(\Phi_i)]^{1/2}\sigma_j(\boldsymbol{\Phi}_i)dW_{jk} \tag{6}$$

where $[2\beta_j(\Phi_i)]^{1/2}\sigma_j(\Phi_i)$ controls the variation of velocity increments for the jth species, $\beta_j(V_j - \mu_j)$ describes the strength with which individual velocities are shifted toward the steady-state mean value, μ_j, and $W_{jk}(t)$ denotes a Wiener process. Here σ_j is the standard deviation of the steady-state instantaneous velocity of particles of the jth species.

Sedimentation can also be studied as an advection-diffusion process [5]

$$\frac{\partial \phi_j}{\partial t} = -\frac{\partial}{\partial x}[\phi_j \mu_j(\boldsymbol{\Phi}_i)] + \frac{\partial}{\partial x}\left[D_j(\boldsymbol{\Phi}_i)\frac{\partial \phi_j}{\partial x}\right] . \tag{7}$$

Generalizing Tory and Ford [8], we write

$$\frac{\partial}{\partial x}\left[D_j(\boldsymbol{\Phi}_j)\frac{\partial \phi_j}{\partial x}\right] = -\frac{\partial}{\partial x}\left[\phi_j \frac{\partial}{\partial x} D_j(\boldsymbol{\Phi}_i)\right] + \frac{\partial^2}{\partial x^2}[D_j(\boldsymbol{\Phi}_i)\phi_j] . \tag{8}$$

When (8) is substituted in (7), the latter has the form of a Fokker-Planck equation that is equivalent to the SDE

$$dX_{jk}(t) = \left[\mu_j(\boldsymbol{\Phi}_i) + \frac{\partial}{\partial x} D_j(\boldsymbol{\Phi}_i)\right] dt + (2D_j(\boldsymbol{\Phi}_i))^{1/2} dW_{jk}(t) \tag{9}$$

where

$$\frac{\partial}{\partial x} D_j(\boldsymbol{\Phi}_i) = \sum_{j=1}^{m} \frac{\partial}{\partial \phi_j} D_j(\boldsymbol{\Phi}_i)\frac{\partial \phi_j}{\partial x} . \tag{10}$$

This reduces to

$$dX_{jk}(t) = \mu_j(\boldsymbol{\Phi}_i)dt \tag{11}$$

when hydrodynamic diffusion can be neglected.

We can generate the N paths of the spheres in parallel. At each time, each sphere in the ith bin ($i = 1, \dots, K$) moves independently according to the values of the parameters determined by $\boldsymbol{\Phi}_i$. However, each $\boldsymbol{\Phi}_i$ is also a random variable that changes with each step as the configuration evolves and the number of spheres in the ith bin changes. Equations (3), (5–6), (9), and (11) are all Pickard–Tory [4] processes. However, to produce a realistic simulation, the dependence of the parameters on $\boldsymbol{\Phi}$ must be known, the fluctuations that result from random initial distributions and stochastic processes must be controlled in a natural way, and concentration gradients must be generated naturally when appropriate.

3 Deterministic simulation of dilute polydisperse suspensions

Table 1. Species distribution

Species	Diameter	d_1^3/d_j^3	Number
1	87.69	1.0000	30,566
2	83.45	1.1603	35,465
3	79.21	1.3568	41,472
4	74.97	1.6002	48,912
5	70.73	1.9056	58,247
6	66.49	2.2939	70,115
7	62.25	2.7953	85,441
8	58.01	3.4592	105,734
9	53.77	4.3374	132,577
10	49.53	5.5494	169,623
11	45.29	7.2584	221,860

At present, only $\mu(\boldsymbol{\Phi})$ is accurately known. The Richardson–Zaki equation

$$\mu(\phi) = u_\infty(1-\phi)^n \tag{12}$$

is supported by many experimental studies. Its generalization to m species is the Masliyah–Lockett–Bassoon equation [1], which yields

$$\mu_k(\boldsymbol{\Phi}_i) = u_{\infty 1}(1-\phi_i)^{3.65}(\delta_k - \sum_{j=i}^{m} \delta_j \phi_{ij}),\ k \geqslant i \tag{13}$$

as the velocity of the kth species in the ith bin. By definition, species i is the largest in the ith bin. Here $u_{\infty 1}$ is the Stokes velocity of the largest species and $\delta_k = d_k^2/d_1^2$. We simulated the sedimentation of 11 species with $\phi_{01} = \phi_{02} = \cdots = \phi_{011} = 0.00682$ and hence $\phi_0 = 0.075$. The diameters and numbers of each species are shown in Table 1. Our deterministic simulation used 1,000,012 spheres with μ_k given by (13) and $u_{\infty 1} = -0.673\,\mathrm{cm/sec}$.

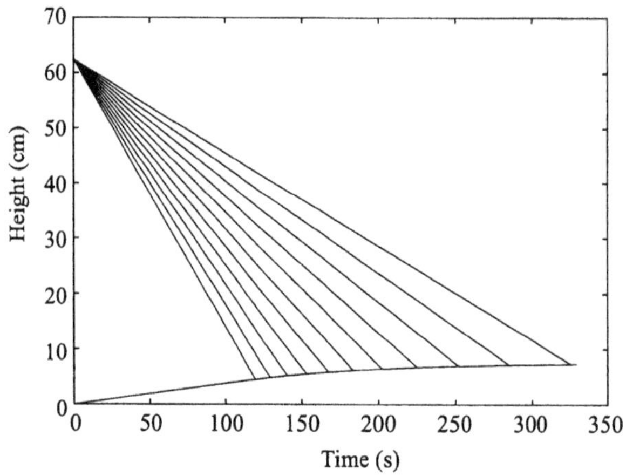

Fig. 1. Trajectories of top spheres for each species

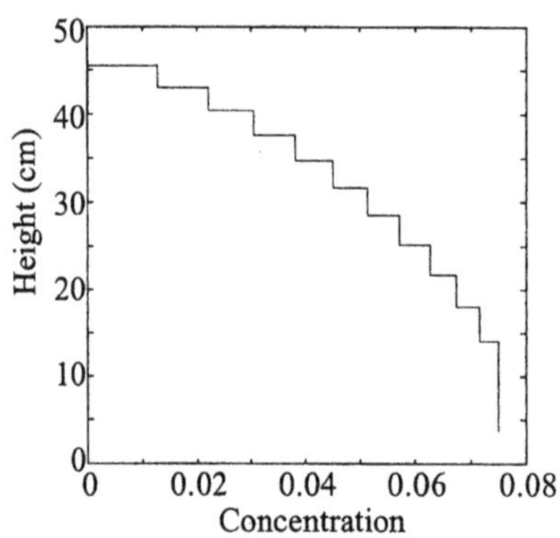

Fig. 2. Solids profile

Figure 1 shows the trajectories of the top spheres of each species. The natural floating bins are the regions delineated by these trajectories. Figure 2 shows the solids profile. It might seem that the steps should be of equal width, but the concentration of the smaller species is increased by the elimination of the largest species from the zone. This increase happens automatically in simulations [8].

4 Simulation of concentration gradients

The principal short-range interaction in the entire path of a sedimenting sphere occurs as it settles into the packed bed and lubrication terms become important. This suggests that a floating personalized asymmetric bin might imitate the actual behaviour when concentration gradients begin to form near the bottom of the container. Thus, we assume that the velocity of a particle at z_i is governed by the number of particles in $[z_i - h, z_i]$, where h is small. As the behaviour of a sphere approaching a flat surface is qualitatively similar to that of a sphere approaching a fixed sphere, we created an artificial layer of bed of thickness h (which, of course, does not appear in any of the results).

A method that can be used for stochastic simulations must not only form a concentration gradient, but also be stable when concentrations fluctuate, either from the random initial distribution or the variation in flux caused by the stochastic terms. These are always troublesome, but especially when $|\phi\mu(\phi)|$ decreases as ϕ increases. We used (11) and (12) with $n = 4.65$ and $\phi = 0.35$ in a deterministic simulation. Since the inflection point occurs at $\phi = 0.354$, this should be an MS-4, i.e., the solution consists of two constant states, ϕ_0 and ϕ_∞, and a concentration gradient [2].

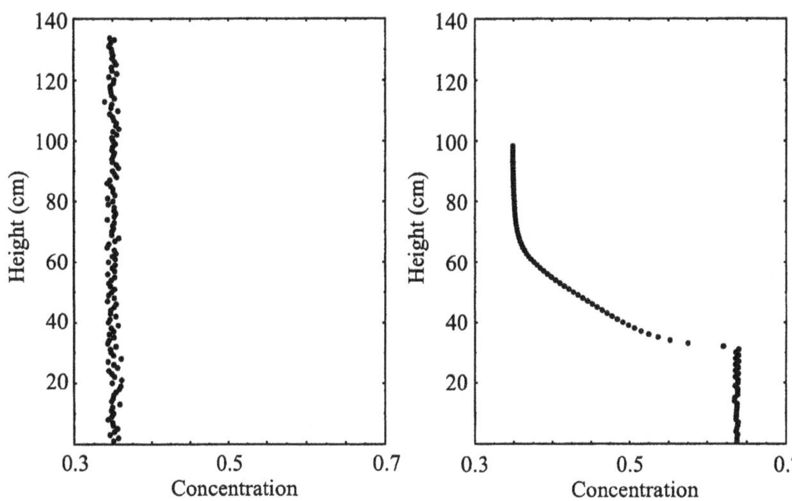

Fig. 3. Formation of a concentration gradient. (**a**) $t = 0$ (left), (**b**) $t = 4000$ (right).

We distributed 40,000 spheres randomly (according to a uniform distribution) over a height of 134 cm. As expected, the number of spheres varied substantially with height as shown in Fig. 3(**a**). Figure 3(**b**) shows that a gradient has formed and the variability has been greatly reduced. The concentration in the upper zone is constant at 0.35 and there is a gradient with a very sharp increase in ϕ (corresponding to a theoretical jump) near the packed bed. Though the predicted jump from 0.35 to 0.354 is too small to register, the overall agreement is excellent.

5 Discussion

As computing power increases, the solution of problems by simulation becomes more and more practical. In the Theory section, we introduced several candidates for simulation. Equation (3) is useful only for simple cases with few spheres, (9) is practical only when $\partial_x D(\boldsymbol{\Phi}) = 0$, and (11) when stochastic terms are negligible. However, (5) and (6) use the steady-state parameters $\mu_j(\boldsymbol{\Phi})$, $\sigma_j(\boldsymbol{\Phi})$, and $\beta_j(\boldsymbol{\Phi})$ of the Markov model to predict the transient behaviour of individual particles and hence of the entire suspension, including shocks, gradients, and upper rarefaction waves [7]. This is much more economical.

Acknowledgments This paper was largely written while E.M. Tory was a visitor at the University of Stuttgart. The invitation from Prof. Dr. W.L. Wendland, financial support from the Collaborative Research Center (Sonderforschungsbereich 404), and helpful discussions with S. Berres and Dr. R. Bürger are gratefully acknowledged.

References

1. Bürger, R., Karlsen, K.H., Tory, E.M. and Wendland, W.L. (2002). Model equations and instability regions for the sedimentation of polydisperse suspensions of spheres. ZAMM (to appear).
2. Bustos, M.C., Concha, F., Bürger, R. and Tory, E.M. (1999). Sedimentation and Thickening. Kluwer Academic Publishers, Dordrecht.
3. Mazur, P. and van Saarloos, W., (1982). Many-sphere hydrodynamic interactions and mobilities in a suspension: Physica A 115: 21–57.
4. Pickard, D.K. and Tory, E.M. (1987). A Markov model for sedimentation: fundamental issues and insights. In MacNeil, I.B. and Umphrey G.J. (Eds.), Advances in the Statistical Sciences, Vol. IV, Stochastic Hydrology, D. Reidel, Dordrecht, pp. 1–25.
5. Tory, E.M. (2000). Stochastic sedimentation and hydrodynamic diffusion. Chem. Eng. J. 80, 81–89.
6. Tory, E.M. and Pickard, D.K. (1977). A three-parameter Markov model for sedimentation. Can. J. Chem. Eng. 55, 655–665.
7. Tory, E.M. and Ford, R.A., (2000). Stochastic simulation of sedimentation. In: M. Rahman and C.A. Brebbia (Eds.), Advances in Fluid Mechanics III. WIT Press, Southampton, UK, pp. 663–672.
8. Tory, E.M. and Ford, R.A. (2002). Simulation of sedimentation of bidisperse suspensions. Int. J. Mineral Process. (submitted).

Iterative Substructering Schemes for Fluid Structure Interaction

Wolfgang A. Wall, Daniel P. Mok, and Ekkehard Ramm

University of Stuttgart, Institute of Structural Mechanics, Pfaffenwaldring 7, D-70550 Stuttgart, Germany

Abstract. This paper discusses partitioned analysis approaches for the transient interaction of incompressible viscous flows and nonlinear flexible structures with large deformations. It is shown that for this kind of coupled systems the commonly used sequential staggered coupling schemes exhibit weak instabilities and may eventually fail, mainly due to inconsistencies in the coupling boundary conditions. As best remedy to this problem subiterations should be invoked which ensure kinematic and dynamic continuity across the fluid-structure interface, thus ensuring stable and accurate numerical solutions even for long-time simulations. For the desirable acceleration of convergence of such iterative substructuring schemes two robust and problem-independent methods are proposed and their efficiency is demonstrated with selected numerical examples.

1 Introduction

Partitioned analysis techniques enjoy great popularity for the solution of multiphysics problems. This is due to their computational superiority over simultaneous, i.e. fully coupled monolithical approaches, as they allow the independent use of suitable discretization methods and modular, optimized analysis software for physically and/or dynamically different partitions. However major drawbacks in terms of accuracy and stability problems can occur if partitioned methods are used inappropriately. It is therefore subject of this paper to analyze and discuss specific problems of partitioned analysis techniques and to introduce possible remedies for the considered class of applications – the transient interaction of incompressible viscous flows and nonlinear flexible structures.

2 Fluid Structure Interaction Environment

Our partitioned fluid structure interaction environment is described in detail in Wall [8] or Wall et al. [7] and is presented here in a comprising overview in figure 1. In this approach a non-overlapping partitioning is employed, where the physical fields fluid and structure are coupled together at the interface Γ, i.e. the wetted structural surface. A third computational field Ω_M, the deforming fluid mesh, is introduced through an ALE description. The single fields are solved by semidiscretization strategies with finite elements and implicit time stepping algorithms.

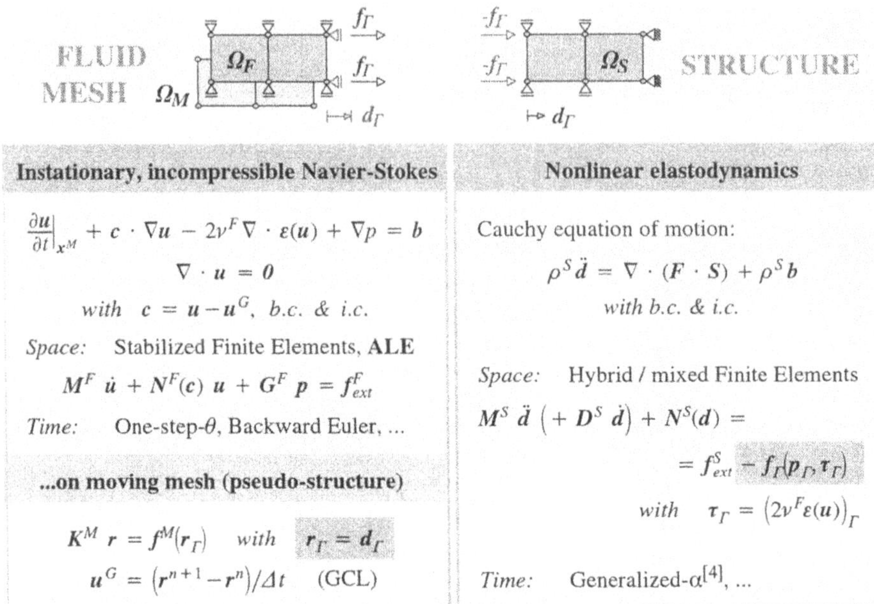

Fig. 1. Nonoverlapping partitioned fluid structure interaction environment

Key requirement for the coupling schemes is to fulfill two coupling conditions: the kinematic and the dynamic continuity across the interface at all times. Kinematic continuity requires that the position of structure and fluid boundary are equal at the interface, while dynamic continuity means that all tractions at the interface are in equilibrium:

$$\boldsymbol{d}_\Gamma(t) \cdot \boldsymbol{n} = \boldsymbol{r}_\Gamma(t) \cdot \boldsymbol{n} \quad \text{and} \quad \boldsymbol{u}_\Gamma(t) = \boldsymbol{u}_\Gamma^G(t) = f(\boldsymbol{r}_\Gamma(t)) \tag{1}$$

$$\boldsymbol{\sigma}_\Gamma^S(t) \cdot \boldsymbol{n} = \boldsymbol{\sigma}_\Gamma^F(t) \cdot \boldsymbol{n} \tag{2}$$

with $\boldsymbol{n}$ denoting the unit normal vector on the interface. Satisfying the kinematic continuity leads to mass conservation at Γ, satisfying the dynamic continuity leads to conservation of linear momentum, and energy conservation finally requires to simultaneously satisfy both continuity equations. In this paper (and in figure 1) only no-slip boundary conditions and sticking grids at the interface are considered.

3 Unified Algorithmic Framework

The partitioned analysis schemes under consideration with synchronous time discretizations in the fluid and structural part (figure 2) can be cast in a unified algorithmic framework. They are discussed in detail in Mok [11]. In the following, $(\cdot)_I$ and $(\cdot)_\Gamma$ denotes variables/coefficients in the interior of a subdomain Ω_j and on the coupling interface Γ, respectively, while a vector without any of the subscripts I and Γ comprises degrees of freedom (DOFs) on the whole subdomain (interior and interface).

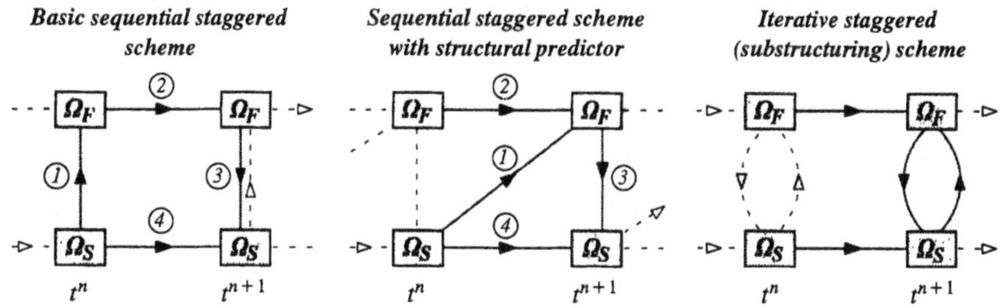

Fig. 2. Synchronous partitioned analysis schemes

In every time interval $[t^n, t^{n+1}]$ the following algorithmic steps have to be performed in order to compute the new coupled solution at t^{n+1}, starting from a known state of motion before and at t^n.

1a. Compute structural predictor $\boldsymbol{d}_{\Gamma,0}^{n+1}$ in Ω_S for the interface displacements at t^{n+1} (see section 3.1). Set $i = 0$.
1b. Pass over $\boldsymbol{d}_{\Gamma,i}^{n+1}$ from the structure Ω_S to the fluid mesh Ω_M.
2a. Solve fluid mesh Ω_M for new positions.

$$\boldsymbol{K}_{II}^M\, \boldsymbol{r}_{I,i+1}^{n+1} = -\boldsymbol{K}_{I\Gamma}^M\, \boldsymbol{r}_{\Gamma,i+1}^{n+1} \tag{3}$$

$$\text{with Dirichlet b.c. } \boldsymbol{r}_{\Gamma,i+1}^{n+1} = \boldsymbol{d}_{\Gamma,i}^{n+1} \tag{4}$$

2b. Derive new grid velocity in compliance with the geometric conservation law (GCL).

$$\boldsymbol{u}_{i+1}^{G,n\to n+1} = \frac{1}{\Delta t}\left(\boldsymbol{r}_{i+1}^{n+1} - \boldsymbol{r}^n\right) \tag{5}$$

2c. Pass over mesh positions and grid velocity from the fluid mesh Ω_M to the fluid Ω_F.
2d. Integrate fluid partition Ω_F on new deforming mesh for new fluid velocity and pressure fields (one-step-θ time integration scheme, fixed-point-like iterations).

$$\begin{aligned}\left(\frac{1}{\Delta t}\boldsymbol{M}^F + \theta \boldsymbol{N}^F(\boldsymbol{c}_{i+1}^{n+1})\right) \boldsymbol{u}_{i+1}^{n+1} + \theta \boldsymbol{G}^F\, \boldsymbol{p}_{i+1}^{n+1} \\ = \theta \boldsymbol{f}_{\text{ext}}^{n+1} + (1-\theta)\boldsymbol{f}_{\text{ext}}^n \\ + \left(\frac{1}{\Delta t}\boldsymbol{M}^F - (1-\theta)\boldsymbol{N}^F(\boldsymbol{c}^n)\right) \boldsymbol{u}^n - (1-\theta)\boldsymbol{G}^F\, \boldsymbol{p}^n\end{aligned} \tag{6}$$

with convective velocity

$$c_{i+1}^{n+1} = u_{i+1}^{n+1} - u_{i+1}^{G,n\to n+1}, \qquad c_{i+1}^{n} = u_{i+1}^{n} - u_{i+1}^{G,n\to n+1} \tag{7}$$

3. Pass over fluid pressure $p_{\Gamma,i+1}^{n+1}$ (and viscous stresses $\tau_{\Gamma,i+1}^{n+1}$) from the fluid Ω_F to the structure Ω_S.

4a. Convert pressure and viscous stresses into a consistent nodal coupling load vector $f_{\Gamma,i+1}^{n+1}\left(p_{\Gamma,i+1}^{n+1}, \tau_{\Gamma,i+1}^{n+1}\right)$ in Ω_S.

4b. Integrate structure partition Ω_S with Neumann b.c. $f_{\Gamma,i+1}^{n+1}$ for new displacements $d_{i+1}^{n+1} := \left\{\tilde{d}_{\Gamma,i+1}^{n+1} \; d_{I,i+1}^{n+1}\right\}^T$ (Generalized-α time integration scheme [4] in predictor-multicorrector formulation with Newton-Raphson equilibrium iterations).

$$\begin{aligned} \frac{1-\alpha_m}{\beta \Delta t^2} M^S\, d_{i+1}^{n+1} + (1-\alpha_f)\, f_{\text{int}}\left(d_{i+1}^{n+1}\right) \\ = -\alpha_f f_{\text{int}}\left(d^n\right) + M^S\, h\left(d^n, \dot{d}^n, \ddot{d}^n\right) \\ + (1-\alpha_f)\left\{\begin{matrix} f_{\Gamma\,\text{ext}}^{n+1} - f_{\Gamma,i+1}^{n+1} \\ f_{I\,\text{ext}}^{n+1} \end{matrix}\right\} + \alpha_f \left\{\begin{matrix} f_{\Gamma\,\text{ext}}^{n} \\ f_{I\,\text{ext}}^{n} \end{matrix}\right\} \end{aligned} \tag{8}$$

with history terms

$$h\left(d^n, \dot{d}^n, \ddot{d}^n\right) := \frac{1-\alpha_m}{\beta \Delta t^2} d^n + \frac{1-\alpha_m}{\beta \Delta t} \dot{d}^n + \left(\frac{1-\alpha_m}{2\beta} - 1\right) \ddot{d}^n \tag{9}$$

5. *For iterative schemes only:* Determine suitable relaxation parameter $\omega_i \in \mathbb{R}^+$ (see sections 3.3 and 3.4).
6. *For iterative schemes only:* Compute relaxed update of predicted interface position.

$$d_{\Gamma,i+1}^{n+1} = \omega_i\, \tilde{d}_{\Gamma,i+1}^{n+1} + (1-\omega_i)\, d_{\Gamma,i}^{n+1} \tag{10}$$

7. *For iterative schemes only:* Check convergence of interface displacements and/or residual. If not converged, then set $i \to i+1$ and go back to step 1b.
8. Proceed to next time step by setting $n \to n+1$.

3.1 Sequential staggered schemes

For schemes *without* iterations, so-called *sequential staggered algorithms*, the quality of the predictor (step 1a above) has a major impact on accuracy and stability of the partitioned method. Simply choosing

$$d_{\Gamma,0}^{n+1} = d_\Gamma^n \tag{11}$$

yields the *basic sequential staggered scheme* dating back to Felippa et al. [2,3]. A far more accurate choice is the *second order predictor* suggested by Piperno [5]:

$$\boldsymbol{d}_{\Gamma,0}^{n+1} = \boldsymbol{d}_{\Gamma}^{n} + \Delta t \left(\frac{3}{2} \dot{\boldsymbol{d}}_{\Gamma}^{n} - \frac{1}{2} \dot{\boldsymbol{d}}_{\Gamma}^{n-1} \right) \tag{12}$$

However, a predictor is never exact, i.e. the explicitly predicted interface position $\boldsymbol{d}_{\Gamma,0}^{n+1}$ used for the time integration of the fluid partition as Dirichlet b.c. will not be the same as the interface position $\tilde{\boldsymbol{d}}_{\Gamma,1}^{n+1}$, which is implicitly computed in the structural partition afterwards using the fluid load $\boldsymbol{f}_{\Gamma,1}^{n+1}$. Thus every synchronous sequential staggered scheme necessarily violates the kinematic continuity condition, thus leading to accuracy reduction and, much worse, weak instability of the numerical solution. Numerical studies (see examples) have further shown that surprisingly those weak instabilities occur much earlier with the more accurate predictor (12) than with the simpler one (11). In such situations one could try to avoid or delay the beginning of the occurence of weak instabilities simply by reducing the time step. However, when incompressible flows are involved, the so-called artificial added mass effect (discussed in detail in Le Tallec et al. [10], Wall [8] and Mok [11]) has an opposite impact on the solution: the smaller the time step gets, the earlier the instability appears; so cutting the time step size is not a suitable solution to the weak instability problem for the systems under consideration. Other possible remedies have also been proposed as there are:

- Introduction of strong viscous damping, *but*: changes the physics!
- Introduction of fluid compressibility, *but*: changes the physics!
- Asynchronous staggered scheme (Lesoinne & Farhat [6]), *but*: fulfils the kinematic and the dynamic continuity condition at different points in time, restricts the structural time integration to the midpoint integration scheme; both facts again cause instabilities (the latter one for nonlinear structural formulations).
- Subiterations, *but*: increase of numerical costs, convergence not automatically ensured.

3.2 Iterative staggered schemes – iterative substructering schemes

Apparently the only appropriate choice, which neither involves artificial changes in the underlying physics nor is restricted to time integration schemes that can become unstable in nonlinear regimes (midpoint rule), are subiterations. These so-called *iterative staggered schemes* must then of course be designed to be as cheap and robust as possible. The iterative scheme described in the above algorithmic framework for fluid structure interaction is due to Le Tallec et al. [10]. It can be interpreted as an *iterative Dirichlet-Neumann substructuring scheme* based on a *nonstationary Richardson iteration.* This becomes obvious from the iterative evolution equation,

$$\boldsymbol{d}_{\Gamma,i+1}^{n+1} = \boldsymbol{d}_{\Gamma,i}^{n+1} + \omega_i \, \boldsymbol{S}_S^{-1} \left(\boldsymbol{f}_{\Gamma \, \mathrm{ext}}^{\mathrm{mod},n+1} - (\boldsymbol{S}_F + \boldsymbol{S}_S) \, \boldsymbol{d}_{\Gamma,i}^{n+1} \right) \tag{13}$$

which is reduced to the degrees of freedom on the interface Γ and can be derived [10,11] by plugging eqns. (3) through (9) into the relaxation equation (10). $\boldsymbol{S}_F$ and $\boldsymbol{S}_S$ denote the Schur complement matrices of the fluid and structural fields, respectively, and $\boldsymbol{f}_{\Gamma \, \mathrm{ext}}^{\mathrm{mod},n+1}$ is the external load vector resulting after static condensation of the DOFs in the interior of the subdomains. In this iterative scheme convergence is accelerated and ensured by the relaxed updates of the interface position. The iteration then converges to the simultaneous solution, exactly fulfilling the required coupling conditions: kinematic and dynamic continuity at every discrete time $t^n, t^{n+1}, \ldots$

However a key question remains: How to choose optimal relaxation parameters ω_i? The commonly used strategy to employ an experimentally (by trial and error) determined fixed parameter is unsatisfactory, because such a parameter is very problem-dependent, in general suboptimal (especially for nonlinear problems), and it requires a careful, time-consuming and difficult determination. In this paper two techniques are proposed which are both robust in the sense that they have problem-independent acceleration properties even for nonlinear systems, and user-friendly in the sense that the relaxation parameters are determined automatically without any user-input being necessary (see Mok [11] for more details).

3.3 Iterative substructering schemes accelerated via the gradient method

The first technique is an acceleration via the application of the *gradient method (method of steepest descent)* to the iterative substructuring scheme. This method also guarantees convergence. In every iteration a relaxation parameter ω_i is computed,

$$\omega_i = \frac{\boldsymbol{g}_i^T \, \boldsymbol{g}_i}{\boldsymbol{g}_i^T \, \boldsymbol{S}_S^{-1} (\boldsymbol{S}_F + \boldsymbol{S}_S) \, \boldsymbol{g}_i} = \frac{\boldsymbol{g}_i^T \, \boldsymbol{g}_i}{\boldsymbol{g}_i^T \, (\boldsymbol{S}_S^{-1} \, \boldsymbol{S}_F \, \boldsymbol{g}_i + \boldsymbol{g}_i)} \tag{14}$$

that is locally optimal with respect to the actual search direction, i.e. the residual

$$\boldsymbol{g}_i = \boldsymbol{S}_S^{-1} \left(\boldsymbol{f}_{\Gamma \, \mathrm{ext}}^{\mathrm{mod},n+1} - (\boldsymbol{S}_F + \boldsymbol{S}_S) \, \boldsymbol{d}_{\Gamma,i}^{n+1} \right) = \tilde{\boldsymbol{d}}_{\Gamma,i+1}^{n+1} - \boldsymbol{d}_{\Gamma,i}^{n+1} \tag{15}$$

The following procedure for evaluating eq. (14) without explicitly computing and storing the Schur complements $\boldsymbol{S}_F$ and $\boldsymbol{S}_S$ has been proposed in Wall et al. [9]. It is based on the fact that applying a [inverted] Schur complement operator to a vector is equivalent to the solution of the respective partition with that vector as Dirichlet [Neumann] b.c. The algorithmic realization of this acceleration scheme can be described by replacing step 5 in the algorithmic framework given above by the following steps, solving a homogeneous auxiliary problem in order to first determine the term $\boldsymbol{S}_S^{-1}\ \boldsymbol{S}_F\ \boldsymbol{g}_i$ and then the relaxation parameter ω_i.

5a. Compute residual $\boldsymbol{g}_i$ and pass it over from the structure Ω_S to the fluid mesh Ω_M.

$$\boldsymbol{g}_i = \tilde{\boldsymbol{d}}_{\Gamma,i+1}^{n+1} - \boldsymbol{d}_{\Gamma,i}^{n+1} \tag{16}$$

5b. Solve fluid mesh Ω_M with Dirichlet b.c. $\hat{\boldsymbol{r}}_\Gamma = \boldsymbol{g}_i$.

$$\boldsymbol{K}_{II}^M\ \hat{\boldsymbol{r}}_I = -\boldsymbol{K}_{I\Gamma}^M\ \hat{\boldsymbol{r}}_\Gamma \tag{17}$$

5c. Compute grid velocity in compliance with the geometric conservation law (GCL).

$$\hat{\boldsymbol{u}}^G = \frac{1}{\Delta t}\left(\hat{\boldsymbol{r}} - \boldsymbol{r}^0\right) \tag{18}$$

5d. Pass over mesh positions and grid velocity from the fluid mesh Ω_M to the fluid Ω_F.
5e. Solve fluid partition Ω_F on this mesh without external loads and history terms.

$$\left(\frac{1}{\Delta t}\boldsymbol{M}^F + \theta\left(\boldsymbol{N}^F(\hat{\boldsymbol{c}}) + \boldsymbol{K}^F\right)\right)\hat{\boldsymbol{u}} + \theta\ \boldsymbol{G}^F\ \hat{\boldsymbol{p}} = \boldsymbol{0} \tag{19}$$

with $\hat{\boldsymbol{c}} = \boldsymbol{u}_{i+1}^{n+1} - \hat{\boldsymbol{u}}^G$
5f. Pass over fluid pressure $\hat{\boldsymbol{p}}_\Gamma$ (and viscous stresses $\hat{\boldsymbol{\tau}}_\Gamma$) from Ω_F to Ω_S.
5g. Convert it into a consistent nodal coupling load vector $\hat{\boldsymbol{f}}_\Gamma\left(\hat{\boldsymbol{p}}_\Gamma, \hat{\boldsymbol{\tau}}_\Gamma\right)$ in Ω_S.
5h. Solve structure partition Ω_S without external loads and history terms.

$$\begin{bmatrix} \boldsymbol{K}_{\Gamma\Gamma\,\text{eff}}^S & \boldsymbol{K}_{\Gamma I\,\text{eff}}^S \\ \boldsymbol{K}_{I\Gamma\,\text{eff}}^S & \boldsymbol{K}_{II\,\text{eff}}^S \end{bmatrix} \begin{Bmatrix} \hat{\boldsymbol{d}}_\Gamma \\ \hat{\boldsymbol{d}}_I \end{Bmatrix} = \begin{Bmatrix} -\hat{\boldsymbol{f}}_\Gamma \\ \boldsymbol{0} \end{Bmatrix} \tag{20}$$

with $\boldsymbol{K}_{\text{eff}}^S = \frac{1-\alpha_m}{\beta\Delta t^2}\boldsymbol{M}^S + (1-\alpha_f)\ \boldsymbol{K}_T^S\left(\boldsymbol{d}_{i+1}^{n+1}\right)$ from step 4b (last N.-R.-Iteration)
5i. Compute relaxation parameter.

$$\omega_i = \frac{\boldsymbol{g}_i^T\ \boldsymbol{g}_i}{\boldsymbol{g}_i^T\left(-\hat{\boldsymbol{d}}_\Gamma + \boldsymbol{g}_i\right)} \tag{21}$$

3.4 Iterative substructering schemes accelerated via the Aitken method

A second technique for explicitly calculating a suitable relaxation parameter is the application of *Aitken's acceleration scheme for vector sequences* according to Irons et al. [1] It can be formulated within the algorithmic framework shown above by again modifying step 5 as follows.

5a. Compute difference between actual and previous interface solution.

$$\Delta \mathbf{d}_{\Gamma,i+1}^{n+1} := \mathbf{d}_{\Gamma,i}^{n+1} - \tilde{\mathbf{d}}_{\Gamma,i+1}^{n+1} \tag{22}$$

5b. Compute Aitken factor (for $i \geqslant 1$ and with $\mu_0^{n+1} = \mu_{i_{\max}}^n$, $\mu_0^1 = 0$).

$$\mu_i^{n+1} = \mu_{i-1}^{n+1} + \left(\mu_{i-1}^{n+1} - 1\right) \frac{\left(\Delta \mathbf{d}_{\Gamma,i}^{n+1} - \Delta \mathbf{d}_{\Gamma,i+1}^{n+1}\right)^T \Delta \mathbf{d}_{\Gamma,i+1}^{n+1}}{\left(\Delta \mathbf{d}_{\Gamma,i}^{n+1} - \Delta \mathbf{d}_{\Gamma,i+1}^{n+1}\right)^2} \tag{23}$$

5c. Compute relaxation parameter.

$$\omega_i = 1 - \mu_i^{n+1} \tag{24}$$

Even though a rigorous analysis of its convergence properties does not exist, numerical studies have shown that the Aitken acceleration for vector sequences applied to the fluid structure interaction problems considered here shows a performance which is at least as good as the acceleration via the gradient method. Furthermore the evaluation of the relaxation parameters via the Aitken method is extremely cheap (in terms of both CPU and memory) and simple to implement.

4 Numerical Examples

4.1 Cavity with oscillating top and flexible bottom plate

A 2D simulation of a square liquid-filled tank with a flexible bottom and harmonically oscillating top plate is analyzed (figure 3). The bottom structure is $h = 0.002$ thick, has a density $\rho^S = 500.0$ and a Young's modulus $E^S = 250.0$. The fluid density is $\rho^F = 1.0$ and its kinematic viscosity is $\nu^F = 0.01$. Stabilized Q1Q1 and Q2 finite elements are used [8,7] to model the fluid and the structure, respectively. Both fields are implicitly integrated in time with $\Delta t = 0.1$.

This example is chosen to be relatively good natured so that the sequential staggered solutions are at least stable for a while. The pressure load histories in figure 4 demonstrate the typical weak instabilities of sequential staggered schemes. The basic sequential staggered scheme produces a solution that is stable for a much longer period compared to the second order

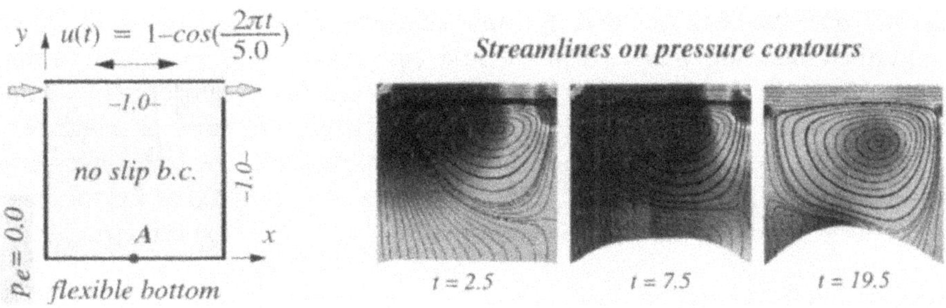

Fig. 3. Cavity with flexible bottom – problem statement, deformed configurations; colored, see page 380

predictor scheme. On the other hand the second order predictor scheme is far more accurate (second order accuracy). Within the stable time range its solution curve is almost undistinguishable from the reference solution generated by the iterative staggered schemes, while the first order accurate basic sequential staggered scheme clearly deviates from the reference solution. The iterative staggered schemes yield a perfectly stable solution. They converged in only $i_{\max} = 2$ to 4 subiterations per time step (with standard convergence criterion: $||\boldsymbol{g}_i||/\sqrt{n_{\text{eq}}} < 10^{-6}$) for both acceleration techniques.

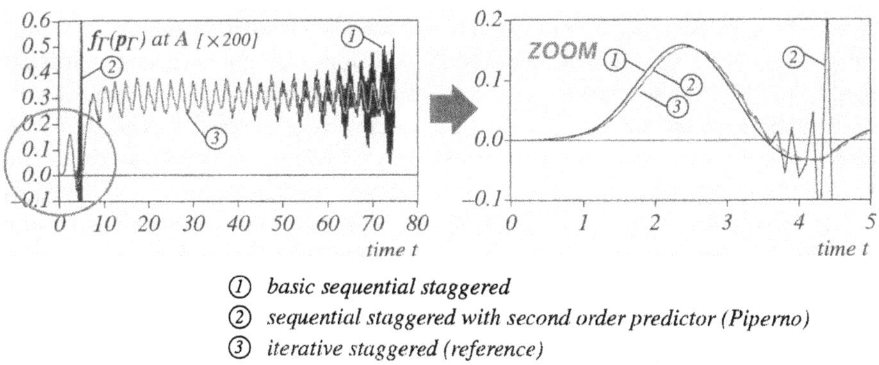

Fig. 4. Cavity with flexible bottom – pressure load histories; colored, see page 380

4.2 Flexible restrictor flap in converging channel

The second example uses realistic geometry and material properties. It is a 2D simulation of a converging channel with a flexible restrictor flap (figures 5 and 6; due to symmetry only one half of the channel is modelled). The material of the flap is a stiff rubber with Young's modulus $E^S = 2.3 \cdot 10^6\,\text{N/m}^2$, density $\rho^S = 1500.0\,\text{kg/m}^3$ and Poisson ratio $\nu^S = 0.45$. The fluid is a silicon oil with density $\rho^F = 956.0\,\mathrm{kg/m^3}$ and dynamic viscosity $\mu^F = 0.145\,\mathrm{kg/(m \cdot s)}$. The prescribed inlet flow velocity's profile and time-history ($\bar{v}(t)$) are given in the figure. The fully developed flow has a Reynolds number of approx. $Re = 100$, i.e. the flow is laminar. Again, stabilized Q1Q1 and Q2 finite elements are used [8,7] to model the fluid and the structure, respectively. Both fields are implicitly time integrated with $\Delta t = 0.1\,\mathrm{s}$ using backward Euler for the flow and Generalized-α with numerical dissipation ($\rho_\infty = 0.6$) for the structure.

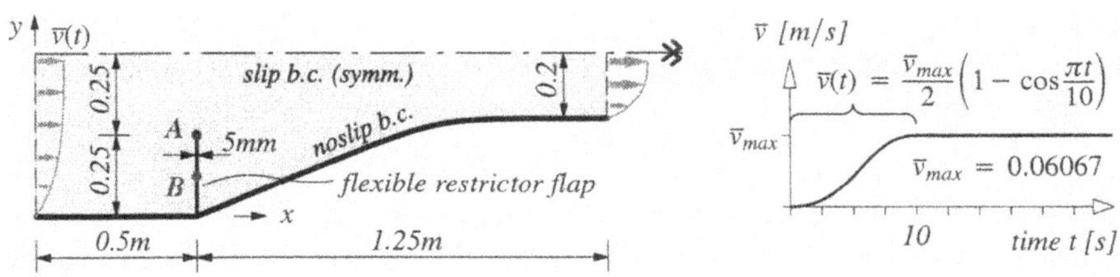

Fig. 5. Restrictor flap in channel – problem statement

In this example the sequential staggered schemes both completely failed to produce stable solutions, the weak instablity already occured after the first few time steps, even when reducing the time step size to $\Delta t = 0.0001\,\mathrm{s}$ (artificial added mass effect!). In contrast, the iterative staggered schemes were again perfectly stable over the whole computed time range. Figure 7 shows the number $i_{\max}$ of subiterations over the partitions required within every time step to yield a converged solution. As convergence criterion $\|\boldsymbol{g}_i\|/\sqrt{n_{\mathrm{eq}}} < 10^{-9}$ was used. A computation using a carefully chosen, "best" fixed relaxation parameter $\omega = 0.125 = \mathrm{const.}$ is compared to the two proposed techniques, where the iterative staggered scheme is accelerated via the gradient and the Aitken method. The unrelaxed algorithm ($\omega = 1.0$) diverges in this case. The results clearly show the suboptimal behaviour of the strategy with fixed relaxation parameter, and the efficient acceleration of convergence of the two proposed schemes.

5 Conclusions

In the present paper partitioned analysis approaches for the transient interaction of incompressible viscous flows and nonlinear flexible structures with large deformations have been discussed. A unified algorithmic framework for partitioned schemes with synchronous time integration in the fluid and structural field has been formulated. It was shown that for this kind of coupled systems the commonly used sequential staggered coupling schemes exhibit weak instabilities and may eventually fail, mainly due to inconsistencies in the coupling boundary conditions. As best remedy to this problem it is recommended to invoke subiterations which ensure kinematic and dynamic continuity across the fluid-structure interface, thus ensuring stable and accurate

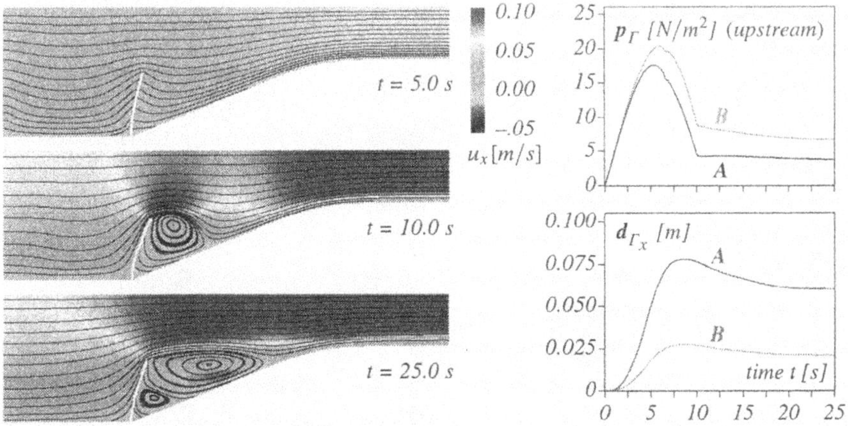

Fig. 6. Restrictor flap in channel – deformed confirgurations (streamlines on horizontal flow velocity); colored, see page 381

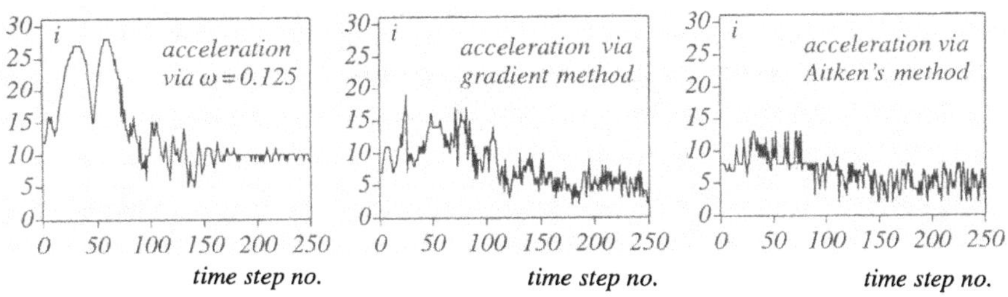

Fig. 7. Restrictor flap in channel – convergence study (number of subiterations i per time step); colored, see page 381

numerical solutions even for long-time simulations. For the desired acceleration of convergence of such iterative staggered or iterative substructuring schemes two techniques have been proposed: acceleration via the gradient method and via Aitken's method for vector sequences. These techniques are both robust in the sense that they have problem-independent acceleration properties even for nonlinear systems, and user-friendly in the sense that the relaxation parameters are determined automatically without any user-input being necessary. The computational efficiency and robustness of the proposed accelerated iterative staggered schemes – even in such complex simulations where the sequential staggered schemes become instable from the very beginning of the computation – has been demonstrated with selected numerical examples.

6 Acknowledgements

The present study is supported by a grant of the Deutsche Forschungsgemeinschaft (DFG) under project B4 of the collaborative research center SFB 404 'Multifield Problems in Continuum Mechanics'. This support is gratefully acknowledged.

References

1. Irons B., Tuck, R.C. (1969) A version of the Aitken accelerator for computer implementation, Int. J. Num. Meth. Eng., **1**, 275-277
2. Felippa, C.A., Park, K.C., DeRuntz, J.A. (1977) Stabilization of staggered solution procedures for fluid-structure interaction analysis, in: T. Belytschko and T.L. Geers (Eds.), Computational Methods for Fluid-Structure Interaction Problems, AMD Vol. 26, ASME, New York, 95-124
3. Felippa, C.A., Park, K.C. (1980) Staggered transient analysis procedures for coupled mechanical systems: Formulation, Comp. Meth. Appl. Mech. Eng., 24, 61-111
4. Chung, J., Hulbert, G.M. (1993) A time integration algorithm for structural dynamics with imporved numerical dissipation: The Generalized-a method, J. Appl. Math., **60**, 371-375
5. Piperno, S. (1997) Explicit/implicit fluid/structure staggered procedures with a structural predictor and fluid subcycling for 2D inviscid aeroelastic simulations, Int. J. Num. Meth. Fluids, **25**, 1207-1226
6. Lesoinne, M., Farhat, C. (1998) A higher-order subiteration free staggered algorithm for nonlinear transient aeroelastic problems, AIAA Journal, **36**(9), 1754-1756
7. Wall, W.A., Ramm, E. (1998) Fluid-structure interaction based upon a stabilized (ALE) finite element method, in: E. Oñate and S. Idelsohn S. (Eds.), Computational Mechanics, Proc. WCCM IV, Buenos Aires, CIMNE, Barcelona
8. Wall, W.A. (1999) Fluid-Struktur-Interaktion mit stabilisierten Finiten Elementen, Ph.D.-Dissertation, Report No. **31**, Institute of Structural Mechanics, University of Stuttgart
9. Wall, W.A., Mok, D.P., Ramm, E. (1999) Partitioned Analysis Approach of the Transient, Coupled Response of Viscous Fluids and Flexible Structures, in: W. Wunderlich (Ed.), Solids, Structures and Coupled Problems in Engineering, Proc. ECCM '99, Munich
10. Le Tallec, P., Mouro, J. (2001) Fluid structure interaction with large structural displacements, Comp. Meth. Appl. Mech. Eng, **190**, 3039-3067
11. Mok, D.P. (2001) Partitionierte Lösungsansätze in der Strukturdynamik und der Fluid-Struktur-Interaktion, Ph.D.-Dissertation, Report No. **36**, Institute of Structural Mechanics, University of Stuttgart

Appendix

International Conference on Multifield Problems

April 8-10, 2002 — University of Stuttgart, Germany

Conference Programme

Monday, 8. April 2002

09:00 – 09:45 Room 57.02	OPENING			
09:45 – 10:30 Room 57.02	**M. Wheeler** *Mathematic and computational modeling of multiphysics couplings*			
10:30 – 11:15 Room 57.02	**H. Gao** *How fast can cracks propagate*			
11:15 – 11:40	COFFEE BREAK			
	Section 1: **Material Modelling and Multiscale Problems** **Room 57.04**	**Section 2:** **Numerical Methods** **Room 57.05**	**Section 3:** **Solid Mechanics** **Room 57.06**	**Section 4:** **Porous Media and Fluid-Structure- Interaction** **Room 57.03**
11:40 – 12:05	**R. Piat** *Micro-cracking in pyrolytic carbon*	**A. Nowakowski** *Performance of some finite elements in numerical simulation of complex incompressible three dimensional flow*	**A. Miranville** *Some generalitions of the Cahn-Hilliard equation*	**B. Albers** *Linear stability of a 1D flow through a porous membrane*
12:05 – 12:30	**H. Stumpf** *Elastic Damage: Thermodynamical microtheory and consistent nonlocal gradient models with numerical applications*	**C. Coclici** *Magnetoplasmadynamic thrusts: plume / tank flow interaction*	**K. Willner** *Thermomechanical coupling in contact problems*	**W.R. Bielski** *Flow of Stokesian fluid through random deformable porous medium*
12:30 – 14:00	LUNCH			

Monday, 8. April 2002

	Section 1: Material Modelling and Multiscale Problems Room 57.04	Section 2: Numerical Methods Room 57.05	Section 3: Solid Mechanics Room 57.06	Section 4: Porous Media and Fluid-Structure- Interaction Room 57.03
14:00 – 14:25	**B. Svendson** *Multifield extensions of crystal plasticity accounting for the effects of local deformation incompatibility of the material behaviour*	**R. Bürger** *Numerical methods for conservation laws with discontinous flux modelling clarifier-thickener units*	**J. Gwinner** *Discretization of some contact problems in one-dimensional thermoelasticity*	**H. Jakobs** *A new method for the numerical approximation of the effects of heterogenities on multiphase flow behavior*
14:25 – 14:50	**Ch. Eck** *A Two-Scale Phase Field Model for Phase Transitions with dendritic Microstructure*	**R. Stark** *Coupled numerical simulation of pneumatic tunneling*	**M. Bach** *Contact problems treated with boundary integral equations*	**W. Ehlers** *A continuum approach for the swelling of charged hydrated media*
14:50 – 15:15	**M Arndt** *Modelling and numerical simulation of martensitic transformation*	**G. Micula** *On some new applications of spline functions*	**W.L. Wendland** *A residual-based error estimator for BEM-discretications of contact problems*	**M. Schanz** *Boundary element formulations for linear poroelastodynamic continua*
15:15 – 15:30	**COFFEE BREAK**			
15:30 – 15:55	**Ch. Miehe** *Computational Homogenization of Inelastic Microstructures at Finite Strains based on Minimization Principles*	**H. Steeb** *Adaptive space-time-Gallerkin methods for volumetrically coupled problems*	**E. Brommundt** *Moving contact with friction at an elastic strip*	**M. Peszynska** *Couplings of models for multiphase flow and transport in porous media with multiple scales*
15:55 – 16:20	**M Lambrecht** *Two-Scale Relaxation Analysis of Inelastic Solids based on First-Order Rank-one Convexification of Incremental Stress potential*	**I.L. Sofronov** *A linear far-field modelling and artifical boundary conditions for plasma flows in MPD-accelerators*	**G. Szefer** *Contact of elastic bodys with negative Poisson ratio*	**K. Wilmanski** *Surface waves on a flat interface: poroelastic medium/ fluid*
16:20 – 16:45	**C. Bayreuther** *Coupling of Homogenization Techniques with Multigrid Solvers for Unstructured Meshes*	**T.-A. Langhoff** *Modelling chemical vapour infiltration: a moving boundary problem*	**V. Loboda** *Contact zone models fao an interface crack in a piezoelectric bimaterial under thermal, electrical and mechanical fields*	**O. Soboleva** *Impure transport in the fractal porous media*
16:45 – 17:00	**COFFEE BREAK**			
17:00 – 17:45 Room 57.02	**C. Glocker**			
19:00	**WELCOME–PARTY IBZ**			

Tuesday, 9. April 2002

09:00 – 09:45 Room 57.02	**H. Matthies** *Strong coupling algorithms*			
09:45 – 10:30 Room 57.02	**St. Müller** *Multiscale folding patterns in compressed elastic films and rigorous derivation of nonlinear plate theory*			
10:30 – 10:45	**COFFEE BREAK**			
	Section 1: Material Modelling and Multiscale Problems Room 57.04	**Section 2: Numerical Methods Room 57.05**	**Section 3: Solid Mechanics Room 57.06**	**Section 4: Porous Media and Fluid-Structure- Interaction Room 57.03**
10:45 – 11:10	**A. Mielke** *Relaxation and Young measures for rate-independent material models*	**D. Braess** *Advantages and limitations of enhanced assumed strains*	**B. Muth** *Collision detection-efficiency comparitions between verlet-neighbourhood and the linked linear list method*	**J. Kubik** *On dynamic couplings in a porous multiphase system*
11:10 – 11:35	**M. Efendiev** *Upper and lower semicontinuity of the stable set for rate-independent process*	**B. Scholz** *The identification procedure of the Cosserat material constants as an inverse problem*	**E.V. Zahariev** *Some new aspects of the kinematic and dynamic analysis of multibody systems*	**W.A. Wall** *Iterative substructuring schems for fluid structure interaction*
11:35 – 11:50	**COFFEE BREAK**			
11:50 – 12:15	**D. Mittenhuber** *Invariant metrics on Lie groups and their application in elastoplasticity*	**B. Wohlmuth** *Optimal multigrid methods for nonconforming spaces*	**C. Zhang** *Transient dynamic response of a cracked piezoelectric solid under impact loading*	**S. Meyen** *Vortex-induced vibrations of a cylinder in a channel flow: Numerical simulation and experiment*
12:15 – 12:40	**K. Hackl** *Relaxed Potentials and microstructures in damage mechanics*	**C. Wieners** *Numerical simulation in solid mechanics*	**J. Jarušek** *On admissable coefficents of friction fort he sovability of contact problems with Coulomb friction for homogenous isotropic material*	**H. Schippers** *Aerodynamic, structural and electromagnetic interaction applied to conformal antennas*
12:40 – 14:00	**LUNCH**			

Tuesday, 9. April 2002

14:00 – 14:45 Room 57.02	**C. Schwab** *Two-scale regularity and sparse grids for homogenization problems*			
14:45 – 15:00	**COFFEE BREAK**			
	Section 1: Material Modelling and Multiscale Problems Room 57.04	**Section 2: Numerical Methods Room 57.05**	**Section 3: Solid Mechanics Room 57.06**	**Section 4: Porous Media and Fluid-Structure- Interaction Room 57.03**
15:00 – 15:25			**B. Hu** *Multi-time scale simulation of impact responses*	**A.N. Guz** *Dynamics of rigid bodies in a compressible viscous fluid (piecewise – homogenous medium model)*
15:25 – 15:50				**E. Tory** *Simulation of sedimentation of monodisperse and polydisperse suspensions*

17:45	**Meeting for the bus tour**
18:00 – 20:00	**GUIDED BUSCITYTOUR IN GERMAN/ENGLISH/RUSSIAN**
20:00 – 22:00	**DINNER IN THE "MARMORSAAL" WITH MUSIC FOR YOUR CONVIENIENCE**

Wednesday, 10. April 2002

09:00 – 09:45 Room 57.02	**D. A. Drew**	***A turbulent dispersion model for particles or bubbles***		
09:45 – 10:00	**C O F F E E B R E A K**			
	Section 1: **Material Modelling and Multiscale Problems** **Room 57.04**	**Section 2:** **Numerical Methods** **Room 57.05**	**Section 3:** **Solid Mechanics** **Room 57.06**	**Section 4:** **Porous Media and Fluid-Structure- Interaction** **Room 57.03**
10:00 – 10:25	**T. Roubicek** *Optimal design of laminated composites*	**T. Merkle** *A robust a-posteriori error estimator for finite element discretizations of elliptic problems in composites*	**D. Knees** *Quasi monotonicity and regularity results for solutions of transmission problems for the Laplace and Lame operators*	**T. Weidl** *Wave guides in acoustics, quantum mechanics and elasticity: Some mathematical issues*
10:25 – 10:50	**A. De Simone** *Multiscale Modeling of advanced material coupling with magnetic or nematic order*	**J. Breuer** *Wavelet approximation of the symmetric boundary integral formulation*	**A. Dimitrov** *Stress singularities in multi-material structures*	**B. Rasuo** *On solving boundary value problem in fluid mechanics by Fourier`s method: Wall interference of transonic wind tunnels*
10:50 – 11:05	**C O F F E E B R E A K**			
11:05 – 11:30	**F. D. Fischer** *Grain Growth and coarsening – A multifield problem in material mechanics*	**J. Aares** *Schwartz analysis for a class of domain decomposition methods for multiscale elliptic problems with high aspect rations*	**V.A. Kovtunenko** *Quasi static propagation of cracks*	**St. Berres** *Kinematical models and particle tracking for polydisperse sedimentation*
11:30 – 11:55	**S. Lurie** *Multiscale modelling in the mechanics of Materials: Cohesion, Interfacial Interactions, Inclusions and Defects*	**B. Khoromskij** *H-metrix approximation in FEM and BEM applications*	**A. Sheveleva** *A plastic strip model for an interface crack in an anisotropic biomaterial under mechanical and thermal fields*	**Y. Lifshitz** *Low-Dimensional models of flow over a body of revolution at high incidence*
11:55 – 12:40 Room 57.02	**P. Papadopoulos**	***Analysis and implementation of dual formulations for the finite element solution of the two-body contact problem***		
12:40 – 14:00	**L U N C H**			

Wednesday, 10. April 2002

	Section 1: **Material Modelling and Multiscale Problems** Room 57.04	Section 2: **Numerical Methods** Room 57.05	Section 3: **Solid Mechanics** Room 57.06	Section 4: **Porous Media and Fluid-Structure- Interaction** Room 57.03
14:00 – 14:25	**K. Kirchgässner** *Dispersiv Dynamics of Waves in Euler-Systems*	**P. Coorevits** *Automation of finite element analysis with multi-criteria methods*	**I.A. Miklashevich** *The generalised space and the solid state deformation description*	**D. Ivanović** *Injection-ejection fluid influence on velocity and vorticity distributions in laminar-turbulent transition of unsteady incompressible boundary layer on porous surface*
14:25 – 14:50	**D. Kuhl** *Interaction of calcium leaching and mechanical damage in concrete*	**A. Sellier** *On the numerical approximation of the derivatives of potentials on a smooth and three-dimensional boundarys by using boundary integral equations only*	**G. Jaiani** *Static and dynamical problems for cusped elastic beams*	**M. Auweter-Kurtz** *Fluid-solid interaction: Adaptive meshes for the anode in magnetoplasmadynamic thrusters*
14:50 – 15:15	**G. Lasko** *Mesomechanical model of plastic strain localization in polycrystals*	**G. Of** *The fast multipole method for the symmetric boundary integral formulation*	**A.V. Chigarev** *Propagation of nonstationary waves in inhomogeneous viscoelastic medium with initial stresses*	**J. Heiermann** *Fluid/ cathode interaction in magnetoplasmadynamic thrusters*
15:15 – 15:30	**COFFEE BREAK**			
15:30 – 15:55		**W. Bangerth** *Adaptive finite element solution of distributed parameter estimation problem*	**R. Wojnar** *Rearrangement of pentagon-heptagon dislocations in grain boundaries*	**T. Schwartzkopff** *Direct simulation of aeroacoustics using heterogenous domain decomposition*
15:55 – 16:20		**I.S. Pop** *Analysis of discritization method for the Richards-equation*	**K. Chelminski** *Mathematical analysis of a poro-plastic model for geomaterials*	**D. Dinkler** *Simultaneous solution to viscous fluid flow interacting with large amplitude structural vibrations*
16:20 – 16:45				**M. Fischer** *Simulation of acoustics-structure-interaction of non-conforming discretizations*
16:45 – 17:00	**COFFEE BREAK**			
17:00 – 17:45 Room 57.02	**D. Arnold**	***From the de Rham complex to colliding black holes: Differential complexes in numerical analysis***		
17:45	**CLOSING SESSION**			

Colored Pictures

Hermann G. Matthies, Jan Steindorf:

Strong Coupling Methods

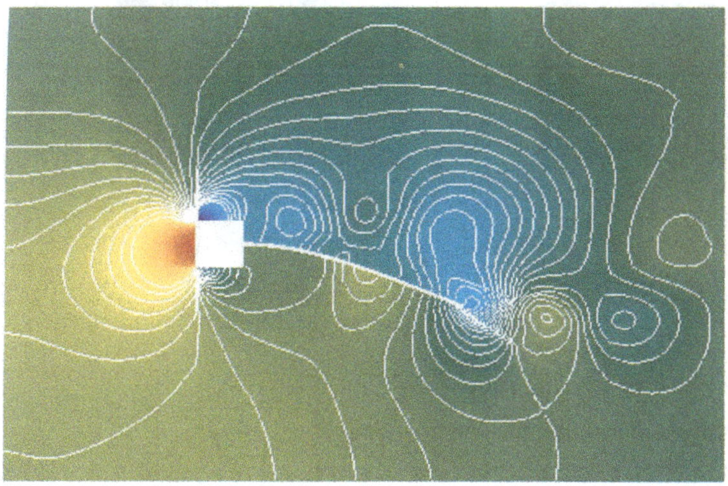

Fig. 1. Vortex Shedding and Pressure Field.

Mary F. Wheeler et al.:

DG for Flow and Reactive Transport in Porous Media

Fig. 2. Pressure Field

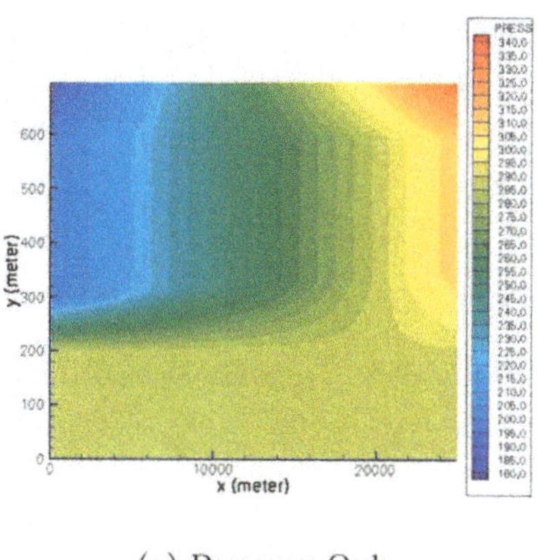

(a) Pressure Only

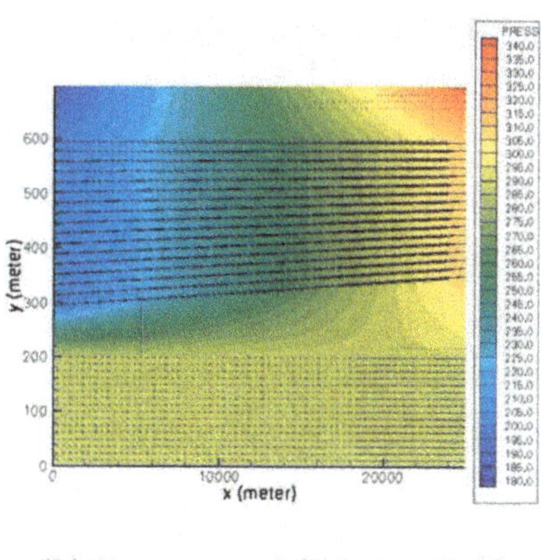

(b) Pressure and Velocity Field

Fig. 3. Concentration of Iodine with Dispersion

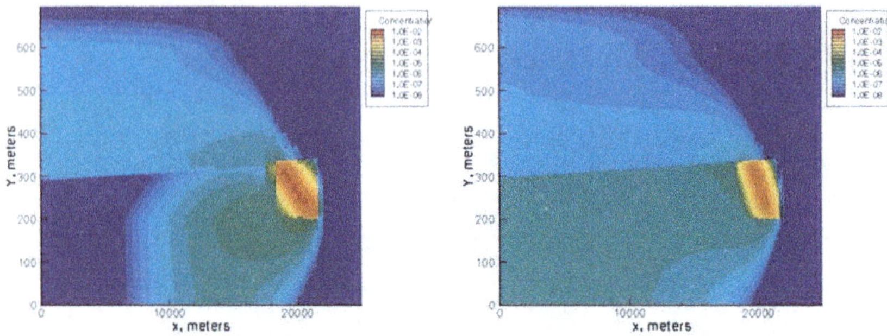

(a) 500k years (b) 1M years

Fig. 4. Concentration of Iodine without Dispersion

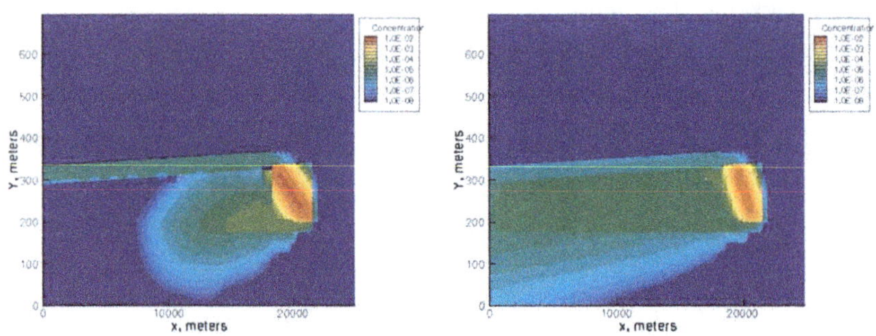

(a) 500k years (b) 1M years

Fig. 5. Slope Limiting the Concentration of Iodine with Dispersion (Time=500k years)

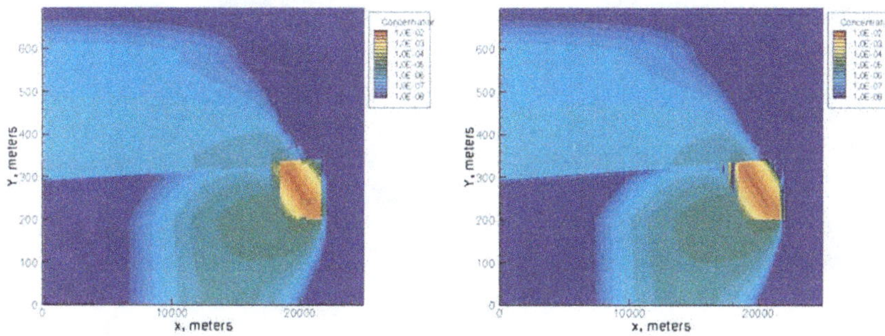

(a) Slope Limited (b) Not Slope Limited

Fig. 6. Concentration of Iodine at Large Times with Dispersion

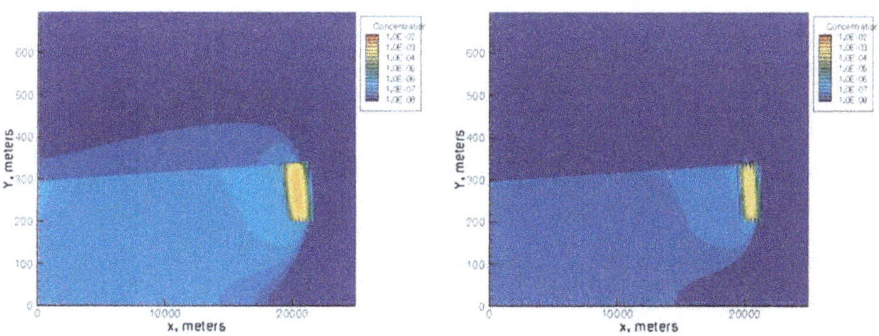

(a) 3M years (b) 5M years

Fig. 7. Concentration of Plutonium

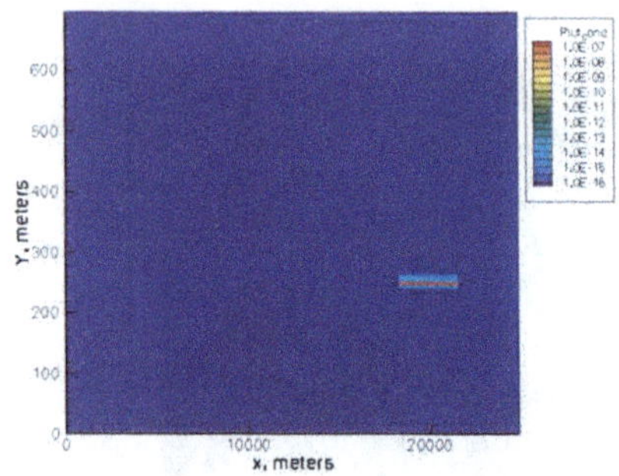

(a) 500k years

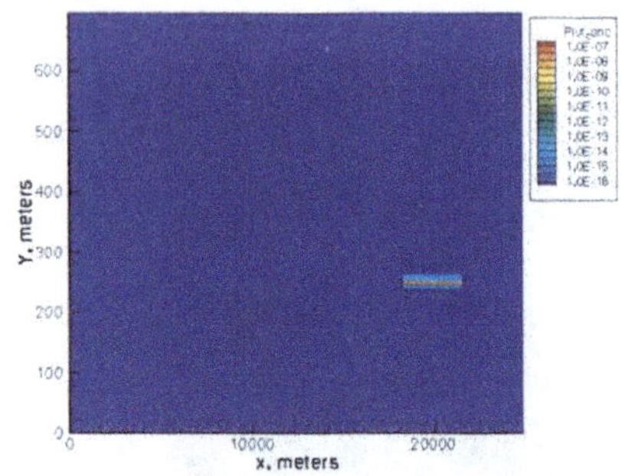

(b) 1M years

R. Bürger and K.H. Karlsen:

Clarifier-thickener units

$$S(x) = \begin{cases} 0.1 + 90(x+1)^2 & \text{for } -1 \le x \le -0.9, \\ 1 & \text{for } -0.9 < x \le -0.5, \\ 0.5 & \text{for } -0.5 < x \le 0, \\ 1 & \text{for } 0 < x \le 0.3, \\ 1 - \frac{0.8}{0.49}(x-0.3)^2 & \text{for } 0.3 < x \le 1 \end{cases}$$

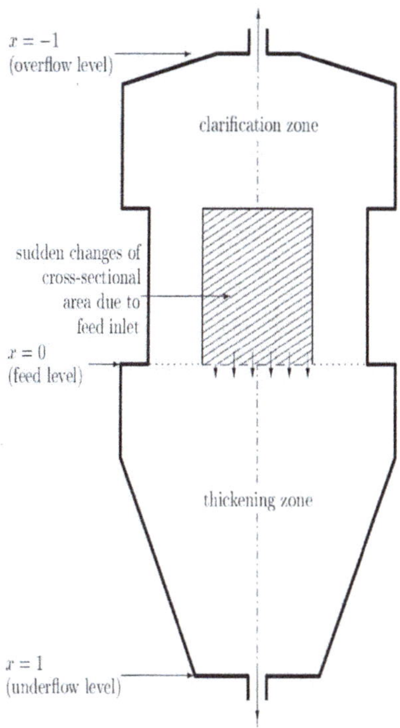

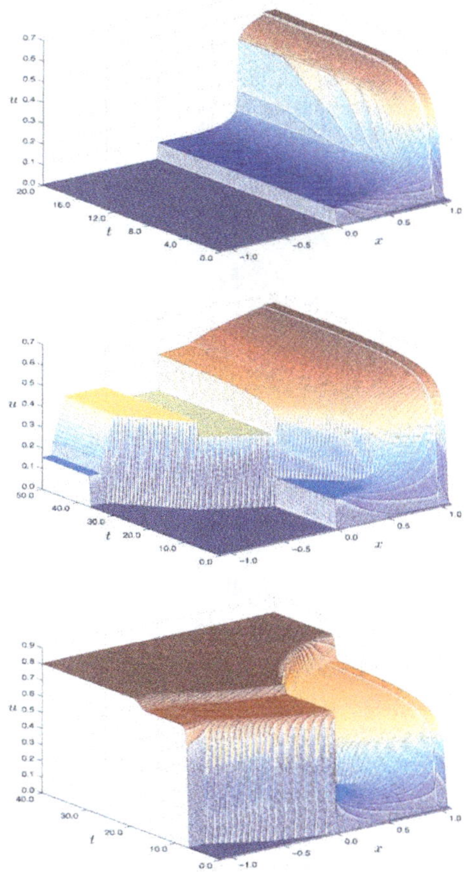

Fig. 1. A clarifier-thickener unit defined by the cross-sectional area function (top left) and simulations of continuous sedimentation processes with $u_F = 0.3$ (right column; top), $u_F = 0.4$ (middle) and $u_F = 0.8$ (bottom)

Iuliu Sorin Pop:

Numerical Analysis for Richards' Equation

Fig. 1. Absolute permeability (left) and the wetting front (right)

Bernd Scholz, Wolfgang Ehlers:

Identification of the Cosserat Material Constants

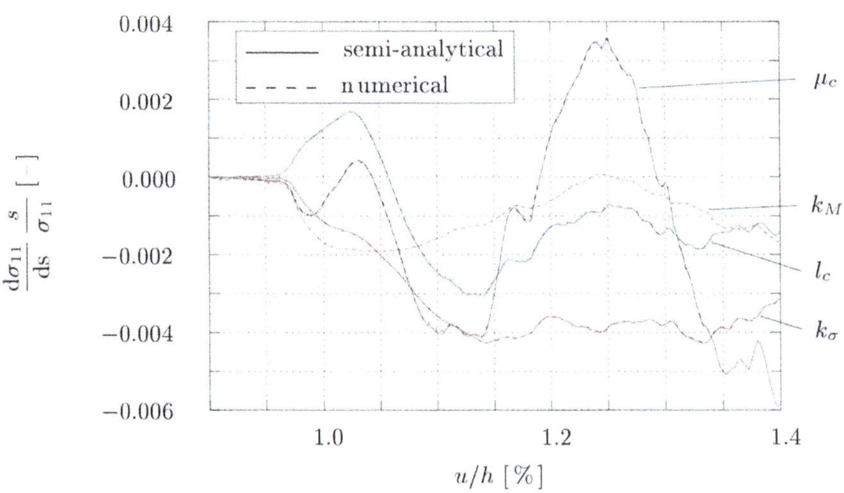

Fig. 3. Comparison betw een semi-analytical and numerical sensitivities.

Bin Hu et al.:

Multi-time Scale Simulation of Longitudinal Impact Responses

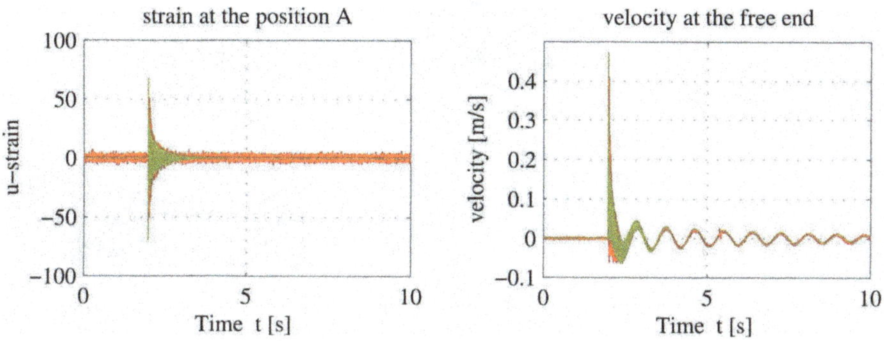

Fig. 2. Comparison between the measured and simulated results

S. Berres, R. Bürger and E.M. Tory:

Mixed systems modeling polydisperse sedimentation

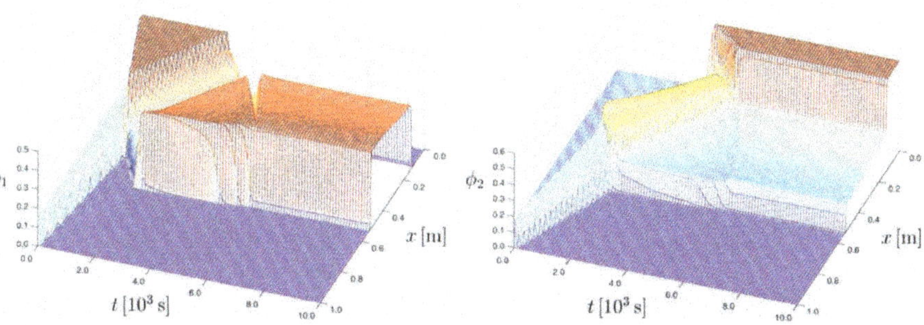

Fig. 1. Simulation of the settling of a bidisperse suspension: concentrations of the larger (left) and the smaller (right) particles.

W. Ehlers et al.:

Swelling of Charged Hydrated Media

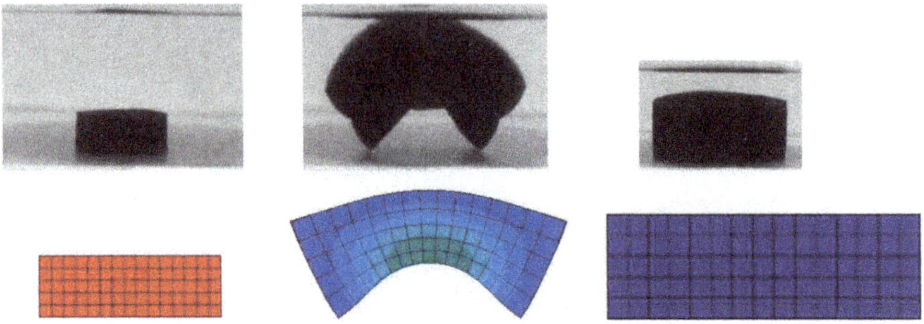

Fig. 1. Qualitative comparison of a free swelling experiment on stoot-colored hydrogel disc (J. M. Huyghe, 1999) with the FEM simulation

Małgorzata Peszyńsk a:

Couplings of models in porous media

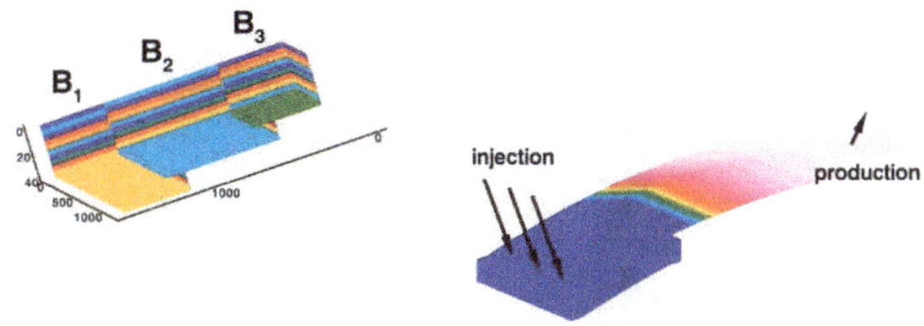

Fig. 1. Left: permeability field and geometry. Right: oil concentration contours at $t = 0$ and well locations indicate that $\Omega^2(t) \supset B_1 \cup B_2$. Gas saturation (not shown) indicates that $\Omega^3(t) \subset B_3$

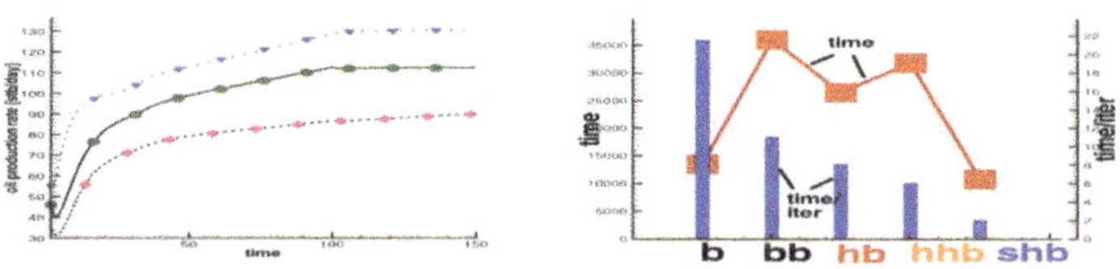

Fig. 2. Left: comparison of the oil production rates obtained with m ultiph ysics (symbols only) and with single model (dotted, solid and dashed lines, respectively). F rom top to bottom: in blodk B1: unsaturated oil; saturated oil but no gas cap; gas cap. Right: efficiency of multiphysics couplings

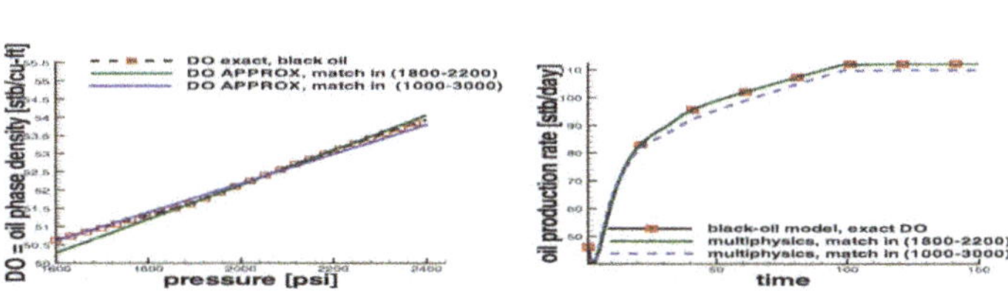

Fig. 3. Sensitivit y to densiy matching. Left: density of oil (exact and approximated using (3)). Right: corresponding well rates

Harmen Schippers et al.:

Aerodynamic, structural and electromagnetic interaction

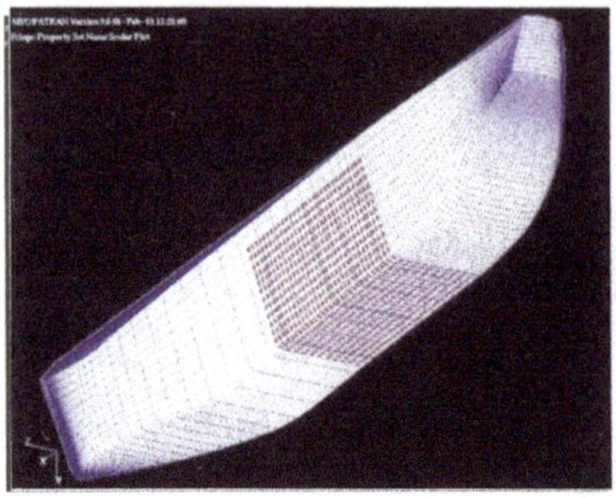

Fig. 1. Artist impression of phased array an tenna on reconnaissance pod

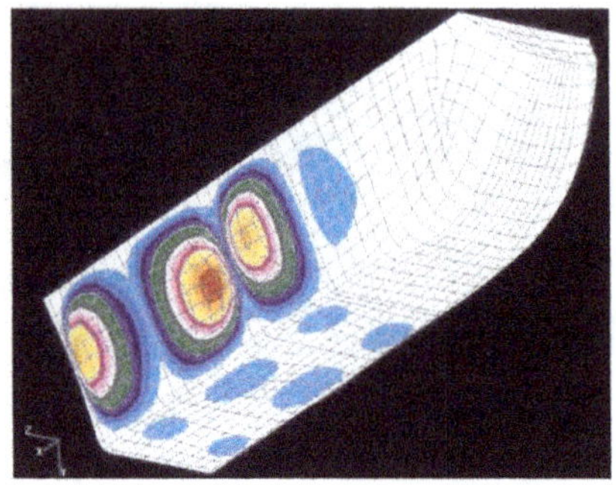

Fig. 3. Vibration mode at 229.6 Hz

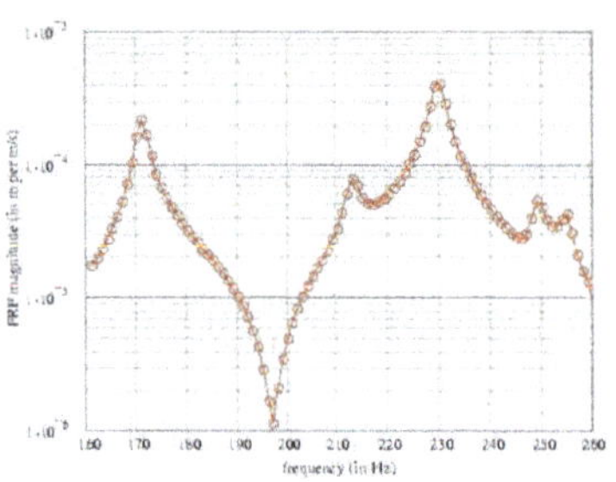

Fig. 4. F requency response function related to gust load $w_g = 1$

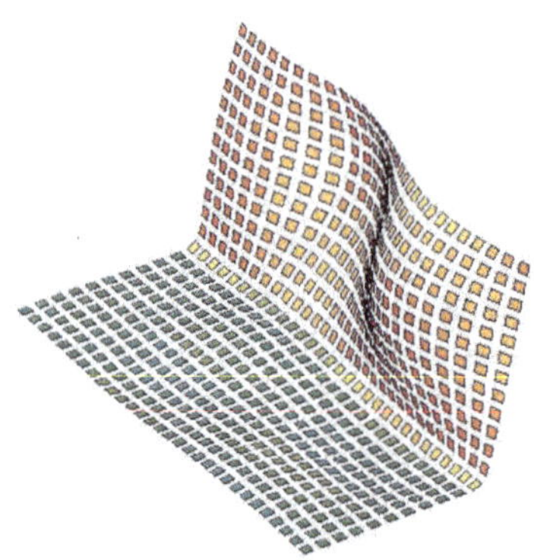

Fig. 5. Deformation of an tenna aperture

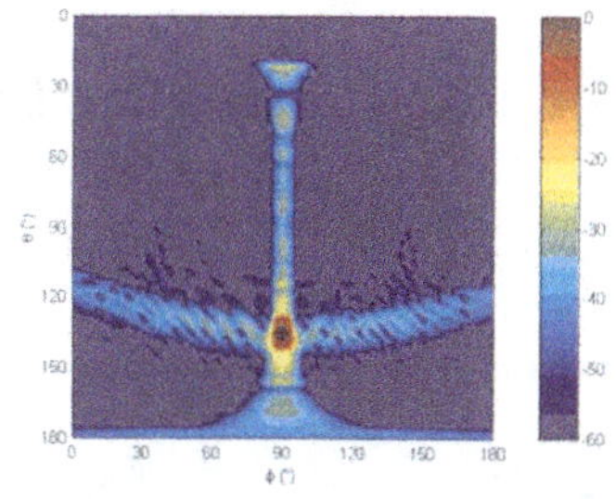

Fig. 9. 3D radiation pattern of the non-deformed SLAR antenna

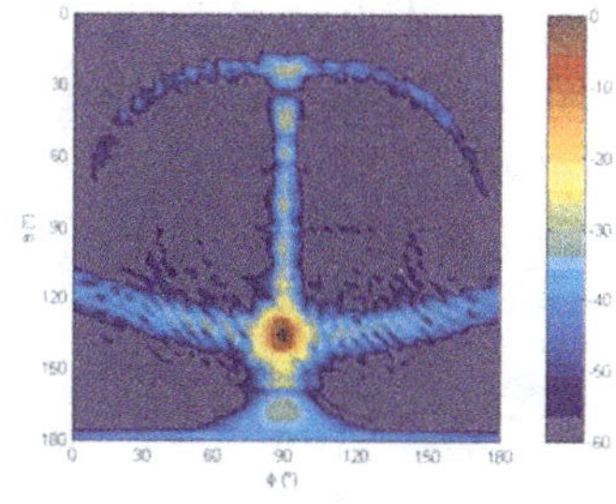

Fig. 10. 3D radiation pattern of the SLAR antenna, surface deformed by vibration mode with resonance frequency 229.6 Hz

Wolfgang A. Wall et al.:

Iterative Substructering Schemes for Fluid Structure Interaction

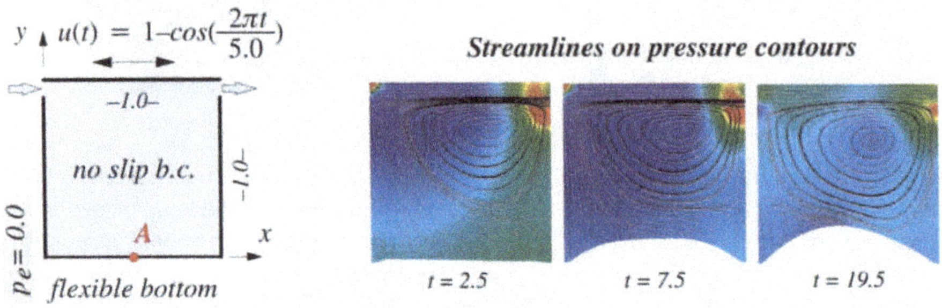

Fig. 3. Ca vity with flexible bottom - problem statemen, deformed configurations

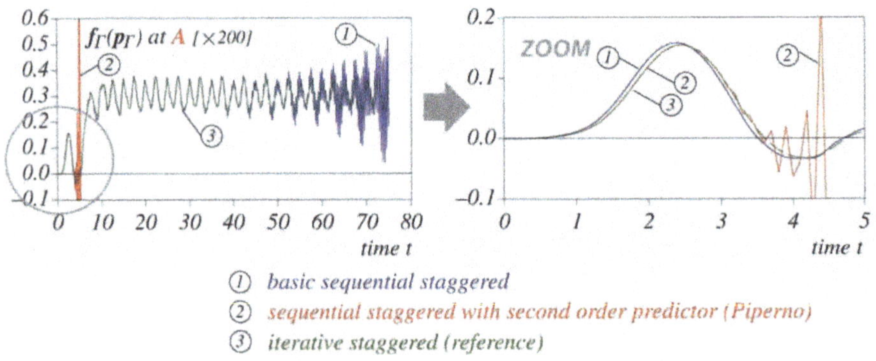

Fig. 4. Ca vity with flexible bottom - pressure load histories

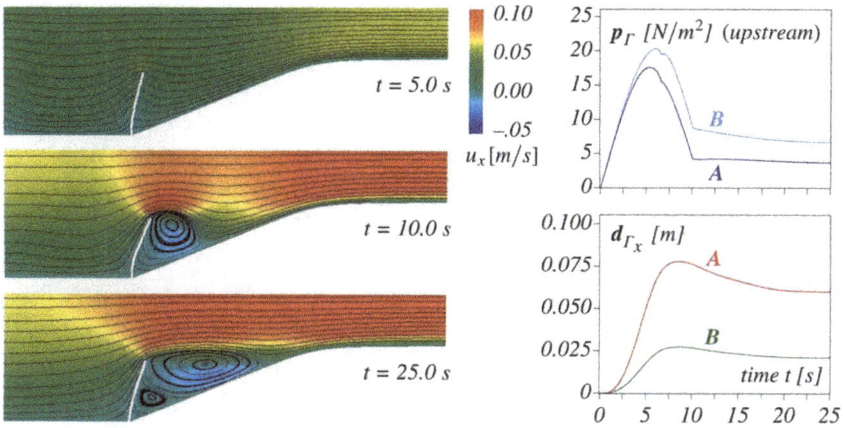

Fig. 6. Restrictor flap in channel – deformed confirgurations (streamlines on horizon tal flow velocity)

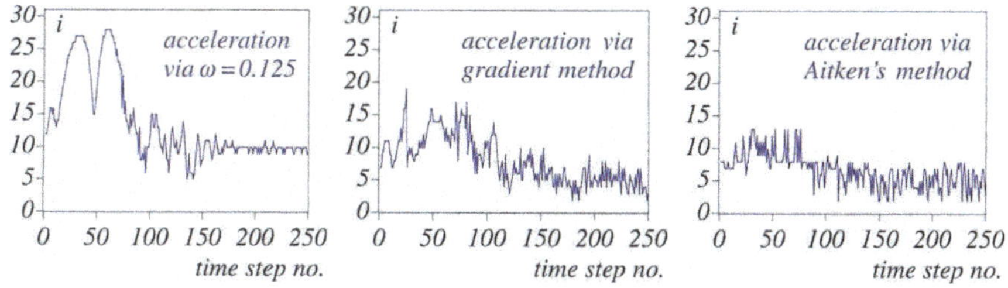

Fig. 7. Restrictor flap in channel – convergence study (number of subiterations i per time step)

Printing: Druckhaus Berlin-Mitte GmbH
Binding: Buchbinderei Stein & Lehmann, Berlin

GPSR Compliance
The European Union's (EU) General Product Safety Regulation (GPSR) is a set of rules that requires consumer products to be safe and our obligations to ensure this.

If you have any concerns about our products, you can contact us on

ProductSafety@springernature.com

In case Publisher is established outside the EU, the EU authorized representative is:

Springer Nature Customer Service Center GmbH
Europaplatz 3
69115 Heidelberg, Germany

www.ingramcontent.com/pod-product-compliance
Lightning Source LLC
Chambersburg PA
CBHW071233230926
51849CB00001B/85
* 9 7 8 3 5 4 0 0 0 6 9 6 1 *